CONSUMER TEXTILES

ANNE FRITZ + JENNIFER CANT

Tamar Holman

OXFORD
UNIVERSITY PRESS

Melbourne

OXFORD UNIVERSITY PRESS AUSTRALIA

Oxford New York Toronto
Delhi Bombay Calcutta Madras Karachi
Pataling Jaya Singapore Hong Kong Tokyo
Nairobi Dar es Salaam Cape Town
Melbourne Auckland
and associated companies in
Berlin Ibadan

OXFORD is a trademark of Oxford University Press

First published 1986
Reprinted with corrections 1987
Reprinted 1988,1989

National Library of Australia
Cataloguing-in-Publication data:

Fritz, Anne, 1946
     Consumer textiles.

     Bibliography.
     Includes index.
     ISBN 0 19 554647 4.

     1. Textile fibres. 2. Textile fabrics. 3. Textile
     industry. I. Cant, Jennifer. II. Title.

677

Designed by Peter Shaw
Illustrated by Guy Holt
Typeset by Post Typesetters, Brisbane
Printed by Impact Printing, Melbourne
Published by Oxford University Press,
253 Normanby Road, South Melbourne

Cover photograph by courtesy of
Grob and Co. Ltd, CH-8810 Horgen, Switzerland.
Grob flat steel healds made of rolled steel wire
have been manufactured since 1891, the year
in which Grob and Co. was established. During the
past four decades, Grob has produced millions of
such healds, and delivered them throughout
the world, where they have proven their endurance
under the most varied circumstances.

Contents

Introduction

This book is for the discerning consumer of textiles, the consumer who wishes to know what properties can be expected in a textile and how these properties can best be exploited. It is for the consumer who chooses with care, who is concerned about quality and price, and about global issues of textiles production.

We have taken a very broad definition of textiles. In this book, textiles refers to textile materials (fibres, yarns, fabrics), and to textile products (those items made from textile materials and those which are used in ways associated with textiles: items such as apparel, furnishings, industrial goods).

An understanding of textiles arises from knowledge of where and why textiles are used, and how the components of textiles combine and are modified to give a textile product. A study of textiles therefore unites climatology, chemistry, physics, engineering, sociology, economics, and, of course, art and design. The consumer of textiles cannot be expected to be expert in all these fields, and no single book can cover them all in any detail. This book, therefore, gives a little of each, sufficient to account for textile properties and to guide consumer choice, and, we trust, to stimulate the student of textiles to further reading.

Throughout the text, research activities and practical exercises are included to illustrate and extend the discussion. The practical activities are in many cases adapted from established industry usage, but simplified so as to make complicated equipment unnecessary. They do not demand a high degree of skill but, carried out with care, they illustrate textile properties much more vividly than could any text.

At the end of most chapters, questions related to consumer problems are answered. These serve not only to solve any problems the reader may have in textile use and care, but also relate the technical information to everyday situations.

Scientific concepts and technical descriptions have been kept as simple as accuracy will permit, and technical jargon has been avoided where possible. The result is an informative and useful guide to textiles, their properties and significance, of interest to the student and consumer alike.

Many people and organizations have helped in the compilation of this book by providing information and illustrations, by reading and commenting on various chapters in the original draft, and by being continued sources of professional support over the years. Our thanks to all these people, but especially to R. Selinger, R. Riach, Professor C. H. Nicholls, Dr M. Pailthorpe, Dr R. Griffiths,

Dr M. Young, Dr N. Johnson, R. Petterson, I. Angliss, P. Williams, and M. Cant. Particular thanks are due to Mr Vaughan Walker B.Sc.Hons., C.Text. A.T.I. of the Scottish College of Textiles, who read the entire book in galley form and saved us from many embarrassing errors and omissions. The project has benefited greatly from his assistance; needless to say any errors and oversights in the published book remain entirely the province of the authors.

Note

Words such as Avril, Kevlar, Nomex, Dacron, Scotchgard, Gore-Tex, Thinsulate and Perlon are proprietary names and should always be spelt with a capital letter.

Textiles in our lives

The use of textiles

How important are textiles in our everyday lives?

Pause a while and reflect on the many USES of textiles that you may encounter on any day.

Some uses of textiles are obvious, such as in clothing. Clothes create a close personal environment, which may be free and comfortable, or so restrictive that it damages your health. The clothing *choices* depend on many factors: climate and weather, available technology, fashion, cultural requirements, the economy, your sex and age, your income and employment, the range of textiles available to you. Clothes appropriate to July in Cooma would be a health risk in Marble Bar in January; the characteristics desirable in a firefighter's uniform are quite different from those needed in a nurse's uniform; the latest fashion releases in Sydney or Melbourne may not appear simultaneously in Tennant Creek; clothes you would wear at a wedding depend on whether you are the bride or the groom, one of the official party, the celebrant, or a guest, and on your religious beliefs and social attitudes.

Another obvious use of textiles is in furnishings and other features of your near environment: the bicycle seat, the car seat, the office chair, the blinds and curtains, the carpet or linoleum, the filters in the air-conditioner. Again, the choices available depend on many factors.

Less well-known uses of textiles are in roadworks, theatre sets, industrial belting, shadehouses.

Textiles therefore protect, contribute to our comfort, have social meaning, and perform important functions in industry and construction.

TRY YOUR HAND

1 Brainstorm for as many uses of textiles as you can find.
2 Select some of the more unusual examples and find out more about them. Try to make use of a wide range of sources of information, such as libraries, manufacturers, and users.

Many of the examples of textiles you discuss in this exercise will be relatively new developments – new materials, or the result of new or improved methods of manufacture.

Within Australia, the range of climates dictates a range of clothing styles.
(Snowy Mountains Authority) (W. Brindle/Australian News and Information Bureau)

Economic importance of textiles

The question at the beginning of this chapter could have been answered in another way: the ECONOMIC, rather than FUNCTIONAL, importance of textiles could have been considered.

You may have heard the expression 'Australia rides on the sheep's back'. While wheat has always been important and the growth of mining has helped broaden Australia's economic base, wool is still a major contributor to Australia's export income. Cotton is an increasingly valuable export crop.

The textile industry is a major employer. Textile factories tend to be quite small, with few staff, but the huge number of these small concerns means that in total they employ a very large number of people. Some, particularly clothing manufacturers, also employ OUTWORKERS, people who work at home for so much an item.

Retailers employ staff to sell textile products in all their forms, and so textiles are indirectly an important source of employment here too.

TRY YOUR HAND

Using *Year Book Australia* and other references in your library, find recent figures for:
1 The dollar value of wool exports.
2 The number of people employed in the textile and clothing industries. How do these figures compare with other products and industries?

If you found a State-by-State breakdown of these figures, you will have noticed that the textile industry is largely concentrated in Victoria, with some industry in New South Wales, and very little in the other States.

Wool is produced in all States, including the Northern Territory. New South Wales produces the most in dollar terms, but the wool industry is important throughout Australia. Cotton requires a hot climate, and is produced in Queensland and New South Wales.

Textiles, therefore, are extremely important in our lives, as they contribute to the nation's wealth, and provide employment for many thousands of people.

The economic importance of textiles extends beyond their manufacture and sale.
(Dale Mann/Retrospect)

Textiles and culture

The range of textiles available to people has developed over the centuries and is a result of the development of TECHNOLOGY. The further back in history you go, the more limited were the choices in textiles. Thus, in any society, the textiles available at any given time are an indicator of the technological development of the society.

Textiles however indicate more than technological development. CULTURE, in addition to meaning 'a trained and refined state of manners, taste and understanding', means 'the stage, form or type of development of a country at a particular time'.

There is a close interconnection between the manufacture and use of textiles and the culture of any group of people; and a culture can be characterized in part by its use of textiles. The study of a culture often includes an analysis of its use of textiles and its ability to manufacture or trade to meet its textile requirements.

TRY YOUR HAND

National Geographic magazine frequently runs articles on the textiles of a particular culture.

1   In groups, select a range of *National Geographic* articles which depict various cultures and their textiles. Depending on the articles each member selects, you may not be able to answer all these questions, but they should still serve for a basis of discussion.

2a   What is the emphasis of the article? Does it deal mainly with clothing, or with textile artefacts?

 b   How are these articles manufactured? By whom? Are the textiles manufactured by simple means, or by sophisticated machinery?

 c   Is there any special meaning attached to the textiles in question?

 d   Have there been any recent changes in the manufacture, use and meaning of these textiles? Why?

 e   Does the article imply that the textiles you studied are a central part of the lives of the people, or not?

3   Share your findings with others in your group and consider the similarities and differences between the cultures you have studied. Consider whether there was any fundamental difference between clothing and textile artefacts.

4a   From your own experience, and that of your friends and family, what can you identify about Australian culture from our textiles?

 b   Do you live in an affluent and/or technologically advanced society? A comparison with the textiles in an earlier period, say the 1850s (gold-rush era), may help here.

 c   Consider (i) your attitudes about what is important, for example about the use of furs and leather; (ii) the activities you are generally involved in. How does this relate to your use of textiles?

 d   Consider community values and activities. Can you see any relation between these answers and what Australia manufactures or imports? What would happen to a manufacturer or importer who ignored these considerations? How does this relate to the supply and production of textiles in Australia?

5   Your answers to Question 4 will have been influenced by your family background. If your class members represent a range of cultural backgrounds, these can be used as a resource for further study.

   a   In small groups, select a culture and identify particular textiles which are associated with it. Be careful to avoid stereotyping that culture: remember that there is great diversity within a culture, and that similarities between cultures are usually greater than the differences.

   b   Compare and contrast (i) the cultures selected by each group; (ii) these cultures and the cultures studied in the *National Geographic* articles; (iii) these cultures and your answers to Question 4.

Textiles perform an important role in most societies. They are an integral part of tradition and ceremony, and as such they function as a source of cultural cohesion.

They contribute to *continuity of a culture:* traditional textiles from the past continue to be used into the future. The wedding gown, christening gown, and ceremonial robes are common examples. Textile *crafts,* such as knitting, embroidery, weaving, and lace making, have strong cultural links.

Within a culture, textiles use has shared meaning. Consider the wearing of hats. Once it was a social requirement for all men and women to wear a hat; now, hats and head coverings are usually worn only as part of a uniform, as required work costume (hard hats in the construction industry, hair coverings for food workers and surgeons), in certain socially prescribed situations, or for reasons of religion.

The Lord Mayor's ceremonial robes have a direct cultural link with medieval England. (Melbourne Times)

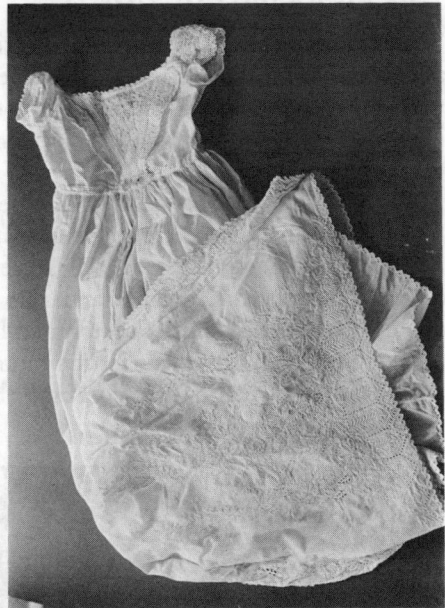

This Ayrshire christening robe, dating from around 1860, is an example of a textile as heirloom. (Dale Mann/Retrospect)

TRY YOUR HAND

List some of the social situations in which people still wear hats. What types of hats are usually worn? Do the hats have any social meaning?

Which sub-cultures in our community usually wear hats or head coverings? Why?

Each of these ideas – textiles as a major contributor to our physical environment; textiles as an important part of our economic life; and textiles as a significant aspect of culture – is explored in greater detail in later chapters.

2 An introduction to fibre chemistry

The molecules of fibres

When you consider the different textiles used for clothing, furnishing, in cars, as geotextiles, it is easy to identify differences in properties between the different textiles. Understanding the reasons for these properties helps a consumer make informed decisions about the selection, use and care of the textile best suited for a given purpose.

Many of the properties of a textile stem from the properties of the textile fibres themselves, so this is a useful starting-point for their discussion.

Fibres may be conveniently classified by the type of molecule they are made of (GENERIC classification), and this is the approach taken in the first part of this book. For example, cotton and linen are composed of *cellulose* molecules, and so have many features in common; wool and silk are composed of *protein,* and have many features in common, but these are different from those of the cellulosic fibres.

Generic classification allows such diverse materials as cotton, linen, rayon, and coir, to be discussed in terms of their common features.
(Dale Mann/Retrospect)

Although these molecules are very different and produce fibres with very different properties, the rules governing the behaviour of the molecules remain the same. An understanding of the properties of these molecules is therefore necessary before we can understand the properties of the fibres.

What is a molecule?

A molecule is a collection of atoms joined together.

The joining together of atoms and molecules is called BONDING. There are several types of bond, and the type formed depends on the properties of the atoms being joined. The type of bonding in a molecule determines many of the properties of the molecule.

Covalent bonds

The atoms join by sharing some of their electrons.

For example, the water molecule is formed by one oxygen atom sharing electrons with two hydrogen atoms. The linkage between the oxygen and each hydrogen is called a COVALENT BOND. It is formed from one electron from the oxygen and one from the hydrogen.

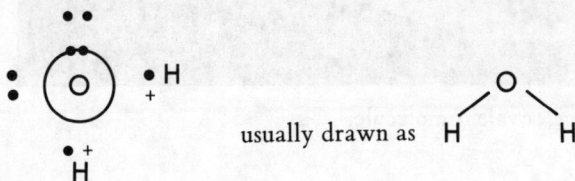

usually drawn as

In carbon dioxide, there are DOUBLE BONDS: each linkage is formed by two electrons from the oxygen and two electrons from the carbon.

$$O=C=O$$

A single bond can bend and stretch and rotate.

Butane: the same molecule.

A double bond can bend and stretch, but it cannot rotate.

trans and *cis* butene: two different molecules.

Double bonds are more easily broken in chemical reactions than are single bonds. A broken double bond becomes a single bond.

Breaking a covalent bond means breaking the link between the atoms: this means changing the nature of the molecule: this is a chemical reaction.

The carbon atom is characterized by an ability to form immense chains and lattices linked by covalent bonds. This unique property is the basis of the vast branch of chemistry known as organic chemistry, of which the chemistry of most textiles is one part.

A diamond is one huge covalent molecule.

The benzene ring

In general, double bonds are far more reactive than single bonds. If a molecule has many *double bonds alternating with single bonds,* the double bonds tend to 'overlap' or 'smear': they form a CONJUGATED system. Conjugated double bonds tend to be even more reactive than isolated double bonds.

In benzene, however, the alternating double and single bonds form a *closed loop* in the ring of six carbon atoms. The double bonds are not localized: their effect is spread over the entire ring, as if, instead of double and single bonds alternating, each carbon atom was joined to the next by one and a half bonds. The resulting AROMATIC system is extraordinarily stable, and resists chemical attack. Benzene rings are found in a number of textile fibres, but are particularly important in dyes.

The benzene ring is somewhere between these forms,

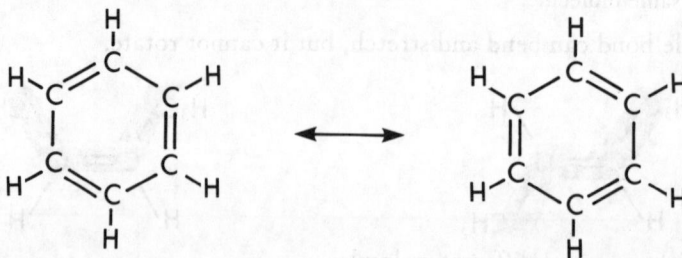

and is usually written like this:

Polarity

The electrons in a covalent bond are not always shared equally. One atom may have a greater attraction for the electrons than does the other: it is more ELECTRONEGATIVE.

When this happens, the electronegative atom ends up with a slight negative charge, and the other atom has a slight positive charge. This makes the molecule POLAR: a pair of covalently bonded atoms with such a charge separation is called a DIPOLE.

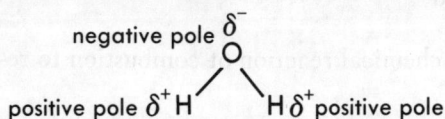

Polarity in the water molecule.

Two atoms important for their electronegativity are *oxygen* and *nitrogen*. *Hydrogen* is ELECTROPOSITIVE: it allows electronegative atoms to pull its electrons away. The carbon atom is effectively neutral.

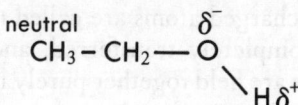

Ethanol (ethyl alcohol).

The hydrogen bond

As described above, covalent bonds can be either neutral or polar. The polarity will be very marked when hydrogen atoms bond to either oxygen or nitrogen. Since positive and negative charges attract each other, a strong electrostatic attraction between these polar molecules is possible.

This attraction can behave like a bond between the atoms: it is called the HYDROGEN BOND.

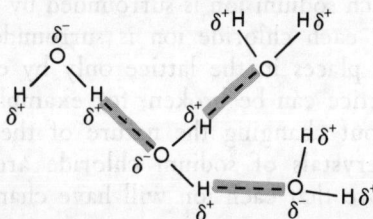

Hydrogen bonding in liquid water.

11

There are important differences between a covalent bond and a hydrogen bond. If a covalent bond is broken, the type, or chemical nature, of the molecule is changed; but a hydrogen bond can be broken and re-formed easily without affecting the nature of the molecule.

This happens, for example, when water evaporates and condenses:

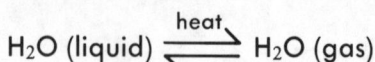

$$H_2O \text{ (liquid)} \underset{\longleftarrow}{\overset{heat}{\longrightarrow}} H_2O \text{ (gas)}$$

reversible reaction

Compare this to the electrolysis (use of electric current to break the covalent bonds) of water:

$$2H_2O \xrightarrow{zap} 2H_2 + O_2$$

breakdown of compound

It takes the violent chemical reaction of combustion to re-form the water.

$$2H_2 + O_2 \longrightarrow 2H_2O + heat$$

combustion

Ionic compounds

Sometimes the difference in electronegativity between atoms is so great that there is complete charge separation: one atom ends up with a whole electron taken from the other. These charged atoms are called IONS.

Because the electron is completely transferred, and not shared, there is no bond between the atoms: they are held together purely by electrostatic attraction.

Sodium is one atom that gives up its electrons extremely easily. Chlorine is extremely electronegative: it forms polar covalent bonds in some of its compounds, but reacts with sodium to form the completely ionic compound we call *salt*, sodium chloride.

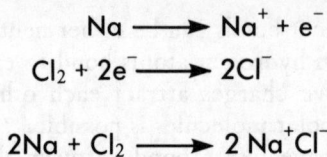

$$Na \longrightarrow Na^+ + e^-$$
$$Cl_2 + 2e^- \longrightarrow 2Cl^-$$

$$2Na + Cl_2 \longrightarrow 2\,Na^+Cl^-$$

In the salt crystal, each sodium ion is surrounded by six chloride (negatively charged chlorine) ions; each chloride ion is surrounded by six sodium ions. They are held in their places in the lattice only by electrostatic attractions. This means that the lattice can be broken, for example by dissolving the salt crystal in water, without changing the nature of the compound. When the water is evaporated, crystals of sodium chloride are recovered unchanged (although the chances are that each ion will have changed nearest neighbours in the lattice).

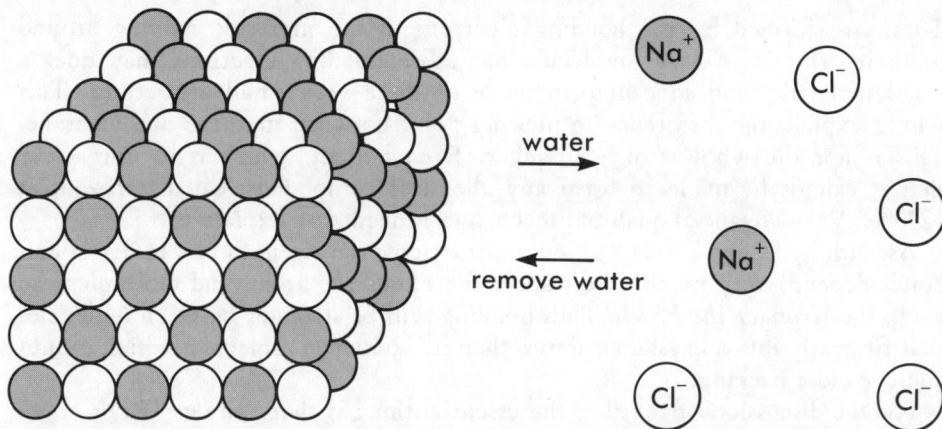

Van der Waals bonding

A substance is a solid if its component molecules are strongly attracted to each other. If the molecules can form hydrogen bonds, these strong electrostatic attractions provide the forces needed. However, many molecules do not contain very polar bonds. Another attractive force is needed for them to be solids.

Non-polar substances tend to have low melting points. Water, which is very polar, melts at 0°C, but methane (CH_4), which is non-polar, melts at −182.5°C.

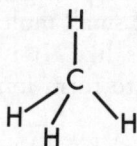

Methane

But methane molecules must attract each other, even if only very weakly, or solid methane could not be formed at all. The much larger, but chemically very similar molecule, eicosane $CH_3(CH_2)_{18}CH_3$, has a melting point of 36.4°C.

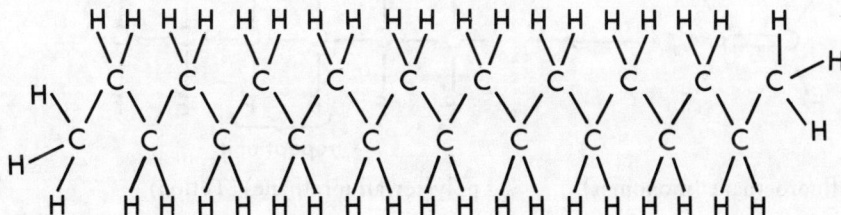

Eicosane

This means that there *is* an attractive force between non-polar molecules, and that this force is stronger in larger molecules. The bonds formed due to this attraction are called VAN DER WAALS BONDS.

The causes of van der Waals bonds are very complex. Part of the explanation seems to be very slight electrostatic attractions caused by very short-lived dipoles.

13

These are formed by the bonding electrons in the molecule moving around in their orbitals. A large molecule has more bonding electrons than does a small molecule, and so can form more of these very small attractions. This would explain the difference in melting point between methane and eicosane. This is not the whole story, however. Even helium, which is so unreactive that it cannot be made to form any chemical compounds, can be frozen, at –270°C. Very advanced quantum mechanics is needed to explain this.

As with hydrogen bonds and ionic attractions, the strength of van der Waals bonds depends very much on distance: the *closer the atoms* and molecules can pack, the *stronger the bonds*. Thus bonding will be stronger between molecules that fit neatly into a crystal structure than it is between molecules which cannot achieve close packing.

For the discussion of textiles, the essential thing is that *van der Waals bonds are weak bonds between non-polar molecules, and the larger the molecule, the more important they become.* Some molecules in textiles are very large indeed, for example, synthetic polymers contain many thousands of atoms, so the existence of van der Waals bonds is very important.

Molecules in textile fibres

The molecules described so far have been small and, in chemical terms at least, very simple.

Molecules in fibres tend to be very large, but their behaviour is governed by the same rules as is that of these small molecules.

Consider what a textile fibre is like. It is long and narrow, and it can be combined with many other fibres to form long, flexible threads, and these can be made different colours.

The molecules in a textile fibre also tend to be long and narrow. Very often, they are composed of very small units (MONOMERS) linked to form long chains. For example, cellulose is simply a vast chain of sugar (glucose) molecules, and the protein of wool is a long chain of amino acid molecules. These long chains of repeated units are called POLYMERS.

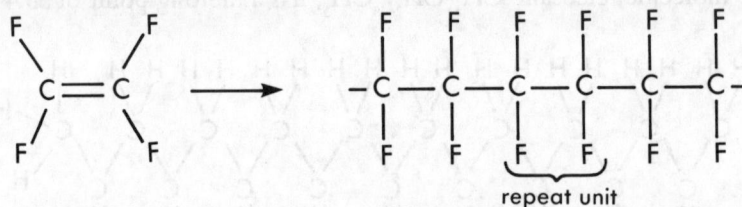

tetrafluoroethane (monomer)   poly(tetrafluorethane) (Teflon)

The properties of the fibre will depend on the nature of these small units, on the length of the chains, and on how the molecular chains are arranged in the fibre.

If the units are small or regular, the molecular chains can line up neatly with one another in the fibre. This leads to a very *ordered arrangement* of molecules in the fibre: it is CRYSTALLINE.

14

A regular structure allows the molecules to pack into regular, ordered, crystalline arrays.

However, if bits and pieces of the molecule stick out, that is, if it has bulky SIDE-GROUPS of atoms that hang off the side of the chain, this makes it harder for the molecule to arrange in an ordered way. As a result, there will be disordered, AMORPHOUS regions in the fibre.

Irregular or bulky molecules pack with difficulty.

The typical textile fibre will have both crystalline and amorphous regions. Their relative proportions will depend on the nature of the molecules themselves, and on the conditions under which they formed the fibre.

In the crystalline regions (CRYSTALLITES), the molecular chains are held together by hydrogen bonds (in the polar molecules) or van der Waals bonds (non-polar), or both. The chains are *tightly packed*, with no spaces between them.

In the amorphous regions there are VOIDS (spaces) between the molecular chains and little inter-chain bonding.

Voids occur in the amorphous regions of the fibre.

How does structure affect fibre properties?

The crystalline parts of the fibre are rigid, but strong. This is because the strength of the many hydrogen or van der Waals bonds has to be overcome before the fibre can be bent or broken in these regions: bending or breaking disrupts the crystal structure.

The amorphous regions are flexible, sometimes elastic, and less strong. This is because between the molecules in these regions there are fewer bonds to be disrupted when the fibre bends, stretches, or breaks.

The looser packing of the molecules in the amorphous regions means there are voids or spaces between the chains. These spaces allow water and dye molecules to penetrate the fibre.

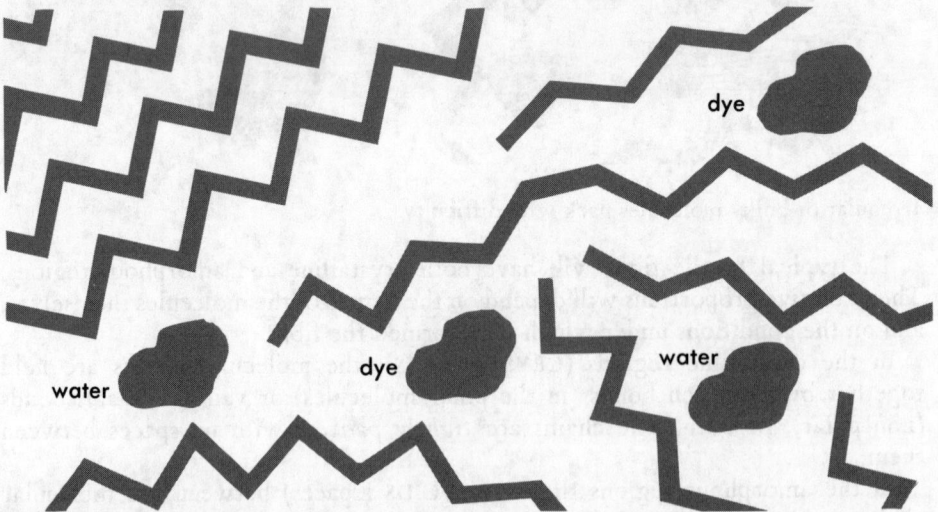
Only in the voids in the amorphous regions is there space to accommodate dye or water molecules.

The more orientated the crystallites in a fibre are, the more they are able to take up the stress of a load; that is, the stronger the fibre. However, if all the crystallites are parallel to the fibre axis, the fibre will be less extensible: it will not stretch as far before it breaks.

A less orientated fibre is more extensible.

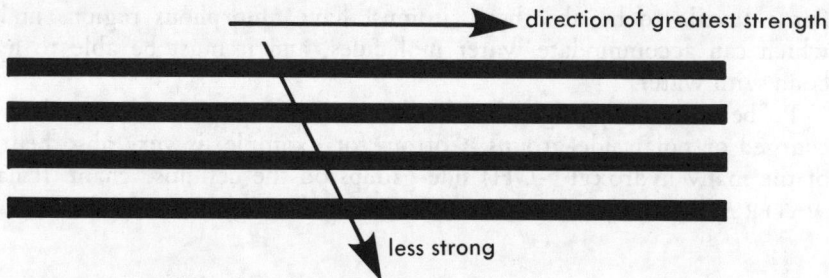

direction of greatest strength

less strong

Crystallites are strongest along their length.

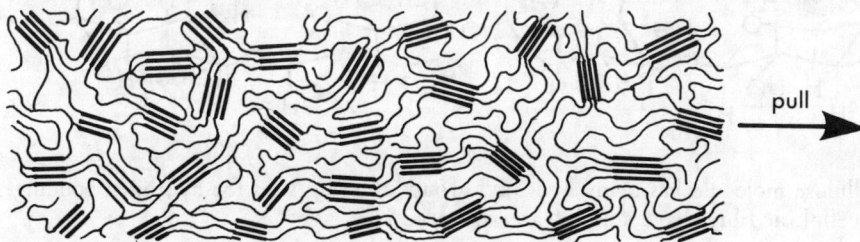

A fibre in which the crystallites are not aligned is more extensible.

pull

less extensible

A highly orientated fibre is strong but inextensible.

The ability of the fibre to accept dyes, and its affinity for water, depend on both the *structure* of the fibre and on the nature of the component units of the fibre molecules. Water is polar, and so will tend to be absorbed into only those fibres which are composed of polar molecules. Dye molecules may enter amorphous regions and lie between the fibre molecules, where they may form bonds with the side-groups of the fibre molecule. The type of dye taken up by the fibre and its fastness within it depend on both the nature of the fibre molecule and the nature of the dye molecule.

The manufacturer of synthetic fibres has some control over the structure of the textile fibre (see Chapter 5 and the discussion of the rayons in Chapter 3). This control is, of course, not possible with the natural fibres.

Treatment of fibres during textile manufacture and finishing depends on the original properties of the fibres, and on what the desired final properties of the textile are.

Water affinity

For a textile to be absorbent, it must have amorphous regions in its fibres which can accommodate water molecules, and it must be able to hydrogen-bond with water.

To be able to hydrogen-bond, the textile fibre molecules need to be have charged or polar side-groups. Cotton, for example, is very absorbent because of the many hydroxyl (–OH) side-groups on the cellulose chain. It has a high WATER AFFINITY, it is HYDROPHILIC (water-loving).

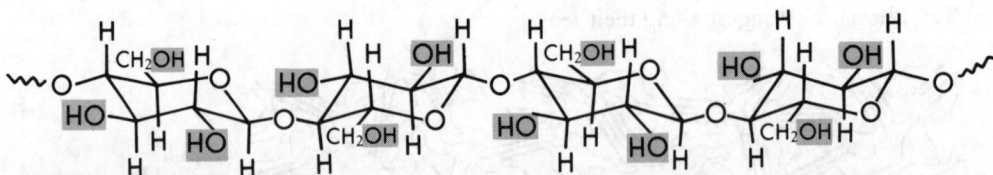

The cellulose molecule has many hydroxyl groups offering sites for hydrogen bonding. Hence cellulosic fibres have a high water affinity.

A textile fibre with few amorphous regions will not be absorbent, and a fibre with few, or no, polar groups in its molecules will not be able to hydrogen-bond. Polyester, for example, is relatively non-polar, has a high proportion of crystalline regions, and is not absorbent. It is described as HYDROPHOBIC, or water-fearing.

Solubility

Solid substances are solid because their component units are strongly attracted to each other. This may be because of (i) hydrogen bonds or other electrostatic attractions, or (ii) their size: in a very large molecule, the cumulative effect of the very many van der Waals bonds is very large. (Metals are a special case, and need not concern us here.)

For a substance to dissolve in a given solvent, the attractive forces within the solid have to be overcome. This requires energy. A system always tends to the lowest energy, and this is usually the most ordered form.

Working in the opposite direction is ENTROPY, or the amount of disorder in a system. A system always tends to maximum entropy. Small units moving at random in a solution have a much higher entropy than do large units, because they have greater freedom of movement. Any substance in solution has a higher entropy than it does as a solid, for the same reason.

This can be summarized as:
 (i) a tendency to minimum energy favours the solid form;
 (ii) tendency to maximum entropy favours solution.

Crystal lattice: low energy, low entropy.   Substance in solution: high energy, high entropy.

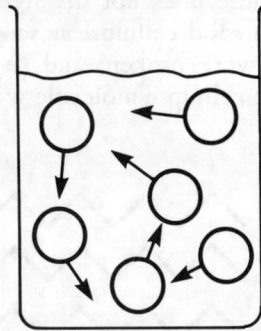

Heating is a good way of increasing the entropy and energy of a system. This is because the heat makes the atoms and molecules move faster (higher kinetic energy) and their movements are more random. Heating generally increases solubility.

The smaller the difference in attractions between: (i) the molecules in the substance for each other; (ii) the molecules in the solvent for each other; and (iii) the molecules of the solvent for the substance to be dissolved; the smaller the gain in energy when the substance dissolves. For this reason, polar solvents dissolve polar and charged substances, non-polar solvents dissolve non-polar substances. Non-polar solvents do not dissolve polar or charged substances, and polar solvents do not dissolve non-polar substances. That is, *like dissolves like*.

Sugar dissolves in water because it is able to form hydrogen bonds with the water molecules. The small gain in energy is more than offset by the large gain in entropy in going from the ordered crystal to the disordered solution.

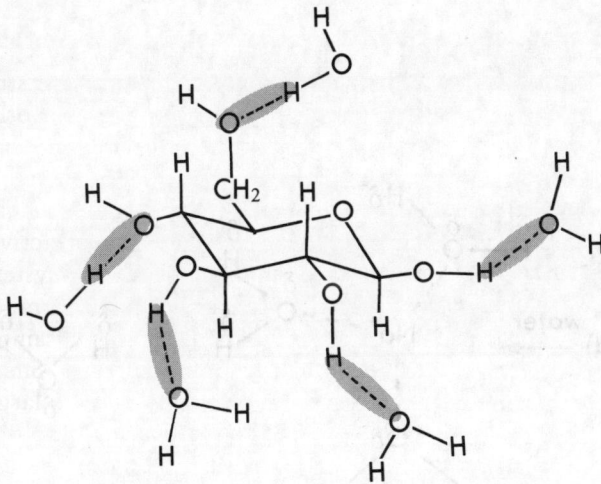

Hydrogen bonding between water and glucose.

19

Cellulose does not dissolve in water because: (i) the number of hydrogen bonds in solid cellulose is so great that there would be a huge gain in energy if they were broken; and (ii) the gain in entropy would be fairly small, as the huge cellulose molecule would not gain much freedom of movement.

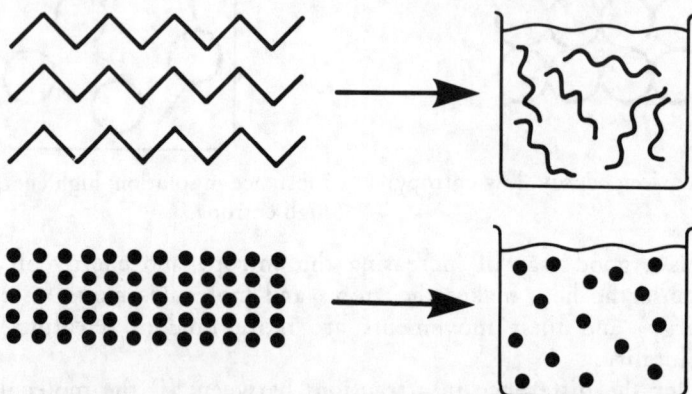

When a large molecule dissolves in water, the freedom of movement gained is small; when small molecules dissolve, the freedom of movement gained is large. Hence small molecules gain more entropy on solution than do large molecules.

Non-polar substances do not dissolve in water because their molecules are not attracted to the water molecules.

Sparingly soluble substances, like some dyes, are borderline: the gain in entropy about balances the gain in energy. The solubility of these sparingly soluble substances can be influenced by adding other substances, such as salt, to the solution. This is because the salt separates into charged ions in solution:

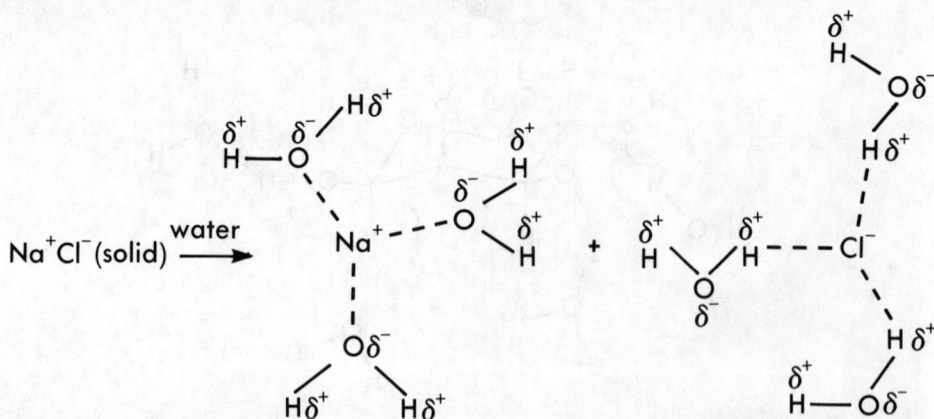

The polar water molecules are strongly attracted to the charged ions. This modifies the attraction of the water molecules for each other, and their attraction for the dye molecules. Adding salt to a solution of dye can force the dye out of solution again.

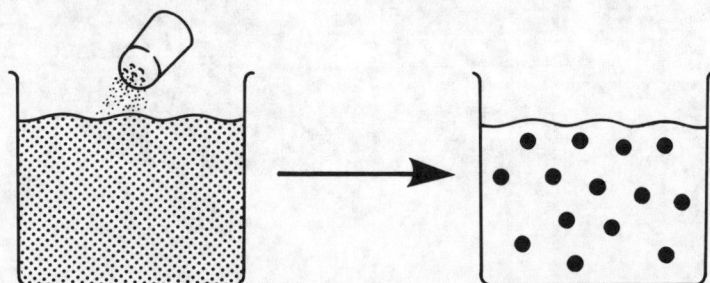

The effect of adding salt to a solution of sparingly soluble dye.

Oxidation and reduction

The terms oxidation and reduction are a pair of chemical opposites which are used as shorthand names for two very common classes of chemical reaction.

OXIDATION means combination with oxygen or removal of hydrogen. Thus when a substance is burnt, it combines with oxygen in the air, and is *oxidized.* An oxidizing agent makes oxygen available for chemical reactions: many bleaches are oxidants.

A vigorous oxidation reaction.
(Dale Mann/Retrospect)

Permanent setting of wool: disulfide bonds are broken by reduction, then reset by oxidation to hold the wool fibre in the desired shape.
(C.S.I.R.O. Division of Textile Industry)

REDUCTION means removal of oxygen, or combination with hydrogen. An example is the reduction of the disulfide linkages in wool treatments:

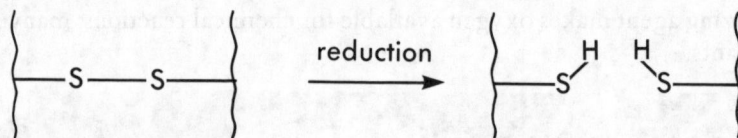

The disulfide linkages are REDUCED.

By definition, when these linkages are re-formed, and the hydrogens are removed, the sulfur is oxidized:

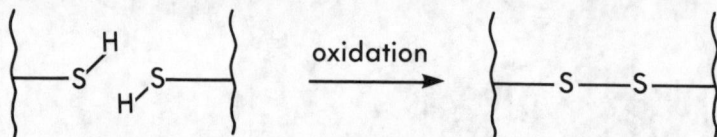

The terminology is not really important: it is used merely because it is convenient. The essential idea is that when these terms are used, they mean that there is a chemical reaction going on.

These ideas may seem very abstract and complex. However, they should make the following discussions of textile fibres clearer, and become clearer themselves in the course of that discussion.

3 Cellulosic fibres

This chapter investigates the fibres of vegetable origin: the CELLULOSIC fibres.

The varieties of cotton, their production, and the special properties of the cotton fibres are discussed. Modifications of these properties are also covered, including crease-resistant and other treatments designed to make cotton conform better to the requirements of the consumer.

Flax and other natural fibres are discussed, as is the regeneration of cellulose. In the study of the acetates, changes to moisture affinity and to heat sensitivity, due to changes in chemical structure and bonding, are explored.

These issues are discussed at the molecular level, to give an in-depth understanding of how such a large range of textile fibre properties can be developed from the cellulose molecule.

Consumer questions which will be answered in the course of this chapter include:

- How do I recognize a real bargain in towels?
- Is it worth paying extra for pure linen tea-towels?
- What type of fibre is most comfortable to wear on a hot, humid day?
- What type of yarn is most suited for a knitted summer top?
- Should I dry-clean a rayon dress labelled 'dry-clean only'?
- What can I do about blackberry juice spilt on an acetate dress?

The cellulose molecule

Cellulose is the basic building-block of all plant structures. It is a long, narrow molecule, made up from glucose rings joined in a chain: it is a naturally occurring polymer. A graphic representation of cellulose looks like this:

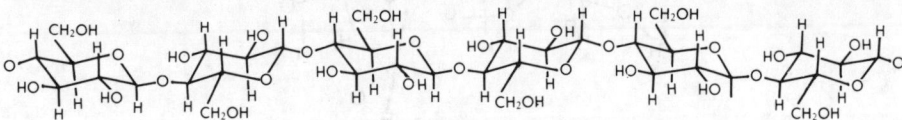

The properties of cellulose in all its forms are dictated by only a few important features.

The shape of the molecule

As can be seen from the structure drawn above, cellulose is like a ribbon – flat, thin, long. This shape means that cellulose molecules can pack neatly into an organized, crystalline system – hence the strength of the stems of plants and the wood from trees.

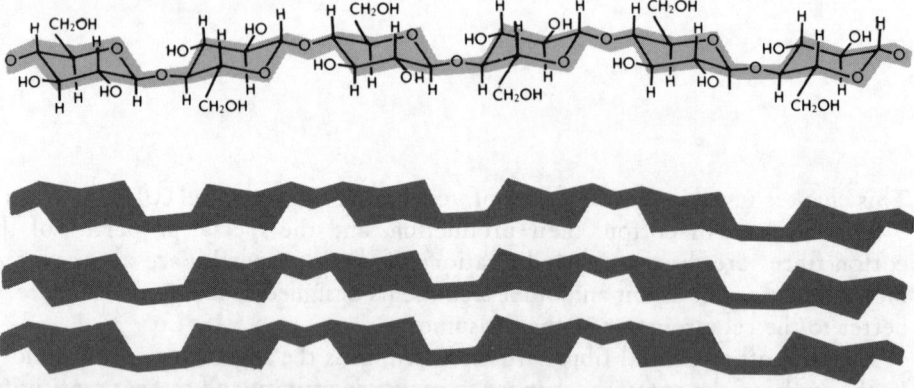

Because of its flat shape, the cellulose molecule can pack neatly into crystals.

The polarity of the molecule

The polarity determines the type of bonding that can occur between cellulose molecules.

Each glucose ring has three hydroxyl groups (–OH) attached to it. Thus the cellulose molecule is polar. Strong hydrogen bonds are formed between adjacent molecules in the fibres, giving rise to strong, crystalline regions.

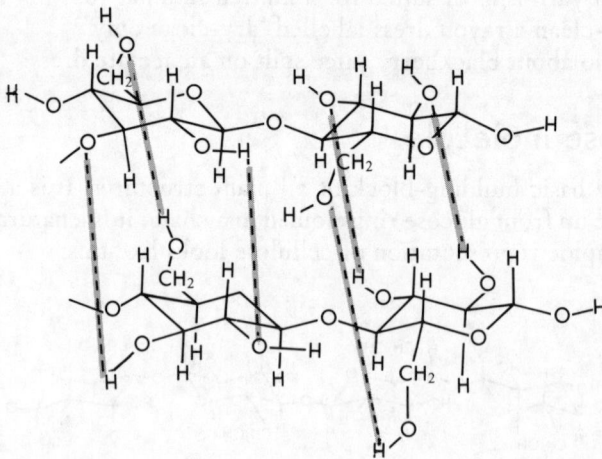

Possible pattern of hydrogen bonding between two cellulose molecules. Note that hydrogen bonding is three-dimensional: a cellulose molecule will hydrogen-bond with all adjacent molecules, those above and below and those on either side.

In the natural fibres, the proportion of crystalline to amorphous regions is determined by the growth patterns of the plant: linen is more crystalline than cotton. In the rayons, the degree of crystallinity depends on the industrial processes in the fibre production.

Affinity for water

The polarity of the cellulose molecule means that it has a high water affinity, as it can form hydrogen bonds with water molecules. Cellulose fibres therefore absorb moisture very readily.

TRY YOUR HAND

You will need:
a flat dish with water
1 piece of household sponge made of cellulose, about 5 cm × 5 cm
1 piece of household sponge made of polyurethane, the same size and thickness
balance

What you do:
Weigh each sponge.
Press one sponge into the dish to absorb as much water as it can.
Carefully lift the soaked sponge onto the balance and record the amount of water it has absorbed.
Do the same with the other sponge.
Now squeeze the sponges very hard, to drain them of the last drop.
Weigh again.
To work out the amount of water held by the molecules of the sponge rather than by the holes, calculate:
(water held) = (squeezed weight) – (dry weight)
1   Which material soaked up more water initially? Why?
2   Which kept more after it was squeezed?
3   If you had to describe one sponge as *hydrophilic* (water-loving), and one as *hydrophobic* (water-hating), which would be which?

When you change your costume after a swim, and don't want to saturate everything else in your beach bag, you wrap it in a cotton towel. The towel absorbs and holds all the water till you get home and hang them both on the line.

Which dries faster on the line, the nylon swimming costume, or the cotton towel? Which holds the water longer? Which one would you call hydrophobic, and which hydrophilic?

TRY YOUR HAND

You will need:
2 pieces of cotton fabric, 10 cm × 10 cm
2 pieces of nylon fabric, 10 cm × 10 cm
water
balance

What you do:

Weigh each piece of fabric.

Wet, wring, and weigh one piece each of the cotton and nylon.

Now roll the wet cotton into the dry nylon, and the wet nylon into the dry cotton, as shown.

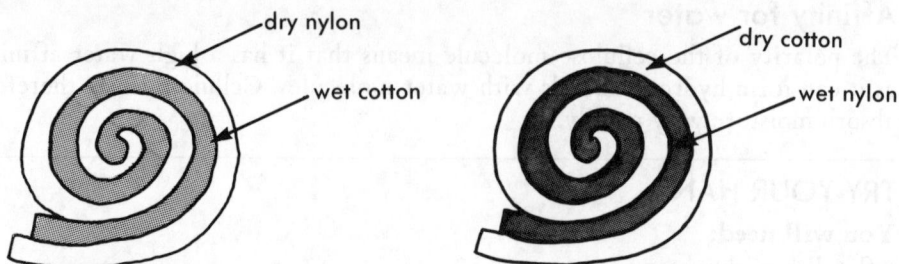

Squeeze, and leave for 5 minutes.

Weigh each of the four pieces.

1. Which fabric absorbed more water on wetting?
2. Which fabric used for wrapping absorbed more water from the piece of fabric it was wrapped around?
3. Which wet fabric lost more water to its wrapper?
4. Which material has a greater affinity for water: cotton or nylon?
5. How could you use this information in your use of textiles?

The organization of molecules

Cellulose materials absorb water readily, because of the polar nature of the molecules. But where does the absorbed water go?

Crystallinity

Cellulose fibres contain both crystalline and amorphous regions.

In the crystalline regions, the cellulose molecules are packed together very tightly. The close packing means that they can form many bonds with adjacent molecules. All this bonding means the crystalline regions are a source of *strength* to the fibre. However, because the molecules are packed so closely, there is no space available to accommodate any other substances – hence there is no absorption of water or any other matter into the crystalline regions. The bonds between the cellulose molecules keep the chains *rigid* and organized, and so the crystalline regions are a source of *stiffness* as well as strength in the fibre.

Amorphous regions

Where did the water go in the experiment? It could not fit into the crystalline regions. The only place it could go is into the *voids* in the amorphous regions. There it has the space between the loosely packed molecules, and it is able to find *anchoring stations*.

For absorption, both space and bonding – or anchoring – are necessary. The polyurethane foam had space, but no bonding sites. The nylon fabric had very little of either. Cotton has both space (amorphous regions) and bonding sites (the polar hydroxyl groups of the cellulose molecule). Hence it absorbs and retains water.

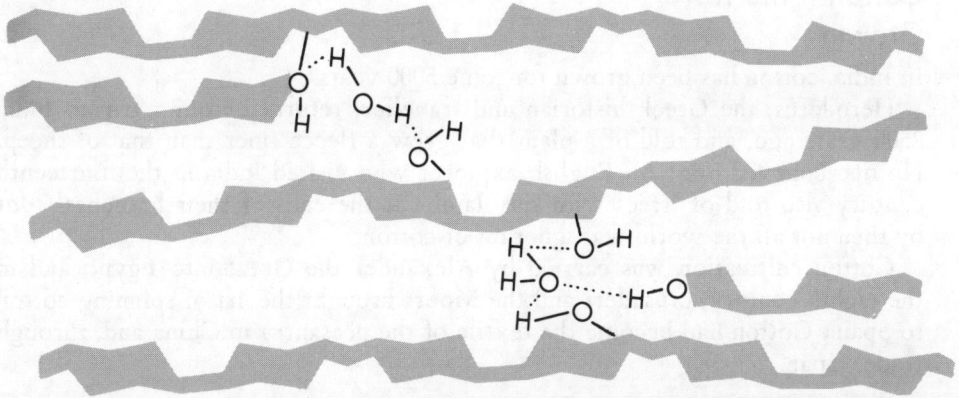

The only place the water can go is into the voids in the amorphous regions of the cellulosic fibre. There, it is able to hydrogen-bond with the hydroxyl groups of the cellulose.

Orientation

Because the molecules are not tightly packed in the amorphous regions, each molecule may move without disturbing its neighbours. This allows flexibility in the fibre. It also means that, if the fibre is to absorb another substance, such as water, then the cellulose molecules in the amorphous regions can move away from each other to accommodate more absorbed molecules. This means, of course, that the amorphous regions increase in size, and the whole fibre *swells*.

When cellulose fibres swell as they absorb water, their diameter grows and their length decreases. This is because the molecules and the crystallites are orientated more towards the axis of the fibre than across it. A randomly orientated fibre would swell equally in all directions.

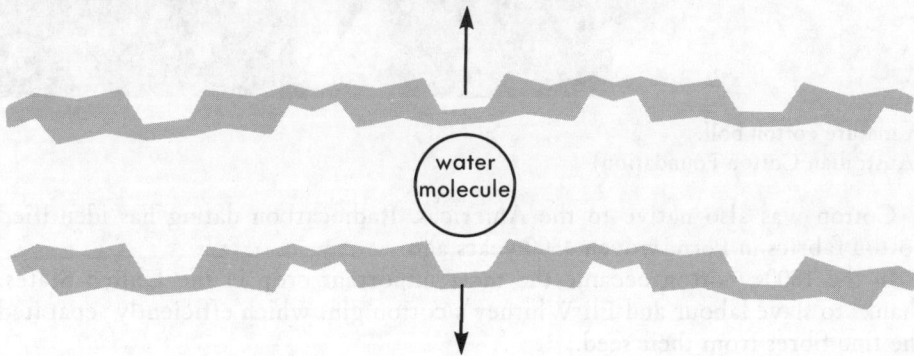

Cellulose chains move apart to accommodate the water molecules. In an orientated fibre, this results in increased width (swelling) but decreased length. In a randomly orientated fibre, swelling would occur in all directions.

Cotton

Cotton – the fibre
Growth

In India, cotton has been grown for some 5000 years.

Herodotus, the Greek historian and traveller, returned from a trip to India 2500 years ago, and told of a plant that grew a fleece finer than that of sheep. No one believed him! An English explorer who visited India in the fourteenth century also told of 'trees with tiny lambs at the ends of their branches'. But by then not all the world was ignorant of cotton.

Cotton cultivation was carried by Alexander the Great into Egypt, and in the eighth century, crusaders and the Moors brought the art of spinning cotton to Spain. Cotton had become the textile of the peasantry in China and, through trade, Japan.

A mature cotton boll.
(Australian Cotton Foundation)

Cotton was also native to the Americas. Radiocarbon dating has identified cotton fabrics in Peru, woven 4500 years ago.

In the 1800s, cotton became the most important crop in the United States, thanks to slave labour and Eli Whitney's cotton gin, which efficiently separated the fine fibres from their seed.

A cotton field ready for harvest looks like a field of low shrubs bursting into white blossom. But it is not flowers that bloom so white: the flower of the cotton plant is small and yellow and fades quickly. The whiteness is from the ripe 'fruit' pods which burst, ready to release their fibre-coated seeds to the wind. Like dandelions, the cotton plant relies on the wind for seed dispersal, and so each seed is covered with long, thin, fluffy fibres.

The more mature or ripe the fruit of the cotton plant, the more mature are the fibres that cover the seeds. Immature, thin fibres cause problems in processing, and are of little value to the grower. Good soil and plenty of water in the growing season help to produce a high quality cotton crop.

In order to ensure that only the most mature cotton is picked, with as little trash as possible, the cotton field is sprayed with defoliants before harvesting. Mechanical harvesters use vacuum suction to pull only the loosest seeds from the most mature bolls. The crop in any one field is therefore harvested over a number of days.

The mechanical harvesting of cotton.
(W. Pederson/Australian News and Information Bureau)

Before the invention of mechanical harvesters, the cotton crop had to be harvested by hand. This was an extremely labour-intensive process.

TRY YOUR HAND

Research the history of cotton and its connection with slavery. A good source of information is the *Encyclopedia of Textiles,* published by *American Fabrics Magazine.*

Morphology

Each cotton fibre which grows on the seed is a *single cell.*

It starts as a bump on the skin covering the seed, and grows rapidly to its full length. It already has its final diameter, but its wall, filled with sap, is still very thin.

The liquid sap in the central canal carries bunches of cellulose molecules from the manufacturing sites in the skin of the seed, and deposits them on the inside of the primary wall of the cotton fibre. The liquid flows in swirls within the long cell, and deposits the cellulose molecules in a spiral path.

The central canal of the cell is called the LUMEN. As the cell wall is built up, layer by spiralling layer, the lumen gradually becomes narrower. The layers are not laid down uniformly: the spirals turn clockwise, and then anticlockwise. They form from twenty to thirty daily concentric growth rings.

When the boll bursts open, the heat from the sun dries out the liquid in the lumen. The fibre collapses into a twisted ribbon shape. The more mature the fibre, the thicker the secondary wall, so mature fibres are almost rod-like: immature fibres are flat and quite twisted. Where the spirals in the secondary wall have reversed direction, convolutions appear in the fibre after collapse.

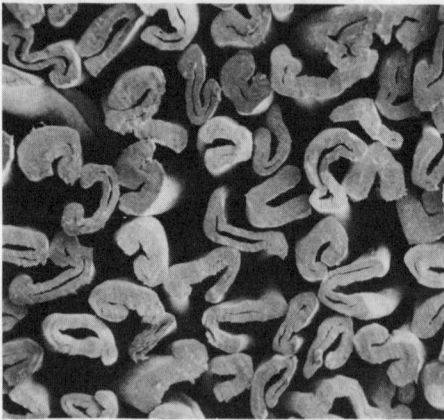

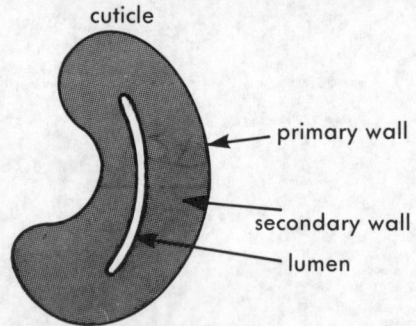

Cross-section of cotton fibres (electron micrograph).
(*Textiles*, a periodical published by Shirley Institute, Manchester, U.K.)

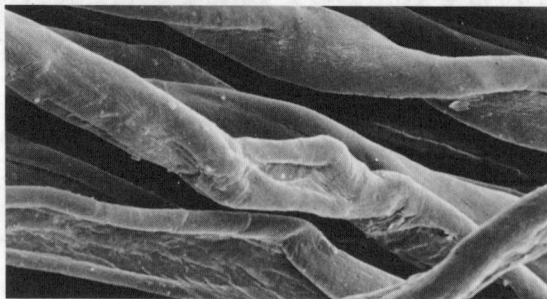

Cotton fibres (electron micrograph).
(*Textiles*, a periodical published by Shirley Institute, Manchester, U.K.)

Microstructure

The cellulose molecules are not deposited around the spiral one by one. Groups of cellulose molecules are manufactured by clusters of active granules on the membranes of the seed's cells. These groups of molecules are in small, long bundles called MICROFIBRILS.

In the microfibril, the many cellulose molecules are aligned parallel and are tightly packed. The microfibrils are laid down in their spiral path to form larger spiralling bundles called FIBRILS. Towards the outer surfaces of the rod-like microfibrils, there is some disorganization. Some molecules stray from the orderly array and form a beard-like fringe, with more spaces between the cellulose chains than in the organized centre.

The fringes of the microfibrils create amorphous regions within the organized fibril. There are also further amorphous regions created between the fibrils as they spiral around the cotton fibre.

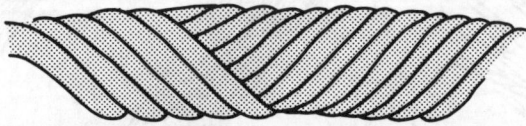

Reversal of fibrils in one layer of the secondary wall.

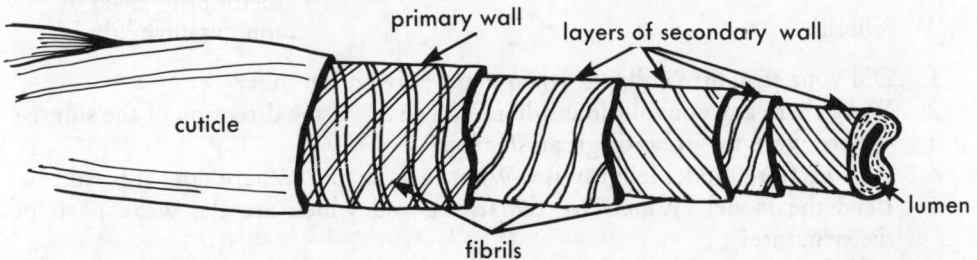

Microstructure of the cotton fibre.

TRY YOUR HAND

A model of the structure of the cotton fibres helps to explain its properties.

You will need:
knitting yarn
craft glue
scissors
plastic film wrap
cardboard tube from centre of roll of plastic wrap or foil, or similar

What you do:
Cover the tube with the plastic.
Cut the yarn into pieces 5–10 cm long. These represent the molecules.

Glue the pieces of yarn into bundles of 10 to 20, as shown in the illustration. These are the microfibrils.

When you have about 50 microfibrils, start to glue them in spiral formation around the tube. Reverse the twist in the spiral about half-way up the tube.

Continue building the microfibrils into fibrils along the fibre, as shown. As you add more and more microfibrils, the fibrils will cover the whole surface layer of the tube. You have just completed one day's growth ring around the cotton fibre.

The next day's growth would occur inside this layer. As this would be difficult to accomplish on the model, build a layer on the outside of the wall, using yarn of a different colour. Remember to vary the direction of your spiralling fibrils.

When the system is dry, remove the central cardboard tube. The space left in the centre is the lumen.

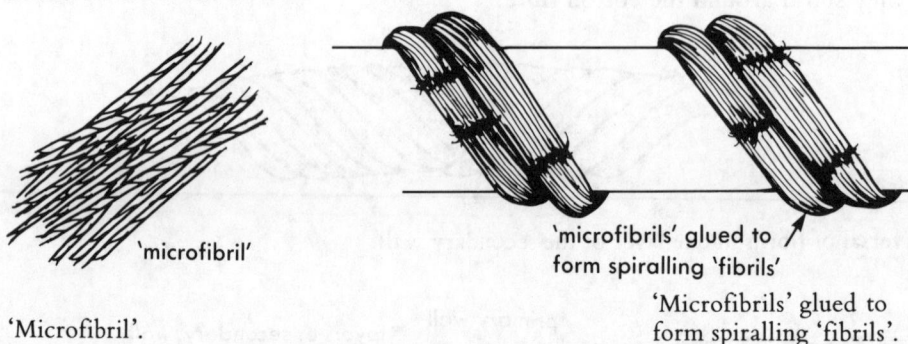

'microfibril'

'microfibrils' glued to
form spiralling 'fibrils'

'Microfibril'.

'Microfibrils' glued to
form spiralling 'fibrils'.

1   Did your structure collapse when you removed the tube?
2   Were there any convolutions where you changed the direction of the spirals?
3   Where were the areas of greatest stress?
4   Pull the fibre model lengthwise. What is its shape? Where does it break?
5   Bend the model. Which are the strong and which are the weak parts of the structure?
6   Which parts would you describe as crystalline, and which as amorphous?

In the model, the fibrils are held together with glue. In a real cotton fibre, there is space – or voids – between the fibrils. The fibrils are linked together by a few molecules which start in one fibril and finish in another.

The fibrillar structure gives strength and flexibility to the fibres. Other types of fibre may be of similar dimensions, but their molecules are not necessarily organized into fibrils.

Orientation

In the cotton fibre, the helical orientation of the fibrils allows for some extension before the fibre breaks.

Because the helix is effectively a spring, the cotton fibre has a small tendency to return to its original shape when the tension is released. This behaviour is called ELASTICITY. Some cellulosic fibres – those which have no helical ordering of the fibrils – are even less elastic than cotton. Other fibres have other, special properties which give them even greater elasticity. Elasticity and resilience (the recovery of a fibre from bending) can be improved by adding special finishes to the fibre. See p. 38.

Types and qualities of cotton

Genetic type

Four different species of cotton plant are grown around the world. Each has its own climatic and cultivational requirements, and each has its characteristic type of fibre.

The longest and finest cotton is called Sea Island cotton, and is grown in the Lesser Antilles island group in the Caribbean. Its fibre length averages 5 cm. The shortest and coarsest cotton is Indian cotton, with fibre lengths as low as 2 cm. The shorter fibre length means that Indian cotton is much more difficult to spin into fine, strong yarns. Egyptian and American cottons are of medium length and fineness.

Within each species, plant breeders are working to develop improved varieties of cotton plant.

Maturity

If the plant has not had time to build up the secondary wall of of the cotton fibre before harvest, the fibres will be immature. Immature fibres are weak, and they are troublesome during manufacture, as they create small tangles (NEPS) during processing. No crop will be of perfectly uniform maturity when it is harvested, but a large percentage of immature fibres will significantly lower the value of the crop.

Grading

Cotton is graded according to fibre length, colour (which can be from white through to light tan), amount of rubbish (TRASH), level of maturity, and strength.

The longest and whitest fibres are given the highest grade, when they are free from trash such as leaf particles, twigs, and sand.

Properties of the cotton fibre

Water affinity

Cotton is a *hydrophilic* fibre. This means that, like all cellulosic fibres, it has a strong affinity for water. Its excellent water absorbency makes it ideal for use in towels and surgical dressings. Because it really absorbs the moisture of perspiration, it is comfortable to wear in hot, humid weather and for active sportswear.

The effect of water

Water molecules are absorbed into the amorphous regions of the cotton fibre and are held there by hydrogen bonds. As the water molecules take up space, they push the cellulose chains apart. As the chains are pushed apart, the amorphous regions increase in size, and the fibre swells in diameter.

Since crystallinity adds to the strength of the fibre, and amorphousness adds to flexibility, this increase in the proportion of amorphous regions should logically decrease the strength and increase the pliability of the moisture-absorbent fibre.

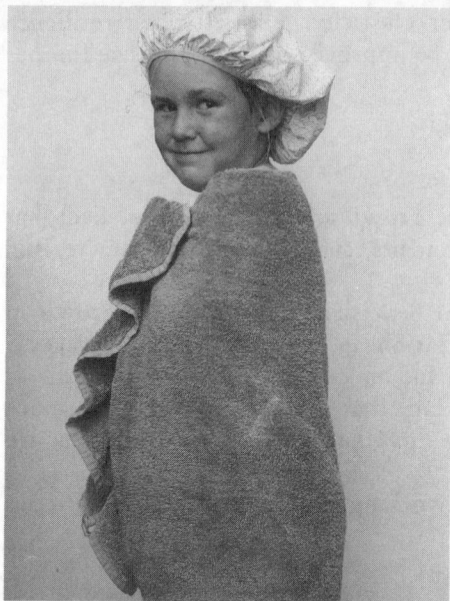

Cotton is an ideal fibre for towels and summer clothes.
(David Bailey)

The strength of the amorphous regions themselves should also decrease, since, as the strength of bonds decreases with increasing distance, the bonding between chains would be very much weakened in the swollen, amorphous regions.

For most fibres this is indeed the case. But cotton is very special.

When the liquid evaporates from the central lumen of the cotton fibre, the fibre collapses to form a long, flat, convoluted ribbon — rather like a sprinkler hose when the tap is turned off. This collapse creates stresses in the cotton fibre – the strength of the convoluted cotton fibre is rather less than the strength of the same fibre in the boll.

When the cotton fibre absorbs moisture, its walls swell, and it returns to something like the shape it had before it collapsed. This aligns the crystallites so *the wet cotton fibre becomes stronger overall*. In other words, wet cotton is stronger than dry cotton.

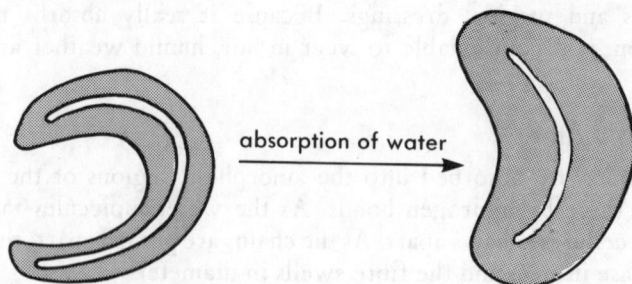

absorption of water

When water is absorbed, the cotton fibre swells. This swelling causes greater alignment of the crystalline regions in the microfibrils, and so increases the strength of the fibre.

34

The effects of chemicals on cotton

TRY YOUR HAND

You will need:

5 pieces of unbleached cotton calico fabric, each 5 cm × 5 cm
dropper
enamelled metal tray to hold fabric samples
microscope
dry-cleaning fluid
acetone or nail polish remover (this is mostly acetone)
household (hypochlorite) bleach (this is an oxidizing bleach)
cuprammonium hydroxide (prepared by adding aqueous ammonia to a solution of copper sulfate, to give a dark blue solution)
concentrated sulfuric acid (H_2SO_4)
concentrated sodium hydroxide (NaOH) (strong alkali)
CAUTION: Concentrated acids and alkalis are highly corrosive. Neutralize spilt acid with sodium bicarbonate. Clean up any spilt chemicals immediately. Chemicals spilt on skin must be washed off immediately, using plenty of cold, running water. Careful, sensible work avoids accidents.

What you do:

Place one drop of dry-cleaning fluid in the centre of a piece of calico.
Repeat for the other four chemicals, using a fresh piece of cloth for each.
Leave for 10 minutes.
Observe and record what happens.

1   Can cotton be dry-cleaned safely?
2   What was the effect of the bleach?
3   What happened with the cuprammonium hydroxide?
4   What did the acid do? Allow the sample to dry. Is the cotton recovered?
5   Did the alkali act in the same way as the acid?
6   Take some fibres treated with alkali and examine them under the microscope at 400× magnification. Compare them with some untreated fibres. Can you see any difference? The effects of alkali are employed in the commercially important process of MERCERIZATION.

What caused the holes in the lab coat?

35

Both the cuprammonium hydroxide and the acid appeared to dissolve the cotton. Look up a chemistry book to find out the formula for the reaction of the acid with the cellulose. Why do you think this is called hydrolysis?

Cuprammonium hydroxide is the only chemical which can dissolve cotton without significantly breaking up the molecules. It acts by breaking all the hydrogen bonds and allowing the chains to separate from each other completely.

TRY YOUR HAND

Dissolve some cotton fibres in a little cuprammonium hydroxide.
Place the solution in a shallow evaporating dish, and allow the solvent to evaporate.
Did you find solid cellulose precipitating out?

Modifications to the properties of the cotton fibre

Mercerization

In 1844, John Mercer discovered a treatment for cotton which was to make his name a household word. He soaked cotton fabric in 17% sodium hydroxide (NaOH; caustic soda) solution, and found that it increased in strength and also in its ability to accommodate dye molecules. Since he was by profession a calico dyer, he was pleased to discover that mercerized cotton needed 30 per cent less dye than unmercerized cotton to achieve a given depth of shade. Later it was found that cotton mercerized under tension was more lustrous than unmercerized cotton, or cotton mercerized without tension.

The saving in dyestuffs is still an important reason for mercerizing cotton goods. Because of the increased strength and lustre imparted by this process, all sewing threads, and most top-quality cotton fabrics, are mercerized.

TRY YOUR HAND

You will need:
cotton yarns removed from a 10 cm long piece of calico or similar cotton fabric
piece of cotton fabric
25% sodium hydroxide (25 g NaOH made up to 100 mL with water). This will be very viscous.
accurate ruler
dropper
CAUTION: Clean up any spills immediately. If any caustic soda (NaOH) is spilt on the skin, wash it off immediately with plenty of cold, running water.

What you do:
Measure the length of the cotton yarns accurately.
Immerse them in the mercerizing solution.
Observe them for 20 seconds, then remove, rinse, and dry without stretching.
Measure their length again. Record your observations.

Using the dropper, apply some parallel streaks of mercerizing hydroxide to the piece of cotton fabric.

1 What was the effect of the hydroxide on the loose threads?
2 How does the reaction of the treated fabric relate to your earlier observations?
3 Does the treated fabric sample remind you of a well-known type of cotton fabric? What is its name? How do you think it is produced? Can you explain the effect?
4 Vary this technique to produce a three-dimensional decorative design on a cotton textile.

The effect of alkali on cotton

The sodium hydroxide *swells* the cotton fibre by altering the arrangement of the molecules in the crystalline regions. As the cellulose chains in the crystallites rotate, they expose more hydroxyl (–OH) groups to the amorphous fringes. More amorphous regions are created.

This has the following results:

(i) The cotton fibre swells, thereby regaining some of its original shape. This relieves some of the stresses that arose when the mature fibre dried out and collapsed, so the swollen fibre is stronger.

(ii) Because the fibre has swollen and has more amorphous regions, it becomes more absorbent – hence it is more efficient in accommodating dye molecules.

(iii) The cylindrical shape changes the normally dull, flat, convoluted cotton into a more lustrous fibre.

(iv) Unless mercerization is carried out under tension, the cotton fibre shrinks considerably in length. This property is utilized in the production of seersucker and plissé effects in cotton fabrics.

Cotton drill, showing some mercerized (cylindrical) and some unmercerized (convoluted) fibres.
(Vivian Robinson)

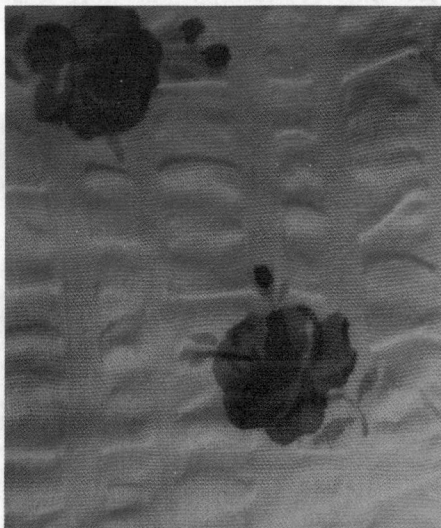

A seersucker made by printing cotton fabric with alkali.
(David Bailey)

Drip-dry

The growing popularity of easy-care synthetic fibres in the second half of this century has encouraged consumers to expect miracles from their textiles.

Garments are expected to be: wrinkle resistant; easy to clean; quick-drying – without creases; ready to wear after laundry – without ironing; comfortable to wear; and, long-lasting.

All these properties were difficult to achieve together in the one garment. DRIP-DRY shirts made of cotton fabric were first produced in the 1960s. (Drip-dry refers to garments which, after laundering, will dry quickly and regain their original shape without ironing.) This behaviour was originally achieved by treating the cotton fabric with formaldehyde.

Formaldehyde has the disadvantages that it is poisonous, and that the cross-linkages tend to break down during storage and use. Today, other, resin-forming, molecules are used, which give fabrics that are practically free of the poisonous formaldehyde.

The pre-resin molecules enter the amorphous regions of the cotton fibre. On heating, they form CROSS-LINKS with other resin molecules, and also link to the hydroxyl groups of the cellulose molecules.

The resin acts as stiffening struts between the polymers of cellulose. When the polymers are pulled apart, the resin resists the force. Hence, wrinkles are less likely to form. If the fibre is bent, the resin acts as a spring, pulling the structure back to its original form. Therefore, resin-treated cotton has both *wrinkle resistance* and good *recovery* from crushing.

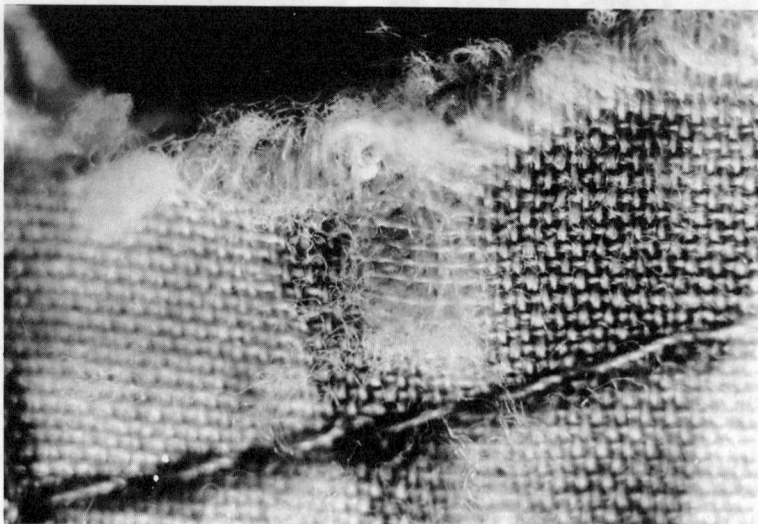

Abrasion damage on a shirt collar.

Cotton fibres treated with cross-linking resin are *less moisture absorbent,* because many of the hydroxyl groups in the cellulose are taken up by the cross-linking resin. The drip-dry effect is thus a combination of *wrinkle recovery* and *fast-drying* properties.

Resin-treated fibres are *less pliable,* because their amorphous spaces are full of resin 'struts'. An additional effect of this stiffening is that the fibre is *more brittle* – more easily affected by anything rubbing against it. This means that drip-dry cottons have *lowered abrasion resistance.*

The points of shirt collars and the edges of the cuffs used to fray and wear out so rapidly that pure cotton drip-dry shirts are no longer marketed. They have been replaced by blended polyester/cotton shirts, as the blended fabric has better abrasion resistance and crease resistance than pure cotton. To improve consumer acceptance further, poly/cotton blends are sometimes also treated with small amounts of resin.

The fragile resin

When the resin is applied to the fabric, it forms cross-links during a heating process called CURING. If not enough heat is used, the cross-links do not form. Too much heat can break the cross-links.

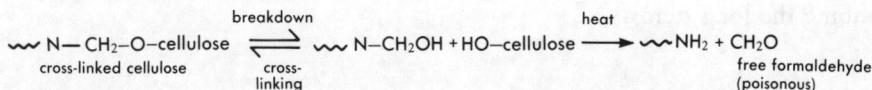

$$\sim\!\!\!\sim N-CH_2-O-cellulose \underset{\text{cross-linking}}{\overset{\text{breakdown}}{\rightleftharpoons}} \sim\!\!\!\sim N-CH_2OH + HO-cellulose \xrightarrow{\text{heat}} \sim\!\!\!\sim NH_2 + CH_2O$$

cross-linked cellulose · free formaldehyde (poisonous)

The breakdown of resin due to excessive heat. The first step is the reverse of the cross-linking reaction; then the resin breaks down in a further, irreversible, step to release formaldehyde.

Steam ironing and hot-water washing combine to break down the resin in a drip-dry garment. *So, the more an easy-care garment is ironed, the less easy-care it becomes.*

Recent solutions to the problem of combining easy-care performance with comfort and durability involve the use of blends of fibres and combinations of finishing processes.

TRY YOUR HAND

You are a clothing manufacturer, and wish to produce an all-cotton dress. This must not give rise to consumer complaints because of either its crushability or its lack of durability. You are given the following information:

a   Cross-linking resins give good wrinkle resistance and crease recovery, but reduce the softness, strength, durability, and moisture absorption of the cotton fibre. Tendency to stain is increased after resin treatment.

b   Mercerization increases the strength, lustre, and moisture absorbency of cotton.

You decide to combine both treatments for a perfect balance of properties.

1   What are the effects of your decision on the consumer?

2   Which process would you carry out first? Why?

3   Design a care label for a dress made from this fabric.

Flax

Growth and morphology

Fibres of flax are obtained from the *stem* of the flax plant, *Linum usitatissimum*: it is a BAST (plant stem) fibre. The long, strong fibres have been spun and woven in Europe since the Stone Age, and it was still the most widely used fibre in Europe until the introduction of cotton from the Middle East only three hundred years ago. Although today it is no longer the most important of the bast fibres, it is still cultivated in many parts of the world.

Long fibres are the most useful for textile purposes. For this reason, the plants are grown close together, so the stalks grow thin and tall. At harvest time, the plants are pulled out roots and all, so as not to cut the useful fibres short. The flax plants are dried in the field, the seeds removed, and the straw prepared for the first stages in production.

Flax fibres are found together in *bundles* around the woody stem of the plant. The woody portion of the stem is first rotted away by soaking it in still water (a process known as RETTING), or by treatment with chemicals. The retting must be carried out with care, or the fibres may be damaged. After retting, the softened woody parts are removed by beating, scraping, and combing the long stems.

Flax plants spread for retting, Buln Buln, 1918.
(Department of Agriculture and Rural Affairs, Victoria)

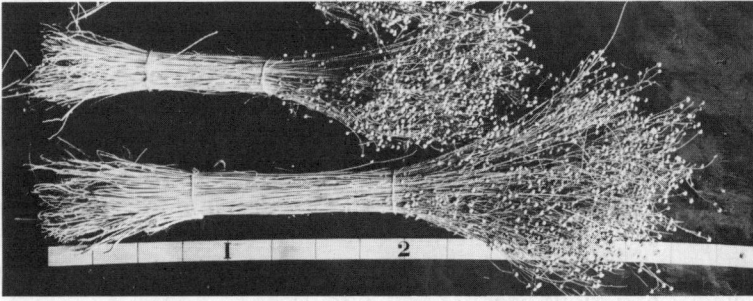

Mature flax plants. The scale is in feet (1 foot = 30 cm approx.).
(Department of Agriculture and Rural Affairs, Victoria)

Scutching flax. The dried stems are crumbled and crushed to release the fibres from the woody stalks. Victoria, about 1920.
(Department of Agriculture and Rural Affairs, Victoria)

Hackling. The scutched fibres are fed through a series of successively finer combs to yield long, regular, parallel fibres. Victoria, about 1920.
(Department of Agriculture and Rural Affairs, Victoria)

Initially, the bundles of flax are fairly thick. If they are spun into yarn at this stage, they give fabric with a characteristic 'homespun' linen appearance. The more often the combing is repeated, the more finely divided the bundles become. The finer the bundles are divided, the finer is the yarn that can be spun from them.

The initial bundles of flax fibres can be up to a metre long. That is why people from the earliest times found them easy to spin into yarns. The ultimate fibrils of flax are less than a centimetre long, and are as fine as the finest cotton fibres. In practice, however, flax is never separated into its ultimate fibrils: bundles of various sizes are spun into yarns.

Properties of flax

Flax consists of bundles of crystalline fibrils held together by an amorphous, absorbent, cellulose-like, complex matrix. As in cotton, the fibrils in each fibre are in a spiral arrangement, with the spirals changing direction at intervals.

The strength and stiffness of the highly crystalline fibrils, the extreme length of the cellulose polymers, and the moisture affinity of the matrix combine to give flax a unique set of properties.

The crystalline fibrils give *strength*, a *stiff* drape (the fabric is not very pliable), and a fairly *high specific gravity* (the fabric is fairly heavy for its thickness). Since flax is often spun into tight, compact yarns, linen fabrics usually are quite *crisp* and have a *heavy drape* for this reason as well. A linen tea-towel feels quite different from a cotton one, and even the finest linen handkerchief has a stiff 'body'.

The lack of flexibility and resilience of the crystalline flax fibre means that pure linen *creases easily,* and is fairly difficult to iron out again. However, the crystallinity means that *high ironing temperatures* can be used, as the cellulose molecules are held firmly together in the crystalline regions, and so are less likely to be disorganized and broken down by the heat. Some fashions have exploited the tendency of linen to crease.

The *amorphous,* cellulose-like *matrix* which holds the fibrils in their bundles *absorbs moisture* very readily. This accounts for the comfort of linen clothes and the absorbency of linen tea-towels.

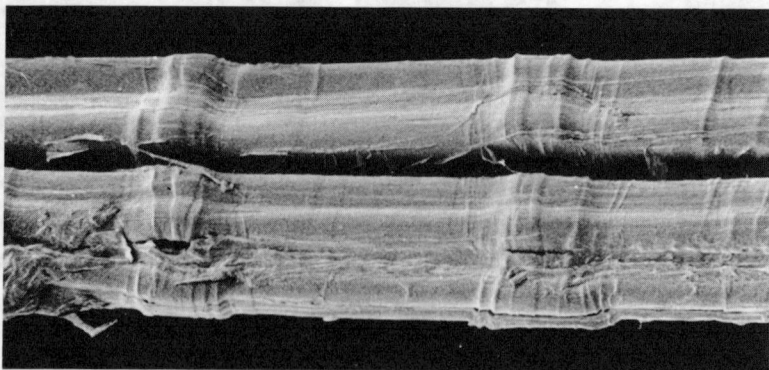

Flax fibres (electron micrograph). The nodes are characteristic of flax.
(*Textiles*, a periodical published by Shirley Institute, Manchester, U.K.)

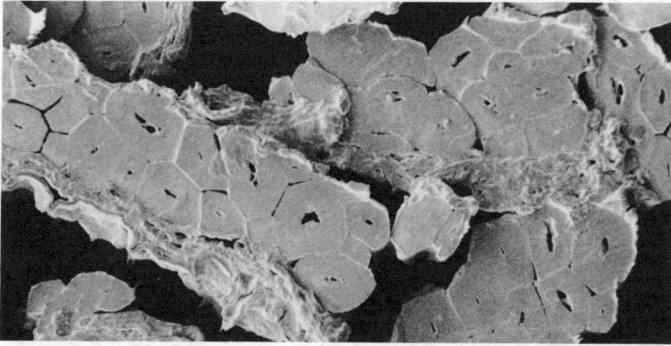

Cross-section of flax fibres (electron micrograph). Individual fibres are held in bundles by a gummy cellulosic matrix. Although flax can be separated into these individual fibres, in practice this is not done.
(*Textiles,* a periodical published by Shirley Institute, Manchester, U.K.)

Strength, due to the crystallinity, polymer length, and the effective holding power of the matrix, allows linen to *withstand repeated laundering* with minimum shedding of fluff and lint. This explains its popularity for table linen and tea-towels.

As in cotton, the spiralling of the fibrils in the fibres means that when flax expands with moisture, it becomes stronger. The increased *wet strength* of flax, as well as the ease with which it can be spun into yarn, made it ideal for one of its traditional uses: in sails.

World trade once depended on sailing fleets, and hence on linen. Historian Geoffrey Blainey has suggested that access to native New Zealand and Norfolk Island flax was one motive for the British colonization of Australia and New Zealand.
(Port of Melbourne Authority)

Linen in the market-place

Traditionally, linen has been a fabric of high status, because of its durability and the comfort of linen clothes. However, today's drip-dry generation of consumers is not always ready to accept the creasing and heavy drape that go with pure linen fabrics. Flax, because of its crystallinity, does not accept easy-care resins very readily. Therefore, linen-like fabrics are often produced from various blends, and some may not contain any flax at all!

TRY YOUR HAND

Go into a fabric shop, and ask to see the 'linens' section. Carefully check the labels on each roll, test each fabric by crushing it in your fist for 5 seconds. Draw up a table of your results, comparing the fabrics for fibre composition, cost, appearance, and crushability/resilience.

1 Could you now tell a pure linen fabric from an imitation one, even if it has no label?
2 Check your new-found skill by examining fabrics used in the clothes in a ready-to-wear shop.
3 Design a garment which uses pure linen to the best advantage, specifying the type of linen fabric which you would use for it.
4 How would you make sure that consumers ignorant of linen imitations become aware of what fabric they are really buying?

Other natural cellulosic fibres

The fibres

Cellulosic fibres which are obtained from the stems of plants are known as *bast fibres*. From its use in clothing, flax is the best-known bast fibre, but

Jute is widely used for sacking, as in sandbags for flood retention and military defence. (Army Public Relations)

the bast fibre produced in the greatest quantity is *jute.*

Most bast fibres are used for cordage or sacking, or for industrial purposes. Hessian and burlap are made from jute.

Other plant fibres include the very coarse *coir,* from coconut husks, which is used to make doormats, and *sisal,* from the leaves of the agave plant.

The annual production of cotton is three times as much as that of all these fibres put together. These natural fibres have to compete with the synthetics for the cordage and industrial market.

Producer countries, such as India, the Soviet Union, and the Philippines, have very active research organizations which work towards keeping natural fibres competitive in the market-place. They aim to improve methods of cultivation and processing, as well as to find ways to reduce costs of production and to expand the areas of use for these fibres.

A comparison of properties

Since all these plant fibres are 70–80 per cent cellulose, and contain similar gums, waxes and pectin (a cellulosic gum), they are chemically the same. Their properties differ mainly by virtue of the differing dimensions and crystallinity of their components.

The following illustrations compare flax and sisal.

Although the fibre *bundles* have the same overall diameter, it appears the fibres making up these bundles are fine in flax, but thick and coarse in sisal. The comparative fineness of the flax means that it is much more supple and flexible. Flax and sisal have similar TENSILE STRENGTH – it takes the same lengthwise pull to break them – but the greater rigidity of sisal makes it far more brittle – it is less abrasion resistant.

This difference in properties shows why flax is suitable for clothing textile applications, whilst sisal is used mainly for cordage.

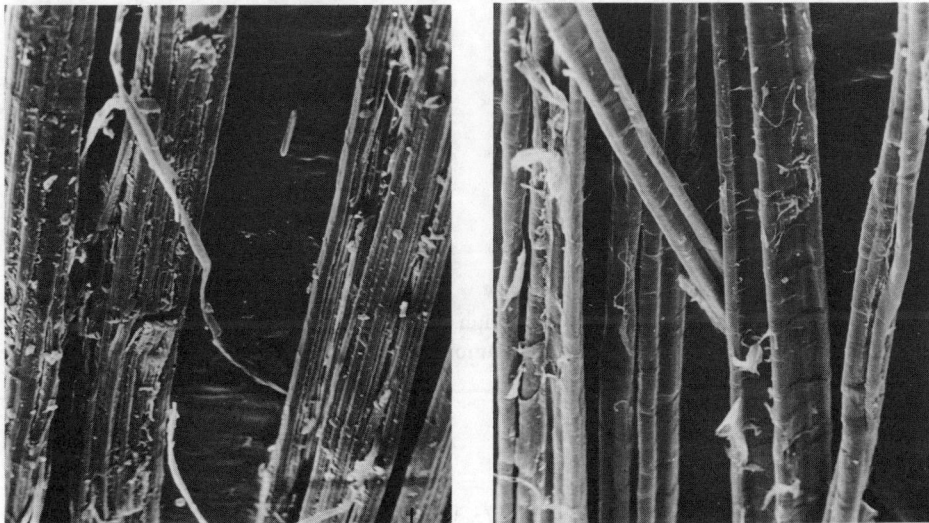

Flax (left) and sisal (right) at the same magnification.
(Vivian Robinson)

Source and uses of plant fibres

Fibre	Type	World production (1978) (million kg)	Major producers	Use and properties
jute	bast	2700	India Bangladesh China Brazil	burlap, hessian bagging cloth carpet backing rope and twine
flax	bast	675	Soviet Union	linen (clothing, household linen)
sisal, henequen	leaf (agave)	480	Tanzania Brazil Mexico	ropes, twine roof insulation backing cloths
hemp	bast	230	India Soviet Union	cordage linen-type weaves
abaca	leaf	90	Philippines	world's most desirable natural cordage – soft and strong
pina	leaf (pineapple)	small, but increasing	Philippines	fine, lustrous clothing fabrics
coir	fruit (coconut)		Sri Lanka India	doormats upholstery stuffing (it is stiff and brittle)

Maize (left) and rush carpeting (right): tough cellulosic floor coverings.
(The Natural Floor Covering Centre, Stanmore, N.S.W.)

The rayons

Regeneration of cellulose

At the end of the nineteenth century, many researchers worked on the idea of taking cellulose and converting it to long, silk-like FILAMENTS. After all, silkworms could do it.

Comte Hilaire de Chardonnet treated cotton with a mixture of nitric and sulfuric acids, converting it to the highly flammable nitrocellulose. Further chemical treatment gave regenerated cellulose in filament form. De Chardonnet exhibited his first 'artificial silk' cloth at the great Paris Exposition of 1889.

In 1883, Sir Joseph Swan had patented a method of making carbon filaments for incandescent lamps, which involved forcing a concentrated cellulose solution through tiny holes (orifices) into a coagulating bath. Chemicals in the bath reacted with the nitrocellulose to give regenerated cellulose filaments. De Chardonnet adapted this method to obtain his artificial silk (NITRO SILK).

Cuprammonium rayon (cupro)

De Chardonnet's method of regenerating cellulose worked, but the highly flammable nitrocellulose was difficult to handle, and the process was expensive. Once the regeneration of cellulose had been shown to be possible, the race was on for chemists to find a cheaper and less troublesome process.

When you were exploring the effects of chemicals on cotton (p. 35), you placed cotton fibres in cuprammonium hydroxide. The cotton fibres dissolved to give a clear blue solution. When the liquid evaporated, the cellulose became visible again as a formless, white deposit on the bottom of the evaporating dish. The molecules of cellulose were not destroyed, they were merely reorganized.

For the cellulose to be useful as a textile fibre, the cuprammonium hydroxide solution of cellulose must be extruded into a coagulating bath so the cellulose can be regenerated as filaments. The difficulty is in getting the solution of cellulose to just the right consistency, so it is liquid enough to go through the holes, and viscous enough to form a good strong filament.

TRY YOUR HAND

You will need:
5 labelled test-tubes, each containing 10 mL cuprammonium hydroxide
5 bunches of cotton fibres, ranging in mass from 0.2 to 2 g
disposable syringes
5 beakers, each containing 200 mL of water and 1 mL sulfuric acid (H_2SO_4)

What you do:
Dissolve a bundle of cotton fibres in each test-tube of cuprammonium hydroxide. Stir to dissolve.
Using the disposable syringes, inject each solution into a separate beaker of dilute acid.
1   Did you encounter any problems in dissolving the cotton?
2   Did you encounter any problems in pushing the solutions through the syringe?
3   Did you have any problems in forming a thread?
4   Which solution formed the strongest thread?
Note: In order to help dissolve the cotton, you could boil it in a 10% solution of alkali (10 g NaOH dissolved in 100 mL water), before you add the fibres to the cuprammonium hydroxide. As you know, the alkali does not destroy the cellulose, but it helps to move the cellulose chains about. The rearranged chains dissolve more readily.

The type of rayon produced in this experiment is called CUPRAMMONIUM RAYON, or CUPRO. It was first produced commercially by J. P. Bemberg Ltd in 1901, using wood pulp. Cuprammonium rayon has since been developed into an excellent, soft, strong fibre. Today the production of Bemberg rayon is about 100 million kg annually, mainly in Japan, the United States, and the Federal Republic of Germany (West Germany).

Viscose rayon: the process

Production of nitro silk and cupro was covered by patents, so their manufacture was controlled by the patent holders. Other large chemical firms also wished to compete on the potentially lucrative artificial silk market. As researchers experimented with a variety of treatments to dissolve and regenerate cellulose, they arrived at a complex set of procedures, which – amazingly – worked. They called the successful fibre VISCOSE RAYON.

The spinning solution may have titanium dioxide (TiO_2) added as a DE-LUSTRANT, to *reduce the shine* of the rayon filaments. Continuous rayon filaments are used to imitate silk. The filaments may also be *cut* to give staple fibres of various lengths. Long staple viscose imitates wool fabrics, and short staple copies natural cotton fibres.

Viscose rayon is *weak,* very *absorbent, easily distorted,* and highly *crushable.* In these ways, it does not resemble any of the natural fibres it seeks to imitate. These undesirable properties arise because, during production, the long cellulose molecules are broken into shorter lengths, and reorganized into a much more amorphous fibre.

Spinning viscose.
(*Bayer Farben Revue* No. 15, 1969, by permission of the editors. *Bayer Farben Revue* is published by Bayer AG, Leverkusen, West Germany.)

The production process for viscose rayon

The process	The effect on the cellulose molecules
Wood chips are purified by boiling and bleaching, and are pressed into sheets of wood pulp.	Some of the lignin and other non-cellulosic materials in the wood decompose, and are washed away. Coloured material in the wood is bleached and the pulp is left as thick sheets of white cellulose.
Sheets of pulp are steeped in 17% (mercerizing strength) alkali (NaOH). They swell.	The cellulose molecules are converted into soda cellulose (a few hydroxyl (–OH) groups are converted to –ONa), and the cellulose chains move further apart. This creates more space in the wood fibrils, thus swelling the sheets of pulp.
The alkali cellulose is shredded into crumbs, and allowed to age. This is needed to help the solubility later on.	Some oxidation occurs during ageing, and some of the cellulose chains break into shorter lengths – the average number of glucose units in the cellulose chain is reduced from 1000 to 350.
The crumbs are churned with a carefully measured quantity of carbon disulfide (CS_2) to produce cellulose xanthate.	The –ONa of soda cellulose is converted to –OCS.SNa, cellulose xanthate. The cellulose xanthate is soluble in alkali.
The orange cellulose xanthate crumbs are dissolved in weak alkali to form the spinning solution.	Bonding between molecular chains of cellulose xanthate is weaker than the bonding between molecules of pure cellulose. In alkaline solution, the xanthate molecules form a complex with (are chemically attracted to) the hydroxide (OH^-) ions. They are thus able to dissolve in the aqueous solution. (This process of chemical attraction, in which several particles – molecules or ions – form a single, complex unit in solution, is called *solvation*.)
The spinning solution is allowed to ripen for 4 to 5 days. At first, the solution is very thick. Then it becomes thinner, and later, more viscous again.	Some of the long molecular chains of cellulose xanthate are broken down further – the final chain length is about 270 glucose units. As the chains become shorter, they become more mobile – hence the solution becomes thinner, more runny. The increase in viscosity occurs because some of the xanthate –OCS.SNa groups are detached from their positions on the glucose rings, and so the chains become less soluble. The viscous solution is now ready to spin.
Extrusion. The viscous spinning solution is extruded under constant pressure through a filter to the round holes of a thimble-shaped spinneret. The holes are usually between 0.05 and 0.1 mm in diameter, and the spinneret is made of precious metal which will not be affected by the corrosive coagulating solution.	As the cellulose molecules are pushed through the spinneret, those nearest the surface of the hole are attracted to the metal. This attraction creates a shear force, which straightens out the tangled molecules as they are pushed through the hole. Those molecules which are straightest end up on the outside of the filament. The centre of the filament consists of molecules which are still quite randomly arranged.

The process	The effect on the cellulose molecules
Coagulation. The filaments are drawn into an acid coagulating bath, the composition of which varies slightly from manufacturer to manufacturer. A typical recipe for the bath is: 10% H_2SO_4 (sulfuric acid); 18% Na_2SO_4 (sodium sulfate); 1% $ZnSO_4$ (zinc sulfate); 2% glucose; 69% water (H_2O).	When the acid of the coagulating bath neutralizes the alkaline sodium hydroxide, the –OCS.SNa of the cellulose xanthate is decomposed to give –OH once more: pure cellulose is regenerated. The first molecules to be regenerated are the straightened molecules on the outside of the filament. These harden into a well-ordered arrangement, to form a skin. The skin slows down the neutralization process for the molecules in the centre of the filament. Eventually these inner molecules harden (or, precipitate) in a disorganized, amorphous form. Thus viscose rayon has a crystalline skin and an amorphous core.
Stretching. During their passage from the spinneret to the take-up rollers, the filaments are under constant tension. They are stretched to about three times their original length.	The stretching orientates the molecules, making them lie more parallel to the direction of the filament axis. At the same time as the filaments become longer, they become thinner as well. The skin of the filaments is quite hard, however, so most of this thinning down is at the expense of the core. This leaves a skin which is too loose for its insides, and so it folds into crenellations. Viewed from the outside, these crenellations appear as stripes or striations along the length of the fibre.
Washing. Some manufacturers use sodium sulfide, Na_2S, to remove all traces of sulfur, and hypochlorite to bleach the filaments. Others wash the filaments thoroughly with water, which removes most of the acid and sulfur from the regenerated cellulose. After washing, the filaments are dried and wound onto bobbins.	All traces of the sodium salts and neutralizing acid are removed, along with any remaining carbon disulfide, CS_2, still lodged between the cellulose molecules. While the filaments are full of water they are swollen, and the cellulose chains are relatively far apart from each other in the amorphous centre. The wet viscose rayon fibre is quite weak. As the filaments are dried and the water molecules are driven out, the chains can move closer again, and link up with hydrogen bonds. The fibre thus increases in strength as it dries.
Special rayons, such as Avril rayon, receive resin cross-linking as an after-treatment, in order to increase their strength, resilience, and crease recovery.	The cross-links form *covalent* bonds which join the cellulose chains. They are not broken by water. Their presence means that the treated fibre resists swelling when wet, and quickly returns to its original form after any distortion.

Viscose rayon is made from cellulose obtained from pulped wood. The cellulose is treated with alkali and carbon disulfide (CS_2), and dissolved in alkali to give a viscous spinning solution. This solution is extruded through a spinneret into an acid coagulating bath, stretched, washed, and dried, to give a soft-cored, hard-skinned, crenellated fibre.

As a result of the ageing and ripening, the cellulose chains in viscose rayon are broken and degraded. The polymers are much shorter than those in cotton, flax, and other natural cellulose fibres, and they are not organized into fibrils. As a result, viscose rayon is weaker than cotton.

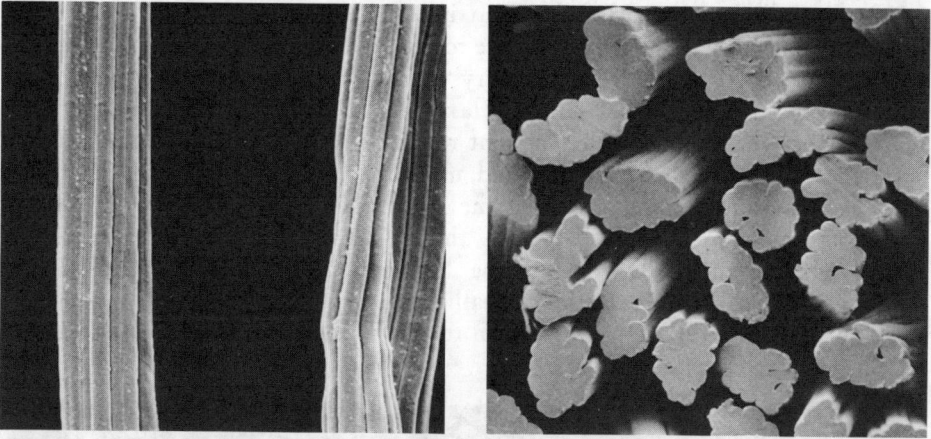

Viscose rayon filament and cross-section (electron micrographs).
(*Textiles*, a periodical published by Shirley Institute, Manchester, U.K.)

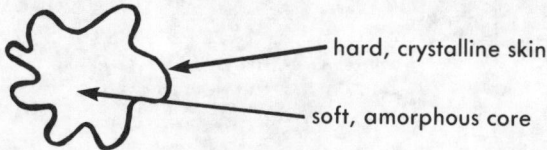

hard, crystalline skin

soft, amorphous core

Cross-section of viscose (schematic).

The shorter polymers reduce the crystallinity, as there are more 'free ends' to disrupt the structure. Fewer cellulose chains are long enough to stretch from one crystalline region to the next. The reduced crystallinity contributes to weakness. The shorter molecular chains also mean that each chain has fewer hydrogen bonds along its length holding it to its neighbours, so the fibres are more easily pulled apart. Length of molecular chains is quoted as DEGREE OF POLYMERIZATION, D.P., the average number of monomer (cellobiose (two linked glucose units) in the cellulosics) units repeated in the chain.

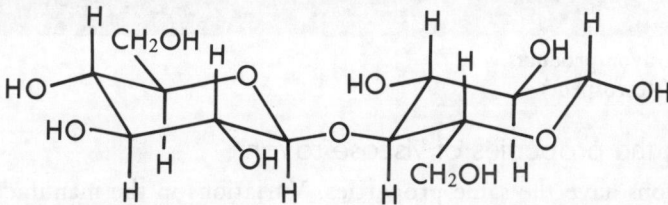

Cellobiose: the cellulose monomer.

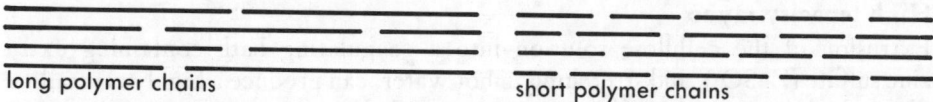

long polymer chains   short polymer chains

Short polymer chains create more voids and disruptions in the crystallites than do long chains, hence fibres composed of short chains tend to be fairly amorphous. As each chain is short, it is not able to form as many bonds to its neighbours as a long chain can, hence fibres containing many short chains tend to be weak.

51

The crystalline regions that do occur in viscose rayon are not organized into fibrils and microfibrils. Some crystalline regions occur in the relatively amorphous core, many more in the relatively crystalline skin. The fibrillar structure is a source of strength in cotton and flax, and its absence means that viscose rayon is the weaker. It also means that rayon *loses strength* on wetting. When a fibrillar fibre swells with absorbed moisture, the crystallites in the fibrils are able to align themselves along the fibre axis, giving increased strength. In rayon, the absorbed water pushes the cellulose chains apart to swell the fibre, reducing the hydrogen-bonding between the chains, but without the strengthening realignment of the crystallites.

Some viscoses are after-treated with resins. This treatment increases the wet strength and the resilience of the fibres.

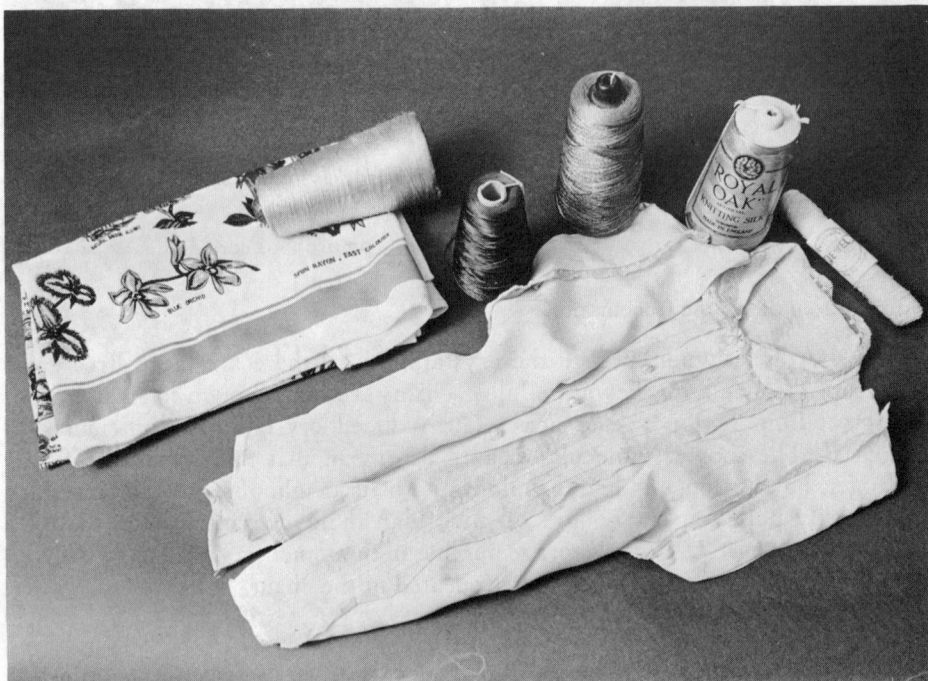

A variety of rayon products.
(Dale Mann/Retrospect)

Changing the properties of viscose rayon

Not all rayons have the same properties. Variations in the manufacturing steps can create remarkable differences in the properties of the resultant fibre.

High tenacity rayon

Extrusion of the cellulose solution into a coagulating bath containing extra zinc sulfate ($ZnSO_4$), and stretching in hot water, can produce a highly crystalline fibre which is strong enough to be used as the belting material for car tyres.

This process allows a much slower solidification. As the fibre remains softer for longer, the molecules have a better chance to become aligned and orientated during stretching. Crystallization occurs uniformly, and so the fibre does not

52

have a distinct skin and core – it is all skin, crystalline throughout, and therefore has a smooth cylindrical shape.

High tenacity rayon has properties typical of a fibre with very few amorphous regions. It is strong along its length (high TENACITY means it can withstand *lengthwise pulls*), but brittle, because it is made rigid by the crystallinity. It does not absorb much water or dye, and does not change strength very much on wetting. It is not suited for apparel purposes.

Polynosic (high wet modulus) rayon

Viscose rayon is highly amorphous, and therefore weak. High tenacity rayon is strong because it is crystalline throughout, but it is also very stiff and non-absorbent.

Researchers have worked towards producing a cellulosic fibre which is stronger than ordinary viscose rayon, but not as rigid as the high tenacity rayon used for tyre cords.

The first step towards higher fibre strength was to prevent the breakdown in length of the cellulose molecules in their passage from wood pulp to spinning solution. This was achieved by eliminating ageing and ripening, and using lower temperatures at all stages of processing. This decreases the solubility of the cellulose, so more carbon disulfide (CS_2) has to be added during xanthation, to make more xanthated sites on each molecule. This increases costs of production.

The combination of strength and flexibility in the natural fibres is due to their fibrillar structure. The organization of the crystallites into microfibrils makes their strength more effective, and the spaces between the microfibrils allow for movement and for the absorption of other molecules.

A fibrillar structure in rayon is achieved firstly by using a less strongly acid coagulating bath. This slows down the process of cellulose regeneration. Secondly, the stretching of the filaments is carried out at a lower temperature and in three distinct phases. This gives the aligned cellulose molecules (crystallites) the chance to organize further into microfibrils. Rayon with this fibrillar structure is called POLYNOSIC.

The crystalline fibrils of polynosic rayon are not affected when water is absorbed into the amorphous regions, so polynosic rayon does not lose strength on wetting as dramatically as does viscose. Hence it is also called high wet strength, or high wet modulus, rayon.

TRY YOUR HAND

Survey a large department store.
1   Where do you find items made from polynosic rayon?
2   Count how many types of items carry the label 'rayon' and how many are labelled 'polynosic'.
3   Did you discover any trade names for polynosic rayons? Are any garments labelled with such trade names?
4   Are sales staff aware of the availability of different types of rayons?
5   In the fabrics section, how do the prices of different rayon yardgoods vary?
6   What do you think are the reasons for the differences in prices?
7   What type of rayon would you use for which purposes?

Properties of rayons, cotton and flax

	Viscose	High tenacity rayon (Tenasco)	Polynosic rayon (Zantrel)	Cotton	Flax
Tenacity	2.5 g/den	3.3 g/den	3.9 g/den	3.2 g/den	5.5 g/den
Wet tenacity	1.4 g/den	2.7 g/den	2.9 g/den	3.5 g/den	6.5 g/den
D.P.	250–270	250–270	250–270	2000–10 000	18 000
Cross-section shape	serrated	round	round	flat	polygonal
Structure	skin and core	'all skin' (uniform across cross-section)	microfibrillar	microfibrillar	microfibrillar

Note: The precise figures for these properties vary between references. The exact figures are not important: what matters is the general trend.

Acetate rayons

From hydrophilic to hydrophobic

Regenerated cellulose is chemically the same as all natural celluloses. These all have many accessible hydroxyl groups in their amorphous areas, and therefore they absorb water readily.

Cellulose can be reacted with acetic acid under certain conditions. The reaction goes as follows:

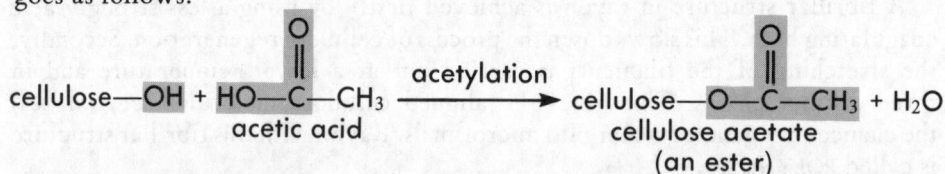

Cellulose is made of glucose rings. Each glucose ring has three free hydroxyl groups. Each of these hydroxyl groups can react with the acetic acid. Therefore, the maximum ACETYLATION that can occur is three acetate groups per glucose unit.

Fully acetylated cellulose is called cellulose TRIACETATE or PRIMARY ACETATE. SECONDARY CELLULOSE ACETATE carries two acetate groups on each glucose ring, on average.

The effects of acetylation

TRY YOUR HAND

You will need:
water
a piece of cellophane

a piece of clear acetate sheet, e.g. from a perfume or chocolate gift box, or an overhead projector sheet

What you do:
Place a few drops of water on each transparent sheet.
1   How did the cellophane react to the water? How did the acetate?
2   Cellophane is actually viscose rayon (without any delustrants) extruded as a sheet. How does this explain its behaviour?
3   How do you explain the different reactions to water of the cellophane and acetate sheets?
4   Suggest some possible uses for acetate where cellophane would be inappropriate.
5   Which material is more expensive?

The hydroxyl group of cellulose is polar because the electronegative oxygen pulls electrons away from the hydrogen. In the acetate group, $-O.CO.CH_3$, there is no hydrogen adjacent to the oxygen, so the oxygen has no source of extra electrons. Therefore it cannot acquire a partial negative charge, and the *acetate group is not polar.*

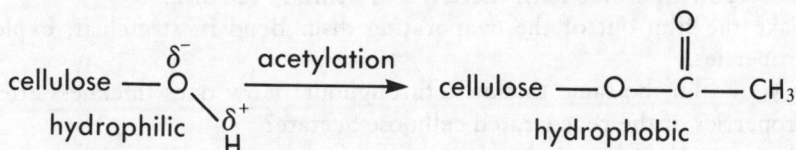

$$\text{cellulose} \overset{\delta^-}{-\!} O \underset{\underset{H}{\overset{\delta^+}{\diagdown}}}{} \quad \xrightarrow{\text{acetylation}} \quad \text{cellulose} -\!O-\!\overset{\overset{\displaystyle O}{\parallel}}{C}-\!CH_3$$

cellulose hydrophilic → cellulose hydrophobic

Acetylation changes *hydrophilic cellulose* into *hydrophobic cellulose acetate.* *Triacetate*, because it has lost all its free hydroxyl groups, is *non-polar* and *hydrophobic.* Because *secondary cellulose acetate* still has one hydroxyl group free on each glucose ring, it is *less hydrophobic* than triacetate, but not as hydrophilic as viscose.

TRY YOUR HAND

Look back to your tests of the effects of chemicals on cotton. Was cotton affected by acetone?

You will need:
cellulose acetate fabric
nail polish remover or acetone (this is a relatively non-polar solvent)

What you do:
Place a few drops of the acetone or nail polish remover on the fabric.
1   What effects can you observe?
2   What is the relevance of your findings to the consumer?
3   Suggest ways in which the sensitivity of acetates to solvents could be used:
    a in the identification of fibres
    b in the manufacture of acetate filaments.

TRY YOUR HAND

CAUTION: Work in a well-ventilated area, away from naked flames.

You will need:
5 g piece of cellulose acetate fabric, cut into small pieces
20 mL acetone in a beaker
stirring rod
evaporating dish
disposable syringe
hairdryer

What you do:
Dissolve the acetate fabric in the acetone, by stirring. You should have a fairly thick liquid.
Pour some of this liquid into the evaporating dish.
Blow across the surface of the liquid with the hairdryer to speed up evaporation of the solvent. Continue until there is a dry film in the dish.
1   Take the film out of the evaporating dish. Bend it, stretch it, explore its properties.
2   Is your film the same thickness throughout? How does thickness affect the properties of the regenerated cellulose acetate?
3   How does the film you have cast differ from a commercial cellulose acetate film? What are the reasons for the differences?
Now place some of the thick acetate solution in the syringe.
Very slowly extrude some of the solution into the path of the hot air from the dryer.
4   Bend, stretch your filament. Explore its properties.
5   How does your acetate compare with the cuprammonium rayon you made earlier? (p. 47)

The second part of this exercise was a small-scale version of the process used industrially to produce cellulose acetate filaments.

This process is called DRY SPINNING, as opposed to the wet-spinning process for viscose rayon. In dry spinning, dry air is used to remove the solvent and to solidify the cellulose acetate fibres.

A thermoplastic fibre

Acetylation, as we have seen, changes the polar cellulose molecule into hydrophobic cellulose acetate. This means that all the hydroxyl groups, which are the source of hydrogen bonding in cellulose, are eliminated. As a result, the cellulose acetate chains are attracted to each other by only the weak, close-range, van der Waals forces.

Cellulose acetate filaments (electron micrograph).
(Vivian Robinson)

TRY YOUR HAND

You will need:
an iron
aluminium foil
5 pieces cellulose acetate fabric, about 5 cm × 10 cm
5 pieces rayon fabric, about 5 cm × 10 cm
stop-watch

What you do:
Fold each piece of fabric in half, and sandwich it inside a folded piece of aluminium foil.

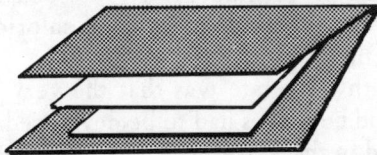

Choose five settings for the iron, from low to very high.
Iron a piece of rayon and a piece of acetate at each of the heats, by pressing the iron on the sandwich for exactly five seconds.
1   How did the rayon react to the increasing temperatures?
2   How did the acetate react?
3   If the iron had a fabric guide on it, what setting did it recommend for acetate fabrics? Why?
Label the ironed fabric samples clearly.
Now wet each sample in hot water, then try to iron out the crease with the iron on a low setting.

4   How effective is this process in removing the crease from rayon? From acetate?

Now repeat this, using a higher heat setting to iron out the crease.

5   Are there any changes in your ability to iron out the creases from rayon? From acetate?

Heat is a form of energy, and vibrates the molecules of the fibres. The more strongly the molecules are bonded to each other in the fibre, the less easily they can vibrate. Van der Waals bonds are much weaker than hydrogen bonds, so a fibre in which the molecules are held together by van der Waals bonds will be affected more by heat.

With the vibrations of heating, the molecules move apart: the intermolecular bonds break, and the molecules can be rearranged easily. As the fibre cools, the molecules move closer together, and new van der Waals bonds are formed between them in their new positions.

This process is called HEAT SETTING. If the heat is increased further, all the molecules of the hydrophobic fibre may start to move about. This is observed as MELTING.

This reaction of fibres to heat is called THERMOPLASTICITY (heat softening).

Heat setting is not possible for fibres which have hydrogen bonds, as these bonds are stronger and therefore harder to break. For polar fibres, water is generally a more effective setting agent than heat.

Generally, *hydrophilic fibres are not thermoplastic.* Cellulose acetates are *hydrophobic and thermoplastic fibres.*

Comparing the acetates

The first acetate produced in large quantities was secondary cellulose acetate. The reason for its commercial success was its ready solubility in acetone.

Secondary cellulose acetate started its commercial career dissolved in acetone and painted on the fabric wings of the aeroplanes used in the First World War. It is usually called simply ACETATE.

Triacetate was harder to manufacture. It needed a less polar solvent still than acetone, such as chloroform or methylene chloride. Chloroform is a dangerous, poisonous solvent.

Another problem with triacetate was that the very hydrophobic fibres were very difficult to dye, and new dyes had to be developed. So, until fairly recently, all the acetate produced in the world was secondary cellulose acetate.

Structure

What are the important differences between acetate and triacetate?

Secondary cellulose acetate is *weak* and has a *low melting point.*

Triacetate molecules are more regular – they form more crystalline structures. Therefore triacetate is also *stronger* and has a *higher melting point* than acetate.

Even though triacetate has a greater number of bulky side-groups than does acetate, it is more crystalline. This is because the side-groups are arranged more regularly along the length of the molecule, and so can organize into a more neatly ordered structure. Because the molecules are able to pack more closely, the van der Waals bonds are stronger and are able to hold the structure tightly together.

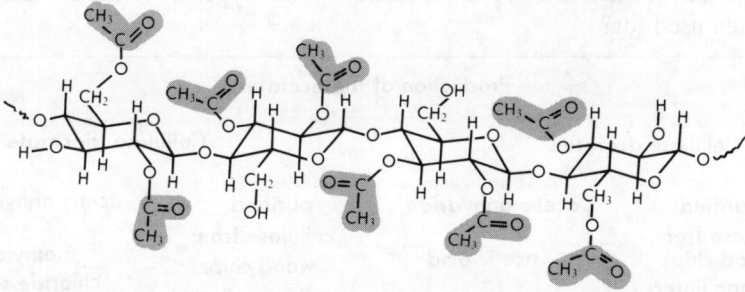

Secondary acetate. Because the acetate groups are positioned randomly, it is difficult for the molecules to pack into an ordered, crystalline structure.

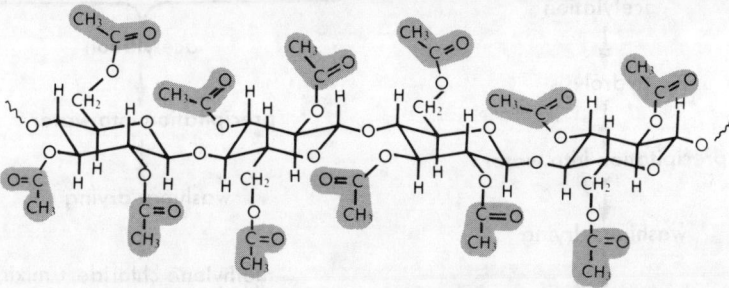

Triacetate. The regular positioning of the acetate groups means that, although they are bulky, they can be packed into an ordered crystalline structure.

When *acetate* is heated, it *decomposes* before it can take on a heat-set form.

Triacetate, on the other hand, can be *heat set,* because of its higher heat stability. Its ability to keep permanently set pleats is an excellent marketing point.

Properties of the acetates

Acetate rayons are *soft, silky* fibres. They have *excellent handle* and *drape* properties, and are used in filament blends in combination with silk and other fibres. Acetate is silkier, and has a better handle, than triacetate.

Acetates are rather weak, and therefore need gentle handling during use and care. Triacetates are a little crisper and more resilient than acetates.

Thermoplasticity allows triacetate to be permanently *heat set* into pleats or other special shapes. However, care must be taken when washing and ironing triacetates: *high temperatures must be avoided at all times.* Acetate cannot be heat set.

The acetates are drip-dry, but tend to feel clammy in hot, humid weather because of their *hydrophobic* nature. Triacetate is more hydrophobic than secondary acetate.

Triacetate is marketed under such names as Arnel, Starnel, Courpleta, Tricel. Secondary cellulose acetate is manufactured under the names of Estacel, Dicel, Celafil, Lansil, Rhodiafil, Aceta, Celanese, Fibroceta, Estera.

Find some advertisements for these names. What types of items are these fibres most often used for?

Production of the acetates

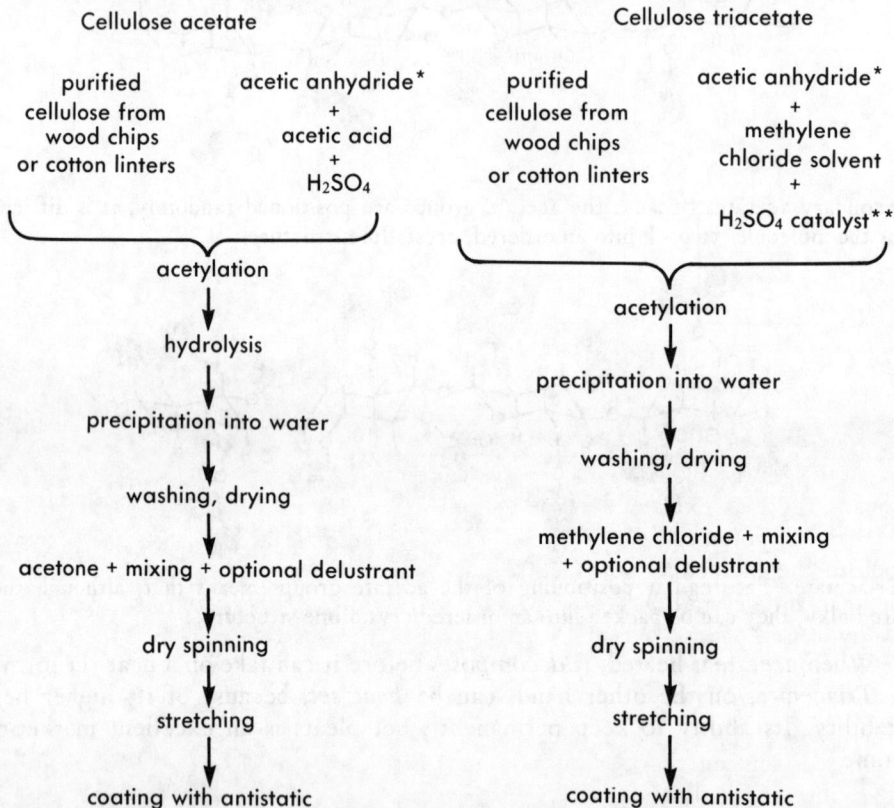

Cellulose acetate

purified cellulose from wood chips or cotton linters	acetic anhydride* + acetic acid + H_2SO_4

↓ acetylation

↓ hydrolysis

↓ precipitation into water

↓ washing, drying

↓ acetone + mixing + optional delustrant

↓ dry spinning

↓ stretching

↓ coating with antistatic

Cellulose triacetate

purified cellulose from wood chips or cotton linters	acetic anhydride* + methylene chloride solvent + H_2SO_4 catalyst**

↓ acetylation

↓ precipitation into water

↓ washing, drying

↓ methylene chloride + mixing + optional delustrant

↓ dry spinning

↓ stretching

↓ coating with antistatic

* acetic anhydride is a more reactive form of acetic acid
** a catalyst is something which helps a chemical reaction along, but which is not itself one of the reactants

The properties of the cellulosic fibres

Measurement of strength

Thus far, we have been describing fibres as either strong or weak. This is too imprecise to be of much use to textile scientists, and it can be very confusing, as when a fibre is described as strong but brittle. Some precise measure of strength is needed.

The ability of a fibre to withstand a *lengthwise pull* is called TENACITY.

Obviously the thickness of a fibre will influence its ability to withstand a lengthwise pull, so before tenacity can be measured, a unit for thickness is needed. By historical accident, the unit used to measure *thickness* (or, more precisely, fineness) is the DENIER. By definition, *9000 metres of one denier fibre weigh one gram.* The standard yarn in women's hosiery is 15 denier.

Tenacity is measured in *gram/denier*, and is the ratio of the mass a fibre can just support without breaking, to the fibre fineness. Thus, a 2 denier fibre with a tenacity of 2.5 g/den can just support a mass of 5 g. If it is loaded with a mass of 5.1 g, it breaks.

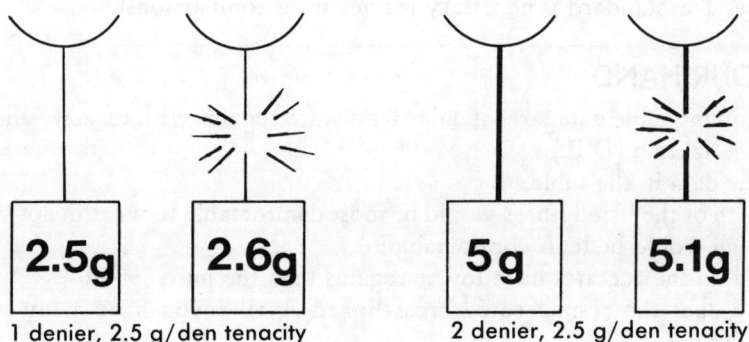

1 denier, 2.5 g/den tenacity   2 denier, 2.5 g/den tenacity

Sources of strength

A fibre is most likely to be high tenacity if: (i) it is highly crystalline, as this means there is extensive bonding between molecules; (ii) it has a high D.P. (degree of polymerization), as this means many more bonds have to be broken before the molecules can slip past one another; (iii) the molecular chains are held in the crystallites by hydrogen bonding rather than van der Waals bonds, as hydrogen bonds are much stronger, and therefore harder to break. A fibrillar structure can increase the fibre's ability to withstand stress.

WET TENACITY will be determined by where the water molecules go in the fibre. If they are absorbed between the molecular chains and force them apart, disrupting the structure, then wet tenacity will be lower than dry. If the fibre is fibrillar, and the water swelling the amorphous regions helps the crystallites align along the fibre axis, then wet tenacity will be greater than dry. If the swelling of the fibre relieves any stresses in its structure, this also contributes to wet strength.

TRY YOUR HAND

The table on p. 62 compares the properties of regenerated cellulose fibres with the chemically identical cotton.
1   How would you explain the differences in tenacity across the table?
2   How would you explain the effects of water on tenacity (wet tenacity)?
3   Which is the most amorphous fibre? the most crystalline?

Water affinity

As in comparing strengths of fibres, some sort of accurate measure is needed when fibres are to be compared for their moisture affinities. 'High' and 'low', or 'hydrophilic' and 'hydrophobic' are too imprecise.

Water affinity is measured as REGAIN. Regain is defined as the amount of water a fibre can absorb, as a percentage of the mass of the fibre:

$$\text{regain } (\%) = \frac{\text{mass of water absorbed}}{\text{mass of dry fibre}} \times 100$$

STANDARD REGAIN is the regain of oven-dried fibres when they are allowed to condition (adjust, return to their normal state) at 20°C and 65% relative humidity. The standard is necessary for accurate comparisons.

TRY YOUR HAND

The following table compares standard moisture regains with tenacity and degree of polymerization (D.P.).
Using the data in the table,

1   Which of the listed fibres would be most comfortable to wear in hot weather?
2   Which would be least comfortable?
3   Why do the acetates have lower regains than the pure celluloses?
4   Why does the resin-treated cross-linked Avril rayon have a lower regain than viscose?
5   Why does viscose have a higher regain than cotton?
6   Why does triacetate have a lower regain than acetate?
7   Why is linen a favoured fabric for use in tea-towels?

Selected properties of cellulosic fibres

Fibre	Tenacity (g/den)		Standard moisture regain (%)	Degree of polymerization (D.P.)
	Dry	Wet		
Cotton	3.2–4.0	3.5–4.5	8.5	2000–10 000
Viscose rayon	2.5	1.4	12.2	250–270
Avril (resin cross-linked rayon)	3.2	2.2	10	250–270
High tenacity rayon	4.0	3.0		250–270
Polynosic rayon	4.5	3.0		500
Cellulose acetate	1.7	0.9	6.5	350–400
Triacetate	1.2	0.8	4.5	350–400
Cuprammonium rayon	2.3	1.2	11	300
Flax	5.5	6.5	10*	18 000

* In flax, some of the moisture is absorbed by the cellulosic cement which holds the half dozen or so cells together to form the fibre bundles.
Note: The precise figures for these properties vary between references. The exact figures are not important: what matters is the general trend.

Electrical resistance

The conduction of electricity in a fibre is usually a function of the water molecules which are absorbed into the fibre. (Exceptions are carbon fibres and metallic fibres: these materials have freely mobile electrons and very high conductivities.)

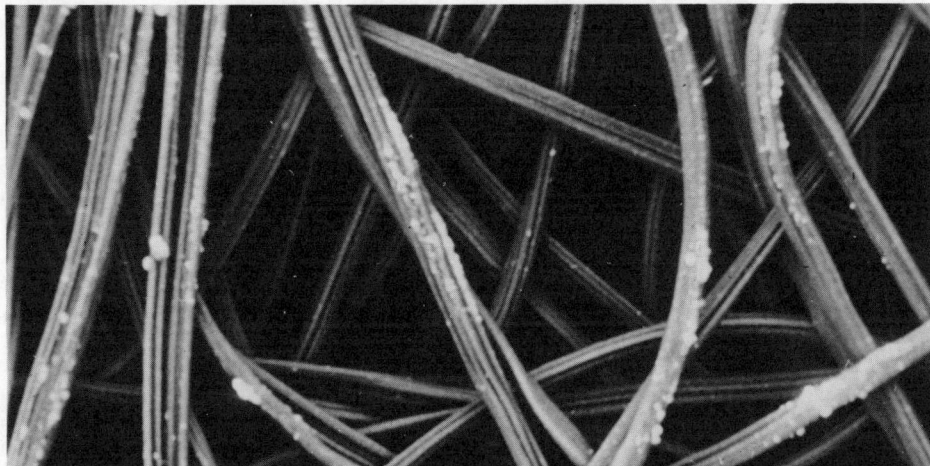

This electron micrograph of a household cleaning cloth shows it is made from viscose filaments. Why would viscose be a suitable fibre for such cloths? (Vivian Robinson)

Because conductivity is a function of the absorbed water, if a fibre has a low moisture regain, it will not conduct electricity effectively.

Friction (rubbing), such as from shuffling the feet on one spot on the carpet, can create a local electric charge by rubbing off electrons. This local charge can build up to quite a large static electric charge. If a person 'earths' this charge, she or he may receive a nasty electric shock.

All textile fibres, apart from carbon and metallic fibres, can develop static electricity in very dry atmospheric conditions.

In less extreme conditions, fibres with *high regain can dissipate charges* because of their absorbed moisture content, and so do not build up static charges. Those fibres with *low regain cannot dissipate charges:* they are insulators. Thus, cellulose acetate and triacetate are good insulators. The pure cellulosic fibres usually do not accumulate static electric charges.

TRY YOUR HAND

You will need:
some paper, cut into very tiny pieces
acetate sheet
cellophane sheet
piece of woollen fabric

What you do:
Rub a corner of the acetate sheet vigorously with the piece of woollen cloth.
Hold the rubbed part of the sheet just above the pieces of paper.
Record your observations.
Repeat with the cellophane sheet.
1   The tiny pieces of paper are attracted by static charges. Which of the two substances built up the most charge?
2   How would you explain this in terms of the composition of the two sheets?

Static charges on clothing are usually only a nuisance, or at most an embarrassment (as when skirts 'creep'). However, near sensitive electronic equipment such as computers, static build-up can create problems.

Biological resistance

Cotton and the other *hydrophilic* celluloses can be *destroyed* by mildew and even by insects. This is because these organisms are able to break the cellulose down and utilize the glucose units for energy. The acetylated, *hydrophobic* fibres have excellent *resistance* to such biological attacks, as mildew and insects are not able to use them for food.

Dyeing

The affinity of a fibre for dyes has a close relation to the chemical and physical structure. The chemical similarity of the pure cellulose fibres means that they can all be dyed with the same sorts of dyes. The cellulose acetates, because they are hydrophobic, cannot be dyed with the water-soluble dyes: instead, a class of dyes called *disperse dyes* is used on these fibres.

The different affinities of different fibres for dyes can be used as a test in fibre identification (Chapter 6).

Dyes and dyeing are discussed in detail in Chapter 8.

Specific gravity

The density of a substance is measured as its mass, in gram, divided by its volume, in millilitre. Thus, 1 mL of water has mass 1 g, so the density of water is 1 g/mL.

A common way of expressing densities is as SPECIFIC GRAVITY, the ratio of the density of a substance to the density of water. Specific gravity has no units, it is just a figure. This figure is the same as the density, that is, the same as the number of grams in one millilitre of the substance.

A substance with a specific gravity (s.g.) of less than 1.0 will float on water; if its specific gravity is greater than 1.0, it will sink.

Crystalline fibres have their molecules packed more closely together than do amorphous fibres. This means, for chemically similar fibres (the same sorts of molecules), amorphous fibres will be lighter than crystalline fibres: their structure contains more voids. Thus a more amorphous fibre has a lower specific gravity.

Drape, or how a fabric behaves when it is hung suspended, or how it flows over the contours of a wearer's body, depends on both the specific gravity and the stiffness of fibres and fabric.

Specific gravities of cellulosic fibres

Flax	Cotton	Viscose	Acetate	Triacetate
1.50	1.54	1.52	1.30	1.32

TRY YOUR HAND

1 Do any of these fibres float on water?
2 Which is the lightest fibre? The heaviest?
3 Which fibre would make the lightest garment?
4 Why are the acetates lighter than the cellulosic fibres?
5 Polyester has a specific gravity of 1.38. What effect does this have on the drape of linen-like fabrics made from polyester fibres?

Elongation

Some substances tend to *stretch* readily when pulled. Ease of extension depends on the ease with which the molecules of the fibre can uncurl or slide past one another. So, amorphous fibres stretch out of shape easily.

The tendency of a garment or other item to stretch out of shape during use or storage will depend in part on the fabric construction, but also on the elongation of the fibres.

Elongation is *not* the same as elasticity: an elastic substance recovers its former shape when the tension is released.

TRY YOUR HAND

From what you know about their structures, which of the following would you expect to be the least extensible? the most extensible? Rank them in order.

   cotton
   viscose
   polynosic rayon
   cellulose acetate
   triacetate

How does the ease of extension of these fibres affect you, as a consumer?

Resilience

RESILIENCE is the ability of a fibre to recover from a small amount of deformation. A simple test of resilience is to take a fabric, crush it in your fist, count to 10, and then release it. The tendency of the fabric to spring back into shape is called resilience. Fabrics which are not resilient tend to crush, sag, bag, and crease.

Resilience of fibres

Fibre	Percentage recovery from small deformation
Cotton	74
Viscose rayon	82
Cellulose acetate	94
Polyester	100

Resilience is measured as the *percentage recovery from a small amount of stretching or bending.*

Which fibre would tend to crease most readily? least readily?

TRY YOUR HAND

Go into a fabric shop.

Try the squeeze test on fabrics made from different fibres. Remember to squeeze each for the same length of time.

Check for any label for treatment, brand-names, and fibre content on your samples.

Grade the fabrics in order of resilience.

1   Did your results correspond with the ranking in the table above?
2   Do you think that any moisture on your palm when you crushed the fabric may have contributed to setting the fibres in a creased form?
3   From their behaviour, do you suspect that any of the fabrics you tested were treated with easy-care resins to improve crease resistance?
4   Which of your samples was made from only one type of fibre? Which were blends?
5   What was the effect of blending on the resilience of the blend?
6   From the table, would you expect the crease resistance of a polyester/ cotton blend to be better than that of pure cotton?
7   Compare a 35/65 polyester/cotton blend to a 65/35 polyester/cotton blend. Is there any difference in the resistance to creasing of the two fabrics?
8   How do these blends compare with pure cotton? With pure polyester?
9   Discuss some advantages of blending fibres to form fabrics with specific properties.

Handle

Softness to the touch is an important sensuous property of textiles. Construction of the *yarn* and of the *fabric* contribute greatly to the final handling quality. However, *some fibres* are much softer than others.

Generally, *amorphous* fibres are *softer* to the touch than are the crystalline fibres. Viscose has a softer handle than cotton, and cellulose acetate in turn has a softer handle than viscose rayon. The soft handle of cellulose acetate has been one of its strongest selling points.

Until recently, the handle of fabrics was assessed purely subjectively. In Japan in 1980, a set of standards was published which seeks to assess the factors which go towards determining final fabric handle. The standards co-ordinate the opinions of textiles experts with the results of various instrumental measurements as the foundation of scientific specifications for this most subjective of textile properties.

Summary of properties of the cellulosics

Properties of the cellulosic fibres

Property	Pure celluloses			Chemically modified celluloses	
	Cotton	Viscose rayon	Polynosic rayon	Acetate rayon	Triacetate
Tenacity (g/den)					
Dry	3.20–4.0 (strong)	2.6 (weak)	3.9 (strong)	1.7 (very weak)	1.2 (very weak)
Wet	3.5–6.5	1.4	2.7	0.9	0.8
Elongation at break (%)					
Dry	9	15	11	25	20
Wet	13	25	16	35	36
Wash care	easy to wash	wash with care	easy to wash	wash with care	wash with care
Standard moisture regain (%)	8.5 hydrophilic	12.2 hydrophilic	11 hydrophilic	6.5 fairly hydrophobic	4.5 hydrophobic
Specific gravity	1.54 heavy fibre	1.52 heavy fibre	1.53 heavy fibre	1.30 light fibre	1.32 light fibre
Reaction to heat	Non-thermoplastic, but yellows on long exposure to heat, when cellulose chains break down	Non-thermoplastic, but yellows on long exposure to heat, when cellulose chains break down	Non-thermoplastic, but yellows on long exposure to heat, when cellulose chains break down	Melting point 230°C with decomposition. Cannot be heat set.	Melting point 300°C, slightly more crystalline, more bonds need to be broken by the heat. Can be heat set.
Handle	Crisp to soft, but not very resilient	Crisp to soft, but not very resilient	Crisp to soft but not very resilient	Very soft, good drape, used in filament blends	Crisper than cellulose acetate, silk-like filaments
Electrical properties	Do not accumulate static, but not insulators – absorb water – high regain	Do not accumulate static, but not insulators – absorb water – high regain	Do not accumulate static, but not insulators – absorb water – high regain	Insulator – static problems – low regain	Good insulator – static problems – very low regain
Dyeing	Direct dyes, azoic, vat, and reactive dyes	Direct dyes, azoic, vat, and reactive dyes	Direct dyes, azoic, vat, and reactive dyes	Disperse dyes used (hydrophobic)	Very difficult to dye. Disperse dyes with carriers used.
Water	Swell in water	Swell in water	Swell in water	Not swollen	Not swollen
Chemical resistance	Attacked by acids	Acids attack rayons even more readily than cotton, because of their amorphous structure	Acids attack rayons even more readily than cotton, because of their amorphous structure	Solvents attack it. (Soluble in 70% acetone, 30% water mixture)	Solvents attack it. (Swells in 100% acetone, but not in 70% acetone)
Biological attack	Moulds and mildew attack them	Moulds and mildew attack them	Moulds and mildew attack them	Not attacked by moulds or mildew	Not attacked by moulds or mildew
Comfort	Absorb perspiration readily. Comfortable to wear in humid weather	Absorb perspiration readily. Comfortable to wear in humid weather	Absorb perspiration readily. Comfortable to wear in humid weather	Feel clammy in humid weather	Feel clammy in humid weather

TRY YOUR HAND

Look at the above table.

1 Why can't the same dyes be used for cellulose acetates and the pure celluloses (cotton and the rayons)?
2 Why is the biological resistance of acetates different from that of cellulose?
3 What properties are affected by the differences in moisture absorption (regain) of the different fibres?
4 How would you launder garments made from each fibre?
5 What kind of garments could you design as a most suitable end-use for each fibre?

Consumer questions answered

Q1 There is a sale of towels. Some are all cotton, some polyester and cotton, others – on special – are made of rayon. Which should I buy if I want them to give long service?

A The rayon towels are more absorbent (higher regain, more amorphous hydrophilic fibre). On the other hand, they will not last as long. The amorphous rayon fibres are weaker, and will break with abrasion. As they break and fall out of the yarns (the fluff or lint collected in the washing machine or tumble dryer), the towels will gradually get thinner and more threadbare. The polyester and cotton blend towels will last longer and dry faster, but they are not as absorbent as all-cotton towelling.

Q2 Is it worth paying the extra money for pure linen tea-towels?

A Linen is a stronger – and a more absorbent – fibre than cotton. Once the smooth starch finish is washed off, linen tea-towels will not only dry the dishes more efficiently, but also they will leave less lint and fluff on the glassware and will last longer.

Q3 The weather is hot and humid. For comfort, should I wear a garment made of cotton, viscose or acetate?

A The fibre which absorbs moisture most readily (has the highest regain) is the one most comfortable in hot weather. Viscose rayon – provided it is not treated with hydrophobic resins – is the best choice for comfort. On the other hand, untreated viscose has a low resilience – it will crease badly, particularly if the garment has a body-fitting style. For a balance between comfort and resistance to crushing, cotton usually gives better value. The hydrophobic acetates are, of course, uncomfortable to wear in hot climates.

Q4 I need to choose between rayon and acetate yarns to make a hand-knitted summer top. Which will be better?

A The knitted structure allows for breeze, so moisture regain is not as important. Acetate is a silky fibre, and lighter in weight than rayon (lower specific gravity). Hence, it is less likely to sag out of shape in use. It will also dry faster after washing: a better choice all around.

Q5 I brought a voluminous rayon jersey dress. Its care label said to dry-clean it. But I know that cellulosic fibres can be washed readily – should I pay for dry-cleaning, or risk laundering the garment?

A   Rayon absorbs 100 per cent of its weight of water when wet. If the large rayon garment is washed and hung out to dry, its wet weight would rapidly pull it out of shape. It could also be easily distorted during spin-drying, because the wet, swollen viscose fibres have very little elasticity. Hot ironing by a careless operator could break down any resin finishes that were applied – small wonder that the manufacturers preferred to avoid some of the risk of complaints by advising consumers to have the garment dry-cleaned.

However, provided they have washable trimmings, such garments can be carefully hand washed, partly dried by rolling in a towel, and then dried flat and smoothed with a cool iron.

Rayon fabrics with a more stable structure than jersey can usually be hot washed and maximum spun without distortion.

Q6   I have accidently ironed a crease into my acetate shirt. What can I do about it?

A   Unfortunately, acetate is a thermoplastic fibre, and can be permanently set or pleated by pressing into shape at a temperature just below softening point. To smooth out a crease, you will have to break the bonds that were set by the iron. For this you will need to heat the acetate to as high a temperature as you used to set the crease in. The higher temperature and pressure may create a shiny patch on the fabric. You must also be careful not to use so much heat that melting or shrinkage occurs.

Acetates must be ironed with caution if they are to be ironed at all.

Since acetate (unlike triacetate) has some –OH groups, it is best to break some of the hydrogen bonds by wetting the fabric. To prevent shine developing, place a thin piece of cotton on the wet crease, and iron carefully, with firm pressure, at moderate heat. Check the results of your pressing and heating by lifting the cotton fabric frequently.

Depending on the temperature at which the crease was set in, you may not be able to eliminate the crease completely.

Q7 I have just spilt some blackberry juice on my new triacetate dress. What should I do?

A Not to worry – triacetate is a hydrophobic fibre, and does not absorb water-soluble dyes (such as the ones found in blackberries). All you need to do is to wash the stain with detergent and plenty of water.

Of course, if your dress had been a hydrophilic fibre – such as cotton or rayon – then you would need to discolour by bleaching any dye which penetrated into the fibres. On a coloured fabric that would naturally lead to a permanent mark.

On acetate, however, the stain is only on the outside of the fibres, and can be carefully washed off. Better do it while the stain is still fresh, all the same!

Q8 My beautiful new rayon shirt looks like a limp rag after the very first wash. Can I do something about it?

A When you bought it, the shirt's fabric probably had a pleasant 'body' due to a resin treatment. If the shirt was washed in hot water or ironed with a hot iron, many of the resin cross-links would have broken down. With the stiffening gone, the rayon fabric regained its characteristic softness. You cannot reconnect the broken cross-links of the resin.

The most practical action is to use some starch – traditional or a plastic fabric filler – in the rinsing water after each wash. Although only an external and temporary treatment, starching will add body to limp fabrics.

Q9 How can I tell an Irish Linen handkerchief from a cotton one?

A There is no way of telling whether it was made in Ireland – other than by looking at the label. Linen handkerchiefs, however, tend to be slightly transparent and to have a fairly crisp handle because they are woven from very fine, even, tightly spun, 'hard' flax yarns.

Q10 Is there an advantage in buying mercerized cotton goods as opposed to unmercerized ones?

A Mercerization is a process through which strength and ease of dyeing are increased. Not all mercerized cotton goods are labelled 'mercerized' – most top quality products are treated in order to achieve high levels of performance and to save on costs of dyestuffs, rather than to sell the consumer on the name.

Q11 I bought a pure cotton shirt with a drip-dry label. It was not as cool to wear as my other pure cotton shirts. Why? Also, I always iron it just lightly before wearing – but now it seems to be losing its drip-dry quality!

A The hydrophobic resin added to your shirt used up some of the polar sites in the amorphous areas of the cotton fibre – so the cotton became stiffer and less moisture absorbent.

Ironing – however lightly – breaks down some of the resin in the fibres. As the cross-links break, drip-dry characteristics are reduced. So the more your easy-care garment is ironed, the less easy-care it becomes!

Q12 What fabric did the Egyptians use for mummy cloth?

A They used *ramie,* a flax-like fibre which grew well in the hot conditions in Egypt. The absorbency of this cellulosic fibre meant that it could be saturated by the embalming chemicals, and so it was resistant to biological attack.

Ramie was also used for the clothes of the nobility and priesthood; the poorer classes wore jute. Ramie has very long fibres, even longer than those of flax, so it was easy to spin. The fibres are, however, brittle, and so ramie is little used today.

4 Protein fibres

This chapter investigates the growth, structure, and properties of wool. Some ways in which these properties may be modified to the benefit of the consumer are also discussed.

Speciality hair fibres – camel, alpaca, cashmere, mohair, angora – are examined and contrasted with wool.

Silk is studied: at the molecular level, to explain why it differs so markedly from the hair fibres; and at the macroscopic level, to highlight features important to the consumer.

Furs and leathers, while not fibres, are important textile materials. As they, too, are protein, their special characteristics are considered here.

In the course of this chapter, the consumer learns the *why?* and the answers to such questions as:

- Which is warmer, wool or acrylic?
- What can I do about a jumper that shrank in the wash?
- How can I remove stains from suede?
- Why are silk garments labelled 'dry-clean only'?
- What is so special about mohair blends?
- Why do some wools irritate the skin?
- How can I remove creases from a suit?

The protein molecule

Protein is the building block of all the textile materials of animal origin: wool and hair fibres, silk, leather, and furs.

These materials may appear to have little in common apart from their animal origin, but at the molecular level they are very similar. This means that despite their many and obvious differences, protein materials will have similar chemical behaviours. They tend to be vulnerable or resistant to the same solvents and chemicals, they react in the same way to heat, they are subject to the same sorts of biological attack – until they are modified in processing.

The peptide linkage

The protein molecule is built up of many smaller units called AMINO ACIDS joined together in a long chain. There are some 26 naturally occurring amino acids, but all have one feature in common: each of these small molecules has a carbon atom which is bonded to both a *basic amino group,* and to an *acidic carboxylic group:*

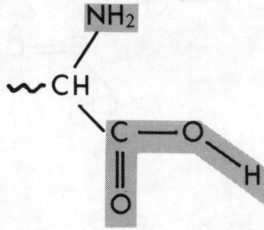

These amino acid ends to the molecules link up in what is called a PEPTIDE LINKAGE: hence the other name commonly used for proteins: POLYPEPTIDE.

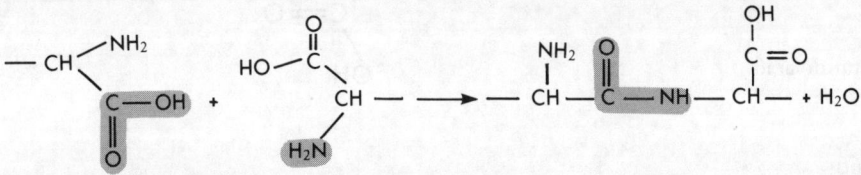

Immense chains of amino acids join up in this way to give the peptide skeleton common to all proteins.

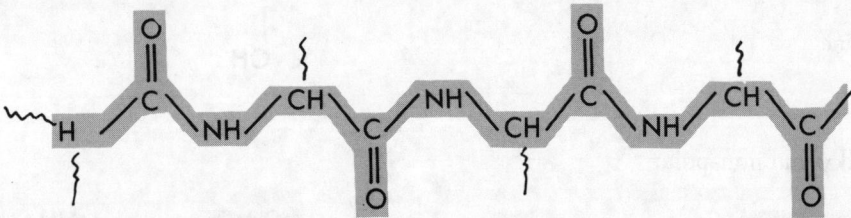

The importance of the side-groups

Differences between the proteins arise because of differences in the side-groups which form the rest of the amino acid molecule. These side-groups may be small, and non-polar

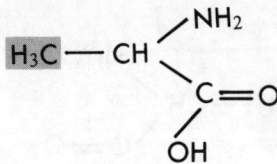

alanine

small and polar

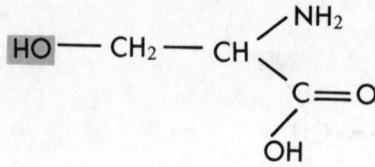

serine

acidic

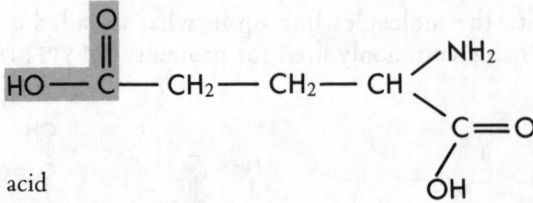

glutamic acid

basic

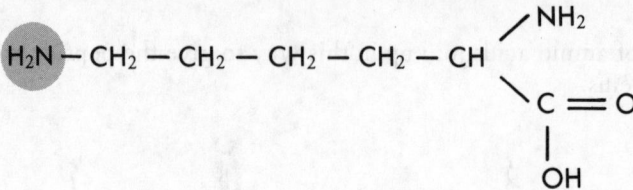

lysine

bulky and non-polar

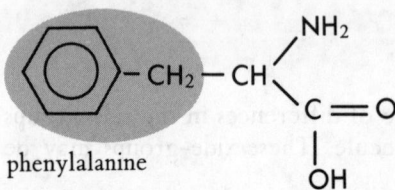

phenylalanine

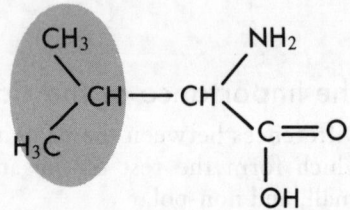

and valine

bulky and polar

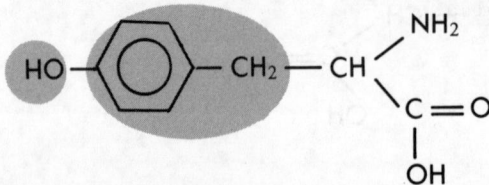

tyrosine

Different proteins are made up of different proportions and arrangements of the many amino acids. A protein that consists largely of amino acids with small side-groups will be able to pack tightly into a crystal structure. If the small side-groups are polar, enabling hydrogen bonds to form, the protein will be even more crystalline. If, on the other hand, the side-groups are bulky, it will be much harder for the protein molecules to pack into a crystal structure.

The arrangement of amino acids, while not random, does not form a simple, repeating pattern. This lack of a simple pattern means that it is even more difficult for proteins with bulky side-groups to pack into a tight crystal structure.

Crystallinity of protein fibres therefore depends very much on which amino acids predominate in their make-up. Silk has mainly small amino acids, and is highly crystalline. Wool has many bulky amino acids, and is relatively amorphous.

The disulfide linkage

One amino acid side-group is of immense importance for another reason. The amino acid *cysteine* ends with a thiol group, –SH. Two cysteine molecules can link their thiol groups to form the more complex amino acid, CYSTINE:

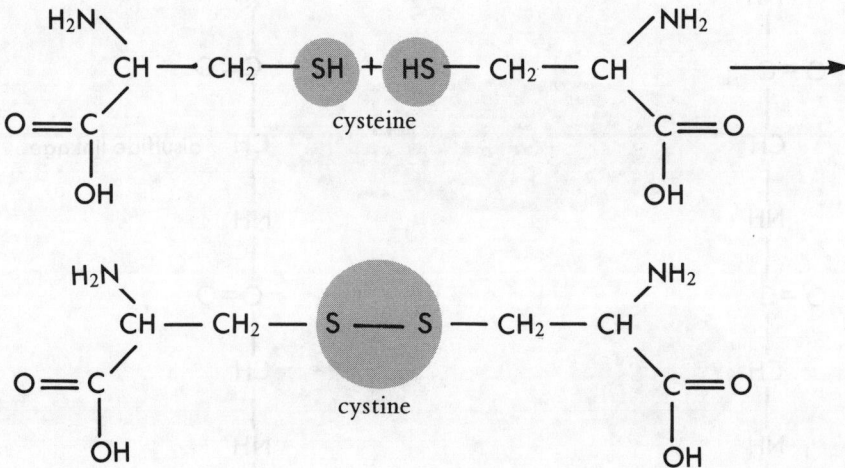

The bond they form is called a DISULFIDE LINKAGE. It is important because the disulfide bonds can form *within* and *between* the protein polymer molecules. They help give structure and stability to the relatively amorphous wool fibre.

Bonding between protein polymer molecules

Four types of bonding occur between molecules in the protein fibres. Always present is the weak *van der Waals* bonding. Its effects are masked by the much stronger *hydrogen bonding* between the various polar side-groups on the amino acids, the *ionic bonds* formed between acidic and basic side-groups (salt formation), and, in proteins containing a large proportion of cysteine, by the cystine *disulfide* links, which are *covalent* bonds.

Disulfide linkages are enormously important in wool, but entirely absent in silk.

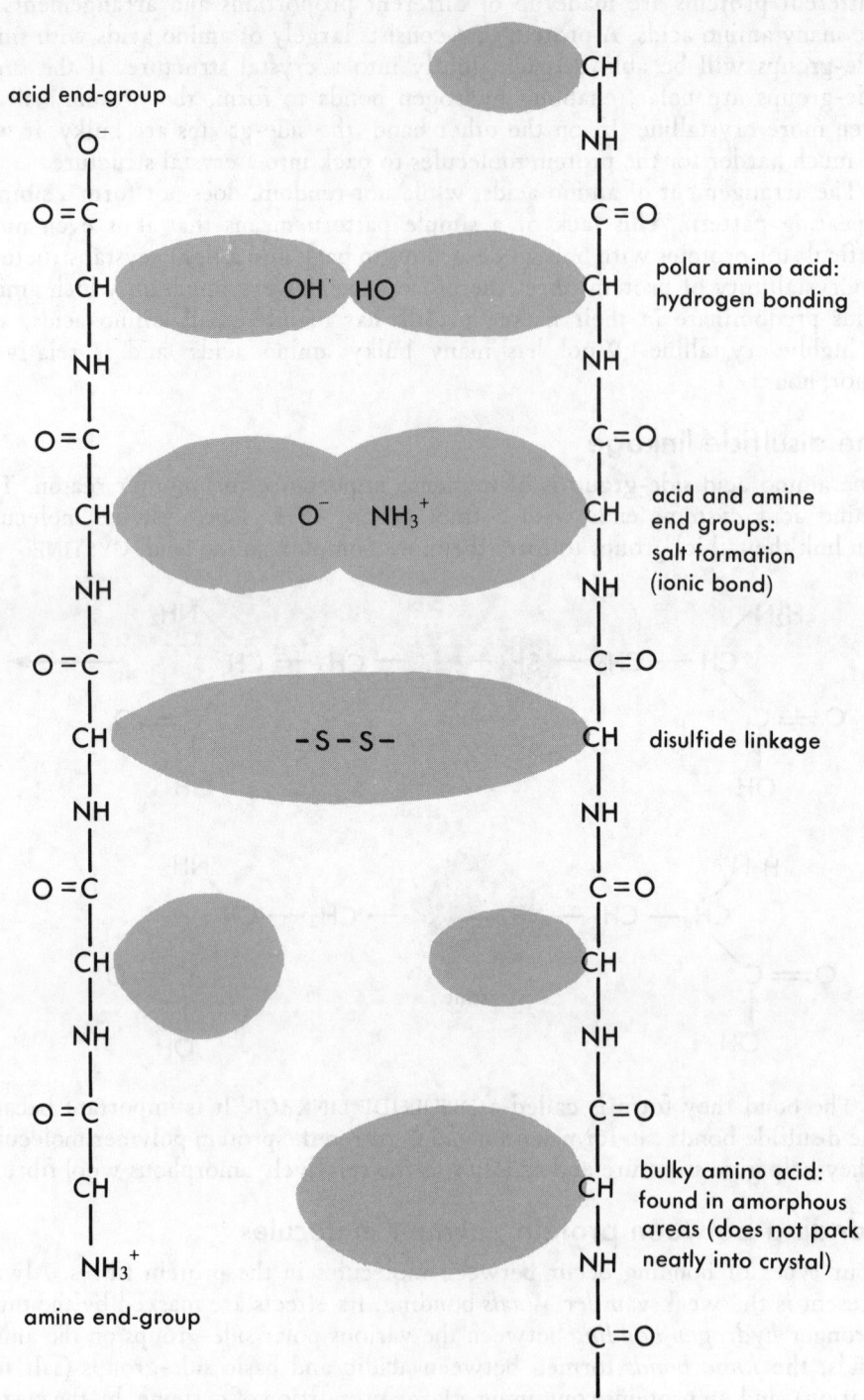

Bonding between protein molecules. The acid and amine groups at the end of each protein chain are available for ionic and hydrogen bonding, and so also contribute to bonding between chains.

Moisture affinity and dyes

The peptide skeleton of the proteins is only slightly polar, so the water affinity of the protein is determined by the amino acid side-groups and the crystallinity. This in turn determines the ability of the fibre to accept dyes. These characteristics are discussed in the examination of each textile material.

Microstructure

The microstructure of the protein textile materials varies enormously, and is discussed in the detailed examination of each material.

It is the difference in microstructure, more than differences in the make-up of the protein molecules themselves, that accounts for the different behaviours and properties of the protein textile materials.

Wool

Wool – the fibre

Growth

Wool is the hair of sheep. Hair from other animals – goats, rabbits, vicuña, alpaca etc. – is also used as textile raw material, but only the long furry covering of sheep is referred to as wool.

Each fibre of wool – or hair – grows out of a depression in the skin called a FOLLICLE. The fibre is formed at the bulbous base of the follicle, where its growth is nourished by many capillary blood vessels. Higher in the tube of the follicle is an opening through which the oil gland – or sebaceous gland – pours its oily sebum. The sebum of sheep is called LANOLIN or wool fat. Lanolin is a valuable by-product of wool processing, and is widely used in the cosmetics and pharmaceuticals industry.

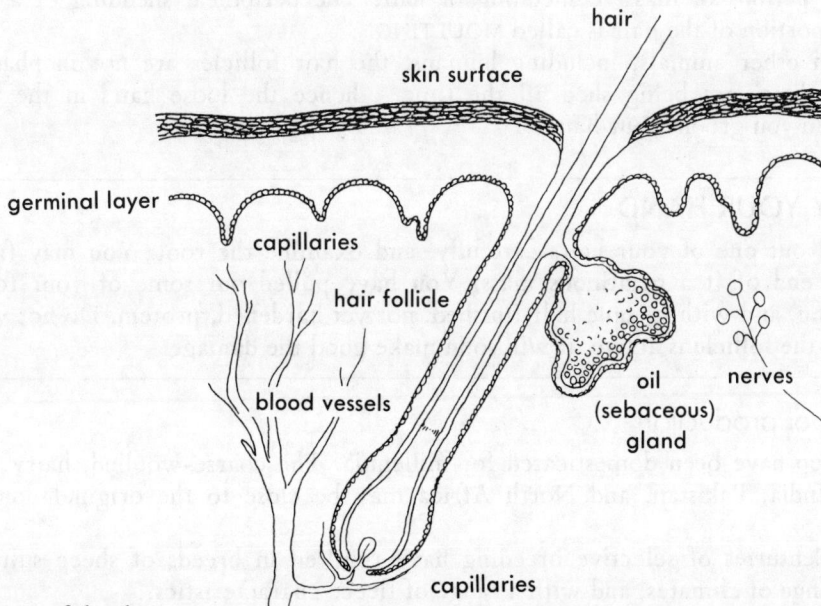

Structure of the skin.

The shape of the follicle determines whether the wool fibre growing out of it will be curly or straight, fine or coarse.

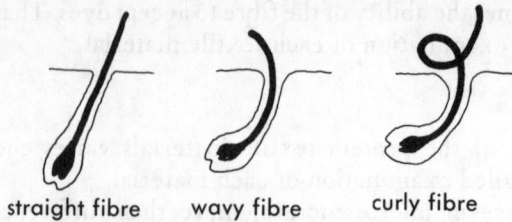

straight fibre   wavy fibre   curly fibre

The shape of the hair follicle determines the shape of the hair fibre.

The cells which make up the wool fibre are produced at the germinal layer of the active bulb of the follicle. Once completed, they are pushed upwards by new cells being produced underneath. About half-way up the follicle they harden and die, permanently keeping whatever shape was impressed on them by the follicle lining. The cells of wool fibre above the skin are dead, with the cell nuclei showing as dark spots in stained slides. The cortical cells are tightly packed together in the wool fibre.

The bulb of each follicle goes through a number of cycles of activity and rest. While the bulb is active, the fibre grows. When the bulb rests, the fibre is held in it for a while. Then as the bulb shrivels, the fibre is shed. Some time later, the follicle starts a new cycle, repeating all the phases of production, rest and regression.

In some animals, the bulbs in the hair follicles grow, rest, and shrivel, in synchrony. This means that the animal goes through periods of hair growth, and periods of massive shedding of hair. The periodical shedding of a large proportion of the hair is called MOULTING.

In other animals, including humans, the hair follicles are not in phase. A few hairs are being shed all the time – hence the loose hairs in the brush when you groom your hair.

TRY YOUR HAND

Pull out one of your hairs carefully, and examine the root. You may find at the end of it a gelatinous mass. You have pulled out some of your follicle lining, and with it some half-finished, not yet hardened, protein. Do not worry – if the follicle is active, it will soon make good the damage.

Wool production

Sheep have been domesticated for millennia. The coarse-woolled, hairy sheep of India, Pakistan, and North Africa may be close to the original domestic sheep.

Centuries of selective breeding have resulted in breeds of sheep suited to a range of climates, and with a range of fleece characteristics.

Berber sheep, Algeria. These hardy sheep are suited to the arid conditions of North Africa. They produce extremely tough carpet wools.
(Australian Wool Corporation)

Merino rams at Concordia stud: superb fine-wool sheep.
(Australian Wool Corporation)

Romney Marsh. This long-wool breed constitutes about 60 per cent of the New Zealand flock.

Lincoln, a long-wool breed unsuited to scrubby pastures.

Australia is the world's biggest producer of fine wool. The Australian Merino, bred from a motley collection of ancestors (including Spanish Merino) since the early nineteenth century, makes up some three-quarters of the Australian flock. It survives particularly well in flat, dry country, where it can graze over a wide area. In drought years, Merino wool is even finer than in years of normal rainfall.

New Zealand and India produce excellent resilient carpet wools. The main breed in New Zealand is the English variety, Romney Marsh. This breed of sheep thrives in the much wetter conditions of New Zealand's mountain pastures. The hairy sheep of India are well suited to that country's hot, dry conditions.

Long-wool sheep, such as the Lincoln, need open country to graze, as their shaggy, hairy coat would pick up burrs and other rubbish in scrubby bushland, and so be unsaleable. Other breeds have been developed to meet other production needs and climatic conditions.

Attempts have been made to produce breeds of sheep suitable for both wool and meat, but this has generally resulted in mediocre returns from both these areas.

Characteristics of selected wools

Sheep breed	Fibre diameter (micron)	Staple length (mm)	Fleece weight (kg)	Uses of wool
Corriedale	28–33	75–125	4.6–6	Medium-weight outer garments, worsteds, light tweeds.
Dorset Down	26–29	50–75	2–3	Fine knitting wool, hosiery.
Drysdale	40+	200–300	5–7	*Speciality carpet wool.* Medullated.
English Leicester	37–40	150–200	5–6	Furnishing fabrics, coatings.
Lincoln	39–41	175–250	5–7	Long, coarse, lustrous – blends well with mohair.
Merino	19–24	65–100	3.5–5	*Speciality fine wool,* for woollen and worsted fabrics. 50 million fibres/sheep
Romney	33–37	125–175	4.5–6	Carpet blends. 15 million fibres/sheep. Comprises 60 per cent of New Zealand flock.
Perendale	31–35	100–150	3.5–5	Low lustre, high bulk, exceptional springiness – shape retention in knitted garments, extra bulk to carpet pile, high insulation factor in blankets.
Coopworth (Leicester and Romney)	35–39	125–175	4.5–6	Heavy apparels and carpets.
New Zealand Halfbred	25–31	75–110	4–5	Nineteenth century Merino and long wool breed (Leicester, Lincoln, Romney). Apparel and fine knitwear.

Source: Wool Research Organization of New Zealand

Types of wool

Some types of sheep have follicles that are active for long periods at a time. These are the long-wool sheep. Other types have many fine follicles, which are active for shorter periods of time. These sheep produce a fleece of short, fine, dense wool. Usually fine wool is curlier – has a finer CRIMP – than the coarse long wools.

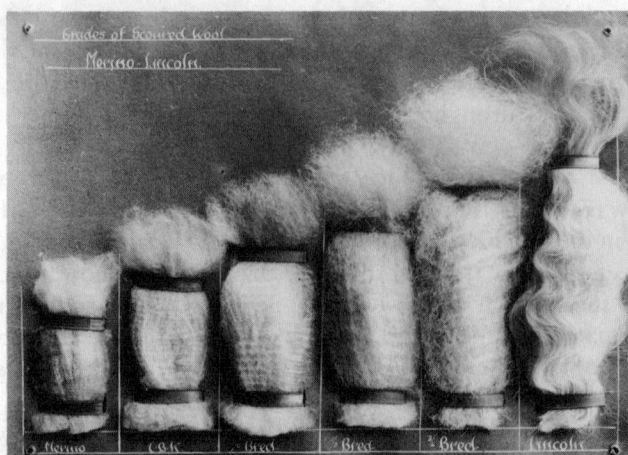

Grades of scoured wool. With increasing Lincoln blood and decreasing Merino, the wool becomes progressively longer and coarser, and has longer waves of crimp.

Wool fibres range from very fine – less than 18 microns (0.0018 cm) in diameter – to more than 40 microns in coarse wools. The fine wools are shorter, 4–12 cm average length, and the coarse wools are longer, up to 30 cm in length. The length of wool fibres depends on the breed of sheep, the part of the fleece from which the wool comes, and, of course, on how often the sheep is shorn. Sheep are usually shorn once a year.

Superfine wool – which has an average diameter less than 17 microns – is so soft that it can be used for babies' underclothes.

Fine wool comes from Merino sheep and from the undercoats of wild sheep and Scottish Blackface. These fibres are fine and short, with small waves of crimp, and are used for most woollen knits and for the finest worsteds.

Medium diameter *wools* are used for blanketing and for coarse woollens.

If more than 10 per cent of coarser diameter fibres are found in even the finest blend of wools, the fabric made from the blend will feel scratchy, and will irritate sensitive skins. The coarser the fibres, the scratchier the wool.

Coarse long wools are tough and resilient. They often look lustrous, due to a canal-like space called MEDULLA along their centre. They are stiff and have very long waves of crimp. Coarse wools are used for tweeds and for carpets. Important coarse wool breeds are Romney Marsh, Crossbred, Lincoln, and Tukidale.

Some tweed wools contain naturally occurring white fibres called KEMP wool. These fibres are coarser to the touch than the rest of the fleece. As they consist mainly of medulla they often have a harsh, flattened cross-section, and they cannot be dyed effectively. They can be used for special decorative effects, but too many kemp fibres reduce the value of the clip.

Shearing is a labour-intensive but seasonal activity, requiring a skilled and mobile labour force.

Skirting the fleece. Wool quality is not uniform over the animal, so immediately after shearing the lower quality outer edges of the fleece are picked out to be baled and sold separately.
(Department of Agriculture and Rural Affairs, Victoria)

The morphology of wool

The wool fibre is very complex, both chemically and structurally. It is not made of a single cell like the cotton fibre, nor from a cemented bundle of cells like flax.

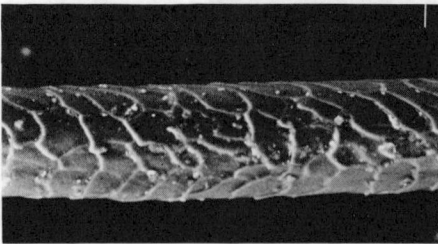

Medium crossbred wool.

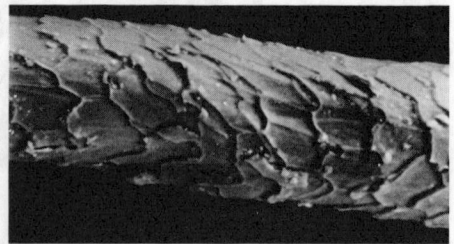

Coarse carpet wool.

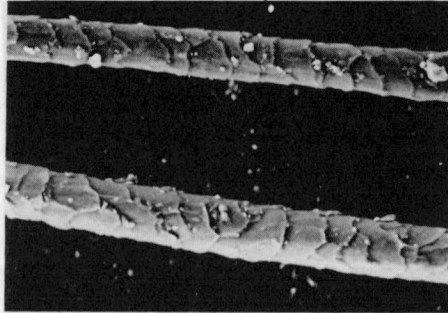

Superfine Merino wool.
(Vivian Robinson)

Its CORTEX – central mass – is made up of two different types of cigar-shaped cells. These cells may be evenly mixed or the fibre may have all the *ortho* cells on one side and all the *para* cells on the other side. This latter occurs in the fine fibres that are produced by curved follicles. Such a fibre is bilateral – has two distinctly different halves – with one half always longer than the other half. This structure creates a helical fibre. Coarser wools have a more uniform distribution of cells, and so they have little or no crimp.

Surrounding the cortex are two to three layers of flattened, overlapping cells. These are the *scales* of wool. They are laid one on the other like tiles, with their edges always pointing in one direction: away from the root of the fibre.

The scales of the wool fibre give it a very important property: its DIRECTIONAL FRICTIONAL EFFECT. This means that it is easier to rub the wool fibre in one direction than the other.

TRY YOUR HAND

Hold a hair from your head (or a friend's, if yours is too short) by one end, without pulling it out. Rub along its length with the finger and thumb of your other hand, towards the root, and from the root towards the end.

Did you find the hair was smoother in one direction than the other? That is, that it was easier to run your fingers one way than the other along the hair? The difference in friction that you felt is the directional frictional effect.

The directional frictional effect in wool is responsible for its ability to FELT, and for the special wool-like handle that synthetic fibres cannot duplicate.

TRY YOUR HAND

You will need:
some wool fibres, cut into 0.5 cm lengths
a microscope with 100× or better magnification
glass slide
coverslip
dropper with water

What you do:
Arrange the wool fibres on the slide so they do not cross over one another.
Now put a drop of water on the fibres, and cover them with the coverslip.
Place the slide under the microscope, and draw what you see at different magnifications.
What do you see?
Are all the fibres the same thickness? Do they all look exactly alike?
What are the lines you see across the fibres?

Microstructure and molecular structure

As described on p. 83, the wool fibre is composed of tightly packed CELLS.

Each of these cells is made up of well-organized *fibrillar* regions. The FIBRILS are made up of MICROFIBRILS. The microfibril is made up of *crystalline* and

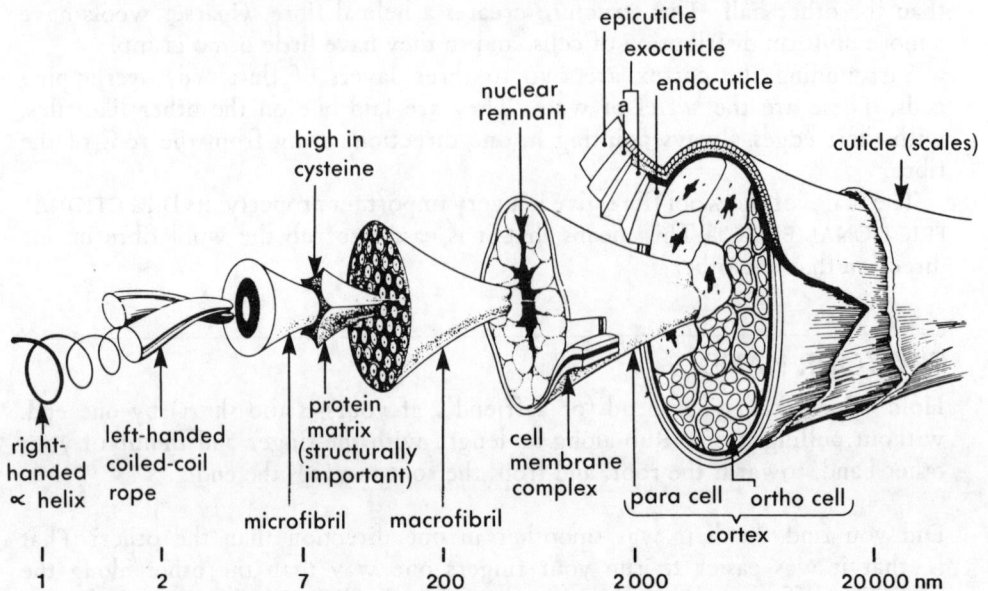

Microstructure of the wool fibre. Ortho cells form the outside of the curve in a crimped fibre.
(adapted from C.S.I.R.O. Division of Protein Chemistry)

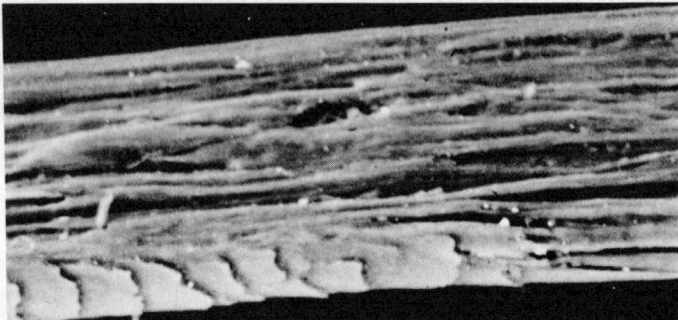

A vertical slice through a wool fibre, showing the complex internal structure (electron micrograph).
(Vivian Robinson)

amorphous arrangements of protein molecules. Each long protein (KERATIN) molecule passes through some crystalline and some amorphous regions.

The crystalline regions of wool protein are organized in a unique way. Three chains of protein molecules wind round each other in a *helical* arrangement. They are held to each other and into the helical shape by van der Waals and hydrogen bonds.

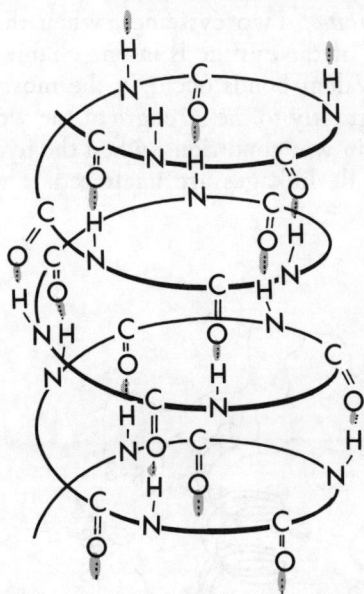

In the crystalline regions of the wool fibre, hydrogen bonds hold the protein molecules in a triple helix.

TRY YOUR HAND

You will need:
ball-and-stick type molecular modelling kit, or
Plasticine and match sticks

What you do:
Make models of 10 or so amino acids, including some of the bulkier ones.
Join them to make a polypeptide chain, remembering to 'lose' one water molecule for each peptide link (p. 73).
Does your polypeptide molecule fall into a helical arrangement?

The helical arrangement allows the crystallites to act as springs. If stretched and released, they snap back into shape. This gives wool its unique resilience.

Wool is not 100 per cent elastic: it can be permanently distorted by excessive extension. During extension, some molecules may slip past their neighbours. The slipped molecules will not return to their original positions. Hence there will be some permanent gain in length.

The effects of extension depend on: (i) amount of extension; (ii) length of time the extension is maintained; and (iii) the time allowed for recovery.

In humid conditions, hydrogen bonds can be broken more readily. This means that wool stretches more easily in humid conditions. However, the same conditions also help recovery and the formation of new hydrogen bonds.

On the whole, wool is a very *amorphous* fibre. This should mean that it is very weak. However, of the 19 amino acids found in wool protein, one is the double-ended *cysteine*. Two cysteines, when they join to form *cystine*, can join so that one half of the cystine is in one chain, and one half in another. The resulting strong covalent bonds occur in the most amorphous parts of the fibre, and so contribute greatly to the *strength* of the wool. They also contribute to the *elasticity* of wool in wet conditions: when the hydrogen bonds are broken by the water, the disulfide linkages are unaltered. They continue to hold the molecular chains in place.

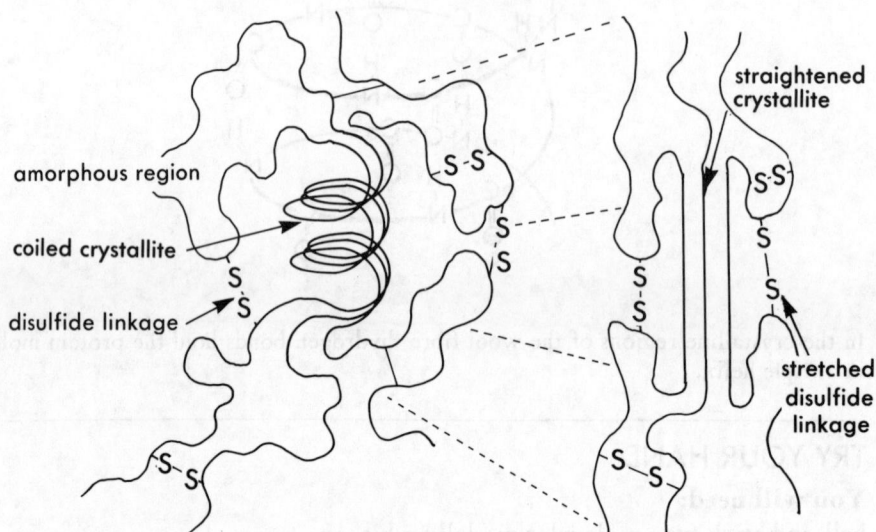

When the wool fibre is placed under tension, the amorphous regions straighten out, and, if the pulling force is great enough, the helical arrangement in the crystallites also straightens out. The disulfide bonds between the molecules are stretched. When the tension is released, the crystallites return to their preferred (lower energy) helical form, and the disulfide bonds help return the amorphous regions to their original arrangement.

The disulfide bond, however, is readily broken by alkalis. This is the reason why wool is so readily damaged by alkaline washing powders and detergents. It is also broken by reducing agents, but this weakness can be put to good use when the properties of the wool fibre are modified (p. 96).

Properties of wool

From considerations of its structure, it can be readily seen that wool is a *soft, weak, resilient* fibre. It is *hydrophilic,* as many of its amino acids are polar, and it has many amorphous regions to accommodate water molecules. It is *sensitive to alkalis,* which break down its disulfide linkages, and so weaken the fibre. Ultraviolet light changes some of the bulky amino acids into coloured products, and so wool *yellows in sunlight.* Bleaching must be done with care, as chlorine bleaches degrade wool and discolour it.

Some of the characteristics of wool are summarized in the table.

Properties of wool

Property	Explanation	Importance to the consumer
Soft	It is amorphous at the molecular level despite complex fibre structure. Finer fibres are softest.	Characteristic handle of wool depends on fibre diameter and the processing the wool undergoes.
Weak	Because of amorphous structure. Strength increases with fibre diameter.	Most wool fibres are suitable for garments. Where strength is paramount, such as in carpets, special coarse wools are used.
Resilient	Covalent disulfide bonds pull the molecular chains back into place after tension is released.	Wool garments do not crease easily, but do not hold pleats well without special treatment. Carpets recover their appearance readily after crushing, as when heavy loads (e.g. furniture) are removed.
Hydrophilic	Polar and ionic groups, and many amorphous regions.	Absorbs moisture from perspiration – comfortable to wear.
Sensitive to alkali	In the presence of the hydroxide ions (OH^-) released by alkalis, the peptide bonds and disulfide bonds hydrolyse (break down) readily.	Great care needed with detergents. Only neutral soaps and non-ionic detergents should be used.
Weakens when wet	As water is absorbed into the wool fibre, its amorphous regions swell and the fibre becomes weaker than it was when dry.	Care must be taken when washing wool garments. They must never by 'hung out to dry', or the weight of the water may permanently stretch the garment out of shape.
Scorches in dry heat	In the presence of moist heat, the molecules in the amorphous regions can break and re-form hydrogen bonds as they rearrange to absorb the heat energy. If dry heat is used, this is not possible, so the molecules themselves tend to break down.	When wool garments are ironed, steam, rather than dry heat, must be used.
Yellows in sunlight, and from contact with chlorine bleaches	Ultraviolet light and chlorine bleaches can degrade (break down) some of the amino acids to yield brown or yellow products.	Wool garments must be dried in the shade, and must *never* be bleached with household bleaches. If wool must be bleached, hydrogen peroxide is relatively safe.

Property	Explanation	Importance to the consumer
Fire resistant	This may be partly because wool normally contains considerable amounts of absorbed water. However, *all* protein materials are relatively fire resistant, so the tendency of wool to char rather than burn, and its self-extinguishing property, appears to be because of its protein nature.	Wool is a fibre recommended for fire-fighting and other uniforms, for upholstery in aeroplanes, and in other high fire hazard situations.
Felts	The scales give rise to a directional friction effect – fibres migrate towards root only.	Care is needed when handling wet wool fabrics: warm, wet, alkaline conditions, combined with strong agitation, produce maximum felting and shrinkage. The ability of wool to felt is put to use in the manufacture of strong nonwoven felt materials.
Warm	The amorphous fibre does not transmit heat readily. Also, crimp and scales can hold still, insulating pockets of air. In addition, wool gives off heat when it absorbs moisture, which adds to its comfort in cold, damp conditions.	Wool is reputed to be warmer than other fibres processed in identical ways.

Felting

TRY YOUR HAND

Take a thin web of wool fibres, and place it inside your shoes.
Wear the shoes as much as possible for a week.
Now remove the web of fibres and examine it.
1   How has the web changed in thickness?
2   In area?
3   Do you think you could make the web felt even more?

The FELTING properties of wool may be useful in making felt cloth from a web of fibres. However, it can also be a great nuisance from the consumer's point of view. A garment which started as size 14 can be reduced to a stiff, unwearable size 8 in a single machine wash.

TRY YOUR HAND

Take a sample of white wool and mix a few black wool fibres with it.
Card your sample to blend the fibres, and form them into a web.

Take a photocopy of your web: the positions of the black fibres should show up clearly.

Place the web between two layers of muslin, roll it up, and felt it by wetting the roll in warm soapy water and pounding it with your hand.

Photocopy again after two minutes of pounding.

Repeat the pounding and photocopying five times.

Check the distribution of the black wool fibres on the succeeding photocopy images.

1 Can you estimate how far a black fibre has moved?
2 Did the fibres move closer to each other, or further away?
3 What has happened to the web of fibres overall? to its area? to its thickness? to its softness and flexibility?
4 Did the felting process stop or slow down at all?
5 Can you explain the changes in the properties of the web by discussing what happened to individual fibres within it?

Felting shrinkage occurs because the scales of the wool fibre give it a directional frictional effect. Directional frictional effects are found only in the hair fibres.

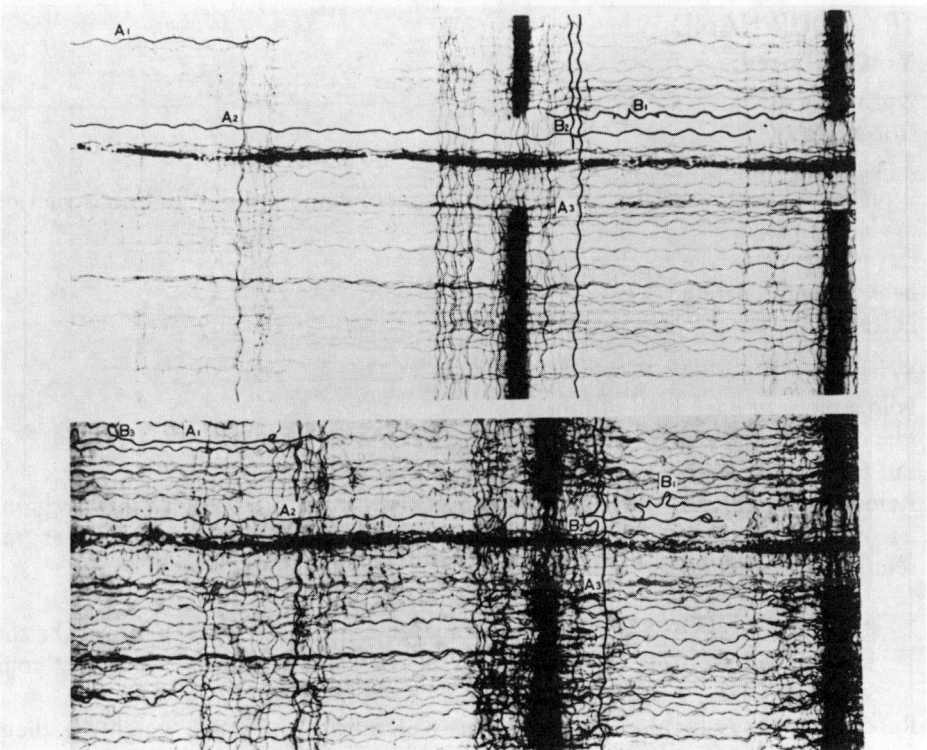

Felting in a tightly constructed two-ply worsted. The upper photograph shows the fabric before washing, the lower is after washing. At the points marked A, the felting has straightened the fibres. At points marked B, the fibres have buckled.
(C.S.I.R.O. Division of Textile Physics)

TRY YOUR HAND

Hold an ear of wheat (or an ear of grass) between the flat palms of your hand.

Rub your palms together.

Does the ear move? Which way does it move?

Now turn the ear over, so it is pointing the other way.

Which way does it travel now?

Can you explain it?

How can you relate fibre movement during the felting of wool to the directional frictional effect of the ear of wheat?

Felting shrinkage can be overcome in two ways. One approach is to eliminate the effect of the scales, by softening them, removing them altogether, or by coating them. The other approach is to anchor the fibres in some way, so that they cannot move. These approaches are discussed in detail on p. 93.

Resilience and elasticity

TRY YOUR HAND

You will need:

2 protractors
stop-watch
3 pieces of cotton fabric, 4 cm × 4 cm
3 pieces of pure wool fabric, 4 cm × 4 cm, of same weight and construction as the cotton
flat board
heavy weight, around 5 kg
kettle

What you do:

Fold a piece of each kind of fabric in two.

Place them side by side, with the flat board on top.

Put the weight on top, and leave them for 5 seconds.

Remove the weight, and check how quickly – if at all – each fabric regains its original flat shape. The best way is to measure the angle of recovery at regular time intervals.

1   Which fabric was more resilient, cotton or wool?

2   Look back to p. 65 for the resilience of cotton and other fibres. On the same scale, wool has a recovery of 99 per cent. Is this borne out by your observations?

Repeat the experiment with fresh pieces of fabric, this time weighting them for 1 minute, and for 10 minutes.

Check the resilience of the two types of fabric.

3   Was the recovery after longer distortion times similar to that after the quick press?

If the fabrics have not lost their crease completely, steam them over a boiling kettle.

Place them on the bench with the crease uppermost, so they form an inverted V.

Leave them to cool and relax.

4   Which showed the best ability to recover from distortion?
5   Why did steaming help crease recovery?
6   How could you apply this knowledge in treating a creased wool jacket?

The RESILIENCE of a fibre is its ability to recover from brief, small distortions. Resilient fibres are needed, for example, for the pile of a carpet, so it does not remain crushed after each step.

ELASTICITY is the fibre's ability to recover from extension. The elasticity of textile materials is especially important in tight-fitting sportswear, where the elbows and knees are repeatedly stretched. If the material is not elastic, it will not return to its original size and shape after each stretching, and the garment will become baggy.

Covalent bonds can stretch, but they are strong. The many disulfide bonds in the amorphous parts of the wool fibre are able to stretch when the fibre is extended. When the fibre is released, the disulfide bonds pull the protein molecules back into their original positions. If there are too few disulfide linkages, as when the fibre has been weakened by alkali, or if the extension is great enough to break some of the covalent bonds, then some polypeptide chains will slide past one another. This will cause a permanent extension of the wool.

The spring-like triple helix of the crystalline portions of the wool protein also helps the fibre regain its original shape after distortion or extension.

Comfort

Comfort is very much a subjective quality. A wearer can decide instantly if a garment is comfortable, but is is very difficult to determine precisely which factors in what proportion are most important in contributing to the sensation of comfort.

Some of the factors which may contribute to comfort are examined below.

Softness

Softness depends partly on the fibre diameter, partly on the cross-sectional shape of the fibre, and partly on the amorphousness of the fibre.

The initial resistance to touch of different fibres can be measured, and is expressed as values of INITIAL MODULUS. The lower the initial modulus, the softer the fibre.

As the table shows, wool is the softest fibre, but its softness can be imitated by some carefully engineered synthetic fibres. Cotton's crisper handle is reflected by its high modulus.

The range of values for the natural fibres is due to their variation in diameter and other properties. Fine wool fibres (e.g. lambswool, and the downy undercoat of wild sheep) and some hair fibres (such as vicuña, alpaca, angora, and mohair) are very much softer than the medium to coarse wools.

Initial modulus of some selected fibres

Fibre	Initial modulus
Wool	0.18–0.24 (depending on type)
Polyester staple (Dacron)	0.20
Acrylic staple (Orlon)	0.34
Cellulose acetate	0.39
Viscose rayon	0.54
Acrylic filament	0.86
Cotton	0.57–1.12 (depending on type)

Loftiness

A fabric should feel full and light, not heavy for its thickness. LOFTINESS depends on a low specific gravity, and on a permanent crimp, which prevents the fibres packing tightly, and so gives bulk. Wool's relatively high density (specific gravity 1.32) is compensated by the fine helical crimp of the finer type fibres.

Moisture absorption

High moisture absorbency allows perspiration to be taken into and through a garment.

This is no problem in the *hydrophilic* cellulosics. Fabrics made from *hydrophobic* materials can be specially designed for porosity, to permit water vapour to pass through (pp. 438–9).

Wool fibres have a *hydrophobic outer layer,* and so wool fabrics resist wetting. However, the inner part of the wool fibre is hydrophilic. Water in *vapour* form is able to pass through the outer layer and be absorbed by the *hydrophilic inner core* of the fibre.

Wool can absorb up to 40 per cent of its own weight of moisture without feeling noticeably damp. In addition, it actually releases heat as it becomes wet, and this may add to the comfort of wool garments in cold, damp conditions.

Wool fibres resist wetting, yet can absorb up to 40 per cent of their mass in water without feeling damp.
(David Bailey)

1   Wool and hair fibres have a hydrophobic outer layer, but absorb water vapour very readily, and give off heat as they do so. In what ways is this advantageous to the animal? Consider all the different weather conditions sheep must endure.

2   In the southern states of Australia, the Bureau of Meteorology issues 'sheep weather alerts' for cold, wet, and windy conditions. Why are these conditions dangerous for sheep? How does this tie in with your answer above?

Modifications to the properties of the wool fibre

Shrink-proofing

Wool has a unique ability to FELT because of the scales that cover each fibre. Methods of shrink-proofing attack the problem by eliminating the effect of scales, or by binding the fibres so they cannot move.

These methods have been so effective that the consumer can take a *Superwash* garment and throw it in the washing-machine, together with other fabrics traditionally regarded as easy-care. Early attempts at shrink-proofing wool were not nearly as satisfactory.

Chlorination

The first method of shrink-proofing, introduced in the 1930s, used chlorine. The chlorination process *softened and degraded the scales* of the wool fibre, and so eliminated the differential friction which caused felting and shrinkage.

Chlorination is an effective anti-felt treatment, and is still widely used. Another method of chemically degrading the scales uses the strong oxidant potassium permanganate, with salt added to confine the oxidative degradation to the surface of the wool fibre.

The use of polymers

More recent methods of shrink-proofing concentrate on the use of polymers to *coat* or *bind* the fibres. This approach aims to eliminate reactions which may damage the fibre or alter its handle.

In the *chlorine/Hercosett* process, mild chlorination prepares the surface of the wool fibres to accept a very thin, uniform coating of polymer. The prepolymer (polymer precursor) is applied to combed wool tops, which are then cured in an oven to develop the cross-links in the polymer coating. If the resin is not applied uniformly, the fabric will not be shrink-proof.

A more recent, improved process, called *Sirolan B.A.P.*, eliminates the need for chlorination. It uses an *acrylic resin* (with polyisocyanate) to cross-link the fibres. This is a 'spot-welding' effect, so it is appropriate for finished fabrics only.

The latest approach is to use *polyurethane* instead of acrylic emulsions. Polyurethane is the resilient, elastic material used to make foam cushions. When it is applied correctly to wool fabrics, it forms *flexible bridges* between the fibres, and prevents their movement during wet treatments.

Since the resin which coats the wool fibres is the substance which comes into contact with the skin of the user, its characteristics are very important in determining the handle of the shrink-proofed fabric. Hydrophobic resins — as all these are – lend some hydrophobic character to the otherwise hydrophilic wool.

Latest directions for research include mixing a variety of resins in order to achieve maximum shrink-proofing with minimum side-effects, and investigation of ways to eliminate the need for costly pre-treatments.

SUPERWASH wool is wool that has been treated with any one of many available processes. Whatever the process, to merit the Superwash label the final product must perform according to stringent standards set down by the International Wool Secretariat.

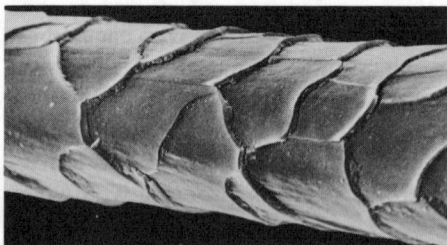

Wool untreated.
(C.S.I.R.O. Division of Textile Industry)

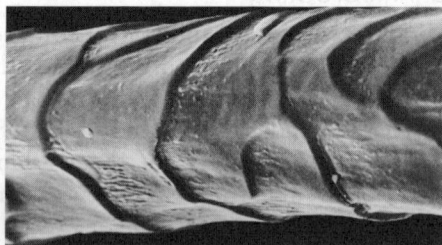

Softened wool scales.
(C.S.I.R.O. Divison of Textile Industry)

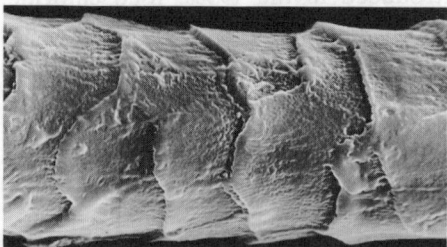

Wool scales coated with polymer.
(C.S.I.R.O. Division of Textile Industry)

Wool fibres bonded to prevent movement.
(C.S.I.R.O. Division of Textile Industry)

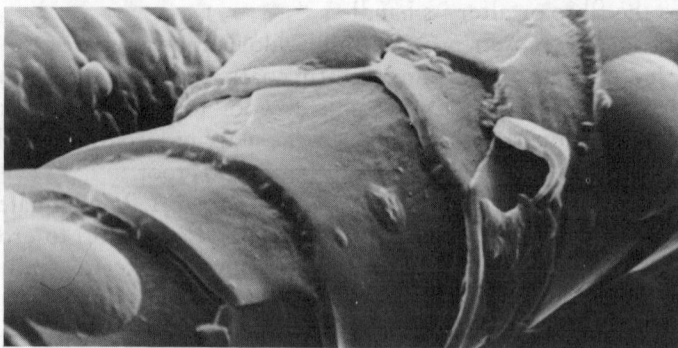

'Stand off' mechanism. The polymer bulges ensure that the wool fibres do not come into close enough contact for their scales to interlock, and so felting does not occur.
(C.S.I.R.O. Division of Textile Physics)

TRY YOUR HAND

You will need:

5 pieces Superwash pure wool knitting yarn, cut in 20 cm lengths
5 pieces pure wool knitting yarn that does not bear the Superwash label, cut in 20 cm lengths
hot soapy water

What you do:

Rub and squeeze the 5 lengths of Superwash wool vigorously in hot soapy water for 5 minutes.
Measure the new lengths of the samples, and calculate their average length.
Repeat this with the samples of the non-Superwash wool.

1   What has happened to the length of the two types of yarn? How does the behaviour of the Superwash wool compare with the other sample?
2   How does the handle of the two types of wool compare? Remember, this may reflect initial fibre fineness: check under the microscope.
3   Repeat the rubbing and pounding of your wool samples. Is there any further change in the yarn lengths?

Permanent setting

Wool is a wonderfully elastic and resilient fibre. For this reason, it is difficult to iron pleats into wool garments. Since the advent of machine-washable wool, the consumer naturally expects that trousers and pleated skirts should keep their creases through the wash. In order to achieve this, the wool fibres must be set permanently into the required form.

TRY YOUR HAND

You will need:

3 pieces of pure wool fabric, 10 cm × 10 cm
a packet of home perming solution with neutralizer
an iron (wool setting)

What you do:

Fold your samples and label them 1 to 3.
Use a dry iron to iron a crease into the first sample.
Wet the second sample and press the fold with the iron until it is quite dry.
Treat your third sample with the permanent waving solution first. Then place a cool iron on the folded edge for 10 minutes. Next apply the neutralizer, press the crease for a further 5 minutes, then rinse the fabric thoroughly and dry by ironing.
Check the effectiveness of each method by wetting your labelled samples in warm water, and hanging them out to dry as shown. Evaluate your samples when dry.
Now use the iron at the wool setting, and try to iron the crease flat again.
First use a dry setting and then a steam one.

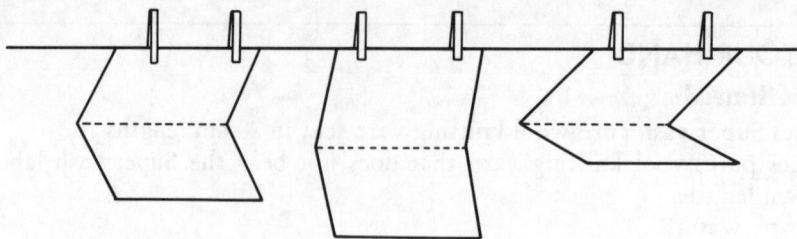

1   Did all three fabrics look and feel the same after setting in of the creases?
2   Which sample lost its set most readily? least readily?
3   Were any of the creases removed completely by ironing? Which?
4   Was the dry iron or the steam iron more effective in removing the creases?
5   What kinds of intermolecular bonds are broken most easily by heat?
6   What kinds of bonds are broken easily by steam?
7   Which are the strongest kinds of bonds between molecules?
8   The wool fibre has van der Waals bonds, hydrogen bonds, covalent disulfide
    bonds and even some ionic bonds serving as bridges between its molecules.
    Discuss which bonds were most likely to be affected by each of the three
    setting processes used.

Wool is a hair fibre, and its setting bears some resemblance to the setting of human hair.

To achieve a permanent change of shape in a wool (or hair) fibre, the *covalent disulfide bonds* need first to be *broken*, and then to be *reset* in a new position, to hold the fibre in its new shape.

Permanent setting lotions contain a strong *reducing agent* (p. 22) which breaks down the disulfide –S–S– bonds of cystine. Once these bonds are broken the fibre can be manoeuvred into a new shape. It is then treated with a *neutralizer,* which oxidizes the –S–H (removes hydrogen) to give new cystine disulfide bonds which hold the molecules of the wool fibre in the new position. Not all the broken disulfide bonds can be joined up again, so permanently set wool (or hair) is *weaker* than the untreated fibre.

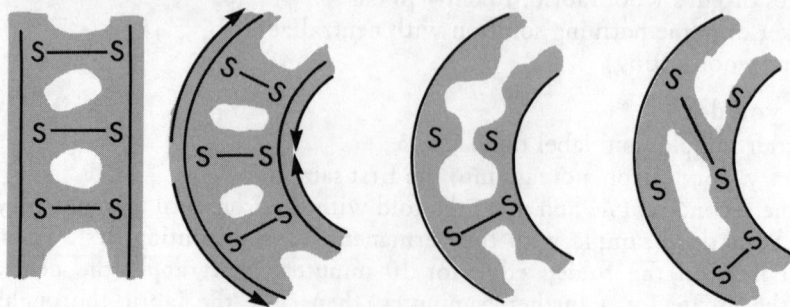

If a straight fibre is bent, the disulfide bonds between the polymer molecule are strained, and so act to pull the fibre back straight again (1,2). However, if the fibre is treated with a reducing agent, some of the disulfide bonds are broken (3). If the fibre in its new position is then treated with an oxidizing agent, some new disulfide bonds are formed, permanently setting the fibre in the new shape (4). Usually, more bonds are broken than reset, so there is some weakening of the fibre.

Extended steaming at a high heat setting can rearrange the disulfide bonds, as well as break hydrogen bonds. This is called STEAM SETTING.

In the *Si-ro-set* process, patented by the C.S.I.RO., wool is treated with a reducing agent (ammonium thioglycollate), pressed, folded, or otherwise arranged in the desired shape, and then *cured* by steam in an oven or autoclave.

In carpet yarns, setting is used to keep a highly twisted yarn in its helical shape. This way, the natural resilience of wool can be further increased.

For some purposes, a wool fabric needs to be set in a *flat* shape. This helps eliminate undesirable *wrinkling* and *creasing*. However, flat-set fabrics can be difficult to shape into a garment – they tend to pucker and strain around shoulders and other curved seams. To overcome this problem, the fabric is treated with the reductant, but is not cured. It is first cut, made up into garments, pressed into shape, and *then* cured to develop the new cross-links. In this way, the fabric will hold the desired shape of the garment.

Such durable press treatments are generally combined with anti-felting polymer treatments to provide even more sophisticated Superwash wool fabrics.

Moth-proofing

Wool and the other protein fibres are readily attacked by the larvae (grubs) of the clothes moth and by carpet beetles. For many years, the only protection the consumer had against the ravages of these creatures was frequent airing and cleaning of wool items, and the use of repulsive-smelling and toxic naphthalene mothballs (or the even worse paradichlorbenzene) in all the cupboards.

Even greater damage occurs in the warehouses where raw wool and untreated wool fabrics are stored from one season to another. The estimated $100 million yearly loss to the industry had been a great incentive to finding a reliable moth-proofing process.

Dieldrin kills the moths and beetles, and has functioned excellently to reduce the common clothes moth to the status of a minor pest. However, because of ecological considerations, dieldrin and related pesticides are falling from favour. They are difficult to remove from textiles, they are not biodegradable, and constant contact with treated wool may possibly cause chronic ill-effects.

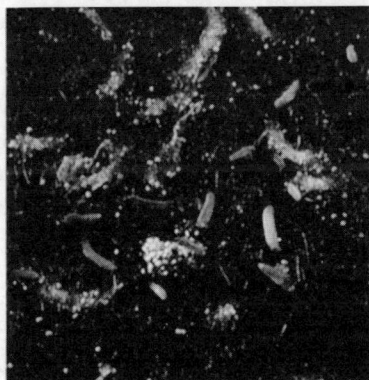

Insect attack on wool.
(*Bayer Farben Revue* No. 17, 1969, by permission of the editors. *Bayer Farben Revue* is published by Bayer AG, Leverkusen, West Germany.)

The most popular moth-proofing agents today are Mitin, Eulan, and the pyrethrins. These poisons are more selective, in that they kill the larvae of the moths but are not dangerous to people.

All wool carpeting and most blanketing and upholstery fabrics are moth-proofed – and today there are not many homes where a suddenly opened cupboard door sends clouds of tiny moths into a flurry.

In a fire, these overalls might save a life. Made from pure new wool, they were specially designed by the Australian Wool Corporation to protect volunteer fire fighters and others, such as farmers, who may confront fire in the normal course of their work. (Australian Wool Corporation)

Speciality hair fibres

Rare and beautiful

In today's textile world, with a plentiful supply of cheap synthetic fibres offering an enormous range of characteristics, wool may be considered to be a luxury fibre. It is comparatively expensive, but it features high on most lists of desirable fibres, for uses as diverse as fine, soft lambswool sweaters, and rugged, resilient carpeting.

In another sense, however, wool is the commonest, and cheapest, *hair* fibre. Sheep are domesticated, bred, and marketed on a huge scale. In contrast, other hair fibres used for textiles are obtained from animals less widely domesticated, and even from animals that are entirely wild. These speciality fibres are even softer, finer, warmer, and lighter, and far more expensive, than the finest wool. Usually they are blended with wool to give special, desired characteristics to the appearance of the fabric.

TRY YOUR HAND

Visit your local wool shop.

Find knitting yarns of the same ply made from wool, acrylic, and blends of speciality hair fibres – alpaca, mohair, angora are all available from time to time, depending on fashion.

For any jumper pattern suited to this ply, estimate the cost of knitting it in each of the yarns you have found.

Look carefully at the labels on the yarns. What information is most important to the consumer? Do you think that more information is needed? If so, what?

Note the countries of manufacture of each type of speciality yarn. Are these the same as the countries of origin of the hair fibres? (See tables, pp. 100, 102, and 103.

Bactrian camel. (Gaye Hamilton, Royal Melbourne Zoo Education Service)

Guanaco. (Royal Melbourne Zoo Education Service)

The camel family

The hair of many members of the camel family is used for exclusive textile fibres. These animals have a coarse outer coat, and a soft, fine, downy undercoat of fur.

Camel hair is obtained from the Bactrian camel of Asia. The stocky, two-humped camel is well adapted to the cold deserts of central Asia, and has dense fur which it moults regularly. Most camels used for hair production are hand sheared.

The *llama*, and its smaller cousin the *alpaca*, have been domesticated for centuries by the Andean Indians of South America. They were beasts of burden

99

Hair fibres from the camel family

Animal	Yield per animal per year (kg)	Total yearly production (kg)	Principal countries of origin	Fibre diameter and length	Notes
Camel	2.5	1 400 000	China Soviet Union Afghanistan	5–40 micron 4–12 cm	Strength similar to wool, but warmer and lighter. Scales not distinct.
Vicuña	0.08–0.5	20 000	Bolivia Peru Chile	7–14 micron 1–6 cm	Very soft, strong, lustrous, resilient. Sensitive to chemicals.
Alpaca	1.5–2		Peru	22–30 micron	Stiffer but lighter than wool, with less tendency to felt. Weakened by bleaching.
Llama	1.5–2	local use	Peru	10–20 micron 8–25 cm	Lighter, warmer than wool.
Guanaco (pelts)			Argentina	18–24 micron 5 cm	Finer than alpaca, but coarser than vicuña. 50 per cent medullated fibres.

Note: All these figures are for the soft undercoats. The guard hairs are usually removed during cleaning and processing, with a loss of up to 50 per cent of the original fibre yield. Production data are compiled from a number of sources, and are 1970s figures.

in the Inca empire, and they still perform that function for the Indians of today. Their fur is shorn every two years. As the animals' grazing conditions may vary over the two year interval between shearings, each fibre in the fleece can vary in thickness from its tip to its root. Alpaca hair is finer than llama hair, but both make *light-weight, warm* textiles. This is because they both have a high proportion of *medullated fibres*. The hollow central canal acts as an insulator by resisting the transfer of heat from one side of the fibre to the other, and also reduces the weight of the fibre.

The *vicuña* and *guanaco* are smaller than the alpaca. They are not domesticated; instead they are hunted for their pelts. The vicuña, which lives in the dry grasslands of the central Andes, is highly valued for its fine, soft, silky fleece,

Angora goats.
(Department of Agriculture and Rural Affairs, Victoria)

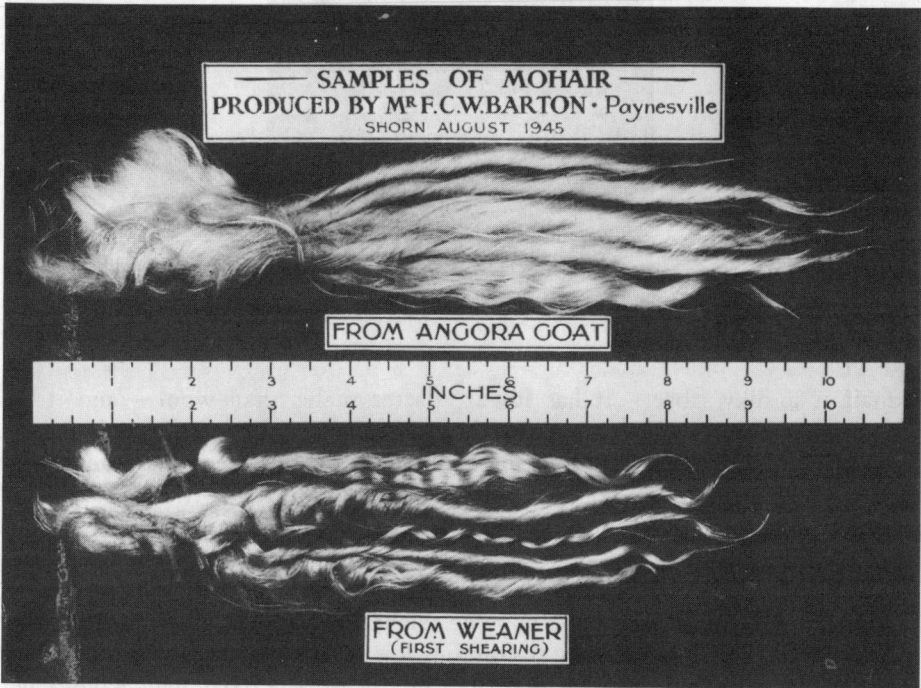

SAMPLES OF MOHAIR
PRODUCED BY Mʳ F.C.W.BARTON · Paynesville
SHORN AUGUST 1945

FROM ANGORA GOAT

INCHES

FROM WEANER
(FIRST SHEARING)

Samples of Australian mohair. (1 inch = 2.5 cm approx.)
(Department of Agriculture and Rural Affairs, Victoria)

which was once reserved for Inca royalty. Today it is protected by a strict hunting quota. The guanaco lives in the southern Andes, where its young are hunted for their soft coats. The pelts are used as fur, and the trimmings are utilized by the textile industry as a source of fibre.

TRY YOUR HAND

Look at the table, and consider:
1   Which fibre would you expect to be most difficult to obtain?
2   Does this – obviously very expensive – fibre have any particular advantages over the other fibres?
3   List the fibres in their order of fineness.
4   What property does the fineness give to textiles made from the fibres?

The goat family

Goats were probably the first animals whose long hairs were used as a textile fibre. Today, two kinds of goats supply speciality hair fibres – the silky, long-haired mohair (angora) and the soft, downy cashmere goat from Kashmir on the India–Pakistan border.

Mohair has been produced in Turkey for thousands of years. Its name derives from the Arabic word *mukhayyar,* which means 'goat hair fabric', and the goat which supplies it is called *angora,* after the district of Ankara in Turkey.

Fragment of a Kashmiri shawl, hand-woven in a paisley design.
(David Bailey)

Mohair is a silky fibre – it has fewer, flatter scales than wool – and it has little tendency to felt. Mohair suitings have a characteristic lustrous appearance.

Cashmere is obtained, by plucking or combing, from the downy undercoat of the Kashmir goats, grazing the dry mountains of Asia. It is a weak and sensitive fibre, but prized for its fineness and softness. In Europe, it first became known through superb hand-woven and embroidered shawls, which were later imitated for the European market by mechanized industry based in Paisley, Scotland. It is from here that the popular oriental paisley design takes its name.

A top quality cashmere sweater contains less than 1 per cent of coarse guard hairs, and needs the yearly yield of 4–6 goats. An overcoat may have 5 per cent coarse hair content, but needs 30–40 animals to supply sufficient fibre for it.

Fine cashmere fibres have no medulla, and are weaker and more easily damaged than wool, but they have outstanding qualities of drape and texture. If natural dark cashmere is bleached, it may be seriously weakened.

Hair fibres from goats

Breed	Yield per animal per year (kg)	Total yearly production (kg)	Principal countries of origin	Fibre diameter and length	Notes
Mohair	10 (5 kg twice per year)	53 000 000	South Africa Turkey United States South America Australia	10–90 micron 10–15 cm	Lustrous, strong, does not felt
Cashmere	0.1–0.5	4 500 000	China Soviet Union Iran	15–20 micron 2.5–10 cm	Scales less distinct than wool, but more pronounced than mohair

Note: Production figures are compiled from a number of sources, and apply to the 1970s and 1980s.

Rabbit

Whilst mohair comes from the angora goat, *angora* is the fur of the angora rabbit. These are domesticated varieties, either French or English, bred for

Rabbit skins displayed ready for auction: before the introduction of myxomatosis. (Department of Agriculture and Rural Affairs, Victoria)

their long, silky fur and their excellent meat. The fur is usually collected during moulting by combing the animals, but the rabbits may be clipped or shorn instead. Because of their high cost, the soft, silky angora fibres are often blended with other fibres.

Before the introduction of myxomatosis to control rabbits in Australia, much felt for men's hats was made from fur from rabbit pelts. This has largely been replaced by wool felts. In the peak years of rabbit hunting in Australia, some 100 000 000 animals were processed each year for export as skins or carcasses.

TRY YOUR HAND

If your area has a hat shop, or a large department store with a hats section, compare a fur felt hat to a wool felt hat. How do they compare for stiffness, resilience, surface finish, price? What fur is used for the fur felt? Ask the retailer for information about the different felts.

Angora rabbit fur

Yield per animal	Total yearly production (1970s)	Principal countries of origin	Fibre length	Notes
0.4 kg total per year, collected four times a year	small scale	France United Kingdom United States Canada	8–9 cm	Silky, lightweight, warm, medullated fibre

Silk

The silkworm

The mature caterpillar of a moth called *Bombyx mori* spins itself a cocoon when it is ready for metamorphosis. It is this cocoon which yields SILK.

The silk fibre is produced as twin *filaments* stuck together by a gummy substance. The silkworm produces this fibre through two glands in its mouth, as it weaves its head about in figures of eight to make the cocoon. The gum which holds the twin filaments together hardens and hold the cocoon in shape. This gum, called SERICIN, can be softened in hot water, and the filaments unwound from the cocoon.

Legend has it that the cocoon of a silkworm pupa fell into the teacup of the Chinese empress Xiling Shi – 4600 years ago – while she was dining in the mulberry orchard. The hot liquid softened the gum that held the threads together. The empress was fascinated, and unravelled the long, shiny filaments.

China became the first manufacturer of silk, and eventually became famous world-wide for the silk fabrics it exported. The secret of silk production was guarded jealously, and the silk route from China to Europe became the most important trading route through Asia.

Some 3000 years after this legendary beginning, the secret monopoly was broken – according to legend – by Buddhist monks who smuggled some silkworm eggs into Korea.

Rich cultural traditions grew up around the production of silk (SERICULTURE) in both China and Japan. These two countries together produce about two-thirds of the world's silk today.

In Europe, silk production dates from the sixth century AD, when the Byzantine Emperor Justinian introduced sericulture into south-eastern Europe. With the rise and spread of Islam in the seventh century, silk culture was introduced into Spain and Sicily. The art of silk cultivation remained when these countries were reconquered, and Italy, and later France, became important centres for silk production.

Cultivated silkworms are fed on the leaves of the mulberry tree. If the caterpillars are fed on other leaves or on artificial food, they produce silk of inferior colour and quality.

In order to produce 1 kg of silk, 104 kg of mulberry leaves need to be eaten by 3000 silkworms. It takes about 5000 silkworms to make a pure silk kimono.

Cocoon of the silkworm.
(*Bayer Farben Revue* No. 15, 1969, by permission of the editors. *Bayer Farben Revue* is published by Bayer AG, Leverkusen, West Germany.)

TRY YOUR HAND

Using books in your library and other references you may have, research the commercial care and breeding of silkworms.

1   How long does it take from hatching of the caterpillar to spinning of the cocoon?
2   What special care is needed by the silkworms?
3   Is sericulture labour intensive or capital intensive?
4   How are the cocoons treated commercially to obtain the silk?

After you have researched the care needed to raise silkworms, you could try raising some yourself.

Obtain some eggs (some pet shops stock them), and divide them into two transparent acetate boxes or shoeboxes. Keep them in a warm place until they hatch.

Feed one box of caterpillars on finely chopped mulberry leaves, the others on other soft leaves; wait and watch how the matured caterpillars spin their cocoons.

Drop a cocoon into boiling soapy water. This kills the pupating caterpillar instantly. Try to unravel the silk: you can find the end of the filaments by lightly brushing the outside of the cocoon.

How does the silk produced by caterpillars on the different diets compare for colour, strength, and texture?

Note: Silkworms cannot survive in the wild. Kill all the caterpillars in the cocoons, when they are pupating and least sensitive, or keep them for an on-going class project.

TUSSAH SILK and WILD SILK are the products of moths called *Antheraea* closely related to the silkworm. Their caterpillars feed on oak leaves, and produce a silk which may be any colour from brown through yellow to green. The wild silkworm moth escapes from the cocoon, and the resulting short fibres are always *spun*.

Although the cocoons of many kinds of moth have been used as a source of textile fibre, only *Bombyx mori* and *Antherea* are of commercial significance.

The microstructure of silk

Silk is a protein fibre. The silk protein, FIBROIN, is made up of mostly very small amino acids, which permit close packing of the molecules in the crystalline regions of the filament. Important among these small amino acids is *serine,* as its hydroxyl end-group allows for hydrogen bonding between the polymer molecules. As a result of this, silk is a highly crystalline fibre.

In the crystalline regions of the silk fibre, the molecules are arranged *parallel,* but not in a helix like keratin. Fibroin differs from keratin also in that there is no cystine in silk, and so there is no disulfide bonding between molecules.

The silk filament itself has no discernable microstructure, and in this, too, it differs from wool and the other natural fibres.

glycine
44%

alanine
27%

serine
polar: 11%

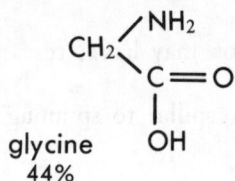

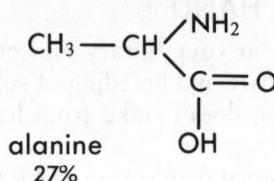

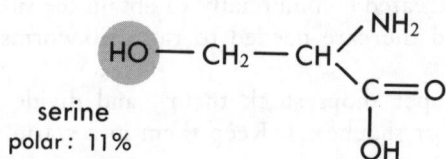

Amino acids important in silk.

TRY YOUR HAND

On the basis of their structures, how would you expect wool and silk to compare for moisture absorbency? strength? resilience?

Transporting silk cocoons, China. Silk production is carried on on a large scale in many rural communities in Asia; in Japan it is often factory-based.
(© Cary Wolinsky, Stock, Boston)

Reeling silk, Chiang Mai, northern Thailand.
(Anna Janca)

The properties of silk

Many of the properties of silk are similar to those of wool. Some of the more important features are listed in the table.

Despite a century of research, no synthetic fibre has yet been engineered to have pure silk's unique combination of handle, drape, lustre, resilience, and comfort.

Properties of silk

Property	Explanation
Stiffer than wool, crisper handle	Highly crystalline fibre resists bending and stretching.
Crush-resistant	Crystal structure resists *small* distortions, so excellent resilience and recovery, provided distorting force is not too great.
Strong	Again, because the fibre is highly crystalline. However, silk has an extension of only 20 per cent at break, compared to wool's 35 per cent: silk is *comparatively inelastic*.
Hydrophilic	Although less moisture-absorbent than wool, silk has a standard regain of 11 per cent: this is quite high. It has few amorphous regions to absorb the water molecules, but it does have some polar groups on its amino acids.
Sensitive to acids and alkalis	In the presence of acids and alkalis, the peptide bonds of the fibroin molecule are hydrolysed. This means great care is needed in washing silks, as ordinary detergents and soaps are alkaline: only neutral soaps and non-ionic detergents should be used. The sensitivity to *acid* means that silk garments should be washed or cleaned after each wearing, as the acids in perspiration can seriously weaken the fibre.
Yellows and is weakened in sunlight and on contact with chlorine bleaches	Although silk does not have many bulky amino acids, those which it does have are readily decomposed by ultraviolet light or the action of chlorine, and yield yellow or brown products.
Burns only slowly	This resistance to burning is a feature of all protein materials. *Weighted* silk (see pp. 108, 436) chars instead burning.

The structure of silk

Silk is extruded from the glands of the silkworm as a continuous filament, about 1000 metres long. Each filament has a triangular cross-section. The three flat sides act as tiny mirrors, and give silk fabrics their special lustre.

To make the silk yarn, 4–8 cocoons, depending on the yarn thickness required, are put into hot soapy water and their filaments reeled together. As soon as one cocoon (or BAVE) is completely unravelled, another is joined in its place, to keep the yarn thickness uniform.

Silk stockings, popular before the advent of nylon, were usually made from twisted silk yarns reeled from 4–6 baves. After 9000 metres was reeled, the hank formed was removed and weighed: the heavier the hank, the thicker the yarn. The yarn thickness was expressed in DENIER, the mass in grams of the 9000 metre hank.

This system of specifying yarn thickness is still in use today: most modern nylon hose are made from 15 denier nylon.

Not all silk is used in its shiny, degummed filament form. If the sericin is not removed, the fibres look creamy yellow and have a duller lustre. They are called RAW SILK.

Sometimes two silkworms will spin their cocoon together. They yield a filament which is irregular in diameter, and which is used for special texture effects. This silk is called DUPION SILK.

Also, the silk growers do not kill every silkworm in its cocoon, as some caterpillars must be kept alive to produce eggs for the next generation. The adult moth has to pierce the cocoon to emerge. The pierced cocoon has only short fibres; other short silk fibres are found from the first brushings of cocoons to find the leading end of the filament, and at the inside end of the cocoon. All these short fibres are called WASTE SILK, but they are *not* wasted. They are spun like any ordinary staple fibre. SPUN SILK (staple silk) yarns look, feel and behave differently from silk filaments.

Silk fabrics – the quality of silks

TRY YOUR HAND

Go to the silks section of a fabric shop.
Identify the range of pure silk fabrics available, and find a comparable range of imitations.

1   Which samples of pure silk are spun silk? Which are filament?
Draw up a table comparing the various fabrics for cost, handle, drape, weight, resilience.

2   Did you find that the fabrics most pleasing to the touch were also the most expensive ones?
3   How did the synthetics compare to the real silk fabrics in price? In feel? In drape? In 'looks'?
4   Was there much difference in the behaviour of spun silk fabrics and filament silk fabrics?
5   What were the care instructions for the various silks and silk imitations?
6   Were the sales staff familiar with the properties of the merchandise?
7   Did you find any fabrics that were marked 'weighted silk'?
8   Were any pure silk fabrics significantly cheaper than the rest? Why do you think they were cheaper? How did they compare in resilience to the others?

Heavy fabrics generally drape better than do light-weight fabrics. Silk is very expensive, so a pure silk fabric with a heavy drape costs a great deal. One way the drape of lighter silk fabrics is improved is by soaking them in salts of tin. The silk can absorb considerable amounts of these metallic salts, which add to its weight, and so a light-weight silk fabric can achieve the drape of a heavy fabric, at less cost.

The metal salts are absorbed into the amorphous regions of the silk fibre, and make the fibre brittle. WEIGHTED SILK crushes more easily and is much less durable than pure silk. All silks are weighted to some extent.

Some silk fabrics – like taffeta – have a special swishing sound, called *scroop*. Although it can be used to distinguish silk taffeta from rayon taffeta, scroop is not necessarily a sign of quality: it is achieved by a special acid treatment of the fabric, which is thought to harden the outside of the fibres.

Like all textiles, silk comes in a range of qualities. In the photographs below fabrics of the same name but different composition – and different price – are compared. In each case, the handle and drape of the more expensive fabrics are superior; but naturally they are difficult to distinguish by untrained eyes and hands.

Most silk garments carry the label 'dry-clean only'. This is not because silk is damaged by water, but because highly twisted yarns will make the fabric crinkle and shrink when wetted. This is due to the slight swelling of the wet fibres, and is a function of the yarn structure rather than of the fibre content. Occasionally, silk may be dyed with dyes which are not washfast; but generally, most silk garments can be hand-laundered with mild agitation. In fact, silk is more resistant to alkali and to heat than wool is. After washing, the damp garment can be ironed under a press cloth till dry.

One problem that the consumer faces with silk is that many dyes are not *lightfast*. Improved dyes and dyeing techniques are going some way to overcoming this, but the consumer needs to be aware of the danger of silks fading.

Pure silk crepe de chine.
(Vivian Robinson)

Polyester crepe de chine.
(Vivian Robinson)

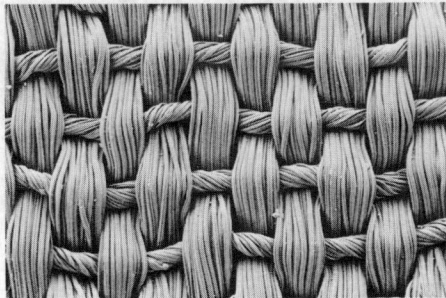

Acetate warp, silk weft crepe de chine. (The highly twisted yarn is the silk weft.)
(Vivian Robinson)

Furs and leather

From hides to clothing

The first protective garment that people ever wore was the natural covering of an animal – its hide. Hides are still used in some societies as hunting trophies or emblems of power – power which was originally associated with the animal itself. To some groups in our own society, furs have a glamour status that bears no relation to their warmth and comfort.

When the hide is stripped from the animal it has small amounts of flesh sticking to it, and is a perishable product. To preserve it, the hide is *scraped* and salted. In this state, however, skins are stiff and not really suitable for use as clothing. Eskimos used to soften hides by chewing them. After each time the hide got wet, it stiffened and had to be chewed again.

Peoples who hunted in forests discovered that if they soaked the hides together with bark from oak trees, then the hides remained permanently softened. This process is called oak TANNING.

Today, chemicals are used to make leathers and furs pliable and water resistant. Different treatments are used to produce different results: chrome tanning (sodium bichromate with glucose and sulfuric acid) is used for most leathers; oil tanning for chamois; alum tanning for white (glove) leathers.

Shadow fox jacket.
(Stephen Dattner, Furrier)

Leather

Leather is used for clothing, upholstery, bags and accessories, and shoes. Most of the leather for these uses is obtained from the hides of cattle. Pigs, goats, lizards, crocodiles, snakes, and even peacocks, yield their hide for speciality leathers in the fashion industry.

Leather is the *skin* or dermis of the animal. Leathers vary enormously in thickness, pliability and even durability. The hide of animals is a natural product which is not uniform even on the animal: it may have scratches and scars, and will vary in thickness and other qualities depending on what part of the animal it is from.

Ths skin of cows is often thick enough to be split into a number of layers. The top layer is called TOP GRAIN, and is the part which carries the characteristic surface pattern of the animal's skin. This is the portion which takes the best finish and which lasts the longest in use.

Only 5 per cent of all cattle hides are suitable for smooth top-grain cowhide in aniline finish (alcohol-soluble dye which *penetrates* the hide). The rest must be *embossed, buffed,* and *corrected*; their colour is applied as a layer to the *surface* of the hide.

The SPLITS are layers taken from deeper towards the flesh side of the hide. There, the elastic collagen fibres which make up the skin's texture are less densely packed than near the outside surface. Split skins are more difficult to finish than top grains, and rough up more easily in use.

The back of the animal yields the best leather, firm and close-grained. The skin on the animal's belly tends to be coarser grained, and can cause problems to the consumer by stretching during use.

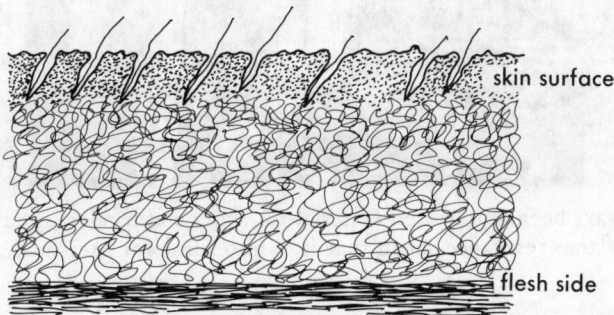

The structure of leather. The collagen fibres are more densely packed near the skin surface.

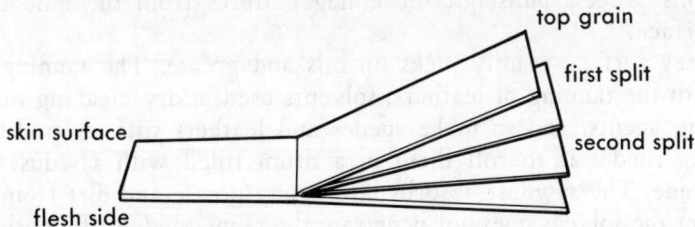

Splitting of thick hide into layers.

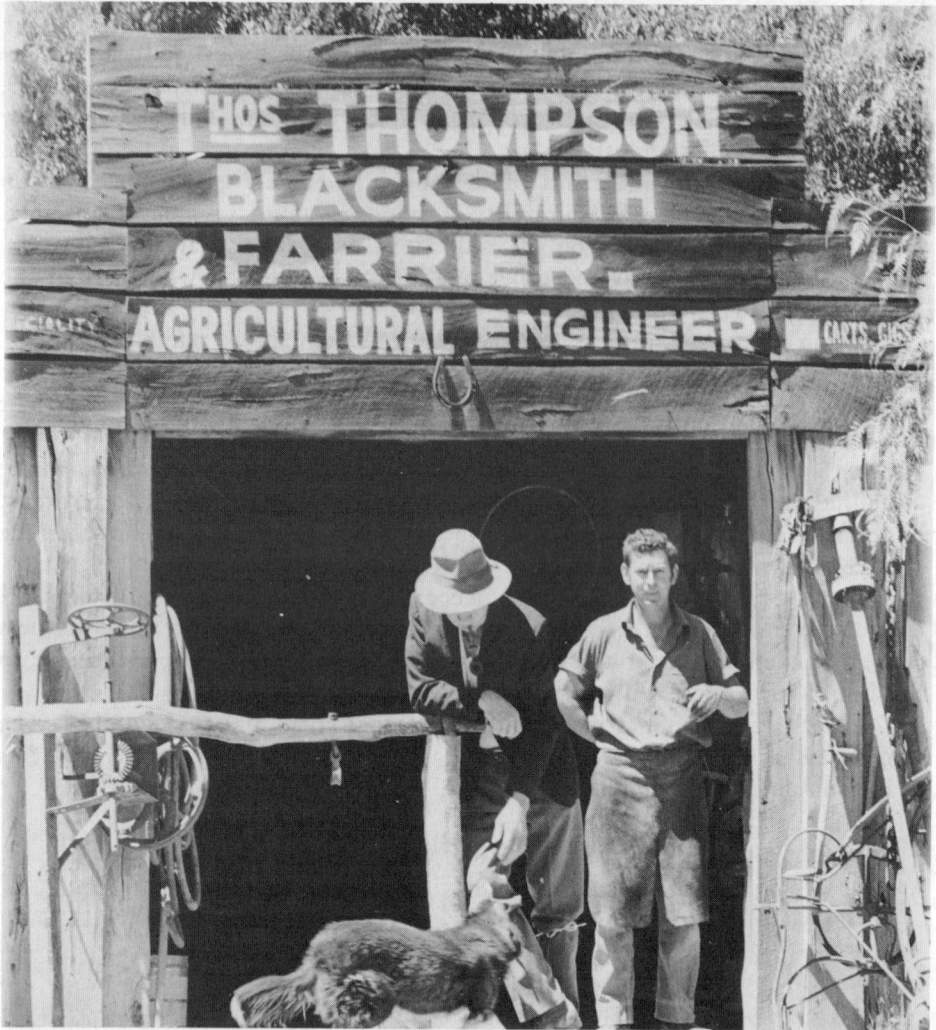

Leather has always been a preferred material for shoes and handbags, but its toughness, pliability, and flame resistance mean that it is also ideal for uses such as blacksmiths' aprons.

Suedes

SUEDES are produced by brushing the flesh side of the skin with emery coated rollers. This process pulls out the collagen fibres from the hide to produce a fuzzy surface.

The fuzzy surface readily picks up oils and grease. The tanning of suedes is similar to the tanning of leathers; solvents used in dry-cleaning may remove the tanning agents, and so make suedes and leathers stiff. A traditional way of cleaning suedes is to roll them in a drum filled with sawdust moistened with benzene. The sawdust absorbs into it any grease and dirt from the suede surface, yet the solvent does not penetrate the tanned hide sufficiently to stiffen it.

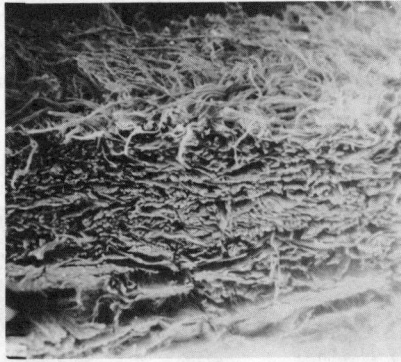

A slice through suede, showing the raised collagen fibres which form the fuzzy surface (Vivian Robinson)

Furs

Because of ecological considerations the hunting of many fur animals has been ruled out by law. Tigers, leopards, cheetahs are protected, together with possums, koalas and most kangaroos.

Although many wild animals are still hunted for their pelts – foxes, racoons, beavers, seals to name just a few – most furs are produced on highly efficient fur farms. Minks, nutrias, Persian lambs, even chinchillas, are carefully bred for the colour and quality of their fur.

Mink has a high status as a fur pelt, but not all minks have the same quality. Winter season male animals always have a lusher coat than females or those skinned in the warmer summer months. The health of the animal can also affect the quality of its fur.

The manufacture of fur garments demands a great deal of skill. The art of the furrier is in matching the natural pelts so that furs of different colour or height do not appear next to each other in the finished garment. Next, the skins must be shaped to produce the final garment form, but without wasting any of the precious pelts. This – in the most expensive minks – is achieved by cutting the skins into strips 1 cm wide and then moving them a little and sewing them together again to give the required pattern shape. In this way a single mink pelt can be made narrower and longer, until it stretches the whole length of the coat. This process (called 'letting out') also emphasizes the central stripe of guard hairs on the pelt.

The pelts of *rabbits* are easily obtainable and inexpensive. They can be dyed and cropped and printed to resemble other furs, and in the past have been sold under names such as lapin, chinchillette, marmink, Australian seal, Baltic leopard and Belgian beaver. Such ambiguous naming is now outlawed because the fur of rabbits sheds readily, and garments made from it cannot be expected to have the warmth or the durability of the pelts it seeks to imitate.

The humble sheepskin is, of course, a fur. It is used for car seat covers, ugg boots, rugged garments, floor rugs. An important use is as *surgical sheepskins*, placed under bed-ridden patients to reduce the risk of bedsores (pressure sores). Surgical sheepskins need to be easy to clean and sterilize, a difficulty that has only recently been overcome. New tanning procedures developed by the C.S.I.R.O. Division of Protein Chemistry permit the sheepskin to be *washed*.

A high quality fur garment can be expected to last for 30–40 years in good condition, provided it is well cared for. Protection from moisture and heat – which produce mildew and rotting – and from insects is important. Furs should not be crushed tightly in a wardrobe, nor folded up, nor kept in plastic bags. Instead of brushing, the garment should be shaken or even lightly beaten to remove dust from between the hairs of the fur.

Furrier at work.
(Anne Fritz)

Consumer questions answered

Q1 Which is really warmer – a wool garment or an identical acrylic one?

A   The thermal resistance of acrylic is quite high – it is not a good conductor of heat. Wool is also a poor conductor of heat. Knitted and woven fabrics, particularly those made from staple fibres, however, depend for their insulative value mostly on small pockets of air trapped in the garment.

Textured acrylic yarns are able to trap many small pockets of air between their fibres. Wool yarns, however, are able to trap their air pockets more effectively, because of the restrictive effect of the scales. Since the air does not move (say from the inside to the outside of the garment), the wool garment is warmer than the acrylic one.

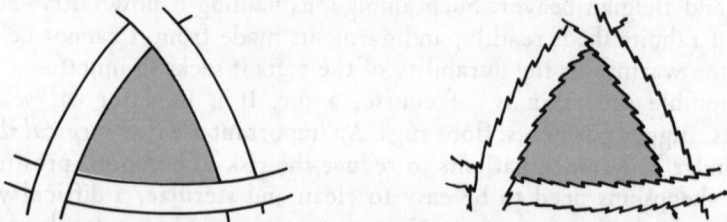

Pockets of air trapped by acrylic (left) and wool (right) of the same fibre diameter.

114

In addition to this, wool absorbs moisture readily, and gives off heat when it becomes wet. Acrylics do not absorb moisture. Although wool is the better insulator, perspiration condenses in an acrylic garment because it prevents free passage of water vapour through it. This means that wool is comfortable to wear, but that acrylics tend to feel clammy.

Q2   I had a pure wool cardigan and put it in the washing machine together with some acrylic and some nylon jumpers. The wool cardigan shrank. What can I do about it?

A   Felting of wool occurs by the interlocking of the scales of the fibres. The soapy water acts as a lubricant, and during the agitation of washing the fibres move relative to one another. As they move, the fabric becomes thicker and smaller, and the fibres move closer so as to leave fewer spaces in the fabric. Because of the one-way direction of the scales on the fibres, each fibre moves in the direction of its root end, and does not move back; hence felting shrinkage is irreversible.

Although the felting shrinkage cannot be reversed, a garment which shrank 5–10 per cent can be stretched out to its original size. This stretching makes use of the easy extensibility of the wool fibres, combined with its ability to be set in the presence of heat and moisture.

To stretch it, the original shape of the garment must be drawn on a board. The garment is then turned inside out, moistened and stretched to fit the drawn shape, and held by pins hammered into it all around. A cloth is placed over the garment (to prevent scorching), and it is ironed with a hot iron until dry. It must be allowed to cool while still stretched – ironing followed by overnight stretching produces good results.

Extreme cases of felting shrinkage cannot be compensated for in this way.

Q3   Should I use home dry-cleaning spray on my suede coat?

A   No. The principle behind the spray dry-cleaning product is to apply to the grease spot a solvent and a fine absorbent powder which will absorb the dissolved grease. Suedes need a coarse absorbent medium such as sawdust, which does not stick to the fine surface fibres of the suede, and a smaller proportion of solvent, so that the tanning agents are not leached out.

Indiscriminate use of dry-cleaning agents could result in stiff, stained suedes and leathers.

Q4   Should I keep a fur coat in a plastic bag?

A   No. Plastic may keep dust and insects out, but it does keep moisture in. In hot, humid weather, mildew and rotting could cause rapid deterioration of the pelts. A calico bag with a few naphthalene flakes – or pest strips – is a better idea.

Q5   Why do some pure wool garments feel scratchy, while others don't?

A   Any garment – whether wool or not – will feel scratchy if it is made of short, stiff fibres which stick out of the cloth. Garments made from very fine wool fibres do not irritate, as any protruding fibres will bend away gently as they come in contact with the skin. Experimentation has shown that if a wool blend contains more than 10 per cent of fibres that are thicker than 30 microns in diameter, then the fabric will irritate the skin. Blends made from ultrafine wool (less than 19 microns) are non-scratchy.

Q6   Why do many silk garments bear the label 'dry-clean only'?

A   Silk is a protein fibre which is much stronger than wool. It has no scales and hence does not felt – there is little reason why silks should not be cleaned by wet processes.

It would appear that some manufacturers feel that an expensive garment is best cared for professionally, to avoid any possible consumer complaints being lodged against them. Such complaints could arise from indiscriminate use of detergents, bleaches or strong washing cycles on delicate fabrics.

If the fabric is made from highly twisted yarns, the manufacturer may consider there is risk of shrinkage as the yarns swell with moisture. It is possible that the dyes used in the garment are not fast to washing; or possibly the threads or other notions do not meet the requirements of hand washing.

In fact, however, many garments labelled 'dry-clean only' can be successfully cleaned by careful and gentle hand washing.

Q7   Why should wools and silks never be soaked in an enzyme stain remover?

A   The stain remover contains *proteolytic enzymes* which are designed to digest the proteins of egg and blood stains. The same enzymes will also attack the proteins of silk and wool, and so weaken the soaked fabric.

Q8   What is so special about mohair suiting fabrics?

A   Mohair is a fine, strong fibre, with a much flatter scale structure than wool has. It has no tendency to felt, and has a special lustre which has aesthetic appeal for some. If well processed, the fabric will keep its appearance for a long time.

Q9   Why are fur felts considered better for hats than wool felts are?

A   Because the fur fibres are generally softer, finer and more pliable than wool, and have no crimp. Because of this, they can be felted closer to one another to give a dense felt which keeps its shape well. Hats made from wool felt stretch out of shape easily.

Q10   What is the difference between lambswool and pure new wool?

A   Although lambswool is pure new wool, it is special because it is the first shearing of a lamb. The fibres are therefore finer than those produced by an adult sheep of the same breed. Lambswool fabrics are generally soft, though they may be less soft than a finest quality Merino wool garment. Under the microscope, lambswool can be identified by each fibre having a pointed tip.

Q11   Whenever I travel, I place my clothes in the suitcase freshly pressed. Nevertheless, my suits get crushed, and my small travel iron will not remove the creases. What should I do?

A   Ironing, with its heat and applied pressure, softens the fibres and dis-organizes the molecular chains. This means they are more easily reset – or creased – after ironing, than if they were not ironed. It is therefore best to pack clothes into the suitcase without pressing them first.

When the clothes are unpacked, they should be allowed to relax: the best method is to hang them on a clothes hanger in a steamy bathroom (*not* under the shower!). Steam will break any hydrogen bonds formed within the hydrophilic fibres during creasing, and so allow them to return to their original set. If the clothes are allowed to hang overnight, this will allow them to reset in their correct shape as new hydrogen bonds form, and so they will be more resistant to further creasing.

5 Synthetic fibres

In this chapter, ideas about polymerization are explored. Synthetic fibres are grouped according to traditional nomenclature, and their properties are investigated through many practical activities. The non-biological fibres – fibreglass, metallic fibres, carbon fibres, asbestos – are also examined. Throughout, the properties of the synthetic fibres are explored and explained in terms of their molecular structure, and their significance to the consumer.

The consumer will find answers to questions such as these:
- How do imitations of natural fibres differ from the genuine product?
- What properties can the consumer expect in a synthetic fibre that were undreamt of in textile materials fifty years ago?
- Should polyester garments be ironed with a steam iron?
- Why do some household sponges work better than others?
- Why do fishermen get so annoyed about knots in their lines?
- What can I do about a quilted jacket that pills?
- What is the best way to care for acrylic jumpers?
- Why won't food stick to a non-stick frypan?
- Why can't I wash fibreglass curtains in the washing machine?
- How could a water-soluble fibre have a use?
- How is power stretch created for foundation garments?

and many other interesting titbits of information.

Polymers

Synthetic polymers

A POLYMER is a vast molecule formed from many small units (MONOMERS) joined together.

The previous two chapters have discussed the natural polymers: cellulose, built up from the cellobiose (double glucose) unit, and the more complex proteins, keratin and fibroin (and collagen, in leathers), built up from amino acids.

This chapter deals with the manufactured polymers: those in which the monomers are chosen and combined in an industrial, rather than living, process.

The synthetic polymers cover a vast range of properties and characteristics. They may be chemically inert and unreactive, like Teflon, or so very hydrophilic that they will dissolve in water, like polyvinyl alcohol. They may be flammable, like acrylics, or fire resistant, like the aramids. They may be strong, resilient, easy to dye, or fragile and difficult to handle.

The properties of the natural polymer systems of wool, silk, cotton, flax, are dictated by the animal or plant, and can be modified only very superficially: by many slow generations of selective breeding, and by treatments during processing. Even triacetate, the least natural of the natural polymers, relies on a plant to produce its cellulose skeleton.

In contrast, synthetic polymers can, in theory at least, be *engineered* to have exactly the desired properties. The *choice of monomer* is the most obvious variable: it is so important that polymers are classified generically by their monomers. It dictates such characteristics as chemical resistance, flammability, ease of dyeing, moisture affinity. The *reaction conditions* under which the polymer is formed, and the *type of chemical reaction* involved, determine the chain length and the pattern of the repeating monomer units. These in turn are crucial to the strength and many other physical properties of the polymer.

Petroleum by-products are the starting material for the manufacture of all synthetic polymers.
(Australian Information Service)

Then, once the polymer has been synthesized, the *method and conditions of extrusion* affect the degree of crystallinity, and thus the strength of the polymer, and many characteristics that are less easily measured: lustre, handle, drape.

A good understanding of performance characteristics and a clear image of her or his requirements are essential if the consumer is to make an informed choice from among the myriad of synthetic textiles on the market. The textiles manufacturer needs a thorough knowledge of how the performance characteristics are achieved in order to design textiles that will meet the consumer's needs, and in order to develop new textiles which will compete successfully in the market-place.

The first appearance

The first synthetic polymer, built up from monomers and *unlike any other material found in nature,* was created by Wallace M. Carothers in 1933.

He started his research in 1928, and for five long years experimented with the sticky, insoluble, difficult-to-handle products that resulted from his combinations of difunctional reagents. By 1938 nylon was commercially produced by Du Pont in the United States.

In 1940, American women were offered the first nylon stockings in a co-ordinated sales campaign. Within a few hours of opening all the shops in all the major cities had sold out their stocks.

Part of this initial success was undoubtedly due to the wartime shortage of silk, but it was also due to the new material promising all the qualities of silk, combined with greater durability.

Silk stockings have now all but disappeared. Since this small beginning, nylon and the other synthetic fibres have increasingly dislodged the natural fibres from their share of the textile market.

Polymerization reactions

There are two main types of polymerization reactions: *condensation* and *addition.* These can be used to form long chain polymers, or complex three-dimensional structures. The three-dimensional structures are sometimes used in plastics technology – to make bakelite, for example – but for textile purposes *linear* polymers are the most suitable.

Condensation polymerization

In a typical CONDENSATION reaction, two different molecules react chemically to yield a third molecule and a molecule of water.

Examples are the reaction of an alcohol and an organic acid to form an ester:

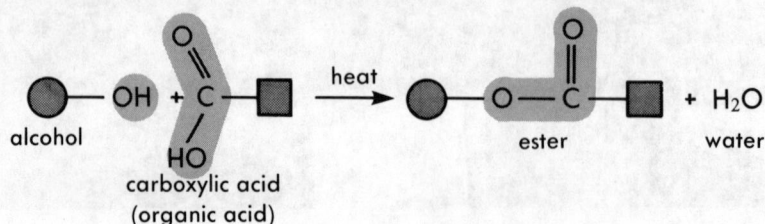

alcohol + carboxylic acid (organic acid) → (heat) ester + H$_2$O (water) and the reaction of an amine and an organic acid to form an amide:

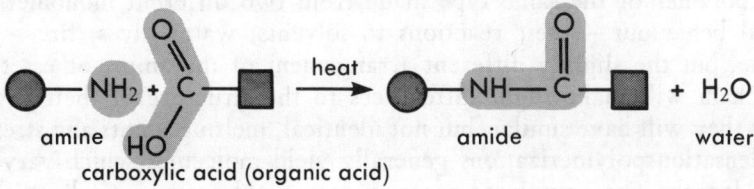

(Does this reaction look familiar? It is identical to the reaction which forms the peptide linkage in proteins.)

If each molecule has only one FUNCTIONAL (reactive) GROUP, the reaction stops there. If, however, the molecules each have two functional groups, the reaction can keep on going, to build up a long polymer chain. If there are three or more functional groups, the polymer can branch in many directions to form a three dimensional structure.

For the linear polymers used in textiles, the monomer may be *one molecule* with *two different functional groups*, or there may be *two monomers*. In that case, one monomer would carry two acid groups, and the other would carry two amine, or two alcohol, groups. The two reagents would be mixed in equal quantities and allowed to react.

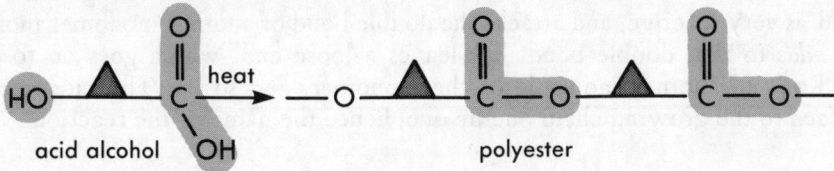

Formation of polyester from a single monomer (schematic).

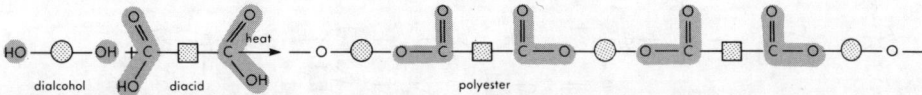

Formation of polyester from two different monomers (schematic).

or

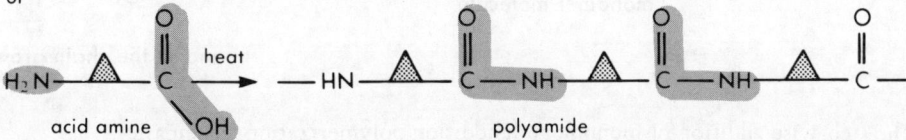

Formation of polyamide from a single monomer (schematic).

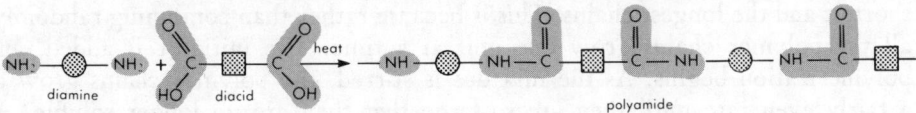

Formation of polyamide from two different monomers (schematic).

A polymer made from only one monomer will have very slight differences from a polymer of the same type made from two different monomers. Their chemical behaviour – their reactions to solvents, water, dyes, fire – will be the same, but the slightly different arrangement of the amine or ester groups in the chain will make slight differences to the structure of their crystalline regions: they will have similar, but not identical, melting points and strengths.

Condensation polymerizations generally yield molecules which vary widely in their length. The completed reaction may yield some very short chains as well as some very long molecular chains. This is because partially polymerized molecules can react at random, and in as few as two steps the molecular chain can grow to four times its original length.

Addition polymerization

ADDITION polymerization occurs quite differently. As mentioned in Chapter 2, a *double bond* is relatively easily broken by chemical reactions. It breaks to become a single bond, but the 'loose ends' of the broken bond must be able to form new bonds with another atom or molecule.

In addition polymerization, an INITIATOR is added to the tank containing the monomer. Each initiator molecule attacks the double bond on a monomer molecule, and breaks it, but leaves one end of the bond 'dangling'. This 'loose end' is very reactive, and attacks the double bond of another monomer molecule. It *adds* to that double bond, but leaves a loose end, which goes on to attack and *add* to another molecule of the monomer, and so on. The monomers are *added* to the growing chain one by one, hence the name of the reaction.

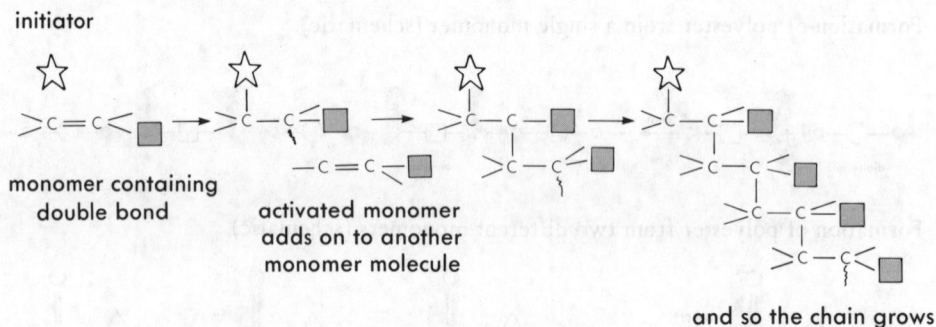

initiator

monomer containing double bond

activated monomer adds on to another monomer molecule

and so the chain grows

The step-wise addition of monomers in addition polymerization (schematic).

The chain length of addition polymerized molecules is far more uniform than that of condensation polymers – there is not as much difference between the shortest and the longest chains. This is because rather than combining randomly, all the polymer chains grow one unit at a time. The initiator is added, and polymerization begins. As the mixture is stirred, the polymer chains grow at a fairly even rate until they are so large that they are no longer soluble. As they precipitate and sink to the bottom of the reaction vessel, they are removed from the polymerization process. Only once – during the termination step of the reaction – do the polymer chains combine with each other.

TRY YOUR HAND

Into the reaction vessel you have put 10 000 monomers, each with a double bond. You have then added 200 molecules of a peroxide initiator.

When the polymerization is complete, and all the polymers have precipitated:

1   How many polymer molecules would you expect to have in the final product?
2   Would all these molecules be the same length?
3   What is the expected average length of the polymer chains?
4   What happened to the initiator molecules?
5   If you wanted shorter molecular chain lengths, would you add more or less initiator to the mix?

Chain length

When two kinds of monomers are mixed in a reaction vessel, then the polymerization reactions occur between those molecules which collide with each other. First single monomers form *dimers,* which then grow to short chains. These short chains then can combine with either more monomers – and so grow very slowly – or with other chains – and so grow rapidly in length.

At the beginning of the reaction, most of the polymer chains are short. At the end of the reaction some polymer chains are very long, and others are short. The *length of polymer chains* is measured as the *degree of polymerization* (D.P.). This is the number of monomers that are joined to make up the polymer.

If most of the molecular chains are long, then the polymer will be strong (compare cotton and viscose rayon among the natural polymers). Short polymer chains yield weak fibres.

Chain length and fibre-forming properties

Degree of polymerization (indicating chain length)	Fibre-forming properties
less than 30	None
30–60	Weak, brittle fibres formed
70–200	Excellent fibres formed
more than 250	Insoluble, infusible polymer (cannot be extruded except in special circumstances)

Copolymerization

A linear polymer is made from repeating units of a monomer. A polymer made from monomer A may have many desirable properties – good strength, suitable melting point – but have poor moisture absorption. The polymer made from monomer B may have excellent water affinity, but be very weak. A polymer which combines the desirable properties of both polyA and polyB would meet the consumer's requirements. How to combine the two?

If A and B have similar reactive groups, they could be mixed before polymerization occurs. They may then form a polymer together: this is called

COPOLYMERIZATION. Usually, such a mix results in a random arrangement of A and B along the polymer chain.

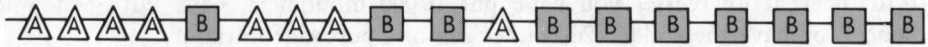

Random copolymer (schematic).

In some cases, the RANDOM COPOLYMER may not perform exactly as desired, or the mixture may not be easy to handle. A more controlled process polymerizes A and B *separately,* to form short chains. These can later be mixed and reacted together to yield BLOCK COPOLYMERS.

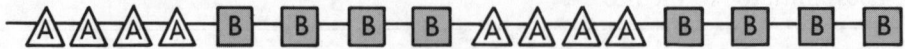

Block copolymer (schematic).

Another method is to treat a completed polymer, for example by irradiation which removes some electrons, and then allow it to react with a different monomer or short-chain polymer. The extra groups are grafted onto the reactive sites created on the main polymer backbone. Such a product is called a GRAFT COPOLYMER. The added side-chains of graft copolymers prevent the main chains from packing very tightly, so they create space within the fibre. They may also form special reactive sites, for the absorption of water for example, or for bonding an inert material such as Teflon to more reactive surfaces. The side-chain graft has been used in the manufacture of elastomeric fibres and in plastics technology, as well as for textile end-uses.

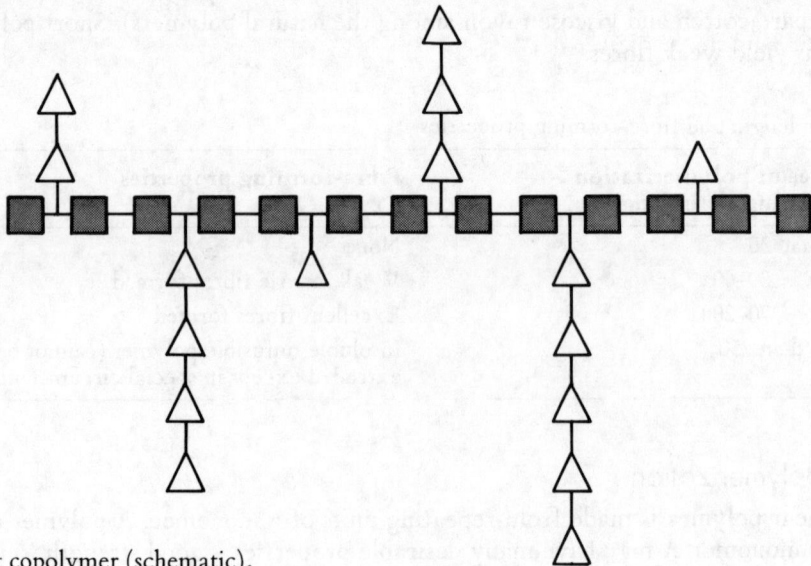

Graft copolymer (schematic).

Addition polymerization is a process eminently suited to the creation of random copolymers. Monomers with a double bond – but otherwise different properties – are mixed in the desired proportions. A certain percentage of branched monomer, and some with polar groups, some linear and non-polar, some inert, whatever is needed, are put into the mixing bowl and emulsified with water.

The initiator is added, the reaction proceeds, and the random copolymer pre-cipitates out. Many acrylics are made by this method.

TRY YOUR HAND

You wish to form a copolymer from two monomers which are suited to addition polymerization. The random copolymer, however, does not quite meet your performance criteria. A block copolymer seems to be the solution. The short chains for the block copolymer are also to be formed by addition reaction. How do you control the reaction to make sure it yields only the length of chain you need?

Bonding between polymer chains

As in the natural polymers, in synthetic polymers there is interchain bonding between the polymer chains. The bonding may be by van der Waals attractions, so important in non-polar molecules, or it may be polar hydrogen bonds, or weaker electrostatic attractions between groups which are only slightly polar, or strong ionic attractions such as between acidic and basic groups, or even covalent bonds. Covalent bonds are not usual between synthetic polymers.

Interchain bonding is often less effective if the chains have large side groups sticking out at random along their length, as these prevent close packing of the molecules. The attraction between polar and charged groups, and the strength of van der Waals bonds, decreases with increasing distance, so anything which prevents close packing weakens the bonding between the polymer chains. On the other hand, specially designed side-groups – such as those of natural or synthetic rubber – can create special effects in the behaviour of polymers.

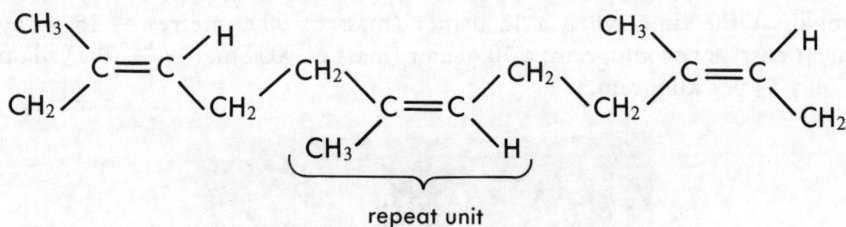

repeat unit

Natural rubber. When rubber is stretched, the methyl ($-CH_3$) side-groups are distorted. The strain is relieved when the rubber snaps back to its original length.

Extrusion

Cross-sectional shape and size

The cross-sectional shape and size of the natural fibres is determined by the genetics of the plant or animal which produces them.

In regenerated natural polymers and synthetics, the *shape* of the cross-section is determined by the spinning process – whether the filament has a hard, crinkled skin and a soft core, or whether it is uniformly crystalline throughout – and by the shape of the orifice through which it is extruded. The size of the orifice determines the fibre *diameter*.

Commencement of extrusion of Nylon 6,6.
(Fibremakers)

Extrusion of a filament involves a lot of stretching: first, to remove the polymer from the spinneret at a constant rate, and then to align the molecules by cold drawing. This means that the final fibre diameter is much smaller than the original orifice. The finer the diameter of the fibre, the softer and more pliable the material will be. A yarn which is made from ten filaments each one unit thick has much softer drape than a yarn made from two filaments of five units each – the strength of the two yarns, however, will be practically the same. It is more expensive to produce fine filaments than coarse ones. For example, at the time when a 15 denier (mass of 9000 metres = 15 g) nylon filament cost $6 per kilogram, a 70 denier (mass of 9000 metres = 70 g) filament cost only $4 per kilogram.

Spinnerets.
(*Bayer Farben Revue* No. 15, 1969, by permission of the editors. *Bayer Farben Revue* is published by Bayer AG, Leverkusen, West Germany.)

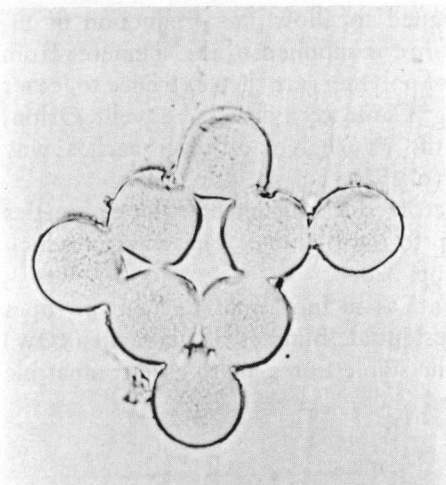

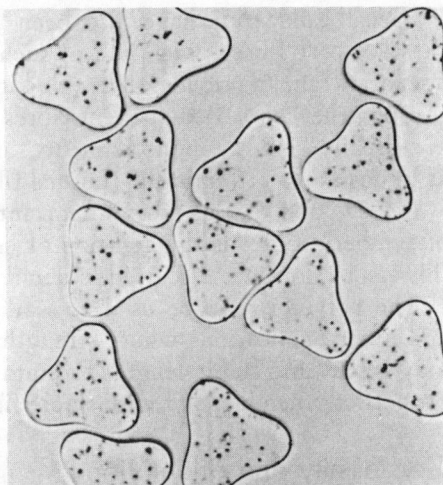

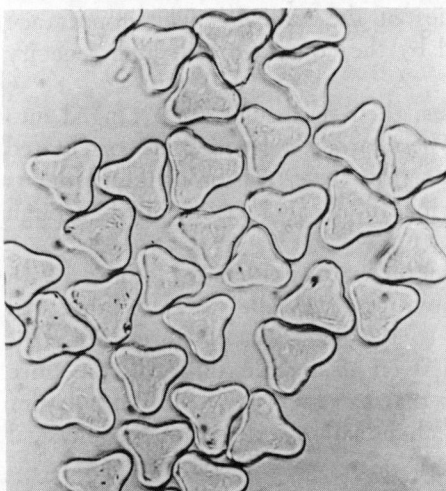

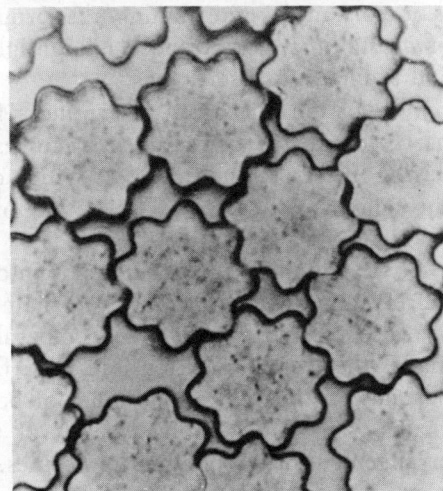

Some cross-sectional shapes in melt-spun synthetic fibres.
(David Bailey)

A fibre with a *hollow*, tubular cross-section is a *better insulator* than a solid fibre. Such fibres are used in blanketing. A fibre with a *trilobal* cross-section is *loftier* – it is harder to pack the fibres together. Such fibres are popular for resilient carpet piles. *Triangular* (delta) cross-sectioned fibres imitate the *lustre* of silk. *Ribbon-like* cross-sections add to *harshness* of touch, but the flat surface will reflect light effectively – as in Lurex, for example. Many other combinations of such ideas have led to a complex range of fibres with varied cross-sections. As the filament is still liquid on extrusion, it is difficult to reproduce the shape of the orifice exactly.

It is very expensive to produce spinnerets with specially shaped orifices. Each hole must be cut into hard tungsten alloy, and to very closely defined size specifications. This cost is quite significant because spinnerets are renewed each 2–10 days of production.

Special spinnerets have also been designed to allow the production of BI-COMPONENT fibres. Each half of such a fibre is supplied to the spinneret from a separate tank of polymer. The two kinds of polymers are then extruded together through the hole. Bicomponent fibres (e.g. Cantrece nylon and Sayelle Orlon) are usually spun as monofilaments, and the two halves differ in such a way as to develop a *crimp* in the finished fibre (p. 151).

Indeed, one of the most important factors that can change the properties of synthetic fibres is the addition of crimp to the filament. The ways in which this can be done are discussed in detail on pp. 320–4.

The polymer may be used as a filament, as in imitation silk, and in ropes and other applications where strength is essential. Bundles of filaments (TOW) may be cut into staple lengths to imitate the staple fibres, or to give compatible properties when blended with staple fibres.

Crystallinity and orientation

In synthetic as well as in natural polymers, the arrangement of the molecules is important in determining the properties of the material. The importance of crystallinity for strength is highlighted by the comparison of high tenacity and ordinary viscose rayons.

During the formation of synthetic fibres, their molecules are arranged into partly organized crystalline structures. These ordered regions are orientated during the stretching processes which occur after the extrusion of the filaments through the spinnerets. The extent of orientation can by controlled through drawing or stretching the fibres.

Some polymers are easier to fit into ordered patterns than are others. With increased drawing, the polymer chains move more parallel to each other and are also orientated more nearly parallel to the direction of fibre axis. This makes the fibre less extensible. The final strength and flexibility of the fibre depends on the kind and number of bonds that can form between the polymer chains. This is a combined result of the chemical structure and the physical arrangement of the polymer molecules.

Additives to polymers

When polymers are solidified into fibre form, they generally have a lustrous or shiny appearance. For some purposes this is not desirable.

In such cases, DELUSTRANTS such as titanium dioxide are added to the liquid polymer and so are included in the finished fibre. The small white particles of titanium dioxide scatter the light falling on the fibre, and make it whiter, duller, and less transparent.

OPTICAL BRIGHTENERS may also be added at the liquid polymer stage, giving long-life 'brightness and whiteness' to the fibre.

ANTISTATIC AGENTS, in the form of small charged or moisture-absorbent molecules, may also be added. These small molecules form hydrophilic sites within the fibre, allowing enough water to be absorbed to dissipate any static charge that may have developed by rubbing of the fibre surface. Their effectiveness is, however, limited.

Some additives change the polymer chemically. Such chemical modifications can alter the polymer's affinity for dyes or for chemicals, and will be discussed under the relevant section for each fibre type.

Polyamides

Petroleum by-products are the starting materials for the manufacture of all synthetic polymers. The basic raw material for nylon is benzene, but alternatives such as phenol and cyclohexane can also be used, depending on economic considerations. From such raw materials the polyamide monomers are synthesized.

Nomenclature

NYLONS are polyamides based on *straight carbon chains,* with acid and amine end groups. The polyamide which is made from the acid amine with six carbon atoms is called *nylon 6.* The polyamide which is made from a diacid with six carbons and a diamine with six carbons is called *nylon 6,6.* Other combinations of reagents are possible, giving different polyamides of different characteristics.

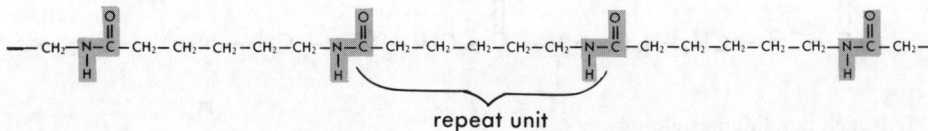

Nylon 6.

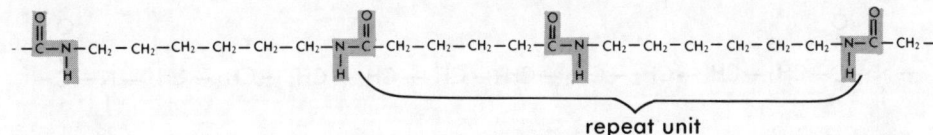

Nylon 6,6.

There is also a newer class of polyamides, which instead of a straight carbon chain contain *aromatic* (i.e. benzene ring) groups. These are called ARAMIDS.

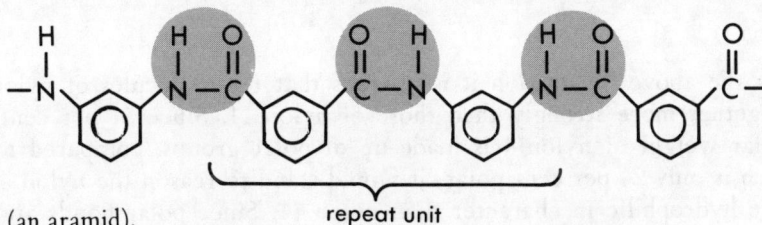

Nomex (an aramid).

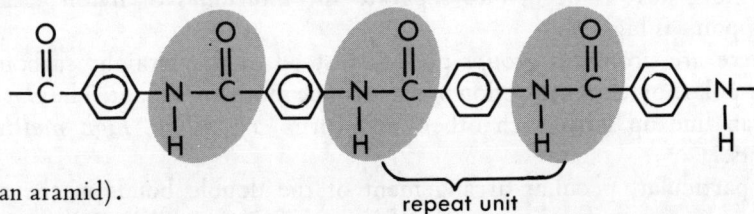

Kevlar (an aramid).

Intermolecular forces

There are two kinds of force which hold the molecules together in polyamides. The *carbon chains* are attracted to each other by *weak, non-polar* van der Waals forces. The *stronger, polar* attraction of hydrogen bonding occurs at the *amide groups*. The nitrogen withdraws some electrical charge from its neighbouring hydrogen, and so a separation of charges occurs. Since nitrogen is not as electronegative as oxygen, the hydrogen bonds of polyamides are not quite as strong as the hydrogen bonds in hydroxyl (–OH) rich cellulose. Nevertheless, the more amide groups in the chain, the more chance there is for hydrogen bonding among the molecules.

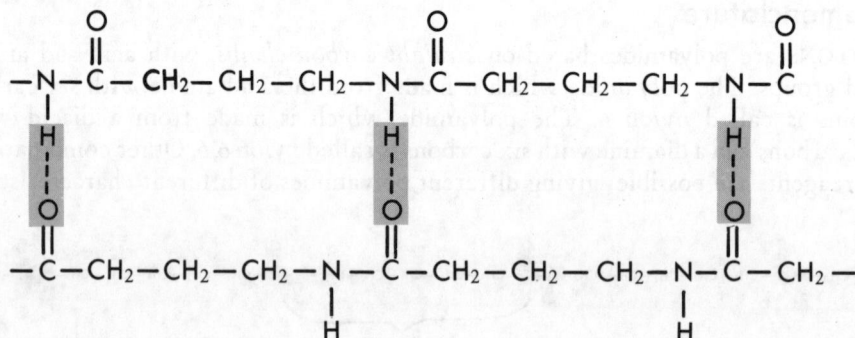

Hydrogen bonding in nylon 4.

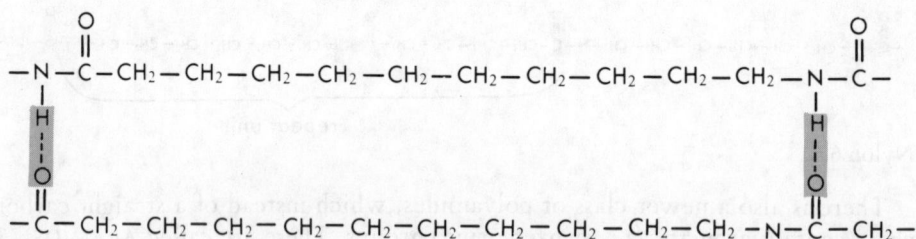

and nylon 11.

From the above illustration it is obvious that the molecules of nylon 4 are held together more strongly than those of nylon 11. Since 51 per cent of the molecular weight of nylon 4 is made up of polar groups, compared to nylon 11 which is only 24 per cent polar, it would stand to reason the nylon 4 would be more hydrophilic in character than nylon 11. Since polar bonds are strong bonds, more heat is needed to separate the molecules of nylon 4: hence its melting point is higher.

If there are *aromatic groups* present instead of the straight carbon chains, the non-polar forces of attraction become much more important. The *flat benzene rings* can line up with each other, and form *crystalline, high melting-point* filaments.

The particular, peculiar arrangement of the double bonds in the aromatic benzene rings gives the rings extraordinary chemical stability. Where normally

Some properties of selected nylons

Name	Melting point	Strength	Moisture regain (per cent)	Comments
Nylon 4	high	low	12 (high)	Decomposes by oxidation (atmospheric attack): hence no commercial success yet.
Nylon 6	220°C	high	4.5 (medium)	
Nylon 6,6	260°C	high	4.5 (medium)	The nylon most commonly used.
Nylon 11	180°C (low)	high	1.2 (low)	Used mainly for brush bristles
Kevlar	above 380°C (very high)	very high	4.5 (medium)	Industrial belting, aircraft reinforcement
Nomex	370°C (very high) Fire resistant	high	4.5 (medium)	Space suits, nightwear, fire-protective clothing

double bonds are weaker than single bonds, and are readily broken, the *benzene ring* is actually *stabilized* by the double bonds. This is why a polymer such as Nomex can be fire resistant.

A modern hot air balloon. The major part of the envelope is polyurethane-coated nylon; the bottom panels are polyester, and heavy heat-set polyester tapes reinforce each gore seam. The scoop skirt is Nomex.
(Kavanagh Balloons Pty Ltd, Wahroongah NSW)

TRY YOUR HAND

Considering the principles affecting melting point, strength and moisture affinity discussed above, answer the following questions:

1 Why do nylon 6 and nylon 6,6 have the same moisture regain?
2 Why does nylon 6,6 have a different melting point from nylon 6?
3 Why do both Kevlar and Nomex have similarly high melting points?
4 Why is Kevlar stronger than Nomex?
5 Why do they both have the same moisture regain as nylon 6 and nylon 6,6?
6 Why would nylon 4 be a very desirable textile fibre if difficulties with its production could be overcome?

Qiana is a new type of nylon fibre (1972). It has been developed by Du Pont, and has had as much as $100 million spent on its development and promotion. It is an expensive, exclusive, silk-like fibre. Its formula has recently been released as the non-aromatic but cyclic polyamide:

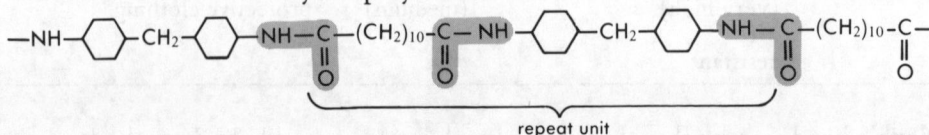

Qiana. The open hexagon represents cyclohexane,
a ring of six carbon atoms with no double bonds.

TRY YOUR HAND

Considering the formula given above, plus what you know about intermolecular bonding:

1 Would you expect Qiana to have higher or lower moisture regain than nylon 6,6?
2 How does that affect its comfort properties for wear?
3 Would you expect its melting point to be high or low?
4 Is it more crystalline than ordinary nylon or less?
5 Is its strength greater or less than that of nylon 6,6? Why?
6 Is it softer and more pliable, or stiffer?
7 If it was put into the market to compete with silk, what kind of cross-section would you expect it to have? Why?
8 How do you expect the properties of Qiana to compare to those of silk?

The following table may help you to check your conclusions.

Comparisons with silk

Polymer name	Specific gravity	Tenacity (g/den)	Moisture regain (per cent)	Melting point
Qiana	1.03	3	2–2.5	275°C
Silk	1.25–1.33	3.5	11	yellows 150°C
Nylon 6,6	1.14	5	4.5	260°C

Melt spinning and cold drawing

Polyamides, like many other synthetic fibres, are extruded by MELT SPINNING. This process involves melting the polymer chips in an inert nitrogen atmosphere – usually by passing them through a heated grid – then carefully filtering the melt and pumping it at constant rate through the holes of the spinneret. The emerging *filaments* are gently cooled with dust-free air. When solidified, they are conditioned – moisturized – by steaming, and wound on a cheese-shaped package.

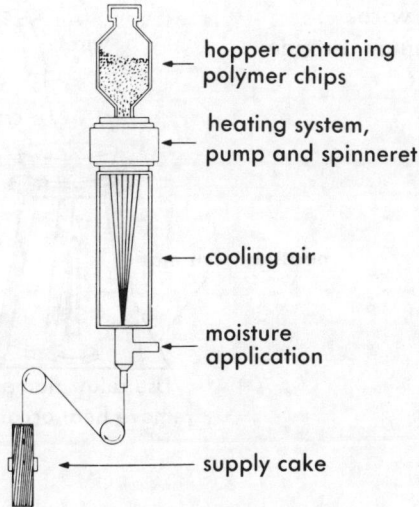

← hopper containing polymer chips

← heating system, pump and spinneret

← cooling air

← moisture application

← supply cake

Melt extrusion (spinning) of nylon. From the supply cake the nylon filaments go to the drawing and texturing processes appropriate to their intended end use. (Fibremakers)

Each spinneret has a number of orifices, and the many filaments emerging from each spinneret are treated as a single *yarn*. Sometimes the products of many spinnerets are combined to form *tow,* which is intended for staple fibre production later.

The conditioned filaments must be DRAWN before they become destabilized by random crystallization during storage. Cold drawing must be carefully and evenly carried out to ensure uniform properties in the final product.

The cold-drawing process aligns the polymer molecules, and makes the fibre more *orientated.* With orientation, the strength, resilience and stiffness of the fibre also increases. Ultrastrong nylon fibres can be created (for uses such as fishing lines) by extra drawing at this stage.

Undrawn and drawn filaments. Drawing pulls the polymer molecules into greater alignment and so increases the strength of the filament.

TRY YOUR HAND

You too can experiment with melt spinning, and find out what variables contribute to the properties of melt-spun fibres.

You will need:

some cut up pieces of nylon fabric,
    preferably shiny
crucible, pipeclay triangle and tripod
Bunsen burner and heat-resistant mat
10 cm copper or other wire
talcum powder or titanium dioxide

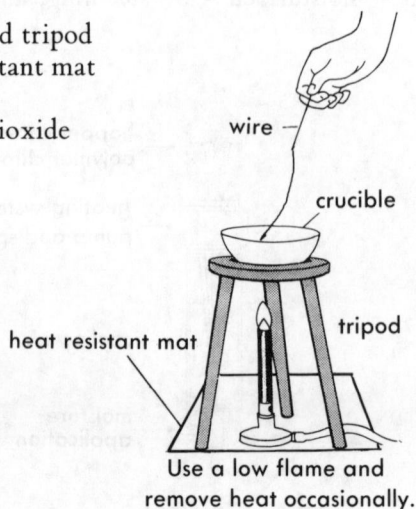

Use a low flame and
remove heat occasionally.

What you do:

Heat the pieces of fabric in the crucible. As soon as the fabric starts to melt, lower the flame so the nylon won't decompose. You will need to reheat it from time to time to keep most of the material melted.

Push the tip of the wire into the molten mass and pull gently upwards. You will find that a fine stream of liquid nylon is pulled away from the crucible, and it solidifies into a filament as it cools.

If you move your hand too fast, the filament will break while it is still liquid. If your hand movement is uneven, the filament will have thick and thin spots. Experiment to produce filaments of different thicknesses.

1   Which are the stronger filaments, thick or thin?
2   Which ones are more pliable?
3   What happens when you stretch the cooled filaments?
4   Are five thin filaments as strong as one very thick one?
5   How could you make a strong but pliable filament yarn?

Add a small amount of talcum powder (or titanium dioxide) to the molten polymer, stir, and extrude a new filament.

6   Can you notice any change?
7   Has the powder affected the strength of the filaments?
8   Was it more difficult to produce even, thin yarns, or thick ones? Would you expect fine filaments to be the same cost per kg to produce as coarse filaments?

The properties of the polyamides

Although there are many chemically different polyamides available, there are only two of major commercial importance. These are nylon 6 and nylon 6,6, of which the latter is produced in somewhat greater quantities. Most nylon producers make both types.

In general, nylon is characterized by high strength, good resilience and abrasion resistance, fairly low moisture regain, and thermoplasticity.

Changes in cross-sectional shape and diameter, amount of delustrant added, extent of cold drawing and even chemical modifications can give nylon a tremendous range of properties.

Strength

The *strength* of nylon can be regulated through the amount of *cold drawing* to which it is subjected. Where the *tenacity* is very *high,* the *extensibility* and *pliability* are *low*. A special feature is that nylon has a very high strength even when knotted. This certainly is a help to the fisherman! As nylon does not absorb much water, its strength is not greatly reduced on wetting.

The high strength of Kevlar allows the manufacture of light-weight and low-bulk bullet resistant vests which can be disguised as part of normal clothing. The vests, consisting of multiple layers of woven Kevlar, can convert a potentially lethal hand-gun attack into a severe bruise. Each successive layer of fabric serves to slow the bullet so that it stops before it can penetrate the body. Although other fabrics can confer a similar protection, many more layers (and so greater bulk) are needed. Bruising from the impact of the bullet may be severe enough to warrant hospitalization, and the light-weight vests do not protect against high-powered rifles or against direct stabbing thrusts with weapons such as ice-picks.
(Du Pont)

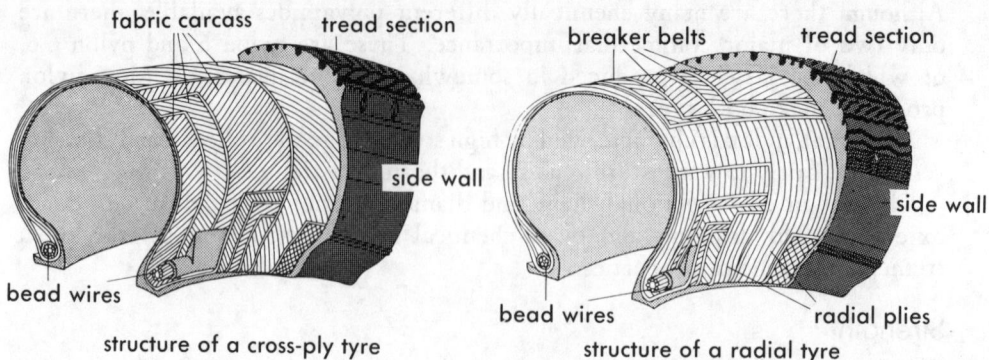

structure of a cross-ply tyre

structure of a radial tyre

High tenacity nylon is used in the fabric carcass of cross-ply tyres and in the radial plies of radial tyres. The tough yarn is highly twisted into cord to improve its fatigue resistance; the cord is then woven into fabric for use in tyre building.
(Fibremakers)

Strength of nylons

Form of nylon 6,6	Tenacity (g/den)	Wet tenacity (g/den)	Elongation at break (per cent)
Standard	4.6–5.8	4.0–5.1	26–32
High strength	up to 9	7.7	20
Staple	4.1–4.5	3.6–4.0	3–40
Knotted	85 per cent of normal strength		

TRY YOUR HAND

1 If standard nylon has tenacity of 5.0 g/den, how strong is a 15 denier yarn from a stocking, i.e. how many grams can it support without breaking?
2 Count how many stitches around the leg of a stocking, and then calculate how much weight you would expect the hose to be able to support. Test your calculations with a suitable experiment!
3 Consider the arrangement of molecules in high strength – crystalline – nylon. Why is the percentage elongation at break of this fibre less than that of standard nylon?
4 What is the percentage loss in strength of nylon on wetting?

Moisture absorption

Moisture is absorbed by nylon at the polar sites provided by the amide groups. The *more crystalline* the fibre, the *less moisture* it absorbs.

Generally, at 65 per cent relative humidity and 20°C, nylon absorbs only 4.2 per cent of its mass in water. As it does not absorb a great deal of water on wetting, it dries rapidly – a useful attribute for laundering.

TRY YOUR HAND

You will need:

mangle or wringer, or a print-making press borrowed from the art department

balance

a piece each of closely woven wool, cotton, and nylon fabrics, cut to weigh exactly 10 g

water containing a drop of detergent

What you do:

Soak each piece of fabric in the water to wet it through (the detergent aids wetting).

Pass each piece through the mangle three times.

Then weigh the fabric pieces, and calculate how much water was absorbed by 1 g of each type of fibre. Expressed as a percentage, this is called the 'percentage imbibition of water'.

1 How significant is the difference between the samples with regard to drying time?

2 How does this relate to the energy required to evaporate absorbed moisture?

Static electricity

The *electrical conductivity* of nylon is *low*. Unless the atmosphere is very humid, nylon fabrics are prone to generate *static electricity*.

   The problem of static may be tackled in three ways. During manufacture, the fibres may be coated with an electrically conductive film of polymer. The fibres may be given 'lifetime' conductivity by charged molecules added at the melt stage of the polymer. These molecules migrate to the surface during wear, and neutralize any charge which may accumulate. If the fibres have not been treated during manufacture, or if these treatments have not proved satisfactory, nylon fabrics may be treated after each laundering by adding anionic fabric conditioners to the rinse.

TRY YOUR HAND

Some people seem to have more difficulty with static than others. Is this because of some characteristic of those people, or does it relate to how they care for their clothes?

Design and carry out an investigation of static and static sufferers. This can be an individual or group project.

You will need to consider what information is relevant to the problem, and what is the best way to gather the information, to analyse it, and to present your conclusions. You will need to set yourself a sensible deadline for the completion of your study.

Points to consider:

1 Who suffers from static?

2 What garments or textiles are involved? What are they made from?

3 Have these textiles received any anti-static treatment? If not, are similar textiles available which have been treated? What are these treatments? Do they work? For how long?
4 What care instructions apply to the textiles? Are these followed by static sufferers? by non-sufferers? Would you expect the care to make any difference to the textile performance?
5 How does the user cope with the problem? What options are available to static sufferers? At what cost?

These are suggestions only: follow any or some or all or none of these in your investigation. A well-planned study should yield some useful advice for static sufferers.

Reaction to heat

Nylon is a *thermoplastic* fibre. The melting point of nylon 6,6 is 250°C, that of nylon 6 is lower at 215°C. Ironing temperatures above 150°C should be avoided for both kinds of nylon, because of danger of the fabric becoming glazed.

Because of their thermoplastic nature, fabrics made from nylon yarns can be *heat set* – either into pleats or into a stable flat form. Such heat setting helps with *easy-care* properties during laundering.

Polyamides which have *aromatic groups* (Kevlar, Nomex) have their molecular chains packed much closer together. This results not only in *higher fibre density*, but also in *stronger forces of attraction* between the molecular chains. When heat energy vibrates the molecules, these stronger attactive forces prevent the chains from separating from each other until much *higher temperatures* (350°C and more) are reached.

Flammability

When a material ignites (catches fire) its molecules react with the oxygen in the air. The molecules break down into their smallest components and release energy in the form of heat.

As they burn, ordinary nylons give off acrid fumes and drip and melt away from the source of the flame. Nylon underwear worn under flammable garments can melt onto the skin and cause severe burns, without actually supporting the fire. After the Falklands war between Britain and Argentina (1982), the British navy banned nylon underwear because of the terrible burns caused in this way.

TRY YOUR HAND

Hold some nylon fibres in tweezers, and slowly bring them close to the heat of a Bunsen flame. Observe their behaviour near the flame, in the flame, after removal. Does nylon support combustion? What odour does its smoke have? After removal from the flame, what is the residue (ash, bead) like? Would you be able to identify nylon by the odour of its burning alone?

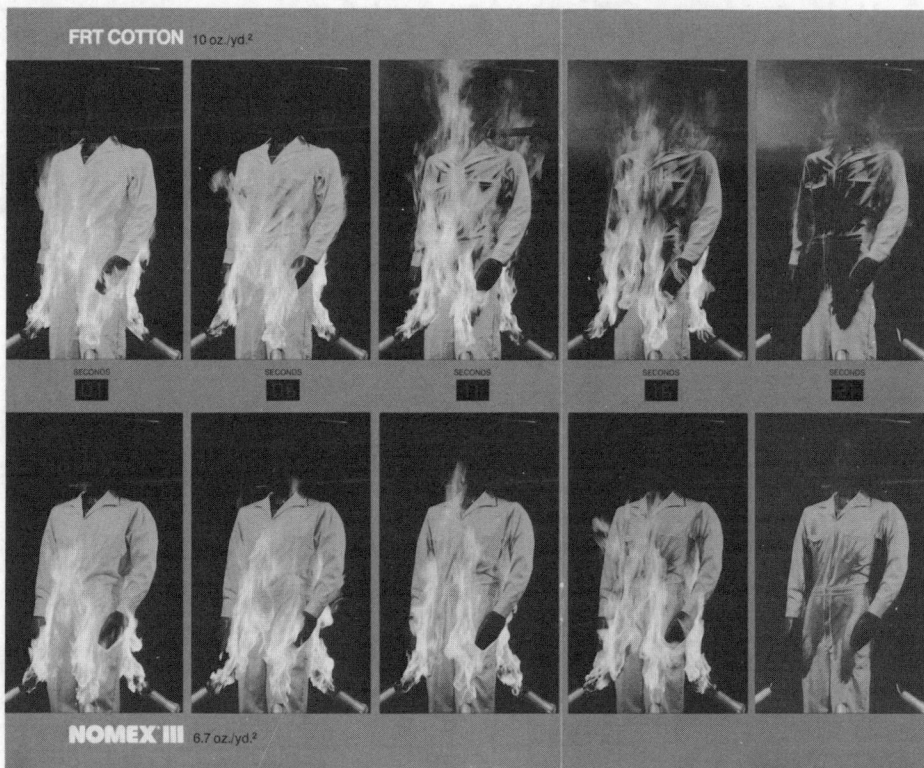

FRT COTTON 10 oz./yd.²

NOMEX III 6.7 oz./yd.²

This comparison between flame-retardant-treated (FRT) cotton and Nomex III shows why the aramid is increasingly being used where there is risk of flash fires. Although the FRT cotton offered some protection, the extent of 'burns' suffered by the manikin when wearing FRT cotton was twice that suffered when the manikin was wearing Nomex III. Untreated cotton, and polyester/cotton blends, gave no protection.
(Du Pont)

Aromatic polyamides (aramids like Kevlar and Nomex) do not ignite or burn because of the very high energies involved in separating their molecules, and because the products of degradation of those aromatic molecules *do not support combustion.*

Drape

The drape of nylon *fibres* depends on *diameter of filaments*, and *extent of crystallinity.*

Qiana is a nylon fibre manufactured as fine filaments to produce a silk-like drape. Antron and Anso are nylon filaments for carpet pile. Because of their thickness and stiffness they have particularly poor drape properties.

The final drape of the *fabric* owes much to features of *yarn* and *fabric construction* as well.

Elasticity and resilience

Nylon is a particularly *resilient* fibre. Fabrics made from nylon recover readily from creasing or wrinkling. If stretched as much as 8 per cent, nylon will

One application in which the strength, resilience, and elasticity of nylon are important. (Herald and Weekly Times)

still have 100 per cent elastic recovery, though it will take some time to return completely to its original dimensions. After an extension of 16 per cent it has 91 per cent recovery.

TRY YOUR HAND

You will need:

5 numbered pieces of nylon fabric, each 10 cm × 10 cm. The fabric should not be too soft or pliable
5 weights, 1 kg to 5 kg
5 pairs of glass sheets 15 cm × 15 cm between which to press the samples
5 rules set vertically upright
clock

What you do:

Fold each numbered sample of fabric in half, and place it between a pair of glass sheets.

Place a weight on each sample, and leave it for 10 minutes.

Remove the weights, and unfold the fabric pieces, placing them pleat upwards as shown.

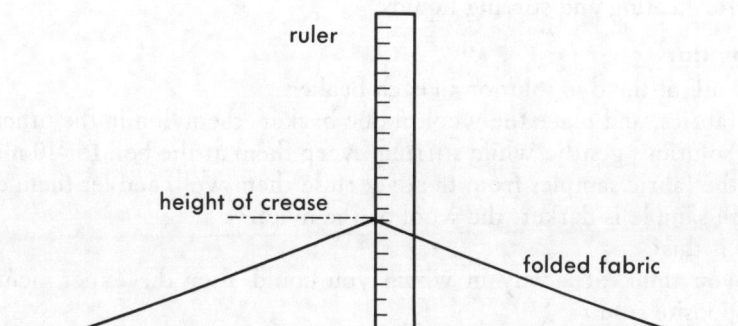

Measure the height of the crease above the baseline at regular intervals. (Conduct a 'dry run' to find an appropriate time interval.)

Graph the height against time to give a 'crease recovery curve'.

1　Under what conditions is the resilience you have observed important?

2　When is it less important?

3　What industrial applications make good use of the strength and elasticity of nylon?

4　What fashion trend of the 1950s relied on the stiffness and resilience of nylon – then a new fibre – for its effect?

Abrasion resistance

The abrasion resistance of nylon is important in many industrial applications – belting, ropes, carpeting. In straight-forward rubbing tests, nylon *withstands abrasion* far better than any other fibre. For that reason, it is often blended with the extremely resilient wool to give excellent carpets.

Dye affinity

Nylons have many amide groups, as well as $-NH_2$ and $-COOH$ end-groups at the ends of each polymer chain. This makes them chemically quite similar to protein fibres.

Wool and other protein fibres are dyed with acid dyes. This involves converting the $-NH_2$ end-groups to $-NH_3^+$ by adding acid (H^+) to the dyebath. These are the dyeing sites to which the negatively charged dye molecule (D^-) is attached.

Nylon can be dyed in the same way, with acid dyes. Since nylon is much more crystalline than wool, dye penetration is more difficult. On the other hand, the molecular chains of the nylon polymer are shorter than those of natural protein fibres. Hence, in a given weight there are more end-groups, and so also more possible dye sites.

TRY YOUR HAND

You will need:

2 beakers, 500 mL capacity
4 g sample of wool fabric
4 g sample of nylon fabric
400 mL of an acid dye solution
facilities for heating and stirring liquids

What you do:

Place 200 mL of the dye solution in each beaker.
Wet the fabrics, and place the wool in one beaker, the nylon in the other.
Heat the solutions gently, while stirring. Keep them at the boil for 10 minutes.
Remove the fabric samples from the dye, rinse them well, and let them dry.

1 Which sample is darker, the wool or the nylon?
2 Why is this?
3 Can you think of a way in which you could alter the experiment to get the opposite result?

Nylon can be dyed with acid dyes because it contains amine ($-NH_2$) end-groups. Like triacetate, it can also be dyed with *disperse dyes,* because it is hydrophobic. Dyes and dyeing are explored in detail in Chapter 8.

Effect of sunlight

If white nylon fabric is exposed to sunlight, it becomes yellow. So while cotton shirts may be hung out in the sun to dry, the same treatment will yellow nylon garments. If a nylon garment is treated with an *optical brightening agent* (Chapter 7), this will make the nylon even more susceptible to damage from light. A nylon garment may appear bright white in sunlight, where the optical brightening agent is active, but quite yellow in shadowy folds.

In spite of this drawback (polyester, for example, is not yellowed by sunlight), nylon is still far more resistant to degradation by sunlight than is the silk it originally sought to replace. After a 16-day controlled test in which silk lost 85 per cent of its original tenacity, nylon, under identical conditions, lost only 50 per cent of its strength.

Biodegradation

TRY YOUR HAND

You will need:

labelled samples of assorted fabrics – cotton, linen, wool, nylon, polyester – 20 cm × 10 cm, 2 pieces for each fabric type

What you do:

Keep a piece of each type of fabric for reference.
Bury each of the remaining pieces of fabric in the earth, near each other, so they are all exposed to the same conditions.

Leave them for six weeks, then dig them up, wash them carefully, dry them, and compare them to the reference pieces.
1 What can you conclude about the susceptibility of the different textile fibres to biological attack?
2 What relation, if any, did you find between the moisture affinity of the fabric (hydrophilic/hydrophobic) and susceptibility to biological attack?

Different nylons for different consumers

As seen earlier in this chapter, nylons can be varied by both physical and chemical means to produce an enormous range of fibres with a wide spectrum of properties. However, if one property is changed (say pliability increased), then it is often at the expense of another, equally desirable property (strength decreased). So in spite of their variability, it is not quite possible to tailor-make synthetic fibres that are perfect for any application.

Nevertheless nylon fibres are used with great effect in many domestic and industrial applications.

The net frame, edge, and tie cords of the camouflage net are made from multifilament nylon coated with polyvinyl chloride. The garniture which provides the actual camouflage is thick polyvinyl chloride film.
(Army Public Relations)

TRY YOUR HAND

Nylon is stong, flexible, resilient. Because of its tenacity, quite strong fabrics can be made from very fine filaments. It is resistant to water, and to salt water. Nylon can be heat set, and it has a high abrasion resistance.

Uses for nylon include parachute fabrics, typewriter ribbons, tyre cords, sheer hosiery, lightweight tents for hikers, spinnaker sails, filter bags for the pottery industry, fishing lines, brush bristles.

1 Which features of nylon make it suitable for the above uses?
2 Can you find other areas where nylon is used?
3 Check each section of a department store for other uses of nylon. Are component fibres always listed clearly? Is nylon often blended with other fibres? Did you find any non-textile uses of nylon?
4 Draw up a table of your observations for each item, noting which properties and characteristics of nylon are important for the particular end-use: tenacity; abrasion resistance; flexibility; softness; thickness; filament or staple; blend (which other fibre?); moisture absorbency; elasticity; resilience; lustre.
   From your table, which features of nylon are most often varied to produce the different properties required by the domestic consumer?

Polyesters

The polymer

Polyesters were the first success in the polymerization experiments of Carothers, but it was not until 1941, well after the commercial launching of nylon, that production techniques for polyester fibres were established.

For polyesters, there is only *one* monomer combination of commercial importance. Terephthalic acid, a *diacid* containing a benzene ring, and ethylene glycol, a small *dialcohol,* are combined to give a long chain POLYESTER.

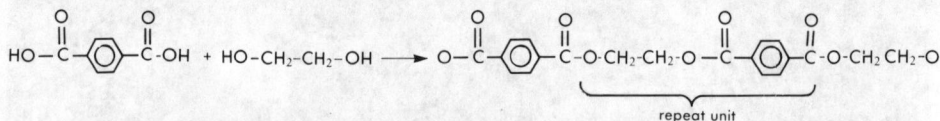

Polyester.

All the well-known brands of polyester – Dacron, Terylene and so on – are chemically identical fibres made from *poly(ethylene terephthalate).*

Only two other polyesters are manufactured for textile use. Kodel, produced by the Eastman Kodak company and originally developed for use in photographic films, contains a dialcohol based on cyclohexane instead of the ethylene glycol. A more recent development is A-Tell, produced by Unitika, which, although reputed to be an excellent textile product, has been slow to gain significant commercial acceptance.

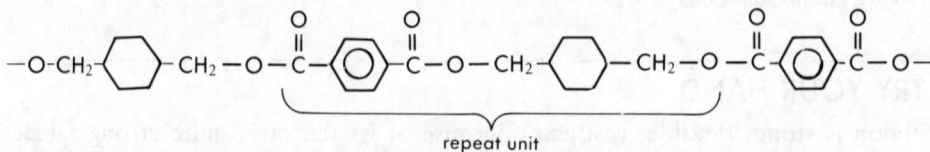

Kodel.

144

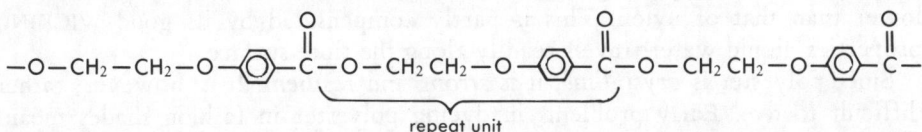

$$-O-CH_2-CH_2-O-\bigcirc-\overset{\overset{\displaystyle O}{\|}}{C}-O-CH_2\ CH_2-O-\bigcirc-\overset{\overset{\displaystyle O}{\|}}{C}-O-CH_2\ CH_2-O-\bigcirc-\overset{\overset{\displaystyle O}{\|}}{C}-$$

$\underbrace{\qquad\qquad\qquad\qquad}_{\text{repeat unit}}$

A-Tell.

Extrusion

As with all other synthetic fibres, the raw materials for polyester are derived from petroleum products. Polymerization takes place with the help of a catalyst.

The polymer is allowed to harden, and is then broken into chips. These chips are then *melt extruded* to produce filaments of the desired cross-sectional shape and size.

Extrusion is followed by *hot drawing,* which increases the tenacity of the filaments to the desired levels.

Properties of the polyester fibre may be modified further by *crimping.* Filaments may be *cut* into staple lengths, to imitate, or blend with, staple fibres.

Characteristics of the polymer

The polyester molecule is totally *hydrophobic:* the ester linkage has no nearby hydrogen to form a polar bond, and both the benzene ring and the carbons of the glycol are completely non-polar. As there can be no hydrogen bonds to hold the molecules together in the fibre, the non-polar van der Waals forces play a most important role.

In the crystalline regions, the benzene rings line up and exert the attractive forces characteristic of aromatic compounds. Although these forces are weak at a distance, at close quarters they can be very strong indeed, and so they hold the polymer chains in a stable structure.

The chains of polyester molecules are remarkably *regular.* This allows the formation of many *crystalline* regions. The high crystallinity in turn gives *strength* and resilience to the fibre, and allows a usefully high *melting point,* in spite of lack of polar bonding.

TRY YOUR HAND

Look back to p. 59, Chapter 3, to the structure of cellulose triacetate.
This is also a totally non-polar polymer.
1   Which would you expect to be the stronger fibre, polyester or triacetate?
2   Which fibre is softer to the touch? Why?
3   Would you expect to use the same dyes for these two fibres? Would water-soluble dyes be appropriate?
4   What is the role of intermolecular spaces in determining the properties of hydrophobic fibres?
5   Which fibre would you expect to be easier to dye? Why?
6   Which has the higher melting point? Why?

As polyester has no polar group, its *affinity for water* is extremely *low,* even lower than that of nylon. This is partly compensated by its good WICKING properties: liquid water travels readily along the fibre surface.

Since polyester is *crystalline,* it is *strong* and *resilient.* It is, however, rather difficult to dye. Early problems in dyeing polyester in fashion shades meant that it was slow in gaining popularity. This problem has been overcome by a new generation of disperse (hydrophobic) dyes, and today there is more polyester than nylon produced in the world.

Comparison of ropes used for towing barges on the Thames

Fibre used for rope	Rope life (days)	Cost per day of life (pence)
Sisal	6	194
Nylon	65	320
Polyester (Terylene)	137	152
Polypropylene	60 (estimated)	68 (estimated)

Comparison of polyester and polyamide

Property	Polyamide (nylon)	Polyester (Terylene)
Moisture regain	4.2%	0.4%
Density (specific gravity)	1.14	1.38
Elasticity: recovery from 8% stretch	100%	80%
Recovery from creasing	good	excellent
Tenacity (normal)	4.5–5.8 g/den	4.5–5.5 g/den
(high tenacity)	8–9 g/den	7–8 g/den
Melting point	215°C–250°C	250°C
Ironing temperature	130°C–150°C	130°C
Resistance to light	yellows	loses strength only very slowly

TRY YOUR HAND

1 Using the figures in the tables, give reasons why polyesters have become successful in the following applications:
curtains
tropical suitings
finely pleated *haute couture* fabrics
sails for boats
'fibrefill' for pillows and quilts
fabrics for ties
dye bags and laundry bags
ropes and nets for fishing.
2 Polyesters have not yet overtaken nylon for use as tyre cord or in carpet pile yarns. Can you think of reasons why polyester performs less well than nylon in these areas?

Blending

Polyesters have low affinity for moisture, and so are suitable for *drip-dry*, easy-care garments. Because of their *hydrophobic* nature, they have great *attraction for oily*, greasy *stains*, which are difficult to remove. They are *resilient* and *crease-resistant*, but tend to feel *uncomfortable* on hot days.

Fabrics made from highly twisted yarns but with a loosely woven structure can overcome this clammy heat problem.

The most practical solution to the disadvantages of polyester for clothing has been to blend it with cotton. The famous *poly/cotton blends* are intimate mixtures of fibres: polyester, of similar diameter to the cotton fibres, is crimped and cut into staple length to match the cotton staple it is to be blended with. Blending can occur at either the cotton blending or the sliver drawing stage (pp. 327, 334).

Pure polyester suiting (left) and polyester/flax suiting (right).
(Vivian Robinson)

In a blend of polyester and cotton, the cotton fibres provide a crisp, cool handle and the comfort of moisture absorbency. Polyester gives the blend excellent *crease recovery* and *drip-dry* properties. Such blends have become extremely popular not only for shirtings, but for sheets and other bed-linen as well. Polyester/cotton blends are also most popular for active sportswear knits.

The major problem with poly/cotton blends has been that of PILLING. The polyester fibres are much stronger than the cotton fibres. With abrasion during use, some of the cotton fibres break. They cannot, however, fall away from the surface of the fabric, because they are 'tied down' by strong fibres of polyester. *Pill resistant* polyester fibres, which are *weaker* than normal, have been developed especially to cope with this problem.

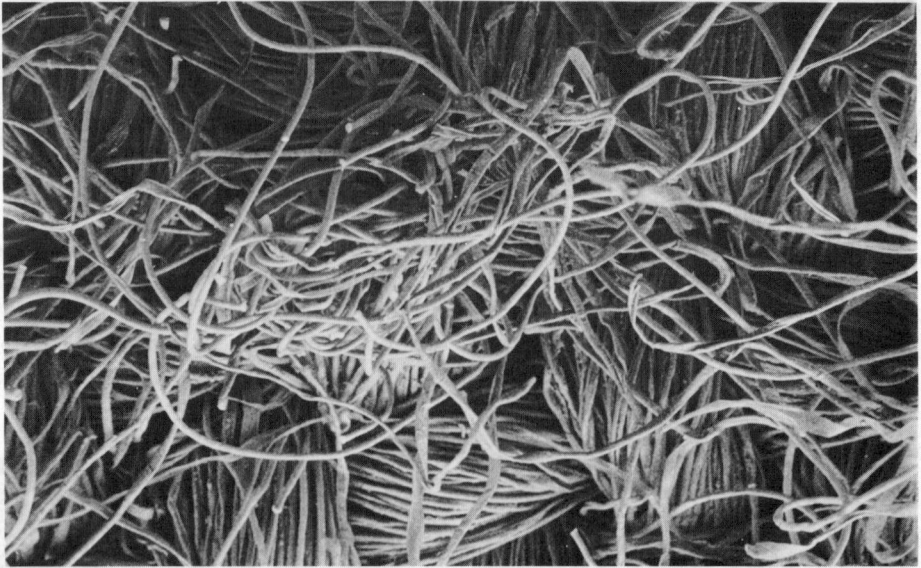

A pill on polyester/cotton blend sheeting.
(Vivian Robinson)

Because of the quite different chemistry of the two fibres, *dyeing* poly/cotton blends to a uniform colour poses problems. Manufacturers may sometimes exploit the different dyeing characteristics of the two fibre types for the subtle 'heather' effect it can give. This is explored further in Chapter 8.

Acrylics

Production

ACRYLIC fibres are made from an *addition polymerization* reaction of ACRYLO-NITRILE (H_2C=CHCN) (vinyl cyanide). When the first acrylic fibre, Orlon, was produced by Du Pont in 1950, it was a pure polyacrylonitrile. Since then, however, many changes have been made. As it is quite easy to copolymerize various vinyl monomers, the variety is great, and all addition polymers that contain at least 85 per cent acrylonitrile are called acrylics.

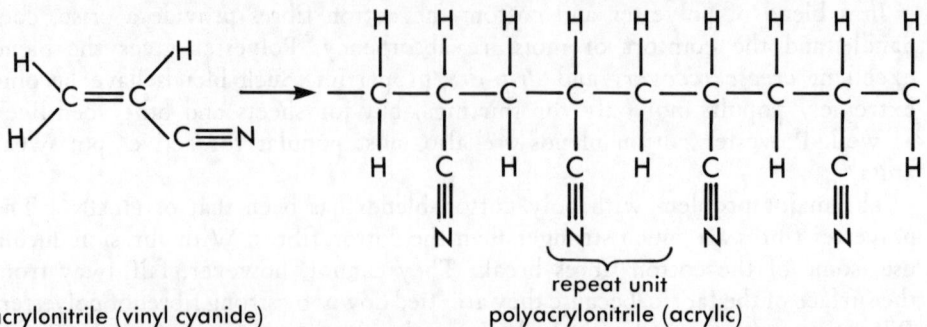

acrylonitrile (vinyl cyanide)

polyacrylonitrile (acrylic)

The formation of acrylonitrile.

148

Acrylonitrile is added to water containing a peroxy catalyst. As the polymer precipitates, it is filtered off, washed, dried, and dissolved in a suitable solvent. The fibres may be wet spun into a coagulating bath, or they may be dry spun, with the solvent evaporated off in hot air.

The solvent usually used for spinning polyacrylonitrile is dimethyl formamide. Other solvents of similar polarity are known, but they are mostly too expensive for commercial applications.

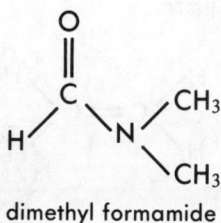

dimethyl formamide

Dimethyl formamide.

Those acrylic fibres which are meant to imitate silk are left as lustrous filaments. Others, which are to imitate wool, are usually delustered, crimped, and cut into lengths to approximate the staple length of wool. The staple fibres can then be bulked by special wet-processing methods (p. 153).

Intermolecular forces

The acrylonitrile monomer is only very slightly polar. In the nylons, the electronegative nitrogen is attached to the electropositive hydrogen, allowing for substantial charge separation and therefore hydrogen bonding. In acrylonitrile, however, the electronegative nitrogen is separated from the nearest hydrogen atom by two neutral carbon atoms. This means that only very slight charge separation (polarity) is possible. Nevertheless, some charge separation does occur, and the molecules of polyacrylonitrile are held together by weak electrostatic forces in the ordered regions of the polymer fibre, as well as by the expected non-polar van der Waals forces.

Pure polyacrylonitrile has its polymer chains packed very close to one another. It has a low moisture regain (1–2 per cent), and is very *difficult to dye.*

The effect of copolymers

The introduction of vinyl monomers with *bulky side chains* opens up the close-packed structure, and allows *disperse dyes* to be more easily accommodated. *Basic monomers,* such as vinyl pyridine, may be added to give attachment sites for *acid dyes,* and so allow wool/acrylic blends to be dyed with the one dye. *Acidic groups* may be added to the monomer mix, to allow the use of *basic dyes.* The moisture regain of acrylic fibres may be varied a little with the amount and type of copolymer used in the manufacture.

vinyl acetate   methyl acrylate

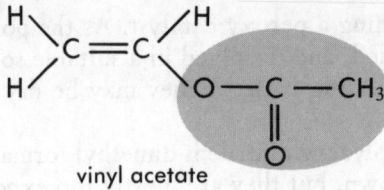

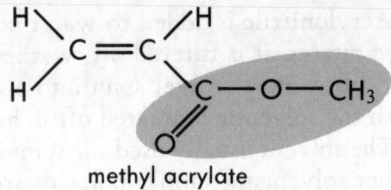

Non-polar but bulky copolymers such as vinyl acetate and methyl acrylate disrupt the packing of the polymer molecules and so create voids within the fibre. This allows dyeing with disperse dyes to take place.

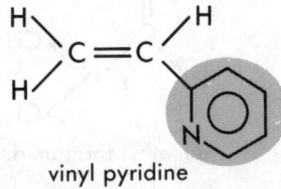

vinyl pyridine

Vinyl pyridine is a basic copolymer which allows the fibre to be dyed with acid dyes. (Pyridine is a benzene ring in which one CH has been replaced by a nitrogen atom. It has a stability to chemical attack similar to that of benzene.)

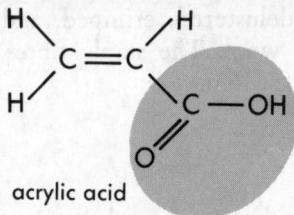

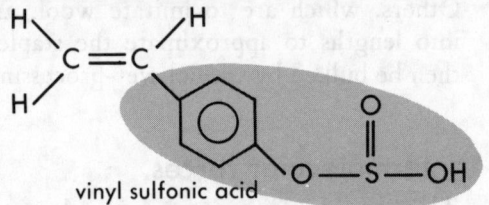

acrylic acid   vinyl sulfonic acid

These acidic copolymers allow the fibre to be dyed with basic dyes. The much smaller acrylic acid would cause less disruption to the polymer packing than would the bulky vinyl sulfonic acid.

Properties and uses

The final properties of acrylic fibres depend on the physical and chemical modifications that were carried out during their manufacture.

Continuous filament acrylics are produced as *high strength, resilient* fibres which are suitable in various thicknesses for double-knit outerwear and for carpet pile yarns. *Staple filament* acrylic fibres are generally softer and more pliable. They are often produced to imitate or to blend with wool fibres. The performance of acrylics in this role is examined and evaluated on pp. 154–6.

The *moisture absorbency* of acrylics is *low* – 1–2 per cent depending on the particular additives used. They all have *excellent sunlight resistance*, and so have gained acceptance as curtaining with long life-expectancy. *Flammability* however is rather *high* and this limits the use of acrylic furnishings for high fire hazard applications. Flammability can be overcome: Acrilan 78 and Creslan 84 are examples of fibres which have been produced with special flame-retardant properties.

TRY YOUR HAND

This flammability experiment imitates the dropping of cigarette ash on a carpet.

You will need:
Bunsen burner with tripod, gauze mat and heat-resistant mat
pair of tongs
stop-watch
1 cm metallic nut
carpet samples, preferably of similar construction, with a range of pile fibre contents

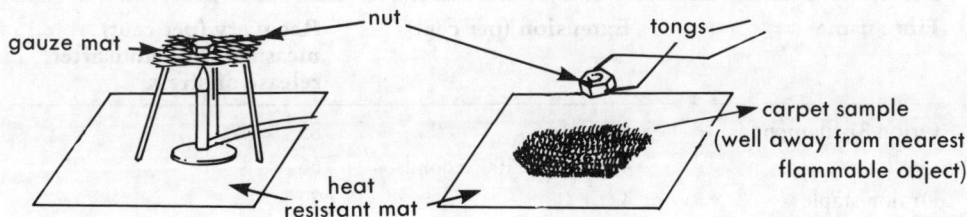

What you do:
Heat the nut until it glows, then, using tongs, place it on a carpet sample for 2 seconds.
Remove the nut and observe the reaction of the pile.
Repeat the experiment for each type of carpet sample, and tabulate your results.
For each fibre type, was the carpet self-extinguishing? Could it be a fire hazard?
Was the mark severe? Permanent?
1   Which fibre performed best?
2   Which performed worst?
3   Did all acrylics behave in the same way? Why?
4   How did the acrylics differ from the other samples in this test?

The *cross-sectional shape* of the fibre varies from dog-bone cross-section to round. *Melt-spun* fibres, which have no solvent to lose from their bulk during fibre formation, have *round* cross-sections. Fibres which are *dry spun* from a solution have collapsed or dumb-bell shaped cross-sections. The more concentrated the initial spinning solution, the less solvent is lost during drying after extrusion, and so the rounder the final cross-section. *Wet-spun* fibres which have been highly stretched have approximately circular cross-sections. *Bicomponent* fibres show their two different halves clearly.

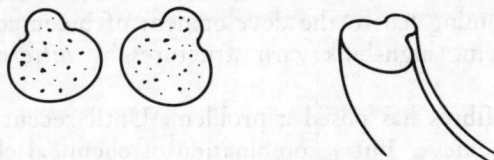

Cross-section of a bicomponent fibre. Because the two parts of Orlon Sayelle have different shrink characteristics, the fibre coils into a helix, creating bulk.

Fibres with round cross-sections are more resilient than the flatter dog-bone shapes, and so are more suitable for carpet pile applications.

Acrylic fibres have quite high resistance to *stretching,* with some variation according to their exact chemical composition and their treatment during manufacture. Although elastic recovery is quite good initially, if the fibre is held extended for some time the fibre will not return to its original length. This can create problems in use: bagging of garments at the elbows and knees, loss of pile height in carpets where furniture weighs it down, stretching of garments hung out to dry after washing.

Resilience of acrylics

Fibre name	Extension (per cent)	Recovery (per cent) measured 1 minute after release of stress
Orlon 81 filament	4	85
	4, held for 100 seconds	66
Dralon staple	4	79
	8	63
Creslan staple	5	40

Note: Orlon 81, a silk-like filament, is no longer in production. Dralon and Creslan are both crimped staples.

TRY YOUR HAND

1   Which of these acrylic fibres is the least resilient? The most resilient?
2   Why does Dralon recover less readily from the greater extension than from the small extension? Why does Orlon 81 recover less readily after the longer period of extension? How does this relate to problems in use?
3   Which of the three fibres would you choose as most suitable for the pile fibre in upholstery velvet?

Acrylic fibres have a *low density* (specific gravity 1.17) and are *poor conductors of heat.* They do not develop *static electricity* quite as readily as polyesters, and this has led to successful applications in light-weight blanketing.

When *heated,* acrylic fibres tend to *discolour,* and then become sticky. Some have *softening points* at about 235°C, and melt at 300°C (Dralon), others *decompose* before melting. In most of their applications, it is not intended that acrylic garments be ironed.

Because acrylics yellow above 150°C, they *cannot be heat set.* Attempts to bypass this shortcoming led to the development of bicomponent fibres and to innovations producing high-bulk yarn structures by *differential stretching* of filaments (p. 153).

Dyeing acrylic fibres has posed a problem. Until recent years deep shades were difficult to achieve. But a combination of chemical changes in the fibre and the use of different dyestuffs has resulted in bright, economical, washfast dyeings with high light fastness.

TRY YOUR HAND

1   In a carpet shop, compare the costs and gradings of carpets made from acrylics with those of similar design made from wool.
    Discuss with the sales person what differences should be expected between these different carpets. Discuss reasons for the popularity of acrylic carpets. If possible, obtain offcuts of the different carpets, and tape them securely in a heavy-traffic area of a corridor or class-room. Try to ensure that all samples are exposed to the same conditions of use.
    Assess the performance of the carpets by whatever criteria seem relevant to you. How does the behaviour compare with the sales pitch?
2   On another shopping trip, compare the handle, look, and feel of curtaining made from acrylic, polyester, cotton, and wool.
    Obtain samples of each, and design your own tests for properties relevant to curtains: durability, resistance to light, flammability, dimensional stability (will the curtain sag?), and any other features that may be important.
    Present your results so that comparisons of the advantages and disadvantages of each fibre can be made. What conclusions can be drawn from your data?
3   In what other major household areas have acrylics gained acceptance?
    What particular properties have contributed to this success?

Imitating natural fibres

Natural fibres are rarely smooth and cylindrical. Wool has a helical spring-like crimp, and cotton is a twisted flat ribbon. In order to make synthetic fibres resemble natural ones, they are treated so as to develop a permanent crimp. Most thermoplastic fibres are heat-set into a crimped shape.

Acrylic fibres, however, *cannot be heat-set* because they yellow at temperatures above 150°C.

When acrylic filaments are hot-stretched, they shrink if allowed to relax in hot water. This property is used for creating special bulked acrylic yarns. Some hot-stretched filaments are blended with unstretched filaments. They are twisted into a yarn, and a fabric is constructed. During hot wet processing the stretched fibres relax and shrink, pulling the other fibres into loops. This produces a bulky, hairy yarn.

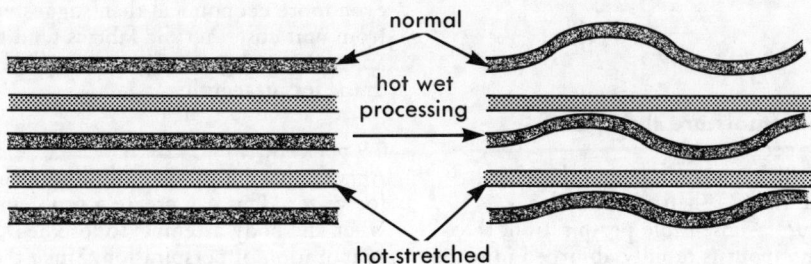

When a blend of hot-stretched and normal fibres is allowed to relax during hot wet processing, the hot-stretched fibres shrink, pulling the unstretched fibres into loops and buckles. This creates a bulky, hairy yarn.

Sometimes the stretched filaments are combined with unstretched ones, and broken into wool-like staple lengths. They are drafted into yarns and twisted into a woollen-type structure. After weaving or knitting is completed, hot wet processing develops the bulk by allowing the hot-stretched filaments to shrink back to their original length.

Orlon Sayelle is a special type of Orlon used to imitate wool. It is produced as a bicomponent fibre. Half of the fibre shrinks during processing, and pulls the bicomponent arrangement into a helical shape.

TRY YOUR HAND

Some acrylics can be processed to imitate the finest cashmere -- others resemble worsteds or crisp fancy yarns.

Make a survey among hand knitting yarns, and list trade names of acrylics against a description of their handle and appearance. This will give you an appreciation of the range of properties of acrylic fibres.

From your investigations it is readily apparent that acrylics are competing with natural fibres in the market-place, and compete particularly effectively with wool. The following table analyses the similarities and differences between staple fibre acrylics and wool. It also gives reasons for the behaviour patterns of each fibre.

Comparison of wool and staple acrylic fibres

Wool	Acrylic
Cost (1980): raw material	
200 cents/kg raw	90 cents/kg staple
Retail price of knitting yarn (4 ply baby yarn)	
$45.00/kg	$9.00/kg
(Australian produced)	(Imported from low labour cost country such as South Africa)
Specific gravity	
Specific gravity is 1.32. Because of its density (heavy weight), fabrics made from wool have a good drape and a rich appearance even in sheer fabrics.	Specific gravity is 1.17. Because acrylic is much lighter than wool, a given weight (e.g. 20 g ball of yarn) will knit into a much larger area of fabric – hence it is even more economical than suggested from unit cost. Acrylic fabrics tend to appear more light-weight than wool and drape less gracefully.
Regain (moisture absorption)	
16 per cent.	0.9 per cent.
Wool is the most moisture-absorbent textile fibre. When the body loses moisture as insensible perspiration, the water vapour is rapidly absorbed into the wool. If the weather is hot and dry, the water evaporates into the atmosphere, thus cooling the wearer. If the weather is cool, evaporation is much slower, and the	Acrylics have a particularly low affinity for water. This can create a problem when the body attempts to lose heat by evaporation of perspiration. Since the water molecules can only escape to the atmosphere *between* the fibres in the fabric and not *through* the fibres (as in wool), many molecules are trapped

Wool	Acrylic

heat of wetting released by wool when it absorbs the moisture compensates for the evaporative loss of heat. Hence wool is *warm*. Because of wool's absorbency, no condensation of the water vapour (emitted by the skin) occurs. Hence, no drops of sweat, and *no clammy feeling* in either hot or cold weather.

between the fabric and the skin. They then condense into liquid sweat which carries heat away from the body by conduction. In cool weather after physical exercise this is registered by the wearer as an unpleasant, cold, *clammy* sensation. In warm weather overheating occurs more readily (since conduction of heat is less efficient than heat loss by evaporation). Also, biological degradation of the *condensed perspiration* may give rise to unpleasant body odour.

Softness

The softness of wool varies greatly with its fineness. Some wools may *irritate* sensitive skins. Fine wool blends which contain more than 10 per cent of 'hard edge' or relatively coarse wool will feel prickly in spite of the generally high quality of the blend. *Reprocessed* and carbonized wools (p. 330) are particularly *scratchy* and irritating for next-to-the-skin wear.

The softness of acrylic fibres can be regulated by varying the diameter of the individual filaments. This of course has other effects, such as on strength and pilling. In general, acrylics are softer than carpet wools but harsher to handle than fine merino or lambswool. If used in sports socks, they may cause *chafing* of the skin.

Drying

Wool is highly moisture-absorbent, and therefore takes a *long time to dry*. Recommended procedure is to remove excess water by blotting between towels (roll in towel and tread on it to squeeze out excess liquid) in order to hasten drying and to reduce possibility of distortion. Wool garments are best dried flat. Low temperature tumble-drying may be used for Superwash wools.

Although acrylics have low moisture regain, they stretch out of shape easily if hung up to dry while dripping wet. This is especially true for bulky garments that can hold appreciable quantities of liquid between fibres within the yarns. Excess water may be removed by blotting or spin-drying. Drying should be either flat or on a low-temperature tumble-dry cycle to prevent heat yellowing. Care must be taken to prevent a build-up of electrostatic charge.

Stains

Wool garments may be *stained* by contact with other coloured items. If stains are not removed in a subsequent wash, the dye molecules which have penetrated the wool fibres must be destroyed by gentle bleaching with peroxide, H_2O_2. Chlorine bleaches cannot be used, as they yellow the wool.

Acrylic fabrics do not generally accept dyes from other fibres – hence *low risk of staining*.

Care, washing

Wool is *sensitive* to *alkalis*, and also tends to *felt* due to the presence of scales. Hence there is a need to use neutral (non-soap) detergents and very *gentle*, brief handling in lukewarm water. Special finishes to prevent felting (e.g. Superwash) have enabled *machine washing* of wool.

Acrylics may be washed in a *machine*, with *ordinary detergent*, spun dry and *tumble-dried*.

Wool	Acrylic

Static electricity

If outside humidity is not too low, the moisture absorption of wool tends to eliminate any problems of static. The mobile water molecules in the wool fibre act as dissipators of any accumulated positive charge.

In almost any atmospheric conditions, acrylics tend to accumulate *static electricity* when rubbed against the skin or other fabrics during wear. This can cause unpleasant crackling and sparking when the static is discharged as the garment is removed. The friction of hot air during tumble drying is particularly effective in creating static charge in acrylic garments. Some of this discomfort can be overcome by the use of *cationic rinse aids* and cautious *brief* drying.

Insulation

Wool is an amorphous fibre and is therefore slow to transmit heat (in the form of vibrational energy). It is an excellent insulator also because its crimped fibres produce lofty yarns which have trapped air spaces. These act as barriers to heat conduction. The air spaces are further immobilized by the projections of the scales of the wool fibres. This makes wool fabrics *outstanding insulators*.

Acrylic fibres are also poor conductors of heat. They can be effectively used in *warm, light-weight* blankets, though their insulative capacity is not as great as that of wool.

Resilience

Wool is a very elastic and resilient fibre. If stretched, it returns to its original shape and size after the removal of the extending force. This elasticity is due to:
(i)   helical crimp of fibre (which acts as a spring).
(ii) helical arrangement of keratin molecules in the crystalline regions of the fibre.
(iii) cystine disulfide links in the amorphous regions. These covalent bonds are distorted when the fibre is stretched, and as the extending force is removed, they return to their original configuration, pulling the rest of the molecules back with them.

Wool's *resilience* and *elasticity* contributes to the excellent *shape retention* of wool garments.

Acrylic fibres in general are not elastic or resilient, though some special types for carpet application are better than others. They have no covalent intermolecular bonds. The weak van der Waals bonds between the molecules are easily broken when the fibre is stretched. The acrylic polymer molecules then slip past one another, and have no way of returning to their original positions. This causes a permanent deformation of the material. Therefore, acrylic garments, especially if tight fitting, will readily *stretch out of shape*.

TRY YOUR HAND

Make a comprehensive list of products which could use wool or a wool-like product. Choose a few items from your list which differ widely. For each product, list the properties required, and indicate whether or not price could be a dominant factor in choice.

Using this information, decide which fibre would be the best choice for each product, and why.

Compare your conclusions with those of other students. Have you always reached the same conclusions? Discuss the factors and priorities that have influenced your choices.

TRY YOUR HAND

Make a careful investigation of the competitive position of acrylics in the marketplace: where they are used, and the comparative costs of acrylics. Visit a department store stocking mainly low-cost goods. In the knitwear section, check all the jumpers (or some other line) for fibre or blend, cost, appearance, handle, and country of origin. Do an identical study on a boutique or 'upmarket' store. What conclusions can you draw from your survey?

Modacrylics and other vinyl fibres

Polymerization and extrusion

MODACRYLICS are acrylic polymers which have been modified by the addition of more than 15 per cent, but less than 65 per cent, of a monomer other than acrylonitrile to the polymerization mix.

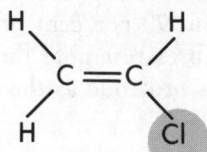

vinyl chloride   vinylidene chloride   vinyl dicyanide

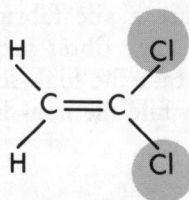

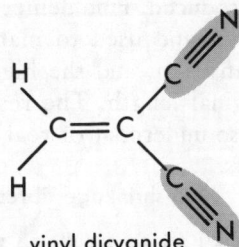

Some copolymers widely used in modacrylics.

These additives generally increase the *chemical and weather resistance* of the polymer. The modacrylics are *flame resistant,* but *shrink* readily at low ironing temperatures. They are excellent *insulators.*

After the polymerization process is completed, the polymer is dissolved in a suitable solvent, dry spun, and hot stretched. The tow is then cut into staple lengths.

Properties and uses

Because of their high *chemical resistance,* modacrylics are used successfully as industrial fabrics, for filters and protective clothing. Due to the high chlorine content, modacrylics based on vinyl chloride and vinylidene chloride *resist burning,* and are often used in blends with acrylics and other fibres to lower their fire hazard.

Properties and uses of modacrylic fibres

Fibre name	Additive	Tenacity (g/den)	Moisture regain (per cent)	Fire resistance	Uses
Dynel (Union Carbide)	vinyl chloride	3	0.4	self-extinguishing	wigs, pile fabrics, industrial fabrics, blankets, air and oil filters, insulation for continental quilts
Teklan (Courtaulds)	vinylidene chloride	3.5–4	1	flame resistant	children's clothing, drapes
Verel (Eastman Kodak)	vinyl chloride	2.8	4	non-flammable	imitation fur fabrics, blended for use in carpets, blended for children's wear and undergarments

Like most other synthetic fibres, modacrylics are relatively easy to care for. Their main disadvantages are their relatively *high cost*, and their *temperature sensitivity*. Modacrylics shrink at 120°C, and so must be washed in warm, rather than hot, water.

Fur-like fabrics

The temperature sensitivity of modacrylics can be exploited to advantage. If the modacrylic tow is *hot stretched* during manufacture, a *high-shrinkage* fibre is produced. Fine denier high-shrinkage fibres are blended with normal, coarser fibres, and used to manufacture a pile fabric. The pile fabric is then heated in an oven, and the high-shrinkage fibres shrink to about 70 per cent of their original length. The result is that the high-shrinkage fibres resemble the soft, dense undercoat of real furs, while the non-shrink fibres protrude as the guard hairs.

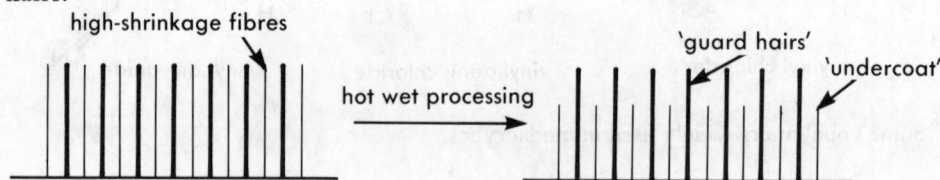

Process for fur-like fabrics from modacrylic fibres.

TRY YOUR HAND

1 Modacrylics are soft, warm, and non-flammable. They would appear to be ideal for children's nightwear. Think of three possible reasons why they have not taken over the nightwear market.

2 What accounts for the success of modacrylics in the fur-like fabrics market? What are the features of real furs that the imitations must compete against?

3 What are the advantages of Dynel wigs over wigs made from natural hair? Go to the wig stand in your local department store, and look closely at how synthetic wigs are made to resemble natural hair. What are the trade-marks for the fibres used? Who manufactures them? What properties are claimed on the label? Look up a textile science reference book to find

out as much as possible about the modacrylic used for the wigs. Compare the fineness of one filament from the wig to the fineness of your own hair. Can you give reasons for any differences?

Other fibres based on the vinyl monomer

When the proportion of additives in the addition polymerization mix is greater than 65 per cent, the fibre is no longer called a modacrylic. Its name will vary according to the chemicals most important in the mix.

Polyvinyl chloride (PVC)

The basic monomer for PVC is vinyl chloride ($H_2C=CHCl$). After polymerization, it is dry spun from a mixture of acetone and carbon disulfide.

The names of PVC *fibres* include Rhovyl, Fibravyl, Thermovyl. *Plastics* made from the same polymer are commonly referred to as *vinyl* or *PVC*. Vinyls are used as furniture covers, car interiors, wipe-off tablecloths, bags, and folders for writing pads.

For textile applications, the most important properties of PVC are its outstanding flame resistance, its *sunlight* and *chemical resistance,* its excellent *heat insulation,* and its *shrinkage when heated.*

PVC fibres are used in blends with wool as a high-shrinkage component. When the fabric is heated, the PVC shrinks, producing a compact fabric with an all-wool surface.

Thermal underwear is also made from PVC fibres. Although its lack of moisture affinity is a disadvantage, it has a unique property. PVC is the only fibre to generate negative static electricity when in contact with the skin. This is claimed by some to be of some benefit to rheumatism sufferers. *Thermolactyl* is an

NOW AVAILABLE IN
LIGHTWEIGHT

Chlorofibre takes 8 seconds

Nature uses the evaporation of body moisture to cool your skin when you're hot, but that evaporation can be a real killer in cold weather. If you want to keep warm, you have to keep dry – and that's where chlorofibre scores over other fibres.
Chlorofibre garments move perspiration away from the skin with uncanny effectiveness. They insulate like nothing else you've ever worn. They wash and dry easily. Most other fibres do not move moisture away, they swell and retain it. Scientific tests prove Chlorofibre moves it in 8 seconds!
The next best is polypropylene in 4 minutes.

(Peter Storm)

159

85/15 blend of the PVC fibre Rhovyl, and Courtelle, an acrylic fibre, and is successfully marketed for thermal underwear with excellent insulation properties. Although PVC fibres are not moisture absorbent, their structure allows some perspiration to be drawn away from the skin by capillary action, so PVC clothing does not cause clammy discomfort: PVC can even be used as a thermal lining for gloves and socks. Care must be taken to use *low temperatures* for washing, otherwise extensive shrinkage may occur.

Some other uses for PVC fibres include *mosquito netting* (where its *flame resistance* is an important property), furnishings, filter cloths, fishing nets, protective clothing, and flying suits.

Saran

Saran is a copolymer of vinyl chloride and vinylidene chloride. It has more chlorine than PVC, and this provides it with even better *resistance to chemicals, weathering and fire*. Also, because of the extra chlorine atoms, it is relatively *heavy fibre* (specific gravity 1.75), which is not an advantage in textile applications. Because of its resistance to solvents, Saran is melt spun.

The uses of Saran have been determined by its range of properties. In pigment dope-dyed form it is used as *shadecloth* for nurseries, for outdoor furniture covers, and for upholstery in car interiors and public transport.

Teflon

Chlorine, bromine and fluorine belong to a group of elements called halogens. The inclusion of halogens in the polymer increases the chemical and other resistance of the resultant fibre, as is shown by the properties of the chlorinated vinyl polymers.

It has been found that *fluorination* is even more effective than chlorination in producing *unreactive fibres.* If all four hydrogens on the original vinyl monomer are replaced by fluorine, addition polymerization will yield polytetrafluoroethylene (PTFE), otherwise known as Teflon.

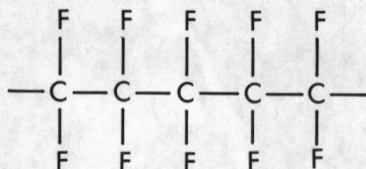

$$
\begin{array}{ccccc}
F & F & F & F & F \\
| & | & | & | & | \\
-C- & C- & C- & C- & C- \\
| & | & | & | & | \\
F & F & F & F & F
\end{array}
$$

Teflon

This polymer is entirely uniform, and electrically neutral – this allows a very close packing of the molecular chains. (Specific gravity is 2.2 – very heavy.) Because of the close packing, there are so many van der Waals forces holding the chains together that it is extremely hard to separate them. Hence Teflon has a *high melting point*: 400°C. Its lack of solubility in all solvents means that special spinning techniques have had to be developed to produce it in fibre form. Its chemical resistance is such that it can be boiled in a mixture of concentrated sulfuric and nitric acids for a day without any loss of strength. Because it is not attracted to any other substances, it has a very low coefficient of friction and a *slippery handle*.

Teflon is used for industrial protective clothing, filtration fabrics, anti-stick bandages. When moulded and used as a plastic, it has great success as low friction rotating points, as a coating for magnetic stirrers for corrosive solutions, and as coating for non-stick frypans.

Polyvinyl alcohol – Vinal, Vinylon, Kuralon, Solvron

Have you ever used PVA wood glue? It is a white emulsion which mixes easily with water. When it dries it forms a clear, hard film. PVA stands for polyvinyl acetate. When PVA is reacted with alkali (NaOH), polyvinyl alcohol is produced.

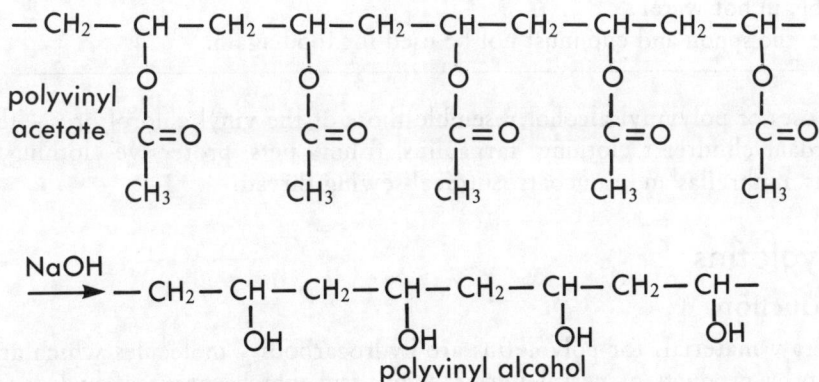

$$-CH_2-\underset{\underset{\underset{\underset{CH_3}{|}}{C=O}}{|}}{CH}-CH_2-\underset{\underset{\underset{\underset{CH_3}{|}}{C=O}}{|}}{CH}-CH_2-\underset{\underset{\underset{\underset{CH_3}{|}}{C=O}}{|}}{CH}-CH_2-\underset{\underset{\underset{\underset{CH_3}{|}}{C=O}}{|}}{CH}-CH_2-\underset{\underset{\underset{\underset{CH_3}{|}}{C=O}}{|}}{CH}-$$

polyvinyl acetate

$$\xrightarrow{\text{NaOH}} -CH_2-\underset{\underset{OH}{|}}{CH}-CH_2-\underset{\underset{OH}{|}}{CH}-CH_2-\underset{\underset{OH}{|}}{CH}-CH_2-\underset{\underset{OH}{|}}{CH}-$$

polyvinyl alcohol

Polyvinyl alcohol has one hydroxyl (–OH) group to two carbon atoms: it is so polar that it is soluble in water.

For special purposes, polyvinyl alcohol is produced as a water-soluble yarn (Solvron or Kuralon), for use as a separator thread between pieces of knit goods such as socks. During hot wet processing the yarn dissolves, and the socks separate.

Polyvinyl alcohol thread soluble in warm or hot water, designed for use as separation threads in sock manufacture.
(Nitivy Co. Ltd, Tokyo, Japan)

In order to make this polymer into an effective textile material for more conventional end-uses, it is cross-linked with formaldehyde. This cross-linking gives it excellent water resistance – indeed, Vinylon has a moisture regain of only 5 per cent.

TRY YOUR HAND

You can make your own polyvinyl alcohol in the following way. Take 1 teaspoon PVA glue and dissolve it in ½ cup methanol to give a clear, thick solution. Add, drop by drop, concentrated sodium hydroxide solution (caution!), and stir. The precipitate is your POVAL (vinyl alcohol polymer). Check if it is soluble in hot water.
Note: the spoon and cup must not be used for food again.

The uses of polyvinyl alcohol resemble those of the vinyl chlorofibres – flame-retardant children's clothing, tarpaulins, fishing nets, protective clothing, uniforms, umbrellas and raincoats, surgical sewing thread.

Polyolefins

Production

The raw materials for polyolefins are hydrocarbons – molecules which are the cheap by-products of petroleum cracking, and which consist of hydrogen and carbon atoms only.

Ethylene can be readily polymerized in an autoclave at high temperature and pressure. The POLYETHYLENE is then melt spun and cold drawn to give filaments.

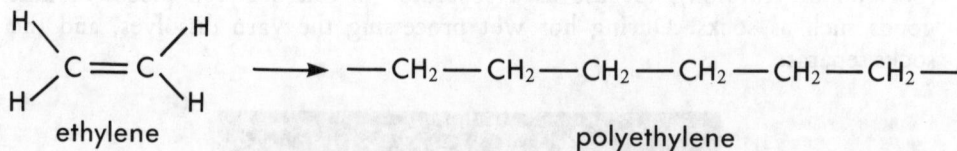

$$H_2C=CH_2 \longrightarrow -CH_2-CH_2-CH_2-CH_2-CH_2-CH_2-$$

ethylene   polyethylene

Because it has *no polar groups,* polyethylene has low friction properties – a slippery, *oily handle.* It also has a *low melting point* – its van der Waals bonds are easily broken during heating. Because its chemical resistance is an advantage, gamma irradiation has been used to cross-link the polymer and so eliminate its heat sensitivity.

Polyethylene was first produced in the 1930s, and was an important plastic during World War II. However, when the process of high pressure polymerization, which was so successful for ethylene, was tried with propylene, the resultant polymer was a greasy mess.

Since propylene is by far the cheapest raw material for synthesizing fibres, much effort was put into developing a method of converting it into an effective polymer. Success came in 1957, when a catalyst developed by Ziegler allowed the bulky side chain of the polymer to be organized in an orderly way.

$$-CH-CH_2-CH-CH_2-CH-CH_2-CH-CH_2-CH-CH_2-CH-CH_2-CH-CH_2-$$

Above the chain, CH₃ groups appear on positions 2, 6, 7 and below on positions 1, 3, 4, 5.

Atactic

$$-CH-CH_2-CH-CH_2-CH-CH_2-CH-CH_2-CH-CH_2-CH-CH_2-CH-CH_2-$$

Syndiotactic

$$-CH-CH_2-CH-CH_2-CH-CH_2-CH-CH_2-CH-CH_2-CH-CH_2-CH-CH_2-$$

Isotactic

With a regular molecular form, the polymer chains are able to pack together tightly, and form fibres suitable for textile applications. The fibres are *melt spun,* and then *cold stretched* to increase the crystallinity. The next process is *annealing,* a final heat treatment which allows the molecules to rearrange themselves and pack together a little better. Annealing further increases the crystallinity of the fibre. The *extent to which the fibres are stretched* during cold drawing controls the *tensile strength* (and elongation at break) of the product.

Stereoregular polymers

A carbon chain is usually represented as a straight line in textbook formulae. In actual fact, however, that 'straight' chain is a zig-zag.

TRY YOUR HAND

You will need:
some models of atoms – the stick and ball variety are the most suitable. If you do not have models, you can try matchsticks stuck into balls of craft clay to make your molecules.

What you do:
First make a model of polyethylene, using up to 24 carbons in the chain. Each carbon atom forms 4 bonds, each bond pointing to the corners of an imaginary tetrahedron.
Note the regularity of the structure of polyethylene.
Make a tracing of the *shadow* of your model by placing it on a sheet of butchers' paper, and lighting it directly from above. Do this twice more so you have three tracings. Cut out the shapes, labelling each 'polyethylene'.

Now replace one of the hydrogen atoms on every second carbon atom by a methyl (–CH₃) group.
Make sure that the methyl groups are randomly arranged. This is an ATACTIC polymer.

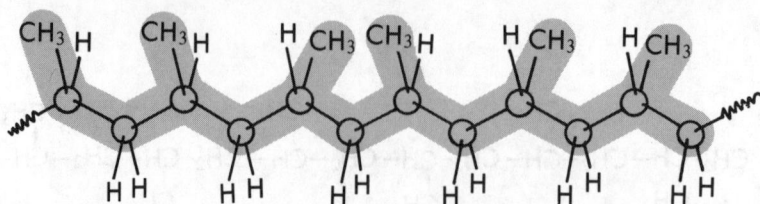

Trace the random polypropylene molecule, making three copies again, and cut it out.
Now rearrange the methyl groups so that they all face in the same direction. This is your ISOTACTIC polymer. Trace and cut three shadow pictures.

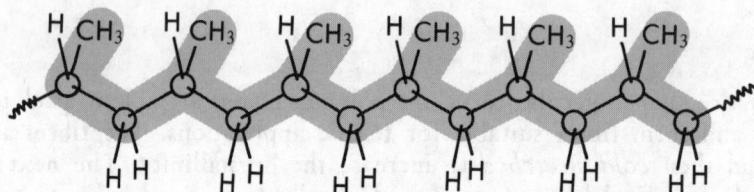

For your SYNDIOTACTIC polypropylene, arrange the methyl groups so that alternately one goes up and the other down. Repeat the process of making three labelled shadow pictures.

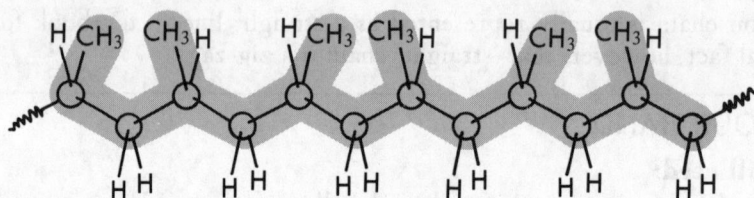

Now for each type of polypropylene, try to pack the molecules as close to one another as possible.
1 Which will cover the largest area? Which the smallest?
2 Would you expect the densities of atactic, syndiotactic and isotactic polypropylenes to be identical? Why?
3 Would you expect them to have the same melting points? If not, which would be higher and which lower? Why?
4 What other differences in the properties of the three STEREOISOMERS (differently arranged but otherwise identical molecules) can you predict?

The polypropylenes

Polypropylene structure	Specific gravity	Melting point
Atactic polypropylene	0.85	semi-liquid
Syndiotactic polypropylene	0.90	150°C
Isotactic polypropylene	0.92	176°C

When the crystalline structure of isotactic polypropylene was investigated, it was found that the molecules were arranged in a helix, with three monomer units (six carbon atoms) forming each turn. This is a very compact arrangement, and allows for close packing of the molecular chains.

If polypropylene is polymerized without a catalyst, the useless atactic polymer is produced. The STEREOREGULAR, isotactic and syndiotactic, POLYMERS are produced with the aid of a *catalyst*.

Three main catalysts are used, each named after its discoverer. These catalysts provide a foundation or mould for the formation of the regular polymers. The monomers align in a particular way on the active surface of the catalyst, and are added one by one to form the stereoregular polymer. The *Ziegler catalyst*, which is used for the formation of most stereoregular polymers, is usually titanium dichloride combined with aluminium triethyl. It is named after Karl Ziegler, who in 1963 received the Nobel Prize for his discovery. The first stereoregular polymers were produced in 1954.

Properties and uses

Polyethylene, because of its *oily handle* and *heat sensitivity*, is not suitable for textile apparel use. Its *hydrophobic* nature and *chemical resistance* make it suitable for car interiors, fishing nets, furnishings, filters and tarpaulins.

Twine made from fibrillated polypropylene film.

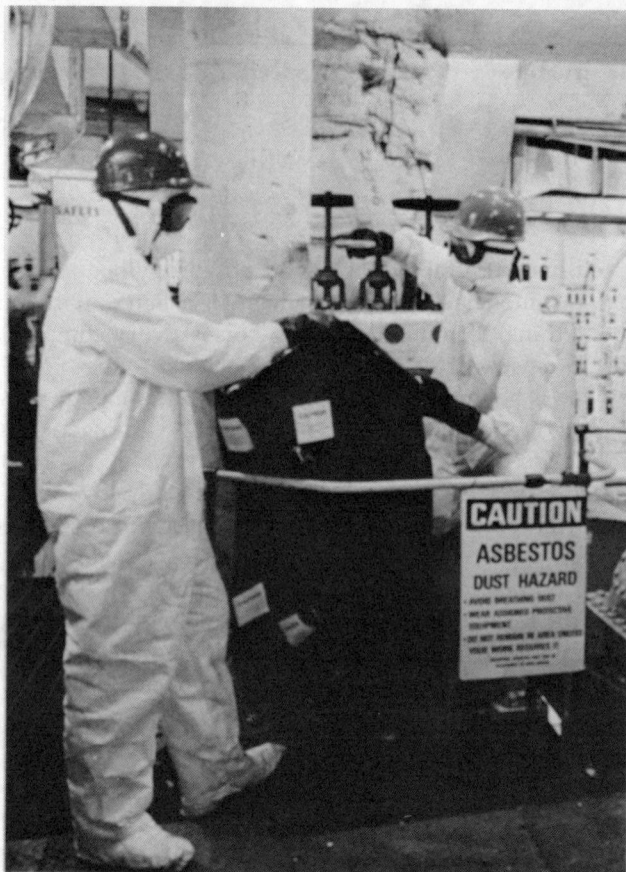

The chemical resistance and low cost of polypropylene make it ideal for disposable protective clothing, for uses where high-risk contaminants could pose a risk during normal laundry procedures.
(Du Pont)

Polypropylene is not only cheaper but is also more versatile. Polypropylene is the *lightest* of all commercial fibres, with excellent *resistance to chemicals, micro-organisms, abrasion* and *light*. Its *low melting point* (165°C) means it cannot be ironed safely. Its lack of affinity for moisture makes it suitable for use in the fishing and similar industries.

Ropes made from polypropylene will float on water (specific gravity is 0.91) and so may be less easily lost at sea than the traditional – if cheaper – manila fishing ropes.

Another fibre gradually being replaced by polypropylene is jute, particularly as sacking and backing material for carpets. To lower costs of manufacture, films of the polymer are extruded, split into strips and then fibrillated and twisted into a yarn.

Indoor/outdoor carpets are successfully produced by the nonwoven method of needling a web of polypropylene filaments, and then shrinking and welding the fibres together in an oven. Most *artificial turf* is also made with polypropylene tape yarns as the pile, to give a resilient abrasion- and weather-resistant product.

Elastomers

Elastic fibres

Elastomeric materials extend when stretched, and return to their original size when the pull is released. Textile materials with very low elasticity are either too rigid and uncomfortable for apparel, or will stretch out of shape during use. Most textile fibres have some elasticity; but a special group known as SPANDEX or ELASTOMERIC fibres have been especially designed to return to their original size immediately the pulling stress is released. Such fibres can be pulled to five or six times their original length before breaking.

TRY YOUR HAND

You will need:
rubber bands of different thicknesses

What you do:
Stretch each rubber band.
1   Did you find significant differences in different rubber bands' resistance to the pulling force from your hands?
2   What factor seems to determine the power of the elastic band?
3   Test the temperature of the stretched and relaxed bands by holding them against your lips.

Light-weight and comfortable garments such as this were an impossibility before the development of elastomeric fibres.
(Magnamail, Brookvale N.S.W.)

When elastomers are extended, some special bonds which hold the molecules in their normal, relaxed shape are distorted. A certain amount of energy is needed to distort each one of these bonds. A 5 cm length of a wide rubber band contains more such bonds than a 5 cm length of a narrow rubber band.

This means the wider elastic needs more force to distort all its bonds than the narrow elastic does. The energy that you put into pulling the elastic is stored up in each of the distorted bonds (it feels warm). When you let go, the stored up energy is released, and pulls the bonds back into their original, relaxed shapes (it now feels quite cold).

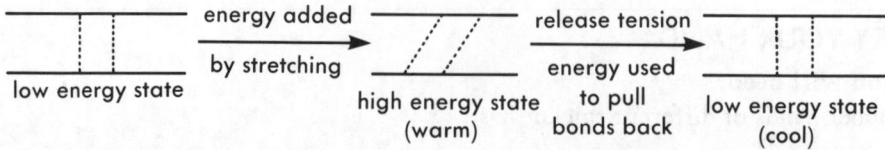

low energy state   energy added by stretching   high energy state (warm)   release tension   energy used to pull bonds back   low energy state (cool)

The force with which an elastomer resists extension and with which it tends to return to its original state thus depends on the number and kinds of bonds that are distorted by the stretching process.

TRY YOUR HAND

Take a curly yarn from a pair of panthyhose, a fine rubber thread (remove the covering of some shirring elastic) and some Lycra thread (from an elastic garment).

Do they all return to their original size after stretching? Which of the three materials do you think has more bonds helping recovery? Which is used most often for figure-controlling garments? Why?

Allowing for variations in thickness, can you estimate which has the greatest resistance to stretching?

Power stretch

Those materials which pull back strongly are called POWER STRETCH elastomers. Those that are elastic, but not as 'powerful', create COMFORT STRETCH when used in apparel. Textured yarns (e.g. the pantyhose thread) provide comfortable ease of movement at places where a lot of extension is regularly needed, such as the knees. Power stretch materials are used for figure control. Lycra has more 'power' than rubber. A figure control garment made of rubber would need to be much thicker than an equally effective one made of Lycra.

Where comfort stretch is needed in apparel.

TRY YOUR HAND

You will need:

some elastic garments and textile materials, such as swimsuit, stretch jeans, knitted cardigan, hat elastic, waistband of underpants, pantyhose

What you do:

Measure each article, and then stretch it as far as you can.

Measure it again while it is stretched. (It may be easiest to make two marks on the article and then measure the distance between them, stretched and unstretched.)

1 Calculate the percentage extension of each item by the following formula:

$$\text{percentage extension} = 100 \times \frac{\text{extended length} - \text{original length}}{\text{original length}}$$

2 Record whether a strong or a weak force was needed to achieve the extension in each case.

3 Suggest the sources of extensibility for each item.

TRY YOUR HAND

You will need:

a range of elastics of different thicknesses and compositions: hat elastic, shirring elastic, elastics from 0.5 cm to 3 cm wide; each piece 10 cm long

a board and drawing pins for mounting elastic samples

S-hooks or fish-hooks

weights from 1 g upwards, with loops so they can be hung on the hooks

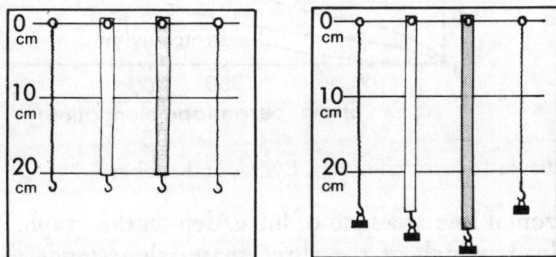

Stretch testing rig.

What you do:

Mount the 10 cm lengths of elastic on the board.

Attach an S-hook to the lower end of each, and mark the board where the lower end of each piece falls.

Now hang a 1 g weight on the elastic, and again mark the position of its lower end.

Continue adding weights, and recording the position of the lower end of each piece of elastic.

On a fresh piece of graph paper for each sample, plot the results of your experiment, with load in gram on the vertical axis and elongation in centimetres on the horizontal axis.

Some elastics elongate readily, even with small loads. You should get a different curve for each sample.

1 What information can you get from your curves?

Remove the weights one by one.

2 Did your samples return to their original lengths along the same curves?

3 The word HYSTERESIS describes a *delay between cause and effect*. Does this apply to your findings?

4 Were all the samples 100 per cent ELASTIC (did they return to their original length), or was there RESIDUAL SET (permanent elongation) in some of them?

The following graph compares the stretch behaviour of Lycra, rubber, and stretch (crimped) nylon.

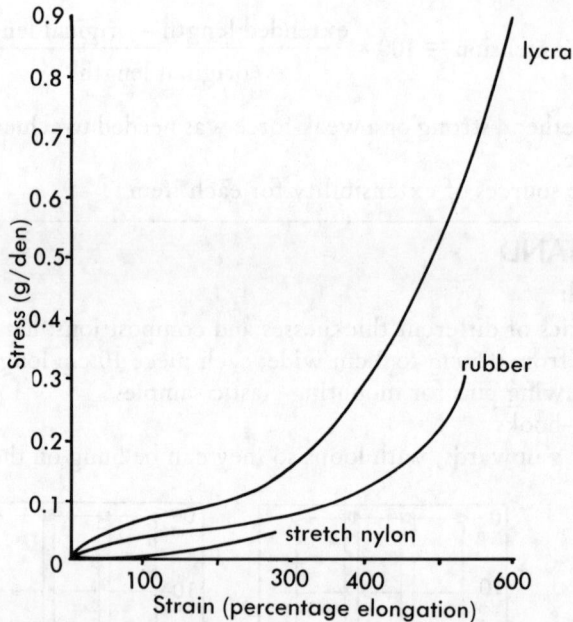

Stress-strain curves for stretch nylon, Lycra, and rubber.

Draw a horizontal line at a load of 0.2 g/den on the graph.

5 At this load, which of the three materials extends the most? Which the least?

6 Which would you expect to be the most effective stretch elastomer?

7 Which would be most suited for comfort stretch? Why?

8 How could comfort stretch be achieved with the stronger elastomer?

The molecular structure of Spandex

Synthetic snap-back fibres are known by the collective name of SPANDEX.

Spandex fibres are *polyurethanes*. They are *block copolymers,* with the urethane groups providing hydrogen-bonded, ordered regions between long, *coiled,* often *amorphous* structures, where there is little bonding between adjacent molecular chains.

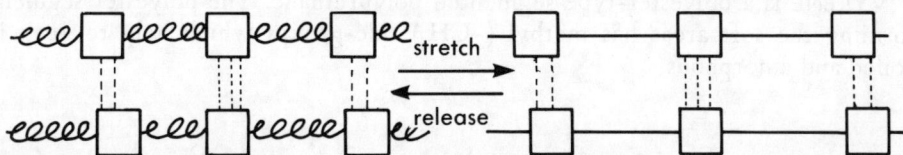

Hydrogen bonds between the 'hard' polyurethane segments provide the bonding between the polymer chains, in both the stretched and the unstretched state.

When the fibre is stretched, the amorphous 'soft' segments uncoil; these, however, contain bulky side-groups which do not allow them to pack in an ordered linear form. When the pulling force is released, the soft segments coil once more to relieve the distortion in the bulky side-groups.

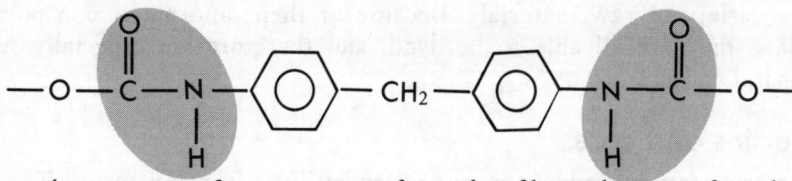

A polyurethane segment from one type of Spandex fibre. The exact formula varies with the manufacturer, but all contain amide ester groups to provide the hydrogen bonding of the hard segments.

When the fibre is extended, the soft segments of the molecule uncoil. The straightened-out length is much greater than the coiled, amorphous length, and so the fibre can be stretched considerably.

The stretched molecules *cannot* slip past each other, because they are held firmly by the hydrogen bonds in the 'hard' urethane segments. If *molecular slippage* occurred, it would cause *irreversible* extension.

The natural state of linear polymers is a coiled, random form. When the molecules are lined up by the drawing and stretching processes in production, *interchain attractions* act to keep them in this *ordered* state. The soft segments of the polyurethanes, however, contain bulky side-groups which make them difficult to pack into a linear arrangement. Not only do the *bulky side-groups prevent close packing,* and so *prevent strong interchain bonding,* they are often *distorted* when the molecules are pulled straight. So, although the amorphous regions straighten out when the fibres are pulled, they return to their natural, random state as soon as the stress is removed.

LYCRA is a polyurethane which has polybutylene ether as the 'soft' segment, with amine groups to provide the hydrogen bonding in the hard, crystalline segment. Lycra is soluble in dimethyl formamide, and is dry spun from its solution.

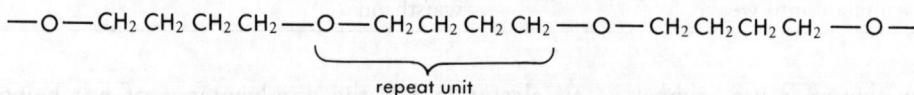

Lycra has a polybutyl ether in its soft segments.

VYRENE is a polyester-type segmented polyurethane. The polyester segment forming the soft areas has methyl ($-CH_3$) side-groups which help to keep it coiled and amorphous.

$$-O-CH-CH_2-O-\overset{\overset{\displaystyle O}{\|}}{C}-CH_2\ CH_2-\overset{\overset{\displaystyle O}{\|}}{C}-O-CH_2-CH_2-O-\overset{\overset{\displaystyle O}{\|}}{C}-CH_2-CH_2-\overset{\overset{\displaystyle O}{\|}}{C}-O-CH-CH_2-O-$$

with CH_3 side groups and labels *repeat unit* / *repeat unit*

In Vyrene, the soft segments are polyesters made from two different dialcohols.

There is a great variety of Spandex yarns manufactured in different ways from a variety of raw materials. Because of their amorphous non-polar soft segments, they are all able to be dyed, and therefore are especially suitable for apparel end-uses.

Properties and care

Elastomeric materials appear in our everyday lives in many forms. The single largest outlet for polyurethanes is elastic foam, used for mattresses, pillows, stuffed toys and carpet underlays.

Elastomers are rarely used in direct contact with the skin. Stretch a rubber band, hold it on the back of your hand and release it. Did it pinch? To allow the elastomer to slide smoothly over the skin, it is usually covered either by fibres (e.g. core-spun yarns) or by another yarn (e.g. covered yarn) or by the knitted or woven structure of cloth.

Elastomers are produced as monofilaments, partly joined multifilaments, or as sheets cut into filaments with square cross-section.

TRY YOUR HAND

Look in your wardrobe for garments which contain elastomers. Can you find examples of all the different forms in which they are produced?
Write more examples in the following table.

Elastomers in use

Type	Examples
square cross-section	bra strap (stay-in-place type)
covered yarn	hat elastic
core spun yarn	'lycra' swimsuit
yarn inlaid into knit	sock top
yarn inlaid into weave	waistband

In consumer use, *rubber* as an elastomer has the disadvantages of not being dyeable, of being readily affected by perspiration, cosmetic oils, perspiration and ageing.

Spandex fibres have mostly overcome these problems. They are *resistant* to acids, alkalis, dry-cleaning solvents, perspiration, oils, insects, mildew, light and ageing. Their one remaining sensitivity is to sodium hypochlorite. This means not only that Spandex garments should *never be bleached* with chlorine bleach, but also that *swimming pool chlorine* can attack such garments if not thoroughly rinsed out. They will have reduced life in areas with a chlorinated water supply.

Spandex garments have *low moisture affinity* – the hydrogen bonds are only in the crystalline part of the molecules – and have low electric conductivity. Reaction to *heat* varies among different brands, but they are generally thermoplastic. This has little importance: because of their resilience, Spandex garments rarely need ironing. Should the occasion arise however, ironing temperatures should not exceed 120°C because of danger of yellowing. Care should be taken also to keep washing temperatures below 60°C, and tumble-drying temperatures below 80°C.

Other textile fibres

Glass

Manufacture

Glass is made from a mixed melt of silica, sand and limestone. The type of glass produced is determined by what other materials are added. The molten liquid is cooled to a clear solid, usually in the form of balls. During the cooling process, the molecules do not form any crystalline structure, but remain completely *amorphous,* disorganized. Despite this, glass has *little extensibility* or *elasticity,* and *tends to snap* readily. Such rigidity makes it very *vulnerable to abrasion.*

TRY YOUR HAND

How can such a rigid fibre be useful for textile purposes?
Take a glass stirring rod, and rotate the central portion of it over a Bunsen flame.
When the glass starts to glow red, pull your right hand away fairly quickly, keeping the left hand part of the heated portion still in the flame.

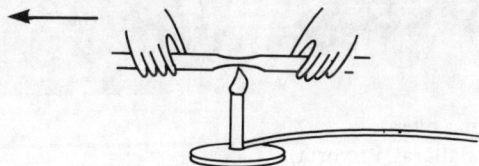

How fine a thread can you pull from the rod?
Repeat the experiment, trying to get glass filaments of very small diameter.
The fine filaments cool quickly. Examine their properties.
1 How does flexibility vary with diameter?
2 What would be the properties of a ten-strand multifilament, compared to a monofilament of the same final weight per unit length?

This experiment is the equivalent of the earliest ways of making glass threads. In fact, the glass wool called 'angel hair' is manufactured by winding threads from heated glass rods around a bicycle wheel. When a thick wool of fine filaments is obtained, cutting across the wool produces bundles of fine, silky angel hair which can be draped prettily to decorate Christmas trees.

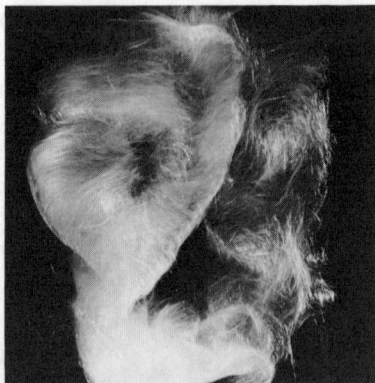

Angel hair: glass fibres of 12 micron diameter.

Glass fibre tissue curing oven.
(Regina Glass Fibre, Ballarat Victoria)

Fibreglass *filaments* are produced by *melt extrusion* and hot stretching. *Staple* fibreglass, such as is used in insulating batts, is made by blowing the hot strands of glass with jets of air.

Glass will not absorb moisture or dyes. The only way to colour it is to introduce into the melt inorganic pigments which can withstand high temperatures. These produce only very pale colours.

174

Stronger colours can be added to glass fibres from the outside, with coloured *resin coating*. Such coloured resins can be applied to the fibreglass fabric as *screen-printed* patterns or as uniform *solid colours*. A process called CORONIZING (introduced in 1964) helps to bond the resin to the glass fibres and also softens the handle of the fabric. Coronizing involves coating the fabric with a light dispersion of silica – this will stop the fibres from slipping over one another – and then heat setting it for 5 to 15 seconds at 650°C. Colouring or printing follows, using coloured latex, which is dried and then bonded to the glass surface. Coronized glass fabrics have a softer, improved handle and better abrasion resistance.

A further improvement has been the introduction of BETA FIBREGLASS. These filaments are about a third as fine as ordinary fibreglass, and therefore fabrics made from them are considerably *softer* and even more *abrasion resistant*. Naturally, the production of *finer filaments* is also *more costly*.

Properties and uses

Glass textiles are outstanding in their *resistance to degradation* by light and by chemical agents. As they are completely *flameproof,* they serve well in curtains as well as in many industrial uses.

Ease of cleaning – generally just sponging or hosing down with water – has made fibreglass drapes popular in hospitals and in offices. Care must be taken, however, that fabrics made of fibreglass are *never* washed in a *washing machine*. During the abrasive action of washing, tiny fragments of glass break away from the fabric, and contaminate subsequent loads. These invisible fine pieces of glass can produce severe skin irritation, and are very difficult to remove once they have lodged in other textiles during the wash.

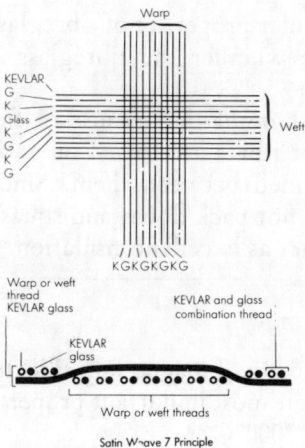

Resin-coated woven glass fibre has long been used for boat hulls. Here, a Kevlar/glass fibre composite fabric is being laminated on a balsa wood base to build a luxury yacht. The composite consists of a satin-weave fabric of Kevlar and glass fibre on a base of glass fibre chopped strand mat.
(Du Pont)

A protective layer of glass fibre and paint being applied to a corroded pipe. The nonwoven glass fibre tissue provides a stable base for the protective layer of paint. The degree of anti-corrosion protection is directly related to the thickness of the paint film. (Regina Glass Fibre, Ballarat Victoria)

Although fibreglass is *strong,* its durability is limited by the *lack* of high *abrasion resistance.* Therefore drapes and curtains made from it must be designed so that edges do not rub repeatedly against windowsills or floors, and so that draughts of air do not whip them out of open windows. If these precautions are observed, fibreglass can give excellent service in domestic as well as industrial applications.

In industry, fibreglass is used as *filter* material, as heat and sound *insulation,* and as *reinforcement* in structural and decorative plastics.

TRY YOUR HAND

Make a list of examples of application of fibreglass materials. Find as many different uses as you can.

Organize your list in a table form, showing the particular properties of fibreglass that make it suitable for that application. Indicate whether the fibreglass is spun, staple, filament, or bulk.

For example, fibreglass batts are such excellent heat *insulators* because of the *rigidity* of the fibres. During manufacture, the short fibres are tossed on each other in such a way that many air spaces are formed between them. Since the fibreglass fibres are both light and stiff, they do not pack down and squash the air pockets. The many *small, still pockets of air* act as excellent insulation.

Asbestos

Asbestos is a mineral fibre. It is a fibrous, crystalline form of magnesium silicate. The mineral fibres are extremely fine and strong. Their most important property is that they are able to withstand temperatures up to 3000°C.

Production

Asbestos is mined. Major producers are Canada, South Africa, and the Soviet Union. Asbestos mined in Western Australia once contributed about 10 per cent of Australia's requirements, but this source is no longer important.

The asbestos fibres are freed from the rock by repeated crushing and sieving. They are then separated from the crushed rock by blowing. Only the longest fibres are used for textile production; the short fibres are used in non-textile products such as building construction materials and thermal insulation, and for specialized applications in electrical installations, combustion chambers, brake linings, and rocket nose cones.

TRY YOUR HAND

Asbestos is a collective name for some 30 fibrous minerals. The most important for commerce is chrysotile.

Research the chemical and crystalline form of this mineral in an encyclopaedia, and discover what other common minerals it is related to.

Compare the fire-resistant properties of asbestos with that of the other related minerals.

Can you find any similarity of uses among the related minerals?

During production and during use, microscopic fibres of asbestos break away from the material and circulate in the air. When inhaled in some quantity these fibres have been shown to produce asbestosis and cancer of the lungs. Currently there is an active campaign to point out the health hazards for workers involved in the manufacture and use of asbestos products, and there are concerted efforts to replace it by synthetic fibres of high heat and fire resistance.

The long asbestos fibres are brittle and smooth, and are therefore very difficult to spin into yarn. For support and ease of handling, they are blended with cotton or rayon, and are then carded and spun in a conventional way.

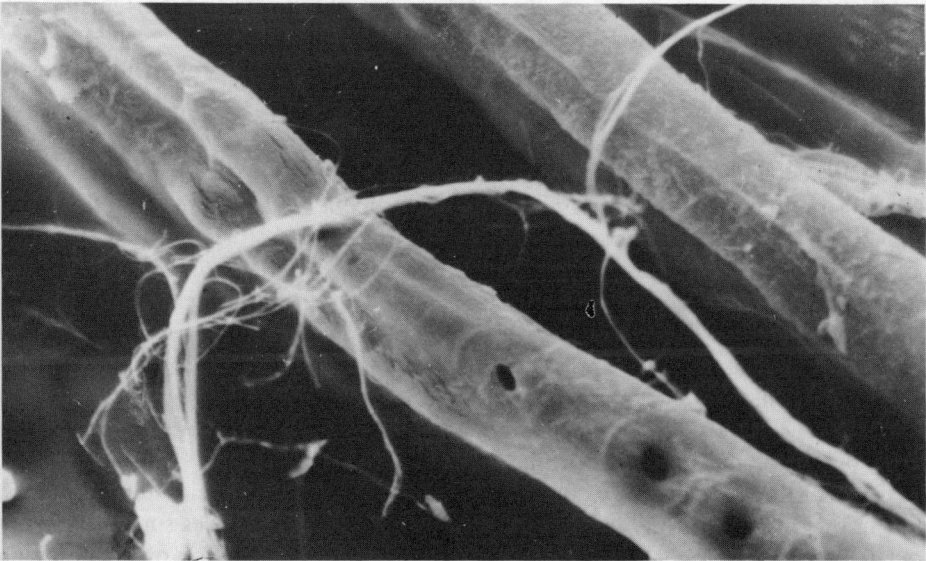

Fine asbestos fibres compared to wood cellulose. The cellulose has a diameter similar to that of coarse wool.
(C.S.I.R.O. Division of Chemical and Wood Technology)

Uses

The Greeks and Romans used asbestos for lampwicks and cremation cloths, but its use was forgotten in Europe by the Middle Ages. It was still used in Asia: in the thirteenth century, Marco Polo brought back from China news of a fabric which did not burn.

Today, asbestos is used for fire blankets, safety clothing, and fireproof curtains, and as insulation material, but its market share is declining because of the hazards associated with its use.

Alginates

Strictly speaking, alginates are not synthetic fibres. They are the calcium salt of *alginic acid,* which is chemically very similar to cellulose: one hydroxyl (–OH) group is replaced by a carboxylic acid (–COOH) group on each ring. Alginic acid is obtained from brown (but not red or green) seaweeds. (The Latin for seaweed is *alga*; hence the name.)

The powdered brown seaweed is treated with caustic soda (sodium hydroxide, NaOH) to extract the soluble alginate. The solution is carefully purified and spun on a viscose machine into an acid calcium chloride coagulating bath.

Alginate fibres have two remarkable properties: they are flameproof, because of the calcium; and they dissolve in soapy water, because they are ionic salts. It is this latter property which gives alginates their commercial importance.

Uses

Delicate *laces* can be embroidered on alginate *supporting cloths.* When the fabric is washed, the backing disappears, leaving the fine lace. During the manufacture of *socks, hosiery,* and other repeat-unit knitted items, alginate yarns can be used as *separating threads* between the knitted units. During wet finishing, the socks or hose separate automatically.

Another important use of alginates is as *thickening agents* for printing pastes used in the dyeing trade.

Carbon fibres

Carbon in the form of *graphite* is such an *excellent conductor of electricity* that it is used as the contact in electric generators and motors.

When used as a textile fibre, carbon *prevents the build-up of static electricity* in specialized, mainly industrial, fabrics.

Carbon can also be included in powder form in the melt of nylons that are to be used in anti-static applications. The small quantity needed minimizes discolouration.

Because of their excellent *strength* and *rigidity,* carbon fibres have their main outlet as *reinforcement* of plastics and metal alloys.

Carbon fibres are produced by *carbonizing* (heating to very high temperatures, but in the absence of oxygen to prevent burning) organic fibres such as acrylics or high-strength regenerated cellulose fibres. Precise control of carbonizing temperatures (up to 3000°C) determines the physical properties of the resultant carbon fibre.

Metal and metallics

The very first fibres ever made artificially were produced by hammering fine wires made from precious metals into long, pliable strips. Threads of real gold decorated hand-woven saris, and the carpets, embroideries and tapestries of many ancient civilizations. Such golden cloths were so expensive that they were the prerogative of royalty. Today fabrics that look like gold can be afforded by anyone.

Metallic threads are produced by sandwiching fine aluminium foil between two layers of plastics. The adhesive used is thermoplastic, and the layers are formed by heating them to about 95°C and passing them through rollers under high pressure.

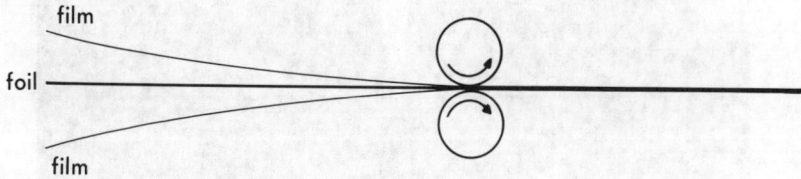

The manufacture of Lurex.

The sandwiched sheets are slit into filaments, usually 0.5 mm wide. At this stage they are silver in colour.

The gold colour which accounts for more than half the metallic thread produced is achieved by mixing a yellow dye into the adhesive. Other colours – bronze, blue and so on – can be added according to the whims of fashion. Sometimes the plastic covering is dyed or printed to achieve a variety of effects.

The plastic most commonly used for metallic yarns is polyester. A more expensive but also more appealing product is made by the vacuum deposition of aluminium on polyester film (Lurex MM). Two layers of metallized film are glued together with coloured adhesive. Because there is no continuous strip of metal, this type of fibre is softer, more pliable and less harsh to the touch than ordinary Lurex.

Metallic thread in a yarn.
(Vivian Robinson)

A traditional use of metals in textiles, Java. An adhesive is applied to the surface of the batik, which is then layered with gold dust or foil.
(Photographic Archives of the Royal Tropical Institute, Amsterdam)

The handle of all metallic yarns is affected by the ribbon-like cross-section of the filaments. The sharp edges of these tend to give a slightly scratchy, harsh handle to Lurex and similar fabrics. Blending with other filaments or fibres during yarn construction can minimize this unpleasant effect.

Another important use of metallic yarns is in carpet manufacture. *Zefstat* is a continuous aluminium strip attached by adhesive to a plastic cover. It can be blended with nylon, acrylic or modacrylic fibres to give a yarn which conducts electricity. About 2 per cent of the weight of the carpet is sufficient to prevent the build-up of static. At this low level of usage, no unattractive glitter can be seen in the finished carpet.

A textile for every purpose

TRY YOUR HAND

1   Here is a list of properties important in textiles. Draw up a chart showing how each of the synthetic fibres studied performs in relation to each property, and indicate its importance to the consumer.

heat sensitivity	biological resistance
strength	build-up of static electricity
resistance to sunlight	density
chemical resistance	moisture affinity
abrasion resistance	grease affinity (oleophilic or not)
resilience	flammability
tendency to pill	reaction to heat (thermoplastic or not)

2   Helen has a large family gathering every Christmas. She usually hires a table and puts it end to end with her own, so everyone can sit comfortably. She cannot buy a cloth long enough to cover the two tables, so she decided to make her own. She bought a suitable length of green linen-like material and hemmed it: it looked great with the red and green table decorations. When she washed the cloth later, she was cross to find that the grease marks did not come out.

What was wrong with her choice of fabric? From the discussion of fibres in this chapter, identify the probable fibre type, and give the reasons for what happened.

Is there a way to remedy the problem?

What fibre types would you advise Helen to choose for her next Christmas table cloth? Give reasons for your choice.

3   George went to the rag bag to get a cloth to wipe up some water he had split on the floor. He was rather puzzled when the fluffy brushed nylon rag did not work very well. Explain the reason for this to George.

4   Mary and Peter are opening a restaurant. They need to choose carpets, chair covers, tablecloths, napkins, and drapes. They wish to choose a patterned carpet so it will not show soiling too easily and, as the restaurant specializes in crêpes Suzette and will be candle-lit, they want non-flammable drapes.

What other considerations should they take into account in their choices? Which textile materials, natural or synthetic, are best suited to meet their requirements at moderate cost?

How would their requirements differ from those for a private home?

The skin of the modern helium-filled airship is a polyester fabric, spray-coated externally with polyurethane impregnated with titanium dioxide, and sealed internally with polyurethane film. The plastics-bodied gondola is reinforced with Kevlar, and the cables which suspend the gondola are also Kevlar. The nose cone is glass-reinforced polyester (GRP). (Design Council/Colin Curwood)

Consumer questions answered

Q1   Should I use a steam iron for ironing polyester shirts?

A   No. Polyesters do not absorb moisture. They have no hydrogen bonds to be broken and reset by steam; therefore steam is of no help in smoothing the fabric.

On the contrary, the temperature at which steam is produced by some irons may soften the synthetic fibre and produce, under pressure, permanent shiny patches on the fabric surface.

Q2   I have seen a variety of scrim fabrics suitable for curtaining; some very exclusive designs in wool, and other attractive but much cheaper designs in acrylics. Does that mean that the wool scrims make a better curtain?

A   No. In spite of its excellent drape and texture, and its heat- and sound-proofing properties, wool is not an ideal fibre for use in curtains – it deteriorates much too rapidly in sunlight.

In this case, the cheaper acrylic, which has excellent sunlight resistance, is a far more durable material. Though they are not as resilient as wool, in a lightweight construction acrylic drapes will not sag out of shape, and will provide excellent heat and sound insulation. They are also resistant to biological attack.

The main disadvantage of acrylic drapes compared to wool is their high flammability. In some situations, this could be a major consideration.

Q3   Why are fishermen always so careful not to put knots in their fishing lines?

A   Fishing lines are made of monofilaments of nylon. When a knot is tied, the filament is made to go around a small circular path. As it is bent into a circle, the part of the filament which is on the inside of the curve is compressed. That part of the filament which is on the outside of the curve is stretched. Draw a curved filament and measure how much longer the outside is than the inside. In a straight filament both sides of course are the same.

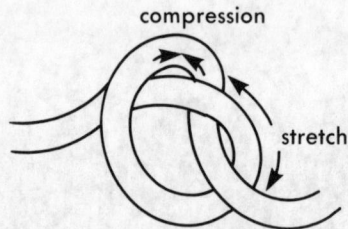

Forces acting in knotted nylon.

When the fishing line is stretched, the portion which is in a knot is already under tension. Added tension will quickly make it reach its breaking point. For example, if a line can support 10 kg before breaking but due to curvature a portion of the knot is already stretched by a force equivalent to 3 kg weight, then a load of 7 kg on the knotted line will break it at the knot. So the knot is the weakest part of the filament.

Q4 I have tried a variety of household sponges. When a wet surface is wiped, some of the sponges leave a trail of water, and some leave the surface dry. How can I tell (without trying each new brand first) which is the better sponge to buy?

A All sponges have holes in them. If these holes are interconnected, then water can pass through the sponge. Sponges which are made from hydrophobic materials – such as polyurethane – are able to hold some liquid water in their holes. Water is held by capillary suction in the smaller holes, but can drain out of the larger cavities.

Sponges made from hydrophilic materials – such as regenerated cellulose – are also able to hold water in their fine capillary spaces. In addition, however, the solid part of the sponge can also absorb water. The molecules of water then penetrate among the molecular chains of cellulose, and the hydrophilic sponge swells.

The water within the material of the sponge is held more firmly than the water in the holes. This is why hydrophilic sponges are able to wipe surfaces dry more effectively than hydrophobic sponges can.

Q5 After three seasons of wear, my Lycra swimming costume wore in places to a thin, transparent web. Why?

A Lycra fabrics are knitted from yarns which are a blend of Lycra and nylon fibres. Chlorine from the swimming pool and in laundry bleach weakens the Lycra, though it leaves nylon unaffected. The weakened Lycra is then destroyed by abrasion during wear, and its crumbs fall out of the fabric, leaving the thin web of nylon behind.

Q6 I have been sewing fibreglass curtains and my hands feel itchy. Why?

A Although fibreglass is strong it is also quite brittle. Many of the fine glass fibres break during handling, and if they stick into the skin they can cause irritation.

Naturally, no fibreglass fabrics should be washed in a washing machine because free fibres that break during the wash can contaminate clothes in the next load.

The same tendency to break can also make the hem of the curtain wear away if allowed to rub against window ledge or carpet.

Q7 I have a quilted jacket filled with polyester wadding. I noticed lately that small white balls of fibre collect on the smooth surface of the coat. What can I do about them?

A The small white balls are pills of the polyester wadding. The fabric covering the wadding is woven loosely enough to allow some polyester fibre ends to push through. These fibre ends then get rubbed against the fabric and against each other, and gradually pull out of the fabric.

Because of the strength of polyester the pills do not break off, but gather and sit on the surface of the fabric.

Preventative measures are the best: the new unpilled garment can be treated with a water-proofing spray. This closes the spaces in the fabric, and provides a smooth surface that does not drag at the fibres of the wadding.

After they occur, the pills can be removed by shaving, brushing, or shearing, and the coat subsequently sprayed to seal the fabric.

Q8 What is the best way to dry an acrylic jumper?

A   Acrylics do not have the resilience of wool. If they are wrung out after washing, the wringing action can distort the garment permanently. If they are hung to drip dry, the weight of water contained in the spaces between the fibres can stretch the jumper out of shape. *Spin drying* puts uniform strain on the garment and does not result in serious distortion, so it is recommended for removing the bulk of the water from the fabric.

The jumper can be dried further by wrapping it in a towel, and stamping on the roll. The pressure of the foot brings the cotton towelling into close contact with the liquid water in the capillary spaces between the acrylic fibres. The water is then absorbed by the hydrophilic cotton fibres. Enough water can be removed in this way to allow the jumper to be hung out to dry without fear of distortion.

Q9 There are many polyester/cotton blends on the market. Why is nylon so rarely blended with cotton?

A   During the manufacturing process, fibres are subjected to a lot of stress and strain. If two different fibres are to be blended evenly, it is important that their mechanical properties be similar.

(Dale Mann/Retrospect)

Spinners find that cotton and polyester work well together during spinning, but the behaviour of nylon is sufficiently different to cause problems in handling blends. Most cotton/nylon combinations are core-spun yarns (p. 347).

In addition to its ability to form uniform blends with cotton, polyester has a better ability to recover from creasing than has nylon. This excellent crease recovery is the most desirable property of polyester when cotton blends are engineered for consumer satisfaction.

Q10 Polyester is often blended with wool for suiting fabrics. Why is wool/nylon the preferred blend for carpets?

A   The pile yarns of carpets are flexed, crushed and abraded during wear. Wool is exceptionally resilient, and can withstand repeated crushing and flexing – it has, however, a very low resistance to abrasion. The abrasion resistance of nylon is very high – higher than that of polyester. Blending even a small amount of nylon with wool produces a significant increase

184

in the abrasion resistance of the blend. Carpets of 80 per cent wool and 20 per cent nylon have excellent performance characters.

Q11   Why do some ropes float on water?

A   Polypropylene has a specific gravity of 0.90. It is lighter than water (s.g. 1.00) and nylon (s.g. 1.14), polyester (s.g. 1.38), or jute (s.g. 1.48). Floating polypropylene ropes are useful for fishermen, because they are not easily lost at sea. There is some danger, however, that they will get tangled up in the propeller of the boat.

Q12   What does the term 'power-net' mean in relation to figure-shaping garments?

A   In the days before elastomeric fibres – not so long ago – corsets were engineered from cotton and whalebone and strings which pulled the garment tight. Even though rubber was available, elastic fabrics made from rubber threads were rather thick, and very hot and uncomfortable. Mostly, rubber was used as elastic panels set into a cooler cotton structure.

Lycra and other elastomeric fibres exert a greater resistance to extension than rubber does, so finer yarns could be knitted into thin yet strong nets. Foundation garments made from such 'power-nets' are light-weight and relatively cool, and allow greater freedom of movement than the cotton garments ever did.

By varying the thickness of the elastomer threads used, garments of a range of figure control 'powers' can be made.

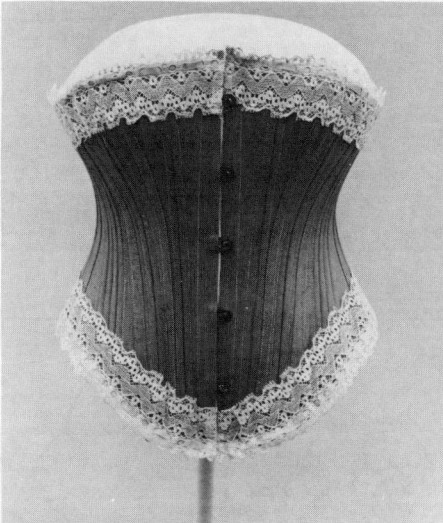

Stays or corsets
Lilac cotton stiffened by close shaped boning, front busk, back lacing
English c. 1890
Purchased by National Gallery of Victoria 1984
(National Gallery of Victoria)

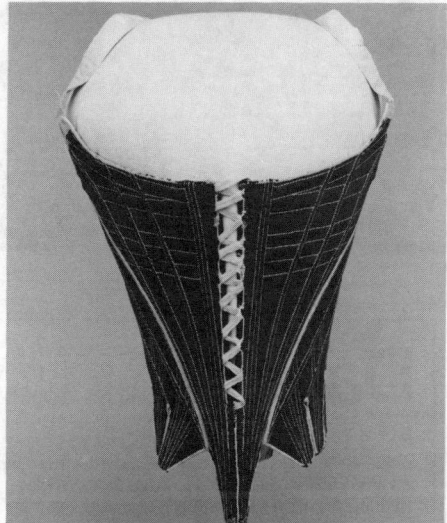

Beige cotton over stiffened foundation, linen lining; closely boned, front busk, back lacing
English 1770–80
Purchased by National Gallery of Victoria 1983
(National Gallery of Victoria)

Figure control was once the province of the mechanical engineer.

Q13 What are 'Hollofil' sleeping bags made of? Are they worth the extra cost?

A 'Hollofil' is a brand name of a polyester fibre which is produced with a tubular cross-section. Air is trapped in the central canal of each fibre – this makes the fibres used for the wadding of the sleeping bags both light and warm.

Hollofil has better thermal insulation properties than ordinary polyester such as 'Dacron fibrefill'. It is also a fair competition to feather filling, but weight for weight it is no match for pure down-filled sleeping bags.

The ultra-fine structure of down traps more air than any synthetic fibre can, and therefore it can insulate most effectively against transfer of heat.

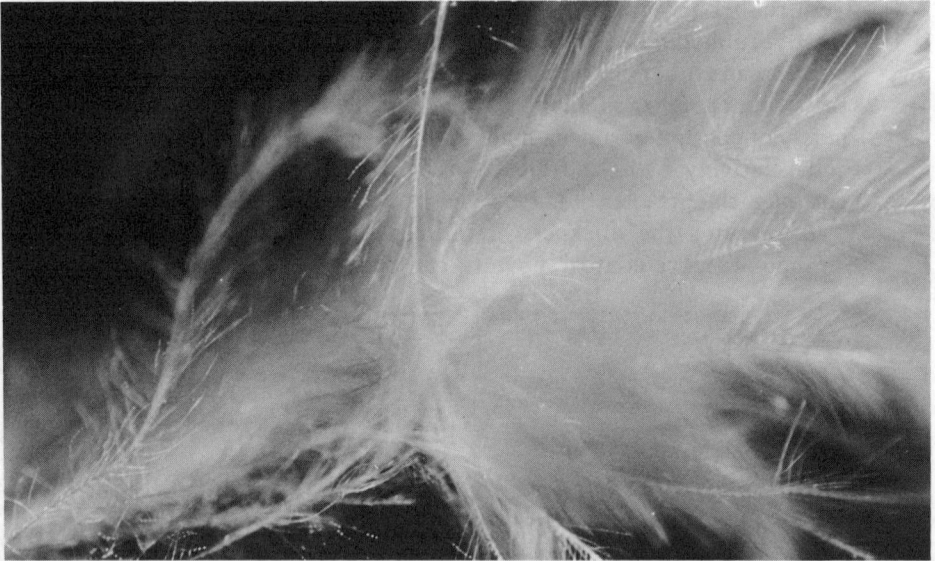

If down is warmer than Hollofil, why would Hollofil be more suited to uses such as this?

(Alp Sports, New Zealand)

186

Q14 In the 'linens' section of a fabric shop I found some rolls of fabric which were marked 'polyester'. They looked like real linen – how can I tell if an unmarked roll is only imitation?

A These days it is becoming increasingly difficult to identify the fibre content of fabrics without resorting to laboratory techniques. Nevertheless, polyester 'linens' are more resilient and crease-resistant than pure linen, and they also tend to be lighter, with a stiffer drape. Pure linen is cool to the touch, while polyester tends to have a warm, slightly oily handle. This latter attribute can be disguised with careful design of yarn and fabric structure.

Q15 What is 'plastic grass' made of?

A Synthetic turf is a tufted carpet made from dope-dyed polypropylene fibrillated pile yarns stitched into a polypropylene backing and held there with a latex adhesive.

6 Fibre identification

Chapters 3–5 highlighted the characteristics of different fibres used for textile applications. With so many different fibres in use, it is often important to find out which particular fibre a textile item is made from.

Is the unlabelled jumper wool or acrylic? Or maybe a blend of nylon and viscose? Can I change its colour by dyeing it at home? Can I wash it or dry-clean it? Will it feel cool, warm or clammy during wear? Will it stretch out of shape, is it strong enough to withstand rubbing at the elbow, is it likely to pill?

Items for sale in shops are required by law to be labelled. The labels carry an indication of fibre content and care instructions. There are nevertheless occasions when both textile consumer and textile manufacturer need to identify unknown textile fibres.

There are some simple, quick tests which can *guide* the experienced person, although they cannot give results that are absolutely certain. For a complete fibre analysis it is important to have more than one test to confirm the identification. Even the best run experiments can give false results on occasion!

When burnt, wool and other protein fibres smell like scorched hair (naturally!), and cotton, rayon and other pure cellulose fibres smell like burnt paper. Cellulose acetate gives off fumes that smell like vinegar. Other smells, other fibres are much harder to distinguish from one another.

There are two main approaches to the positive identification of fibres: the use of microscopy, and various combinations of chemical tests. Careful selection of these techniques can identify all fibres. In this book, however, we will investigate only the more commonly used textile fibres, and leave identification of esoteric samples to the professional textile analyst.

Microscopy

The microscope is useful for the immediate identification of those fibres which have characteristic surface textures. Usually a longitudinal view under about 300× magnification is satisfactory.

Preparation of samples

For *longitudinal* sections it is important not to have too many fibres criss-crossing over each other. At higher magnifications the depth of focus is rather

yarn sample or cut fibres

needles or dissecting needles

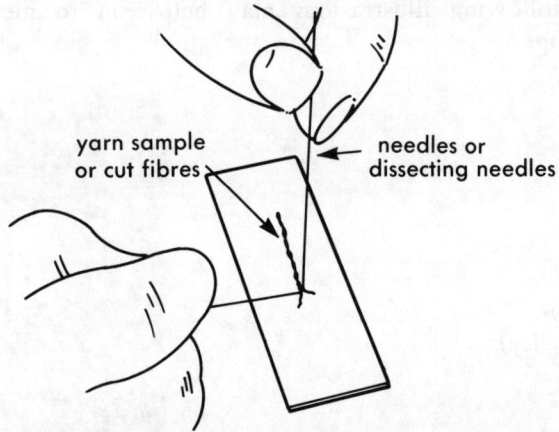

Gently tease the fibres apart so they form a single layer. (Remember to examine both the warp and the weft.)

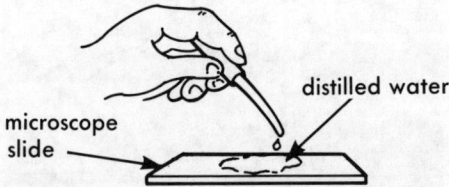

microscope slide

distilled water

cover slip

Preparation of a microscope slide.
Place a drop of distilled water on the slide.

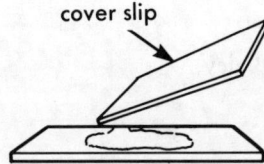

Place a cover slip on the slide.
It is now ready for examination.

shallow, and so a single fibre may move in and out of focus if it weaves up and down too much. Ideally, a *single layer* of *closely spaced* short lengths of fibre is most effective for identification.

For *cross-sectional* examination of fibres it is important to have the section *thin* enough to allow light to pass through the fibres for clear visibility. The spaces between the fibres can be made dark by smearing a mixture of ink and glycerine jelly over the prepared section and then wiping the fibres clean. More sophisticated cross-sectioning methods involve placing a bundle of fibres in a self-setting resin. The fibres and resin are sliced thinly together for accurate cross-sectional views.

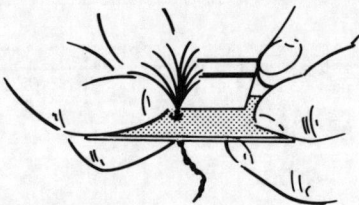

Slicing across a fibre bundle to provide thin samples for a cross-sectional view.

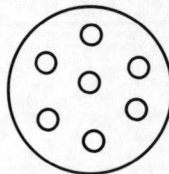

Fibres set in resin.   Jammed fibres.

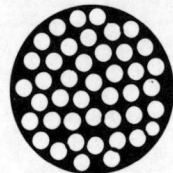

The following illustrations may help you to identify fibres under the microscope.

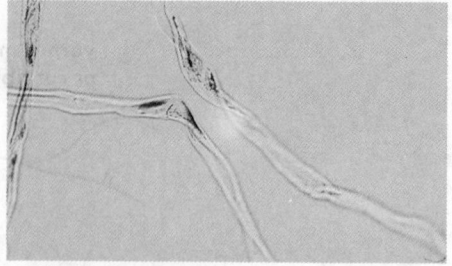

Cotton 40×.
(David Bailey)

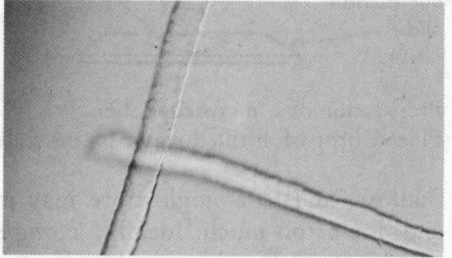

Sisal 40×.
(David Bailey)

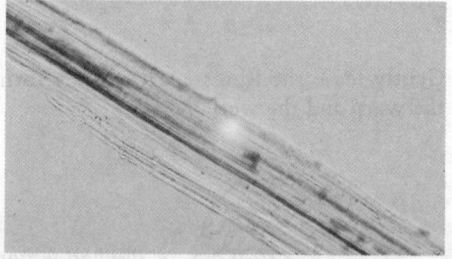

Wool 40×. Note the clearly
defined scale structure.
(David Bailey)

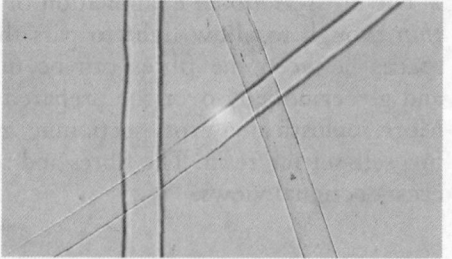

Mohair 40×. Note that there are
very few scales, and that those present
are very small.
(David Bailey)

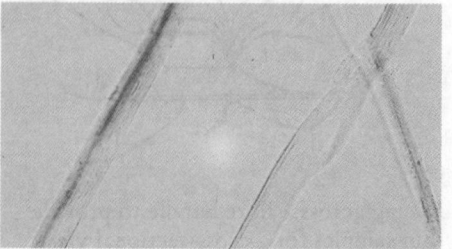

Silk fibre 40×. Silk is much less uniform
than the synthetic filaments.
(David Bailey)

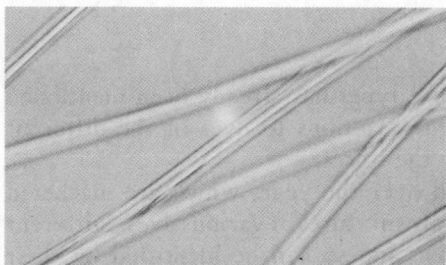

Viscose 40×. The characteristic striations are not present on cupro and other rayons, and some synthetics are also striated, so presence or absence of striations alone is not sufficient evidence for identification. (David Bailey)

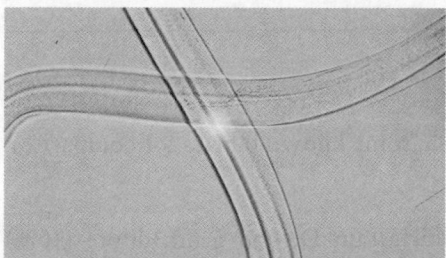

Dynel (vinyl chloride/ acrylonitrile copolymer) 40×. (David Bailey)

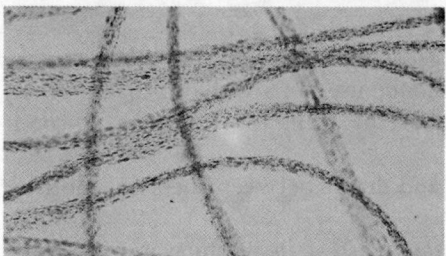

Orlon (acrylic) 40×. The dark patches are delustrant or pigment. (David Bailey)

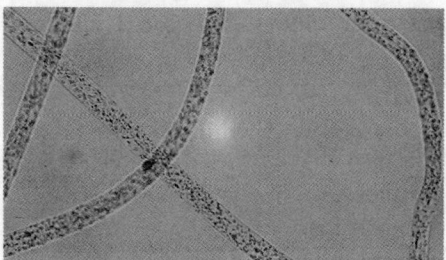

Acrilan (modacrylic) 40×. Again, the dark patches are additives to the polymer, and so do not aid identification. (David Bailey)

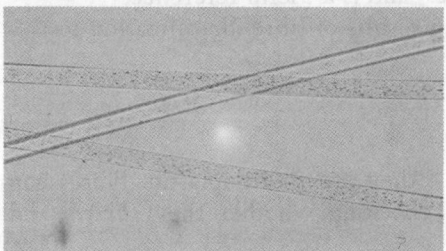

Nylon 40×. The dark speckles are additives. (David Bailey)

191

Many synthetic fibres are difficult to distinguish from one another under the microscope. This is because they are often simple cylinders in form. Some may show a few striations, but this is not enough for positive identification.

Diagnostic staining

Ease or difficulty of dyeing the various fibre types has already been mentioned in the previous few chapters. The different reactions to dyes of the different fibres can be employed as a test to identify the fibres.

A variety of stains have been developed over the years which are marketed by the Shirley Institute (U.K.). These stains are mixtures of various dyes. Different fibres absorb different amounts of each dye and so can be identified by their resultant colours.

TRY YOUR HAND

You will need:

samples of unknown fabrics, and, if you wish, some known fabrics for comparison
Shirlastain A
Shirlastain E
mixture of 15 mL Shirlastain A, 5 mL Shirlastain D, and 5 mL dilute (10%) sulfuric acid

What you do:

Clean the unknown fabric or fibres in a degreasing solvent such as ether.
(*CAUTION:* extremely flammable, and must be used in a well-ventilated area.)
Shirlastain A: dip the sample into a dish of the stain for 1 minute at room temperature.
Rinse it thoroughly under running water, and dry it between sheets of blotting paper.
Shirlastain E: Boil the sample for 2 minutes, wash it in warm water, and dry it.
Mixture of Shirlastain A and D: Boil the sample in the stain for 2 minutes, wash it in warm water, and dry it.

The table indicates the colours that will be obtained with white or pale fabrics that have not received resin treatment.

Note that resin treatments and other finishes are likely to interfere with the colours obtained.

The *Journal of the Textile Institute,* November 1961, contains a coloured illustration of the staining of different fibres, and is a useful reference.

The stains are a useful way of confirming results of fibre identification tests.

Chemical tests

The burning test

Many synthetic fibres melt when heated – they are thermoplastic. Fibres can be readily identified as *thermoplastic* by checking whether they *shrivel* and

Fibre	Shirlastain A (cold)	Shirlastain E (2 min. at boil)	Shirlastain A and D mixture (2 min. at boil)
Cotton	pale purple		
Mercerized cotton	mauve		
Flax	brownish purple		
Ramie	lavender		
Sisal and jute	golden brown		
Raw silk	very dark brown		
Silk	brownish orange		
Wool	yellow		
Nylon	cream to yellow	brown	very dark green/black
Viscose	bluish pink		
Cuprammonium rayon	bright blue		
Acetate	greenish yellow	persimmon	emerald green
Triacetate	off-white	light brown	slightly greenish yellow
Terylene	off-white	bright yellow	light grey
Acrilan		olive green	dark purple
Orlon 42		bright red	light blue
Courtelle		yellow	light green
Dynel		light brown	bright green
Fibravyl (PVC)	light brown	light grey	

melt in the heat near an open flame. The fibre melt solidifies to a *hard bead* of plastic on cooling. *Non-thermoplastic* fibres leave *ash,* the characteristics of which can aid identification.

Careful observation of the burning process, the odour, and the residue, can help in identifying fibres.

TRY YOUR HAND

You will need:
3 cm × 1 cm pieces of various fabrics of known fibre content, including wool fabric, cotton, viscose
tongs or tweezers
Bunsen burner
small piece of meat
hair fibres
sliver of wood

What you do:

Hold one of the fabric samples with the tongs near the flame of the burner.
Move it closer *very slowly,* then allow it to enter the flame.
Work *very slowly* and observe carefully.

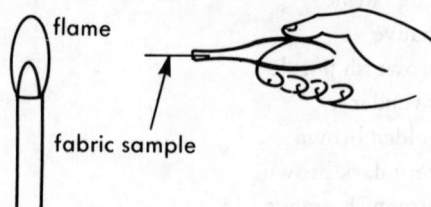

flame

fabric sample

Note whether fibres shrink away from the flame, whether they melt, how they burn, what the odour of the smoke is, and the nature of the *ash* or *bead* (small melted droplet which cools to a hard glassy ball) formed.

Those fibres which seem to shrink away from the heat near the flame, and which melt even before they are burnt, are *thermoplastic* fibres. They usually leave small hard beads where the melting took place. Synthetic fibres are usually thermoplastic. *No* natural fibre is thermoplastic.

Repeat the test with other fabrics.

Those fibres which do not melt on heating are non-thermoplastic and include the *cellulose* and *protein* fibres as well as some special synthetic fibres.

Protein fibres can be told apart from the cellulosics by the odour of their smoke. Try to identify other fibres by doing the following tests.

Burn the meat at the flame. Note the odour. Burn a few of the hair fibres. Burn the small piece of wool fabric.

Can you smell the resemblance? What do all these materials have in common?

Now burn the paper. Burn the wood. Burn the cotton fabric, then the viscose fabric.

Can you smell the resemblance? What do these materials have in common?
Can you add more fibres or other materials to this list?

You might like to use the microscope to distinguish between the fibres in each odour group.

Test more of the fabric samples for their thermoplasticity, the colour of the flame when they are placed into the Bunsen flame, whether they stop burning when they are removed from the flame, and the odour and nature of the residue (what is left after burning).

Draw up a table of your results, listing the fibre type, its behaviour near the flame, its behaviour in the flame, its behaviour after removal from the flame, the odour of the smoke, whether an ash or bead is formed, and the nature of the ash or bead.

With experience, you may be able to separate cellulosic (plant and rayon) fibres, protein (animal) fibres, and the thermoplastic (synthetic) fibres into their correct groups. Accurate identification is almost impossible, however, because many products are subjected to finishing processes which may distort the results.

Chemical analysis

To distinguish between the many fibres which are cylindrical in shape and are thermoplastic, *chemical analysis* is needed.

Acrylics and polyamides have nitrogen in their molecules; PVC has chlorine; polyesters, acetates, and polyolefins have neither; modacrylics have both. A chemical test to identify these elements is thus a guide to the nature of the fibre.

Another group of tests uses the ability of particular solvents to dissolve particular types of fibres. For example, nylons dissolve in 50 per cent hydrochloric acid and in metacresol; cellulose acetate dissolves in acetone.

These tests can be combined in a step-wise fashion to form a 'decision tree' to provide a quick identification of fibres.

Decision tree for fibre identification

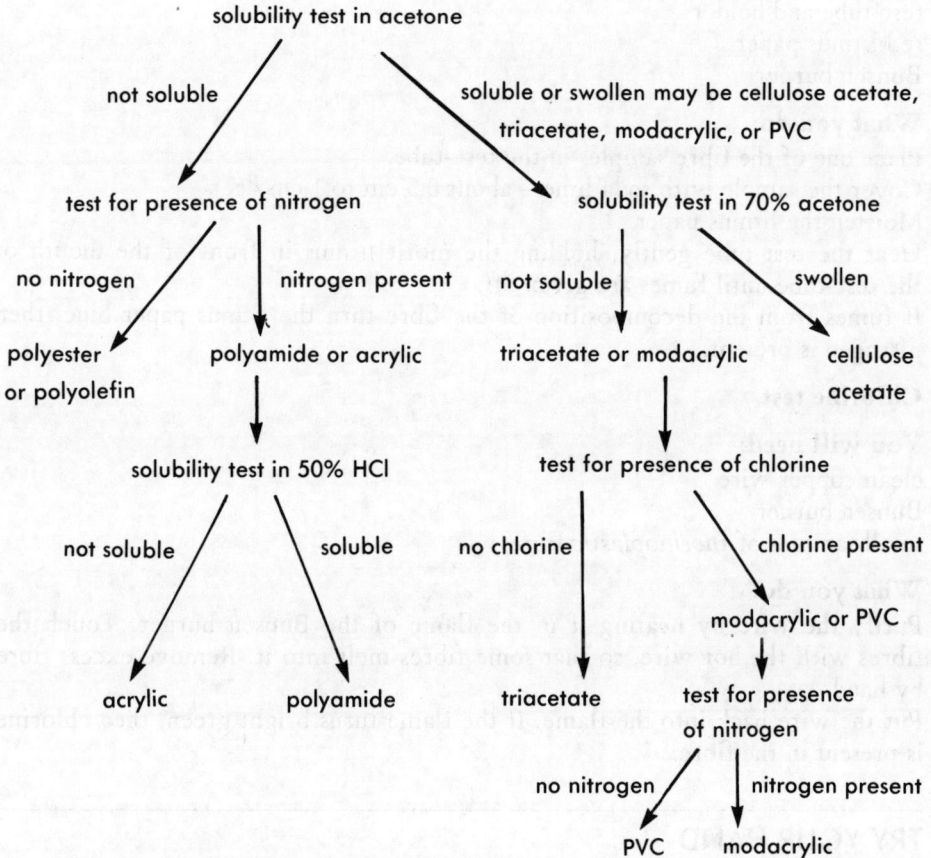

Decision tree for identification of fibres.

TRY YOUR HAND

Note: In all tests involving solvents, care is essential. Many solvents are highly flammable, and many can be dangerous if excessive amounts are inhaled. Always work in a well-ventilated room, away from naked flames. Clean up spills immediately, and wash any chemicals off the skin with plenty of running water, and soap.

Nitrogen test

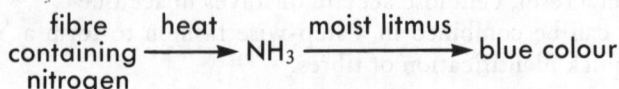

$$\text{fibre containing nitrogen} \xrightarrow{\text{heat}} NH_3 \xrightarrow{\text{moist litmus}} \text{blue colour}$$

You will need:
small samples of fibre or fabric, each about 3 cm × 1 cm
anhydrous soda lime
test-tube and holder
red litmus paper
Bunsen burner

What you do:
Place one of the fibre samples in the test-tube.
Cover the sample with soda lime – about 0.5 cm to 1 cm deep.
Moisten the litmus paper.
Heat the test-tube gently, holding the moist litmus in front of the mouth of the test-tube until fumes are given off.
If fumes from the decomposition of the fibre turn the litmus paper blue, then nitrogen is present.

Chlorine test

You will need:
clean copper wire
Bunsen burner
small samples of *thermoplastic* fibres

What you do:
Purify the wire by heating it in the flame of the Bunsen burner. Touch the fibres with the hot wire, so that some fibres melt into it. Remove excess fibre by hand.
Put the wire back into the flame. If the flame turns bright green, then chlorine is present in the fibre.

TRY YOUR HAND

Now you are familiar with the tests for chlorine and nitrogen, you are ready to identify some unknown fibres.

You will need:

the equipment and materials for the chlorine and nitrogen tests
pure acetone
70% acetone
50% hydrochloric acid (HCl)
small samples of unknown thermoplastic textiles

What you do:

Using only a *few fibres* at a time for the tests, and observing very carefully, start at the top of the decision tree and work through.

Thus, start by trying to dissolve the fibres in *acetone*. If the fibres are not soluble, they may be polyester, polyamide, or acrylic. To distinguish between these, test for nitrogen.

With very careful testing and observations, you can be confident of the accuracy of your results.

Tests to distinguish between polyesters and polyolefins are not suited to class work, as they involve boiling the samples in rather nasty and flammable solvents. They will therefore be allowed to remain indistinguishable.

7 Light, vision, and colour

In this chapter, we touch on the cultural significance of colour, and its importance in our lives.

The nature of light, and the fascinating problem of the human perception of light and colour, are explored in greater detail. These are areas of great concern to textile designers.

The interactions of light and objects are then examined, as we learn why many things are coloured and some are not. The problems of mixing and matching colours are investigated.

In this chapter, we find the answers to such questions as

- Why do notions which matched a fabric in the shop not match it now I have made up the garment?
- What are primary colours, and what is their significance?
- Why are some molecules coloured?
- Can a detergent really wash 'whiter than white'?
- How and why do colours fade?

and many more intriguing ideas.

Colour and culture

Why do we add colour to our textiles?

Colour is something we tend to take for granted. We seem to have always known what colour is: we had to learn to *name* the colours, but the idea of colour itself is so basic that it did not have to be learnt.

Colour plays a very important part in all our lives. It has much cultural significance: black for mourning, red for blood and war in Western society; white for mourning and red for good luck in Chinese society. White in Western society may mean purity – the white wedding dress – but it also has overtones of death and horror.

The English language is rich in colourful words: 'blue blood' for nobility; 'yellow' for a coward; 'greenhorn' for inexperience. Can you find another ten expressions which use colour in similar ways?

Different cultures attach importance to different colours. Some cultures do not differentiate between tones of a colour: in one Indian language, there is only one word for red; nothing like pink, tomato red, pillarbox red and all the other shades. In Vietnamese, blue and green tones are not distinguished:

it is thought that because of the amount of these colours in the environment they are played down in favour of other colours which are special. On the other hand, the Inuit (Eskimos) have more than ten words to describe various shades of the colour white. Can you think why it would be important in their everyday lives to be able to specify different shades of white?

TRY YOUR HAND

1   Many names of colours are associated with natural phenomena: snow white, canary yellow, tomato red. Write down ten variations of the colours red, yellow and blue, using objects from nature to describe exact shades.
2   Look at the colour samplers put out by a paint company. What proportion of the names of colours uses references to plants and other ideas associated with nature?

Colour in decoration
Colour schemes

Gradually, images associated with certain colours have come to give the colours meanings. Fire burns with the colours of red and orange. By association, these are called warm colours. Similarly, blues – sea, sky – are called cool, distant colours; greens – grass, trees – are restful; yellows – sunshine – cheerful. There is no implication that a red room is actually higher in temperature than a blue one. The 'warmth' is a perception, a mental image, in a variety of sensations registered by the body.

Psychiatric studies have established that coloured environments can produce physiological effects on patients. Red surroundings can increase heart rates and help cheer up mildly depressed patients; dark purple was found to be depressing – in fact, it was a colour favoured by schizophrenic patients during painting exercises. While these examples cannot be extended into generalizations, they do show that there may be some real foundation to beliefs held by many interior decorators.

TRY YOUR HAND

1   Look up various colour schemes in decorator magazines. What concepts are associated with various colours? Where are strong, raw colours, like reds, blues, yellows, used? pastels? What are MONOCHROMATIC colour schemes? Make a list of terms decorators use for colours and check their meanings in a dictionary.
2   During the 1960s, fashion colours for men's and women's clothing were bright and strong. Employment levels were high, the economies of Western countries were buoyant, the future looked rosy. In the late 1970s and early 1980s, oil prices were rising and many economies were depressed, unemployment was rising, and fashion colours were more subdued. Periods of 'boom and bust' have occurred frequently in history. Choose several such periods and research which colours were fashionable at the time. (Women's and children's clothes may be easier than men's, but you could research

Male attire in the nineteenth century: sombre, 'sober' colours.

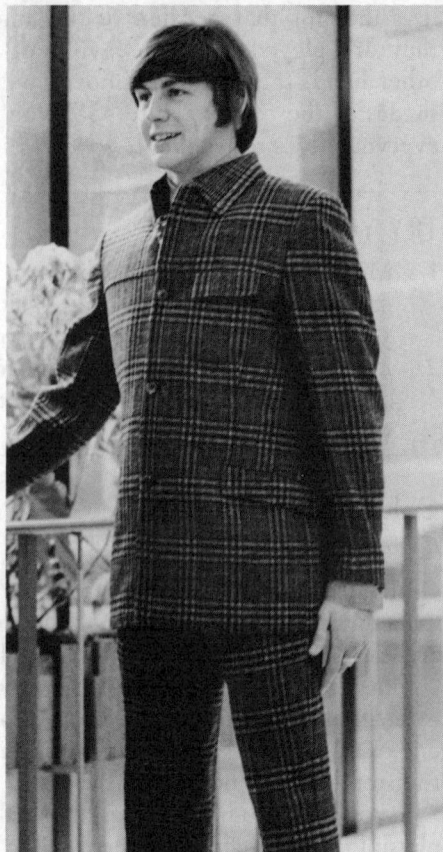

1966.
(*The Age*)

the social climate when men's clothes became so drab in the nineteenth century.) Did you find any pattern, or were fashion colours unrelated to social climate?

Lighting

Lighting is an essential factor in the colours which surround us. Different light sources may have different effects on our perceptions of colour, and on how we feel about our surroundings. The *measurable* differences in the quality of light from different sources will be explored on pp. 213–14.

TRY YOUR HAND

You will need:
some coloured light globes
a small furnished room
a selection of objects of different colours
some friends who are unfamiliar with the room
an evening

What you do:
Illuminate the room with one of the coloured lights.
Ask your friends to come and describe the 'atmosphere' or 'feel' of the room.
Ask them to describe the colours of the objects in the room.
Send them out again, change the colour of the light, and provide some different coloured objects.
Repeat, taking careful note of your friends' comments.

1 Did different coloured lights evoke the same emotional responses in all the observers?
2 What emotional response did yellow light cause? red? blue? green?
3 Which light changed colour perception the most?
4 Which light would be easiest to live with?

It is important to have some white, or neutral, or very soft pastel colour in an interior decorating scheme. This is because of the way in which the eye perceives colour. Edwin Land (inventor of Polaroid film) showed recently that an observer takes in the whole field of view, and at each boundary makes comparisons between neighbouring colours. The lightest colour in the field of view is perceived as white, and the other colours are compensated accordingly.

This compensation process accounts for the fact that we are able to identify a colour accurately, in sunlight, under fluorescent light, and under incandescent tungsten light. Sometimes we are even able to identify colours correctly under more strongly coloured lights. In other words, the eye and brain compensate for the prevailing conditions.

The compensation system depends on the presence of some white (or near-white) as a reference point. Imagine an evening scene in a room with only mushroom and pink tonings, and no white at all: the eye moves restlessly in search of a settling focus, and the observer can feel quite uncomfortable if not already familiar with the scene.

The language that interior decorators and artists use to describe colours and colour schemes is quite different from the language of the colour scientist. Tones and tints, shades, harmonies and colour triads, complementary or pastel colour schemes, employ concepts that are not necessarily measurable quantities. The colour scientists have their own set of terms which they use to specify exact colours and qualities of light.

What is light?

Put simply, light is a form of energy in transmission. The source of the light – sun, flames, tungsten lamp or whatever – consists of atoms which have absorbed excess energy. Their electrons are not in their most stable configuration: they are EXCITED. An excited atom is unstable, and can use the excess energy in chemical reaction with other atoms, or the electrons can drop back to their lowest energy state (ground state) and *emit the energy as light*.

The energy is emitted as discrete packets called PHOTONS. The possible energy values of photons form a continuum from very high through to very low. The exact value depends on the WAVELENGTH: a long wavelength is low energy; a short wavelength is high energy. The inverse of wavelength is frequency:

a low frequency is low energy; a high frequency is high energy. In the entire spectrum of possible energies, a small band of wavelengths, from 300 to 700 nm (0.000 000 3 to 0.000 000 7 metres) is visible to us as light.

A narrow band of energies (or wavelengths) from the visible spectrum appears to our eyes as a remarkably pure colour. For example, the yellow of sodium street lights is due to the emission spectrum of sodium, which is very intense at two wavelengths close to 592 nm. It contains other wavelengths as well – a touch of green and red – but these are so faint that they are not seen.

Generally, we do not see the colours of the spectrum in isolation. The green colour obtained in the test for chlorine (p. 196) is the emission spectrum of copper: it contains red and green and many other colours. The intense blue-white of mercury street lamps consists of red and blue amongst other colours.

How we see

Light rays enter the eye through the PUPIL, which controls the amount of light let through. The light is first focused by the CORNEA, and then the LENS, so a sharp image falls on the retina. It is the RETINA which actually detects the light.

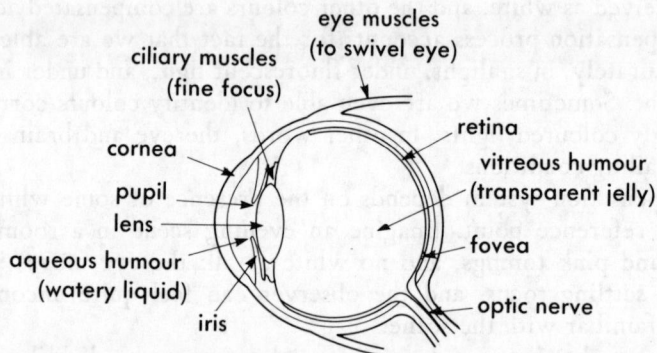

Structure of the eye (simplified).

The retina consists of a mosaic of light-sensitive cells called *rods* and *cones*. These translate the incoming light into nerve signals, which the brain then interprets: this *interpretation* is 'seeing'.

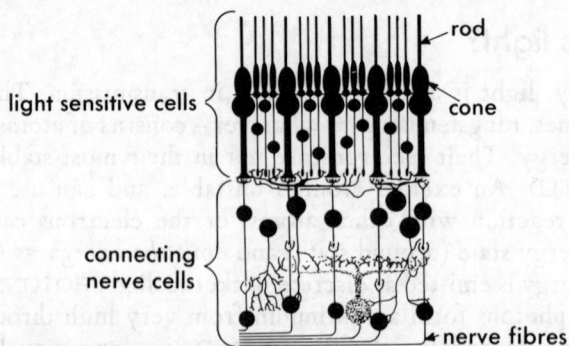

Structure of the retina (simplified).

RODS are very sensitive to small amounts of light, and are responsible for night vision. Their sensitivity is due to a pigment called RHODOPSIN, which undergoes a reversible chemical reaction in light to generate the nerve signal. The regeneration of rhodopsin is not instantaneous, so it is readily bleached out in strong light: hence the temporary 'blindness' caused by a bright light shone in the eyes. Rhodopsin is most sensitive to light in the green part of the spectrum, but the image the cones send to the brain is in black and white: the *rods have no colour sense*. Rods, although they are so sensitive to light, cannot resolve fine detail.

CONES are responsible for colour vision. The seven million or so cones are distributed all over the retina except at the blind spot where the OPTIC NERVE (which carries all the vision information) leaves the eye. The cones are particularly concentrated in a thin part of the retina called the FOEVAL PIT, where there are no rods. Each cone cell terminates in a brush of nerve endings which interconnect with neighbouring cone cells, so that groups of cones operate in conjunction. The cones in the foeval pit provide acute (fine detail) vision as well as colour vision.

Perception of colour

Cones contain a photosensitive (light sensitive) pigment, and it is this pigment's interaction with light that provides the nerve impulse that tells the brain to see colour.

According to the *Young-Helmholtz theory,* which provides the best explanation so far of colour vision, the eye contains three types of cone, which are responsible for our ability to see all the colours. One type of cone is very sensitive to red light, slightly sensitive to green, and not sensitive to blue at all. Another type of cone is very sensitive to green, slightly sensitive to blue, but not sensitive to red. The third type responds strongly to blue, very slightly to green and does not react to red at all.

If pure red light falls on the retina, the 'red' cones will undergo a chemical transformation, whilst the 'blue' and 'green' cones, which do not react to red light, are unaffected. The chemical reaction in the 'red' cone produces a nerve signal to the brain, which is interpreted as 'red'. This is the perception of colour.

When white light, that is light containing equal amounts of all the wavelengths of the visible spectrum, enters the eye, all three types of cone are stimulated, and we perceive 'white'. If light consisting only of equal amounts of red, green and blue enters the eye, then again all three types of cone are stimulated, and again we perceive white. The brain cannot tell the difference.

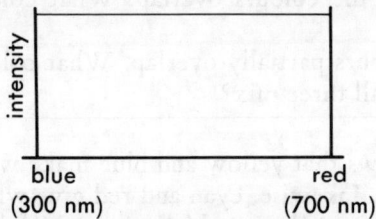

White light composed of equal amounts of all wavelengths of the visible spectrum.

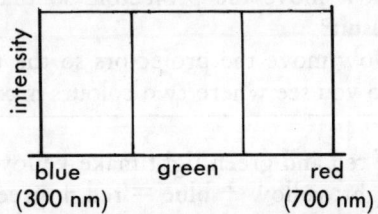

Light composed of equal amounts of red, green, and blue: perceived as white.

So, how do we see colours other than red, green, and blue? If pure spectral *yellow* light enters the eye, it stimulates *both* the 'red' and the 'green' cones, and the brain interprets this mixed signal as 'yellow'. The same mixed signal can be obtained by an appropriate mix of red and green light, and so the brain will see that mix as yellow too.

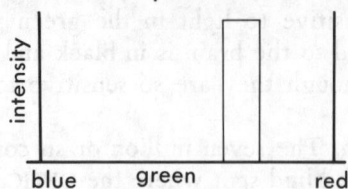

intensity

blue   green   red

intensity

blue   green   red

Yellow light stimulates both the red and the green receptor.

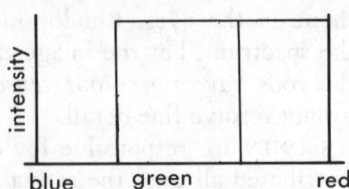

A combination of red and green light is perceived as yellow.

Thus, all colours are seen through appropriate stimulation of the three types of cone, and all colours will be perceived through an appropriate combination of red, green and blue light.

Because red, green, and blue light can be *added* together to give white or any other colour, but cannot themselves be formed from other colours of light, they are called ADDITIVE, or LIGHT, PRIMARIES.

Colour television exploits this mechanism of colour perception. The screen of the colour television contains substances called phosphors. On stimulation, one type of phosphor glows red, another green, and the third blue. The television image is composed of tiny dots of these three colours of light: the brain supplies the rest.

When *two light primaries* are mixed in equal proportion, the SECONDARY colours are obtained: red + green = yellow; green + blue = cyan (a greenish blue); red + blue = magenta (a purplish red).

TRY YOUR HAND
You will need:
3 slide projectors or overhead projectors
thick cellophane or filters in the additive (light) primary colours of red, green, and blue
white wall or screen

What you do:
Project each of the colours onto the white wall or screen.
Now move the projectors so that two of the colours overlap. What colours result?
Now move the projectors so the three colours partially overlap. What colours do you see where two colours mix? where all three mix?

If red and green light make yellow, it follows that yellow and blue make white light: yellow + blue = red + green + blue. Likewise, cyan and red give white, and magenta and green give white. Any two colours which give white light on mixing are called COMPLEMENTARY COLOURS.

Tricks of perception

Vision is not simply a matter of light reaching the eye. It is also a process of *interpretation* by the brain of the nerve impulses from the eye. In some circumstances, the brain will register information which is not there. These effects are often exploited in art and design.

Positive afterimages

TRY YOUR HAND

Accustom your eyes to the light in a dimly but evenly lit room.
Strike a match, and stare at its flame for a second, then blow it out.
What do you see now?
If you saw bright flames dancing in front of your eyes, it is because the receptors in your eyes over-reacted to the sudden strong stimulus, and the brain continued to register messages after the stimulus was gone.

Negative afterimages

TRY YOUR HAND

Stare fixedly (that is, without moving your eyes) at the illustration for 15 seconds.

Now shift your gaze to a light-coloured blank wall.
What can you see?
Are the colours of the squares reversed?
When you stared at the black square with the white centre, the light from the white centre used up a lot of visual pigment where its image fell on the retina. The light from the black square used up much less pigment, so the rods and cones where its image occurred were well rested.
As you shifted your gaze to the neutral background, the same amount of light fell on the whole retina. The rested receptors responded to the stimulus readily, and sent a 'bright' message to the brain. However, the other receptors had used up much of their pigment and so were less able to generate signals to the brain: the message they sent was 'dark'.
Stare fixedly at a brightly coloured object for 15 seconds, then shift your gaze to a white or neutral background.
What do you see?

If you saw a bright image in the complementary colours, it is because this time it is the colour receptors which are overloaded. If the object was, say, bright green, and the green receptors tired, they will be less able to register the green component of the white light from the neutral background. Thus the message reaching the brain will be (white – green), or magenta.

Afterimages are not, however, just a matter of signals from the eyes misleading the brain. They are also a function of how the brain interprets visual information, and so the above explanations are incomplete.

Simultaneous contrast

TRY YOUR HAND

Take a uniformly coloured, but not too bright, surface. A pale green is ideal. Take a sheet of black paper and a sheet of white paper, and cut holes in them to form a lattice.

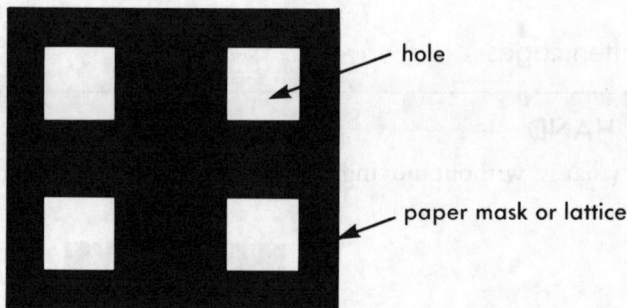

hole

paper mask or lattice

Place the sheets on the coloured surface.

Did you find the colour framed by the black paper appeared darker than the colour framed by the white paper?

This is a good example of the *brain* providing information that is not supplied by the eyes. For some reason, not yet understood, the brain 'borrows' information about colour density from surrounding areas.

Try the exercise again, using coloured lattices. The green should appear to change colour.

Coloured objects

So far we have discussed only the interaction of light and the eye: we have not considered where the colour comes from in the first place. Why, for example, does a ripe tomato appear red in white light?

As red is an additive (light) primary, it follows that only red light reaches the eye from the tomato. All other colours which could stimulate the blue and green cones must be absorbed by the tomato. Likewise, a lemon appears yellow, because only yellow light (or red and green combined) reaches the eye. The blue is absorbed by the lemon, or, to put it another way, blue is *subtracted* from the white light.

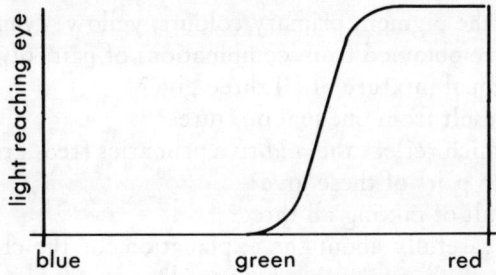

The tomato can appear red because only red light reaches the eye. That means that the fruit is absorbing or *subtracting* the green and blue components from the light.

The colours of objects which are not themselves sources of light are thus *subtractive colours*. When the light primaries (red, green, blue) are *subtracted* from white light we are left with the SUBTRACTIVE PRIMARIES (or PIGMENT PRIMARIES) yellow, cyan, and magenta. These are the primaries of your art classes.

Colours of non-luminous objects are obtained from mixtures of these three subtractive primaries. Thus, if a yellow pigment, which subtracts blue, is mixed with cyan, which subtracts red, the only colour left to be reflected and seen is green.

It follows that, since yellow subtracts blue, cyan subtracts red, and magenta subtracts green, *if the three subtractive primaries are mixed,* all the light primaries are subtracted: *no light is left* to be reflected to our eyes: we see *black*.

TRY YOUR HAND

You will need:

blue ink
yellow chalk
coloured paints

What you do:

Dip the yellow chalk into the blue ink, and then break it in half. What colour has it become?
Here is an explanation of what has happened.

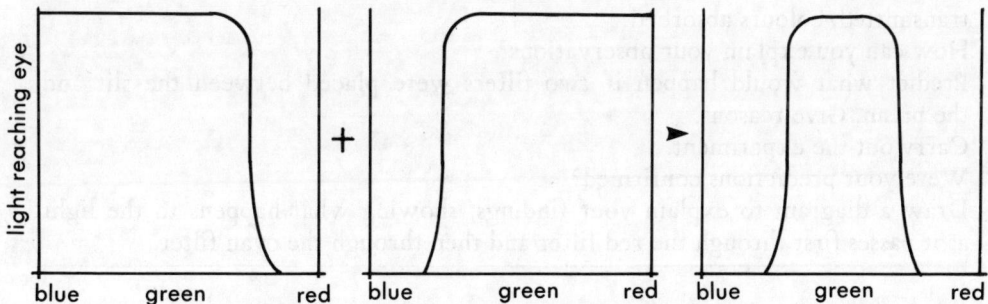

Blue ink absorbs red light, leaving cyan.

Yellow chalk absorbs blue light: remaining green and red is perceived as yellow.

Only parts of light *not* absorbed is green.

Now mix paints of the pigment primary colours, yellow, cyan, magenta.
1   What colours are obtained from combinations of pairs of these colours?
2   What does an equal mixture of all three give?
3   What colours result from unequal mixtures?
Now, mix paints which reflect the *additive* primaries (red, green, blue).
4   What colours do pairs of these give?
5   What is the result of mixing all three?
6   Thinking very carefully about the explanation for the chalk dipped in ink, how would you explain what has happened?

Subtractive primaries are employed in colour printing. Inks or pigments of black, yellow, cyan and magenta are applied in a pattern of tiny dots to build up the colour picture.

TRY YOUR HAND

You will need:

3 high density filters or strongly coloured cellophane in the additive primary colours (red, green, blue)
3 high density filters or strongly coloured cellophane in the subtractive primary colours (magenta, cyan, yellow)
overhead projector
white wall or screen
piece of black cardboard with slit 0.2 cm × 5 cm
glass prism to fit over slit

What you do:

Cover the overhead projector with the cardboard, place the prism over the slit, and project the rainbow onto the screen.
Now place one of the filters between the slit and the prism, and check to see which colours are being transmitted. (If you are using cellophane, you may need to use a number of layers to see the effect clearly.)
Which colours are absorbed by the pigments of the red filter? By the pigments of the cyan filter?
Examine all six filters, and note your results in a table: colour of filter, colours transmitted, colours absorbed.
How can you explain your observations?
Predict what would happen if *two* filters were placed between the slit and the prism. Give reasons.
Carry out the experiment.
Were your predictions confirmed?
Draw a diagram to explain your findings, showing what happens to the light as it passes first through the red filter and then through the cyan filter.

The reflection of light

Until now we have considered light that is *transmitted,* as through a coloured filter, and light that is *absorbed,* as by a pigment.

Highly reflective trilobal nylon 6,6 is combined with dull nylon 6,6 filaments to give the self-stripe effect in these warp knitted garments.
(Fibremakers)

When we see an object that is not itself a light source, we see light that is *reflected:* light that bounces off the object to be detected by our eyes.

Light is always reflected from a surface at the angle with which it strikes the surface: angle of incidence equals angle of reflection. This means that a smooth surface, such as highly polished metal, reflects light uniformly. An irregular surface will scatter the reflected light. Thus a textile material made of yarns in which the fibres are organized to lie parallel will have a smooth surface, and will appear shiny. A textile with a bumpy or irregular surface will disperse light and appear dull or matt. The cross-sections of synthetic filaments are engineered to give the desired light reflectance properties.

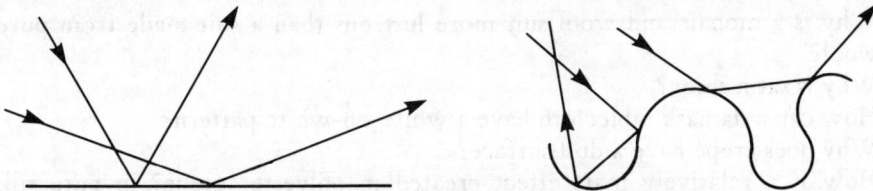

Patterns of reflection from smooth and rough surfaces.

Mirrored glass reflects light from the glass surface and from the metal coating on the back of the glass. A one-way mirror allows some light to pass between the metal grains. These effects are similar to the reflectance of light from a synthetic fibre.

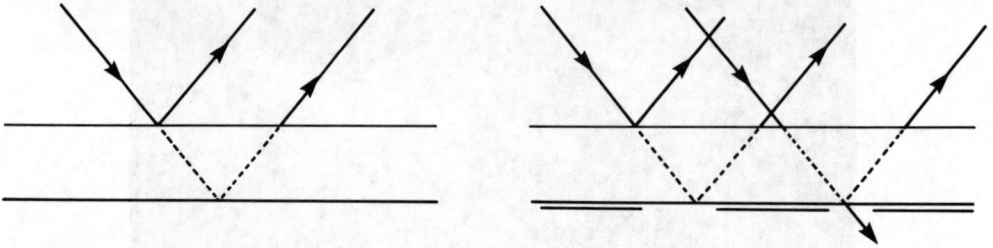

Reflectance from a fully silvered and a half-silvered mirror.

The typical synthetic fibre is extruded as a smooth-surface filament, which does not absorb much light. It is therefore quite shiny: light is reflected from the smooth inner and outer surfaces. To overcome this effect, DELUSTRANT is added. Grains of white pigment are added to the polymer mix before extrusion, and serve to scatter the light. The filament becomes matt. The more delustrant added, the more effective is the scattering of the light.

Obviously, if a coloured pigment is added, it will absorb some of the incident light, and reflect only coloured light.

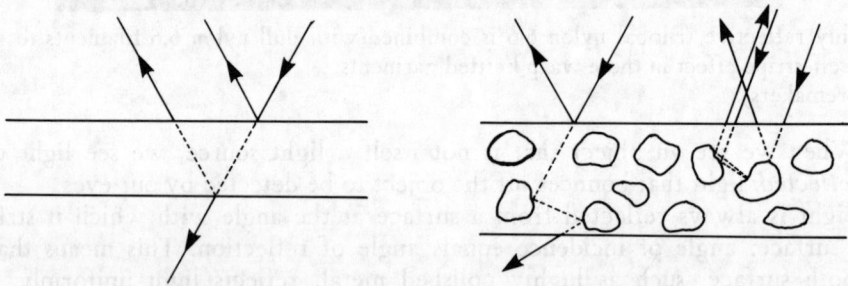

Reflectance from a lustrous and a delustered filament.

TRY YOUR HAND

Investigate the structure of various textiles (with magnifying glass and microscope).

1   Why is a mohair and wool suit more lustrous than a suit made from pure wool?
2   Why is satin shiny?
3   How can a damask tablecloth have a white-on-white pattern?
4   Why does crepe have a dull surface?
5   How is a relatively matt effect created in polyester crepe? in pure silk georgette?

Colours of all kinds

The eye sees white when it receives equal amounts of red, green and blue light. It sees pink when it receives slightly more red: pink is white plus a little red. It sees yellow from a combination of green plus red, and orange from some green and more red.

TRY YOUR HAND

Match the following spectra to the names provided. The graphs show the proportions of red, green and blue reaching the eye.

dark green, light green, aqua, dark grey, sky blue, pale yellow, mustard, brown, bright pink.

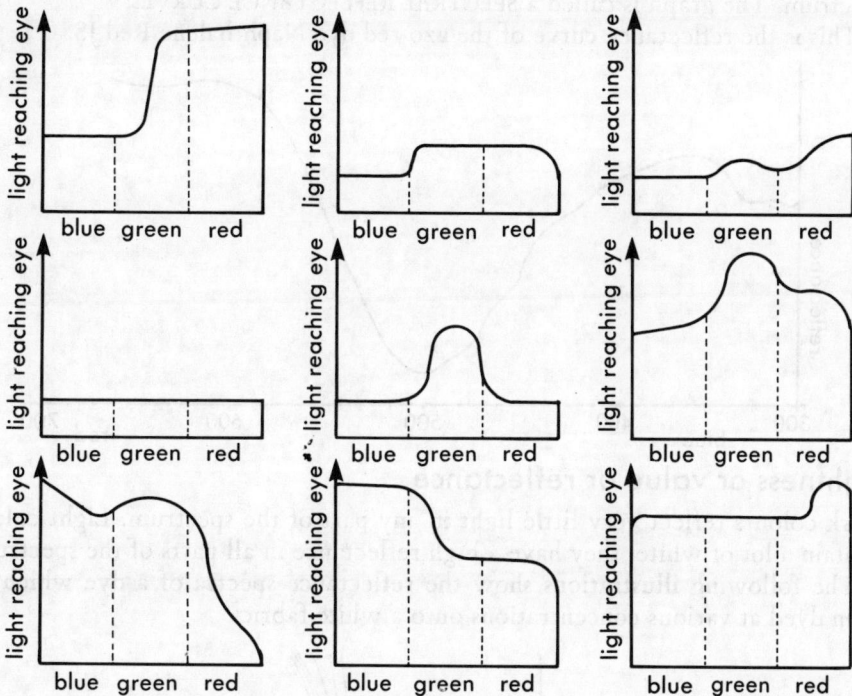

It is possible to take samples of coloured materials and, with an instrument called a SPECTROPHOTOMETER, measure exactly how much light the sample reflects in each portion of the spectrum.

The spectrophotometer divides white light into very narrow wavelength bands, and compares the light from the light source with the light reflected from the sample. The results are expressed as PERCENTAGE REFLECTANCE for each wavelength band. For example, at 500 nm (green), a green sample may reflect 80 per cent of the light falling on it, but the same sample at 650 nm (red) will reflect only 5 per cent of the light. (Remember, green pigments absorb red and blue.)

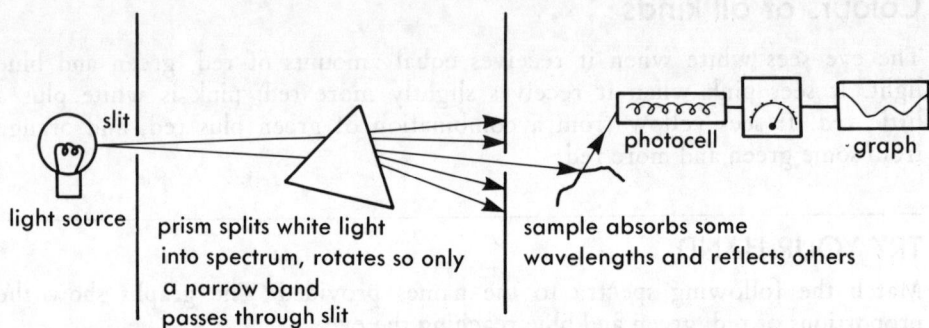

The spectrophotometer.

Percentage reflectance is usually graphed for all the wavelengths in the visible spectrum. The graph is called a SPECTRAL REFLECTANCE CURVE.

This is the reflectance curve of the azo red dye Naphthalene Red JS.

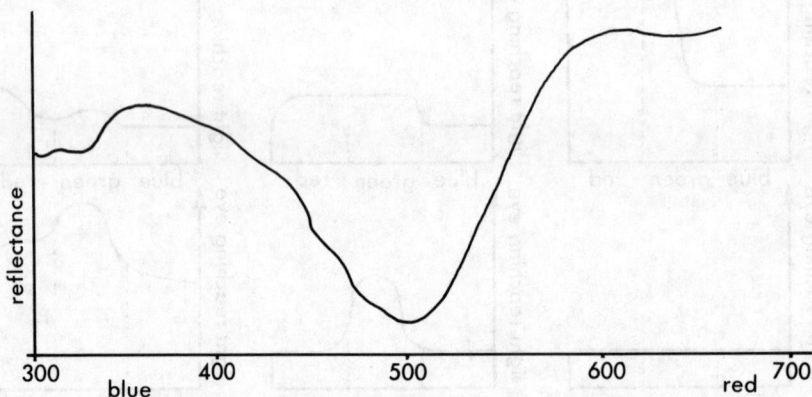

Lightness or value or reflectance

Dark colours reflect very little light in any part of the spectrum. Light colours contain a lot of white: they have a high reflectance in all parts of the spectrum.

The following illustrations show the reflectance spectra of a dye which has been dyed at various concentrations onto a white fabric.

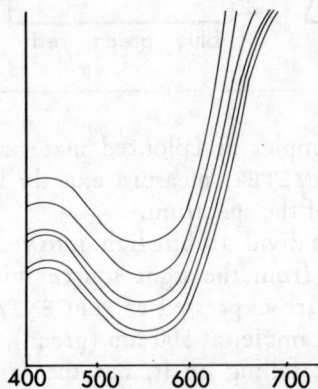

Reflectance spectra for Reactofix Red HE8B dyed on white cotton fabric.

The effect of the light source

A sample can reflect only the light which falls on it, so the colour the eye perceives depends on the spectral quality of both the light source and the pigment. Thus, if red predominates in the spectrum of the light source, the sample will reflect more red light.

TRY YOUR HAND

You will need:

a camera loaded with colour film (print or slide)
a large blue and white plate or tray packed with brightly coloured fruits: ripe tomato, green capsicum, lemon, orange

What you do:

Photograph the arrangement under different light conditions: in a darkened room lit by incandescent (tungsten filament) light; by fluorescent light; in sunlight; in shadow; on an overcast day.

Compare the shades of white (the plate) obtained under the different light conditions. Observe any changes to the other colours.

To explore the effects further, take a photograph showing the inside of a tungsten-lit room with a sunlit scene visible through the window. Take any other photographs you can where two separately lit scenes are recorded on the one frame.

The film can record only the light which reaches it: it cannot adjust like the human eye. The differences shown in this experiment are therefore differences in the quality of the light. (Different films may give slightly different results, due to different pigments and processing, but the same *type* of effect will be observed for all films.)

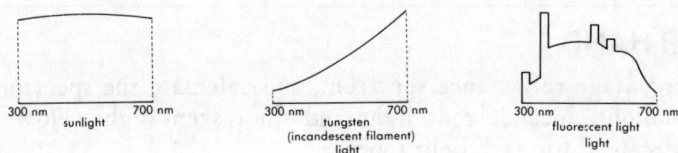

Spectral composition of sunlight, tungsten (incandescent filament) light, and fluorescent light.

1 Describe the colour of the light from each illuminant you used.
2 If you put on make-up under one light source, would you expect your appearance to remain the same under another light source? (This applies to theatrical make-up as well.)
  Experiment with putting on make-up in candlelight and appearance in bright sunlight. Compare with the effect of make-up applied in sunlight and appearance in candlelight. Can you explain the effects?
3 How do your observations relate to matching colours in textiles, for example matching buttons to fabric? How would you ensure a good match in all light conditions?

The perceived colour of an object depends on the actual amounts of each wavelength of light reflected from it.

At any wavelength,

reflected light = incident light × percentage reflectance

Thus the effective spectrum of an object can be obtained by measuring the height of the emission (incident light) curve and the height of the percentage reflectance curve at a given wavelength, and multiplying the two values: this is repeated at small (say 10 nm) intervals to yield a smooth curve.

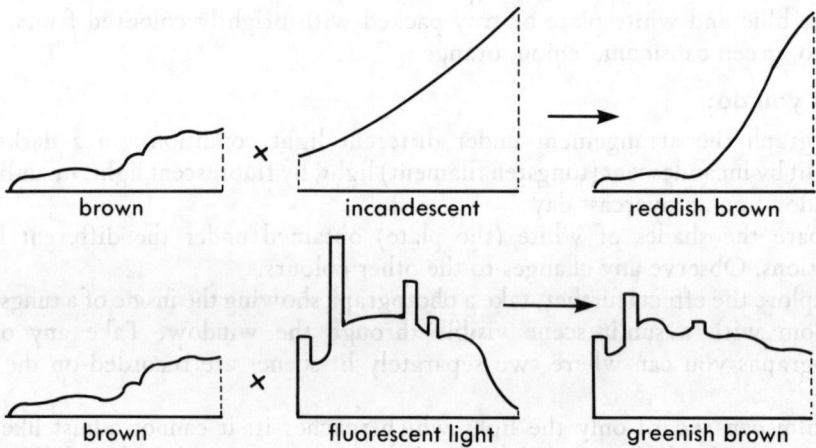

The effect of different light sources on a reflectance spectrum and perceived colour.

TRY YOUR HAND

Invent a percentage reflectance spectrum, and calculate the spectrum it would yield for sunlight, incandescent light and fluorescent light. How would the colour be perceived for each light source?

Ideal white light contains equal amounts of light of all wavelengths. As we have seen, most sources of light are able to produce only approximations to this ideal white light.

Sunlight is the closest to the ideal, but it is not always available. Skylight is often used instead, but its colour varies with the location, season, time of day, weather, and direction of the compass. For reliable colour matching, a reproducible light source is essential.

The Commission Internationale de l'Eclairage (CIE) in 1931 specified three international standard illuminant sources. Source A is a tungsten filament lamp glowing at a temperature of 2854 K (2581°C). Source B, an approximation to sunlight, and Source C, an approximation to skylight, are derived from Source A by passing its light through special filters.

Colour specification and measurement

Visual systems

Colour matching

The simplest way to specify a colour is to send a colour chip (a piece of fabric, paper or other item) to the colourist with the instructions to 'match it'. There is a whole range of colour cards or colour chip booklets available which help the colourist in visually identifying and specifying colours. The colourist (paint or dye manufacturer) then proceeds to mix pigments until the product matches the sample.

Matching is judged by comparison of the product and sample under standard illumination.

Usually, the dyes used in the original sample will not be the same as those used for the match. This could result in shades of, say, green which appear the same under one illumination, but different under a different light source. Matching which does not hold under all light conditions is called METAMERIC MATCHING.

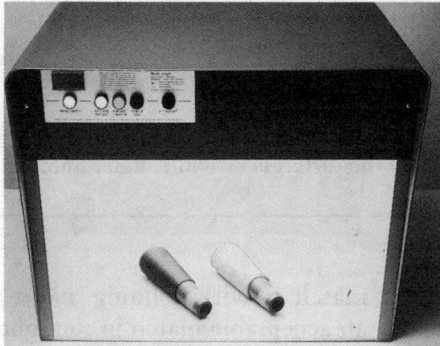

The colour matching booth allows comparison of samples under several standard light sources.
(Instrumental Colour Systems Limited, Kennetside Park Industrial Estate, Newbury, Berkshire RG14 5TE, England)

In some circumstances, a reasonable colour match could be a matter of life and death – but he would change his scarlet beret for a less conspicuous helmet or fatigue cap.
(Army Public Relations)

Metamerism

Metamerism can occur when two coloured samples have similar but not identical reflection spectra. Under one light source, the similar portions of the spectra may be emphasized, and the samples will match; another light source may emphasize the differences, and create a mismatch.

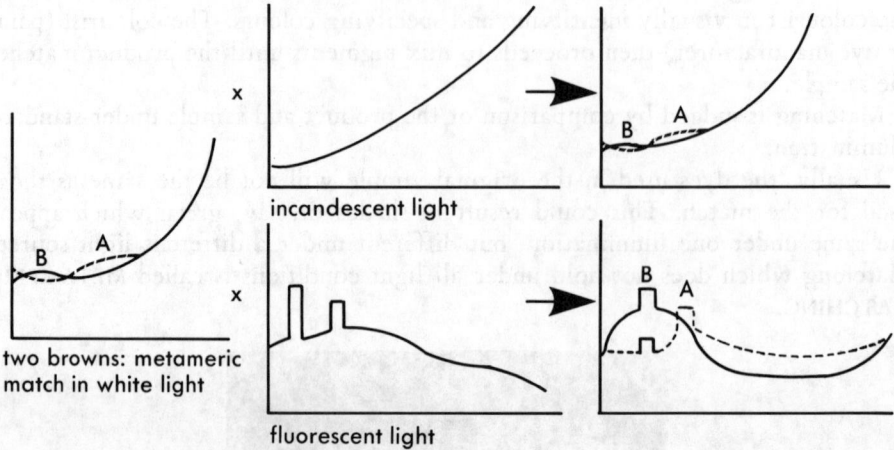

two browns: metameric match in white light

incandescent light

fluorescent light

Metamerism. If two samples have similar but not identical reflectance spectra, one light source may exaggerate the differences while under another they are insignificant.

TRY YOUR HAND

Pick a coloured lace, and match it with a lining and a dress fabric. If they match in the shop, are they an acceptable match in sunlight? in candlelight? Why do you think matching is important?

Problems with visual matching: colour blindness

Visual colour matching is a simple and quick process, and tests the product according to the real needs of its use. If the colours cannot be distinguished visually, the differences are not important, even if the colours can be differentiated by a sensitive instrument.

Not all people have perfect colour vision. Colour vision depends on effective colour receptors. If a person's retina lacks effective receptors for one of the primary colours, then he (it is a sex-linked problem) will have difficulty in recognizing all the colours of the spectrum. The most common form of COLOUR BLINDNESS is the inability to distinguish between red and green. A DICHROMAT is able to match colours to his satisfaction by mixing just two primaries, instead of all three, but the match would not be acceptable to a person with normal colour vision.

A variety of tests exist to check for deficiencies in colour vision. Persons with defective colour vision are not suited for occupations requiring colour matching.

Colour specification: the Munsell system

The Munsell system of colour specification is a visual colour matching system which attempts to sort, classify, and numerically identify all possible colours. It is one of many systems that have been developed over the years, and is widely used in the textile industry.

Munsell divided the achromatic scale from black to white into ten equal steps of subjective brightness. This scale is called VALUE. White, the lightest, is given a value of 10; black a value of 0. This column of values forms the axis of a colour solid, and the colours are arranged around it in a circle.

The colour circle is divided into five principal HUES: red (R), yellow (Y), green (G), blue (B), and purple (P). With the intermediate colours of RY, YG, GB, BP, and PR, this gives ten segments around the circle, each with ten further subdivisions.

The colour circle is arranged as pages attached to the central axis. The outermost edges of the page are the most highly coloured; the ones closest to the central axis are the least saturated. The distance from the axis measures the CHROMA, or saturation or purity, of the colour: the higher the chroma, the purer or more saturated the colour is. Grey has a chroma of 0.

The Munsell system allowed the production of a colour atlas with 960 patterns of 40 hues arranged in a series of charts.

A person who wishes to specify a colour for matching therefore does not need to supply a sample. All that is needed is the numbers and letters for hue, value and chroma for the correct chip in the Munsell colour atlas: the colourist is then able to match the dyed product accordingly.

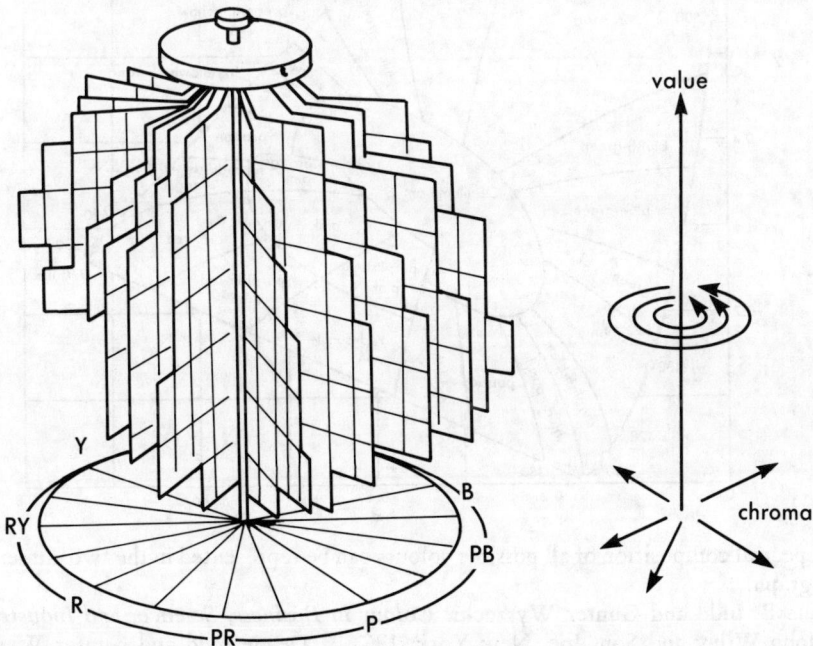

Munsell colour atlas.

The CIE system

Because there are subtle individual variations in colour vision, and because the eye can be fooled, a more reliable method of colour specification is needed.

In 1931, the CIE established an instrumental system of colour specification. This uses a hypothethical standard observer, and three standardized colours of red, green and blue to represent the three visual colour receptors. An instrument called a TRISTIMULUS COLORIMETER determines how much of each of the three additive primary colours is needed to match any colour exactly.

The three components, labelled X, Y, and Z, are called the TRICHROMATIC CO-ORDINATES. Their values, x, y, and z, called TRISTIMULUS VALUES, are obtained mathematically from the measurements of the colorimeter.

$$x = \frac{\text{amount of X}}{\text{total amounts of X, Y, and Z}}$$

$$\text{or } x = \frac{X}{X + Y + Z}$$

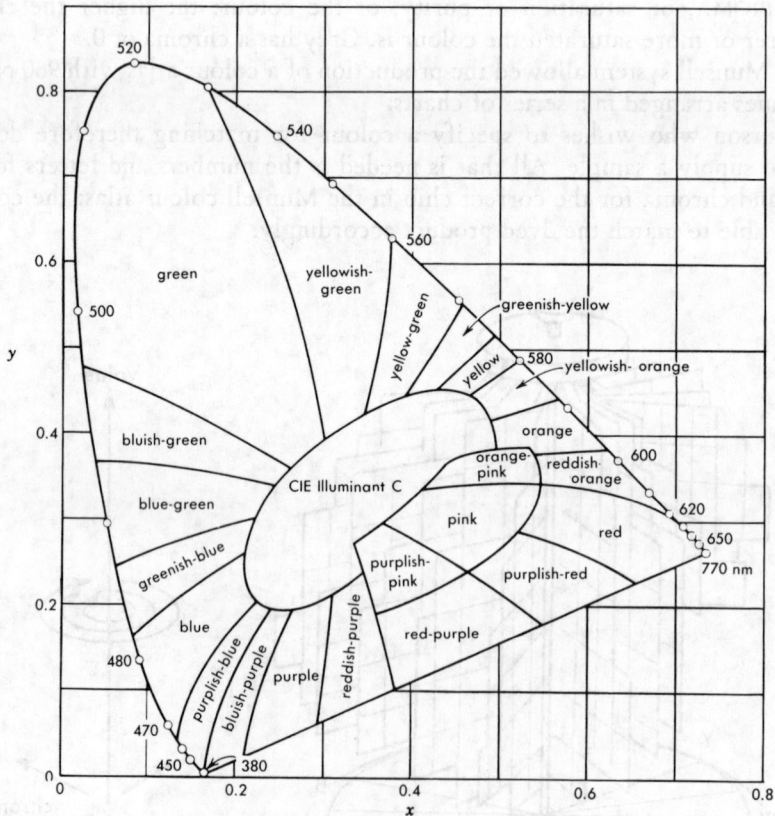

The spectral composition of all possible colours can be represented in the two–dimensional CIE graph.

(Deane B. Judd and Gunter Wyszecki. *Colour in Business, Science, and Industry*, 2nd Ed. John Wiley and Sons Inc. New York, 1963. © Deane Judd and Gunter Wyszecki, 1963. Reproduced by permission of John Wiley and Sons Inc.)

For a standard white light, the three components are used in equal amounts: $\frac{1}{3} X + \frac{1}{3} Y + \frac{1}{3} Z = 1$ (or , $x + y + z = 1$)

For a coloured light or sample, the proportions of X, Y, and Z will vary. For example, for a shade of green, the proportions may be: $0.18X + 0.69Y + 0.13Z = 1$.

The sum of the primaries must always add up to the total amount of light available:

$$xX + yY + zZ = 1$$

so we can say

$$z = 1 - (x + y)$$

In this way, we can always calculate z, provided we know x and y. Because only *two variables,* x and y, are needed to define any colour completely, all colours can be plotted on a two-dimensional graph system.

On the graph, y is the vertical axis and x is the horizontal axis. The standard illuminant light source occurs at $0.33X + 0.33Y + 0.33Z$, that is at $x = 0.33$ and $y = 0.33$. The colours of the spectrum, which are saturated colours with no white in them, form the outer limits of the space within which all colours are defined. Thus the highest point on the graph, the point with the highest y value, represents the purest spectral green. Red is found at the furthest x value, and blue where both x and y are nearly zero.

Near the illuminant the colours are less saturated, less pure, than near the spectral boundaries of the graph. The EXCITATION PURITY of the colour is defined as the ratio

$$\frac{OC}{OS} \times 100\%$$

This corresponds to the idea of chroma on the Munsell system.

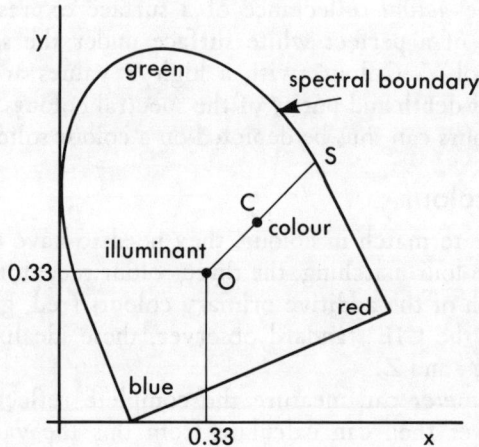

Although the graph defines the proportions of the additive primaries composing a particular colour, it does not define how bright the colour is. Another variable is needed to define the luminosity or BRIGHTNESS.

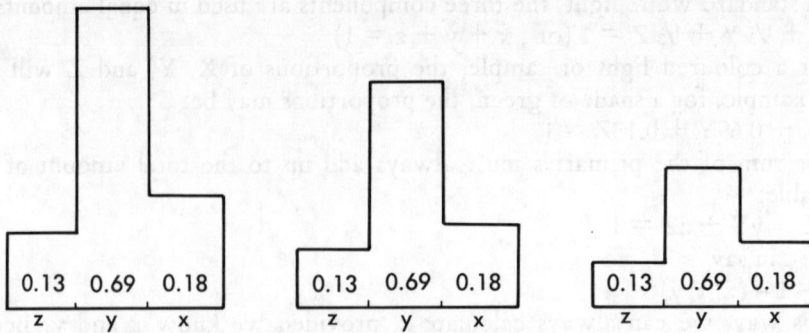

These are the same colour (have the same tristimulus values) but have different luminosities.

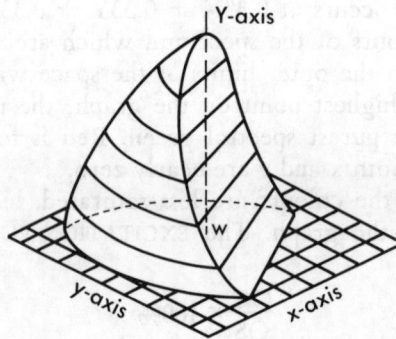

Brightness is represented by a third axis, Y, added to the CIE graph.

Brightness is the *actual* reflectance of a surface expressed as a percentage of the reflectance of a perfect white surface under the same illumination. It is given the symbol Y. Colours with a high Y values are pastels, and so do not have the bright depth and purity of the spectral colours.

All possible colours can thus be depicted on a colour solid.

Matching the colour

If two samples are to match in colour, they need to have the same reflectance spectra. In visual colour matching, the three colour receptors in the eye evaluate the amount of each of the additive primary colours (red, green, blue) reflected by the sample. In the CIE standard observer, these idealized colour receptors correspond to X, Y, and Z.

A *spectrophotometer* can measure the complete reflectance spectrum of a sample. A computer then can calculate from this the values of x, y, and Y which correspond to the reflectance spectrum.

The CIE system corresponds both to the real reflectance spectra of coloured objects, and to the way in which the human eye perceives colour. Also, it is well-suited to transforming a *colour into numbers* which can be handled by a computer. There are a number of other, slightly different, systems which also translate colours into numbers.

220

For the dyer and colourist, a numerical definition of the colour to be matched is insufficient. They need to know what dye or pigment combination will produce the same reflectance curve on another sample.

Pigments *subtract* colour from light. The more pigment, the more light is absorbed or subtracted. Thus, if two pigments are mixed, both subtract light. The total light subtracted can be calculated by *adding* the absorption spectra.

TRY YOUR HAND

Here are the absorption spectra of two different dyes.

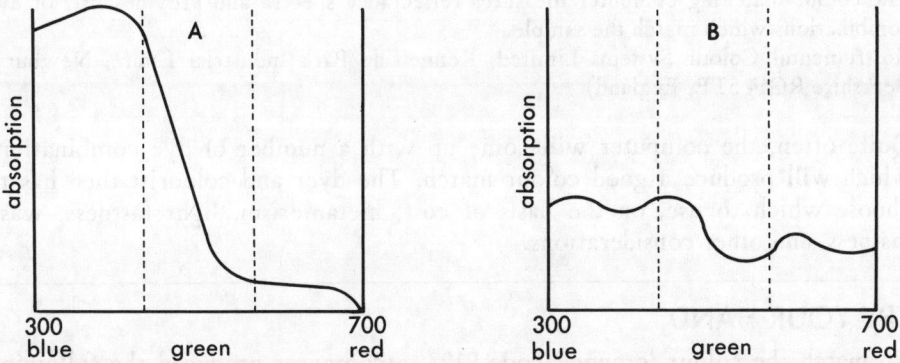

Add them; that is add the heights of the curves for each wavelength. (Height of curve A at 450 nm plus height of curve B at 450 nm equals absorption of mixture at 450 nm.) Adding heights at 50 nm intervals will give a complete absorption spectrum for the dye mixture.

Now place a mirror at the top of the curve. What you see is the reciprocal of the absorption spectrum: the reflectance spectrum (light not absorbed must be reflected).

Predict the colour of the dye mixture.

The colour-matching computer stores data about the reflection/absorption spectra of different dyes, and from these data it calculates how much of dyes A and B (and if necessary C and D) is needed to reproduce the spectrum of the sample.

If a perfect match is not possible, the computer calculates how much metamerism can be expected under different lighting conditions.

TRY YOUR HAND

Using acrylic paints or water colours, see how many different combinations of paints you can use to produce an acceptable match of a colour of your choice.

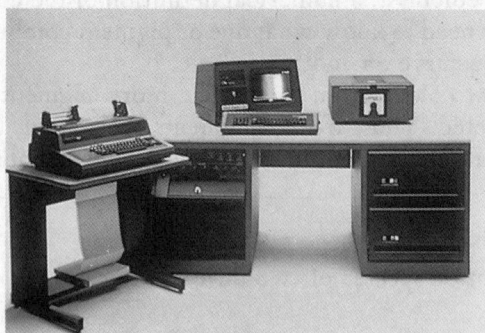

The colour-matching computer measures reflectance spectra and provides lists of dye combinations which match the sample.
(Instrumental Colour Systems Limited, Kennetside Park Industrial Estate, Newbury, Berkshire RG14 5TE, England)

Quite often, the computer will come up with a number of dye combinations which will produce a good colour match. The dyer and colourist then has to choose which to use, on the basis of cost, metamerism, light fastness, wash fastness, and other considerations.

TRY YOUR HAND

To match the colour 'orange', code 9123, a computer produced the following options:

	Concentration	Dye name	Relative cost	Performance
i	0.0021%	Turquoise HA		A = 0.0
	0.2626%	Red HE3B	34.43	B = 1.3
	1.5969%	Yellow HE4R		C = 1.9
ii	0.1124%	Red HE8B		A = 0.1
	3.1033%	Yellow HE4R	64.96	B = 1.2
	0.4948%	B Orange HER		C = 1.0
iii	0.2599%	Red HE3B		A = 0.0
	1.5973%	Yellow HE4R	34.39	B = 1.6
	0.0013%	Navy Blue RE		C = 2.0

A = wash fastness, B = light fastness, C = metamerism. The lower the value, the better the performance of the formula.

1 Are any dyes used more than once? Which ones?
2 Why is option (ii) more expensive than (i) or (iii)?
3 Which option has the best performance values?
4 Which combination would you use? Why?

Coloured molecules

Dyes and pigments are coloured because their molecules are able to absorb some colours of light and reflect others.

Light is a form of energy. The energy is inversely related to wavelength: long wavelength, low energy; short wavelength, high energy. If a molecule selectively absorbs some wavelengths of light, and not others, it means the molecule is selectively absorbing some energies.

The simplest spectrum

The simplest spectrum is that of an isolated atom.

An atom consists of a central positively charged nucleus surrounded by negatively charged electrons. The negative electrons occupy regions of space called ORBITALS: they are attracted to the positive nucleus, and so are at their lowest energy in the orbitals closest to the nucleus.

For any atom, the orbitals have *fixed energy levels*. Each orbital can accommodate only a certain number of electrons. It is possible for an electron to absorb energy in the form of light, and make a jump from a low energy orbital into a higher energy orbital. As the orbitals correspond to fixed energy levels, only energy corresponding to the energy leap (or excitation) can be absorbed. Thus the atom absorbs only specific wavelengths of light.

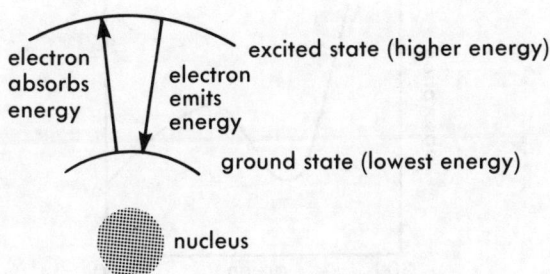

electron absorbs energy

electron emits energy

excited state (higher energy)

ground state (lowest energy)

nucleus

Mechanism of absorption and emission of light in an isolated atom (schematic).

An example is the sodium atom. In a sodium street light, an electric discharge through the sodium gas excites the most loosely held electron so it goes into a higher energy orbital. The excited atom then loses the energy by *emitting* it as yellow light of wavelength 592 nm. If white light is shone through sodium gas, the atoms will *absorb* light of this wavelength and the electrons will be excited.

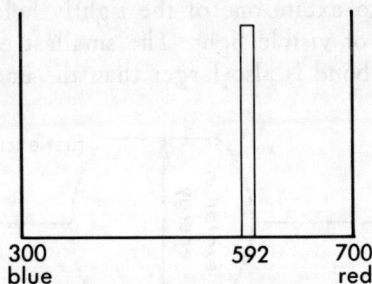

300
blue

592

700
red

The principal spectral line of sodium.

Metal compounds

Metal compounds are used to colour glass. The intense red of blood is due to the presence of iron, and the colours of gems are due to metal compounds.

In these coloured metal compounds, usually *more* than one electron is able to be excited, and usually there is more than one energy jump that corresponds to a wavelength of light in the visible spectrum. This means that several transitions can occur: there can be several wavelengths of light absorbed.

In the gaseous sodium atom, the absorption occurs in a very narrow band of the spectrum. This is because the atoms are isolated. In a metal compound, the metal atom or ion is surrounded by other atoms: these all have electrons which interact with the electrons of the metal atom. The atoms are vibrating constantly in the molecule or crystal; all molecules and crystals vibrate. These vibrations mean that the distances between atoms are changing all the time, and so the interactions between them change. All this means that the possible energy levels of the electrons are not fixed: the exact amount of energy needed for a particular transition will vary from instant to instant because of these interactions. Thus the *electrons will absorb a broad band of wavelengths of light.*

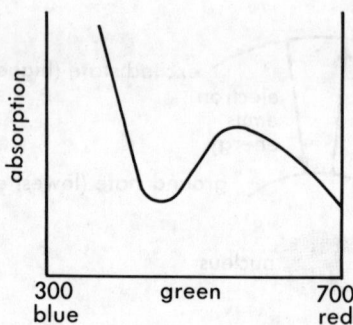

The absorption spectrum of the titanium ion in water. What colour would you predict the solution to be?

Dye molecules

With molecules, the situation is more complex.

A covalent molecule does not have the loosely held electrons of a gaseous sodium atom or the metal ion in a coloured metal compound. The electrons are all either tightly held by their atoms, or involved in bonding between atoms.

The energy needed to excite one of the tightly held electrons is enormous, far beyond the energy of visible light. The smallest energy jump possible for an electron in a single bond is also larger than the energy provided by visible light.

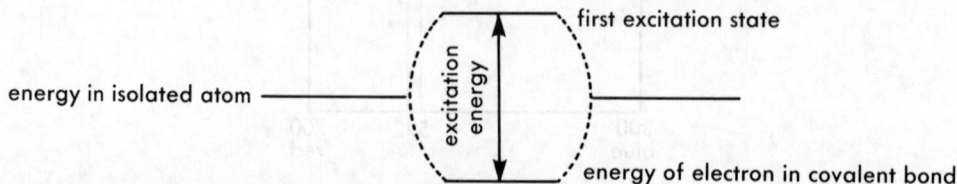

The transition available to an electron in a double bond.

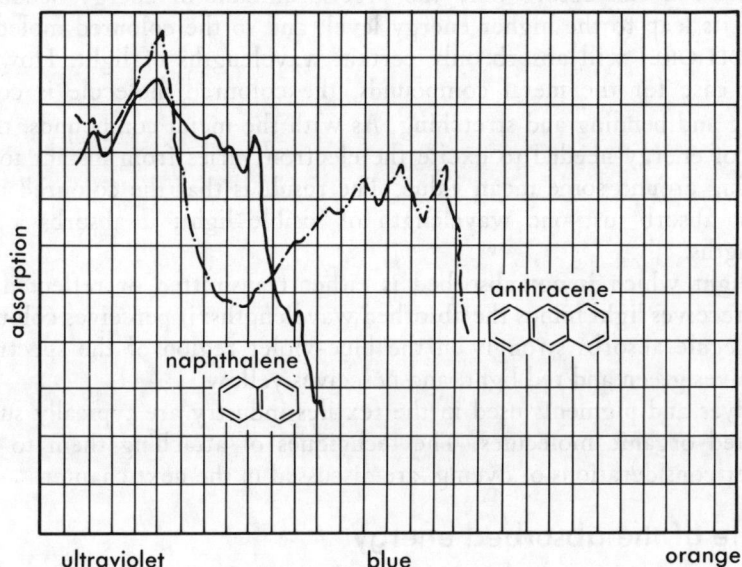

absorption

ultraviolet — blue — orange

naphthalene

anthracene

Absorption spectra of naphthalene and anthracene. Naphthalene, with five conjugated double bonds, absorbs strongly in the ultraviolet region of the spectrum. Anthracene, with seven conjugated double bonds, is edging into the visible spectrum. (Note: here, and in all discussion of coloured molecules, structures are drawn showing the double bonds. This is for easier visualization of conjugation systems.)

The electrons in a double bond are not held as tightly, and so they are able to make their smallest energy leap by absorbing high energy ultraviolet light.

If a molecule has a system of *alternating double and single bonds,* the bonds form a CONJUGATED system: the double bonds overlap or 'smear'. The electrons are less tightly held still, and so molecules with a large conjugated bond system absorb strongly in the near (lower energy) ultraviolet region.

If, to a molecule with such a conjugated system, we add an extremely polar group, this will change the electron distribution and energy of the entire molecule. The energy now required to excite an electron in the conjugated double bond system will correspond to a wavelength of *visible light.* This added polar group is called an AUXOCHROME. The nature of the auxochrome determines its effect on the electron structure of the whole molecule, so different auxochromes will cause the molecule to absorb different colours of light.

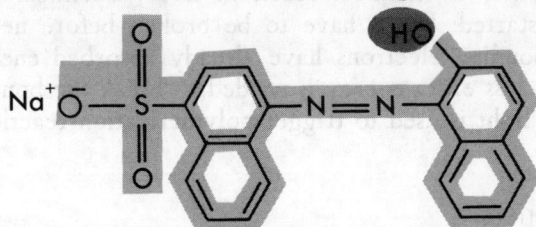

C.I. Acid Red 88. The chromophore (conjugated bond system absorbing in the visible spectrum) is indicated by light tone; the auxochrome (in this molecule a hydroxyl group) is indicated by heavy tone. The spectrum of this dye is shown on page 263.

225

An electron can absorb only the precise amount of energy needed for it to make its leap to the higher energy level, and so the coloured molecule (the CHROMOPHORE) will absorb only certain wavelengths of light. However, as was the case for the metal compounds, the coloured molecule is constantly vibrating and bending and stretching. As with the metal compounds, the exact amount of energy needed to excite the electron varies from instant to instant, fluctuating around some mean value. The result is that the coloured molecule does not absorb just one wavelength of visible light: it absorbs a band of wavelengths.

The light which is not absorbed is either transmitted or reflected, and so the eye receives light minus the absorbed wavelengths: it perceives colour. Thus, if a molecule absorbs strongly in the blue-violet region of the spectrum, the eye receives green and red light, and perceives yellow.

The dyes and pigments used in the textiles industry are typically such large conjugated organic molecules. The techniques of attaching them to textiles, and other considerations of dyeing, are discussed in the next chapter.

The fate of the absorbed energy

What happens to the energy the molecule has absorbed?

If the electrons of a coloured substance stayed at their higher energy levels, the molecule would be able to absorb only so many photons, or packets of light energy, and then no more: once all the electrons were excited, there would be no more left to absorb the light energy.

This means that the molecule must be able to get rid of the energy in some way.

The molecule may simply re-emit a photon of the exact same wavelength as it absorbed. This would be indistinguishable from reflection. As no substance has quite total absorption at any wavelength, this probably does occur, but it cannot be the main mechanism for disposing of the excess energy. If it was, every substance would appear white.

As already mentioned, molecules and crystal lattices are in constant vibration. The energies involved in the bending and stretching and rotating of bonds are much lower than the energies of excitation of electrons. These molecular movements are registered as *heat,* and occur in the infrared region of the spectrum. An excited electron may shed its energy and return to the ground (lowest energy) state by transferring the excitation energy to the molecular skeleton as vibrational energy: the light is degraded into heat.

Another possibility is chemical reaction. Every chemical reaction needs a 'kick' to get it started: bonds have to be broken before new bonds can be formed. If the bonding electrons have already absorbed energy in the form of light, so much less extra energy is needed to break the bonds. This is useful when ultraviolet light is used to trigger polymerization reactions, but it is not desirable in a dye.

Optical brighteners

Some detergents are claimed to wash 'whiter than white'. Some white or pastel textiles glow under ultraviolet (black) light.

Both these effects are possible because special dye molecules in the fabric or the detergent *absorb ultraviolet* light when their electrons are excited, but *re-emit the light* at a lower energy, *in the visible region,* usually in the blue or blue/green region.

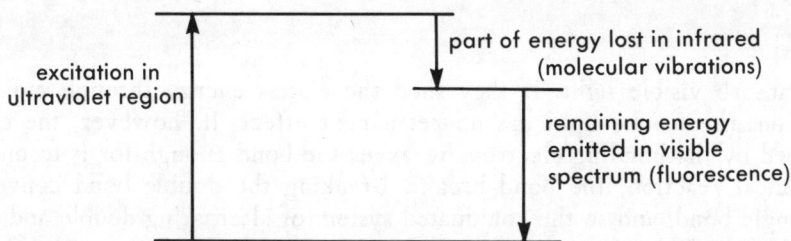

How optical brightening agents work.

The process of emitting visible light after or during ultraviolet stimulation is called FLUORESCENCE. If there is a long delay between absorption of the ultraviolet and emission of the visible light, the phenomenon is called PHOSPHORESCENCE.

Sunlight contains a considerable amount of ultraviolet light. So does fluorescent light. In fact, the *fluorescent light* actually emits a lot of ultraviolet light but very little visible light: the inside of the tube is coated with materials called phosphors which absorb the ultraviolet and re-emit it as visible light: they *fluoresce*. Uncoated tubes produce 'black' light, the invisible ultraviolet.

Tungsten lamps, with their glowing incandescent filaments, are a source of visible and infrared light, and heat, but they do not produce any ultraviolet.

TRY YOUR HAND

You will need:

a darkened room
a 'black' light (uncoated fluorescent tube)
a fluorescent light
a tungsten light
2 pieces of unbleached calico or similar white fabric
'ultrabright' detergent

What you do:

Soak one piece of the calico in the 'ultrabright' detergent and dry it.
Compare the two samples under the different light sources.

1 Which glows under the black light?
2 What are the differences under fluorescent light?
3 How do the samples compare under the tungsten light?
4 Do you think detergent manufacturers are right to claim 'brighter than white' effects for their products?
5 Have you ever noticed shirts or other clothes glowing in black light illumination?

The dyes which produce the 'whiter than white' effect are called OPTICAL BRIGHTENERS. Brilliant whites can be achieved with their use, but treated fabrics may be sensitized to light: degradation and yellowing due to light may be accelerated. The result may be a shirt which appears brilliant white in direct sunlight, but which appears yellow in the folds where it is in shadow.

Fading

Dyes absorb visible light. If they shed the excess energy through molecular vibrations (heat), the light has no permanent effect. If, however, the energy absorbed by the bonding electrons weakens the bond enough for it to undergo a chemical reaction, the bond breaks. Breaking the double bond converts it to a single bond, and so the conjugated system of alternating double and single bonds is disturbed.

When the bond breaks and the conjugated system is disrupted, the energies of the electrons are altered so they no longer absorb visible light. The molecule becomes *colourless*. This process is seen as the LIGHT FADING of dyes.

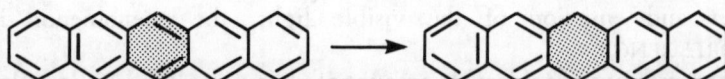

Take, for example, the blue molecule pentacene. If the double bonds in the middle benzene ring are converted to single bonds, the result is effectively two linked naphthalene molecules. The conjugated system is disrupted, and so tetrahydropentacene would absorb in the ultraviolet with a spectrum very similar to that of naphthalene.

The absorption of light only *weakens* the bond: it only *increases the chance* that it will undergo a chemical reaction. Whether or not the reaction takes place depends on other factors.

The *shape of the molecule* may make it impossible for other molecules to come near to the weakened bond to react with it. The *products* from the chemical reaction may be *less stable* than the original molecule, and so there is no energy saving to drive the reaction. Although the bond is weakened, it may *not be weakened enough* to allow a particular reaction to proceed.

All these factors vary between the different dye molecules, and so different molecules have different tendencies to undergo PHOTODEGRADATION. The dyes which react most readily fade the fastest.

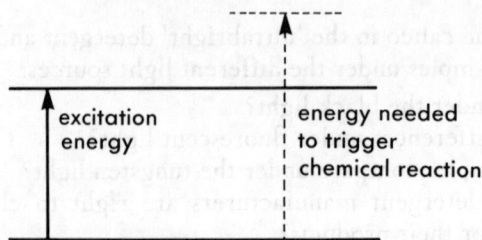

When a bonding electron has already absorbed some energy, much less extra energy is needed to break the bond (chemical reaction).

TRY YOUR HAND

You will need:

15 cm × 8 cm strips of assorted denims and other blue fabrics
large piece of cardboard, and cardboard strips

What you do:

Staple or pin the fabric samples to the board. Do not use glue, as it may discolour them.
Cover half of each sample with the cardboard strips.

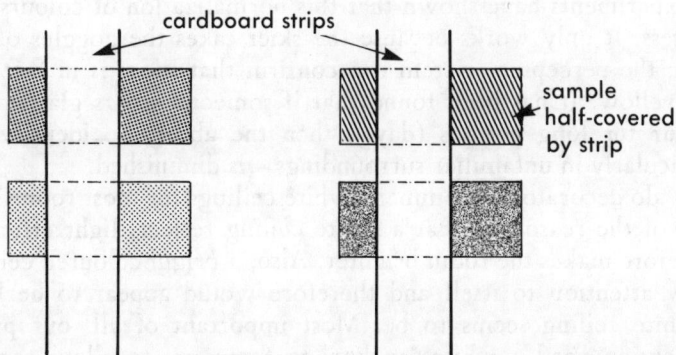

cardboard strips

sample
half-covered
by strip

Expose the display to sunlight for two weeks. If possible, do not keep the samples behind glass, as glass cuts out some of the ultraviolet light.

At the end of the fortnight, remove the protective cardboard strips, and estimate the degree of fading that has occurred.

A more precise test would use standard samples of blue dyes (a 'blue scale') for comparison. It gives each dye a rating, from 8 for lightfast, to 1 for very little light tolerance.

1   What fabric applications require dyes with a high light fastness?
2   List textile applications where light fastness is of little importance.

Consumer questions answered

Q1   I have a room which I call my happy room; it is all a bright yellow – walls, ceiling, furniture, rugs. When I look in the door, the room appears to be perpetually bathed in sunshine. Yet inside the room my complexion becomes pale and sickly looking. Why?

A   When white light (whether from sun or from artificial lighting) is reflected from the walls of the room, most of the blue is absorbed out of it by the paint. This is what makes the room appear yellow. $(Y = G + R)$

Now, rosy cheeks appear red because they absorb more blue out of the light falling on them than does the cream colour of the rest of the skin. $(R = W - (B + G))$

Since the yellow light in the room has no blue, the rosy cheeks reflect much the same colours as the rest of the face and so they appear yellow, pale and sickly. Note – contrary to expectations, red lighting tends to 'wash' colours out even more than yellow does. Can you explain why?

Q2 How is it possible to recognize the colour of various objects even when we wear coloured skiing glasses?

A This process of recognition is related to the comparative way in which all colours are perceived.

The cones register relative amounts of red, green and blue light that impinge on the retina from the different parts of the scene. The lightest colour with a relatively even distribution of the primaries is interpreted as white (even though it appears yellow through the anti-glare skiing glasses). The brain then interprets the rest of the colours relative to this 'white' colour.

Experiments have shown that this normalization of colours is a learnt process. It only works because the skier takes the goggles off now and then: the perceptual system can confirm that snow is in fact white and not yellow. It has been found that if someone wears glasses of a given colour for long periods (days), then the ability to identify colours – particularly in unfamiliar surroundings – is diminished.

Q3 Why do decorators recommend white ceilings for most rooms?

A One of the reasons is that a white ceiling reflects light efficiently, and therefore makes the room brighter. Also, a *bright* coloured ceiling would draw attention to itself and therefore would appear to be lower than a white ceiling seems to be. Most important of all, our processes of perception need a white 'anchor' to focus on, to allow correct inter-pretation of the rest of the colours in the room. Without any white – especially in an artificial environment – the perceptual process is hin-dered, and the viewer can feel uncomfortable.

Q4 Why is it important to check the colours of fabrics and accessories in daylight as well as in the store?

A The light which falls on the surface of a fabric affects the light which is reflected from it. The reddish-yellow light from an incandescent globe can make a cream and a pink fabric appear the same. Daylight, with its more even colour distribution, can soon show up the difference. Metamerism is often a problem when notions such as zippers, buttons and sewing threads are chosen to match a fabric.

Q5 I went dancing, and the dance floor was dimly lit by some purplish 'black light' fluorescent tubes. The strangest parts of people's clothes glowed in the dark – sometimes the whole shirt, sometimes the piping of a skirt, or even just the interlining of a sheer dress. What creates such a weird effect?

A The 'black light' tubes radiate mainly ultraviolet light, which is invisible to the human eye – this is why the dance floor appeared so dark. Some fabrics are dyed with optical brightening agents. These dyes absorb ultra-violet light, and emit some of the absorbed energy as visible light. This is why they appear to glow.

Since not all fabrics are so treated – optical brightening agents are mainly used for whites, not coloureds – the total effect of light emission can look quite strange.

Q6 How can some detergent manufacturers claim that their product makes clothes 'whiter than white'?

A   These detergents contain optical brightening agents, which are absorbed into the fabric. The optical brightening agents absorb invisible ultraviolet radiation, and emit visible radiation – so, in fact, in sunlight, more visible light is reflected from treated fabric than from a standard pure white surface.

Q7   I have a blue rayon scarf. During ironing it becomes purple where the iron has touched it. Later the blue colour returns. How can this be?

A   The dye used in the scarf appears to be one which can exist in two forms – a high energy purple state and a low energy blue state. When heated by the iron, the dye molecule flips into the high energy purple state, thus using up some of the iron's heat energy. During cooling it loses energy, and flops back into its blue state.

Q8   I found some lovely salmon-coloured corduroy in a dress shop – it matched my room perfectly. I bought the fabric and made it up into curtains. The curtains have faded rapidly at the edges where direct sunlight reaches them. I have never had such a problem before – why now?

A   Some dyes have excellent resistance to degradation by light; others don't.
     Dyers choose dyes for fabrics according to the kind of use that is expected for them. Curtain fabrics are always carefully designed for light fastness; for outerwear, wash fastness is more important than light fastness. For this reason, it is never advisable to buy fabric intended for outerwear and make curtains from it!

8 Dyeing

In this chapter you will find the answers to the following questions:
- Why do some fabrics bleed dye during washing?
- Why are some fibres easy to dye at home, whereas others remain white in the craft dyeing liquor?
- Which bleach is best for which fibres?
- How can I do heat transfer prints?
- Why do some dyes rub off onto hands and other fabrics?
- Why do fruit juices stain?
- How, apart from colour matching, are computers used in the dyehouse?

and much more interesting and useful information.

A note on safety

Many of the exercises in this chapter ask you to work with dyes. Please treat the dyes with as much caution as you would give to any other chemicals.

Rubber gloves should be worn at all times, to prevent the dyes being absorbed through the skin. Protective clothing – smock or plastic apron – is recommended to prevent damage to clothes.

Utensils used in dyeing must be used for that purpose only. On no account must they be used for food after they have been used for dyeing.

All work should be performed in a well-ventilated area. Avoid inhalation of dyes.

Dyes and pigments

Textile fibres have a natural colour. Usually, this natural colour is grey, or white, or shades of yellow and brown.

The task of the dyer and colourist is to change this natural colour into the shade demanded by the consumer. Often, the task is to apply colour to a fabric in a pattern.

Dyes

A DYE is a *soluble* colour that is applied from a solution. It *penetrates* and *combines with* the matter being dyed. This means that a dark fabric or fibre cannot be dyed a lighter colour. As a result, any natural colour of the textile fibre must be bleached before the dyeing process can be performed. In some cases, bleaching may present problems: alpaca, for example, was until recently available only in its natural colours, as there was no satisfactory bleaching process for the fibre. It is now available in a wide range of colours, for a price.

Pigments

PIGMENTS are *insoluble* coloured substances and are applied to the *surface* of the matter to be coloured, and attached by some form of binding material.

Because pigments are applied to the surface, they obliterate any colour already there. Thus a dark material may be made a lighter colour by the application of pigment.

Traditional dyeing: a woman in the Hebrides dyes wool with crotal (lichen) by boiling the wool and lichens together.
(Harris Tweed Association)

Traditional use of pigments: Fijian women colouring tapa cloth. Pigments used are a black clay and a red colour extracted by boiling from auru bark and kogona leaves.
(Public Relations Office, Fiji)

Sources of dyes and pigments

Many colourants were once obtained from *living matter*. Cochineal was obtained from an insect, purple was extracted from a sea snail, and an enormous range of colours was extracted from a vast range of plants.

Plant-derived dyes are still widely used by craft workers.

A number of colourants were based on naturally occurring mineral compounds. These were mainly used as pigments: an example is the use of ochres in bark painting.

These days, most colourants are synthesized and manufactured. Apart from a few specialized uses, such as colouring glass fibres, mineral colourants are rarely used. Most colourants are *synthetic organic* (that is, carbon-based) compounds. They are manufactured from raw materials which are ultimately derived from coal and petroleum products: from materials which were once living matter.

Indigo (left) and madder: two historically important sources of dye.
(*Bayer Farben Revue* No. 20, 1971, by permission of the editors. *Bayer Farben Revue* is published by Bayer AG, Leverkusen, West Germany.)

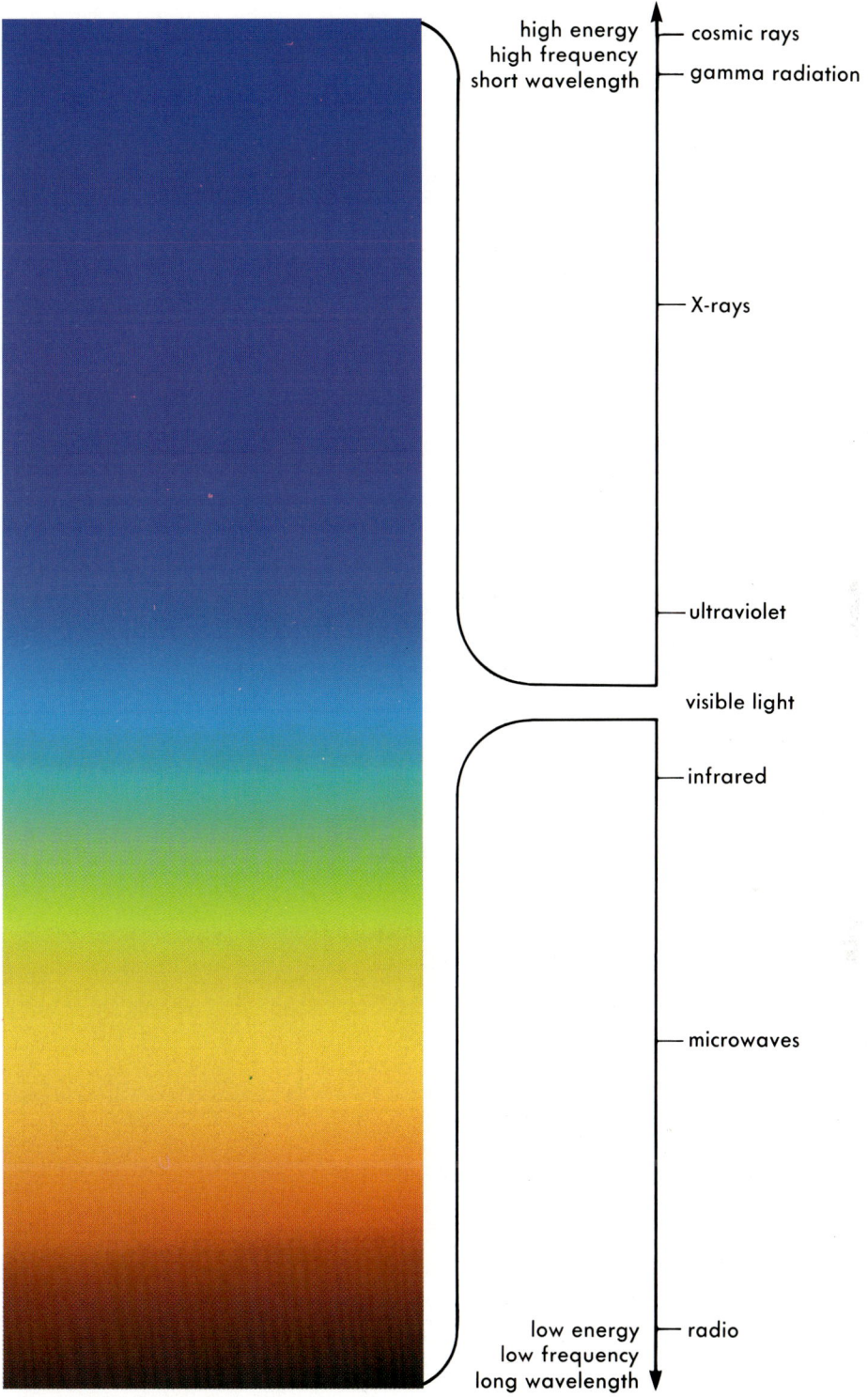

high energy
high frequency
short wavelength

— cosmic rays

— gamma radiation

— X-rays

— ultraviolet

visible light

— infrared

— microwaves

low energy
low frequency
long wavelength

— radio

The electromagnetic spectrum.
(Dale Mann/Retrospect)

Additive colour mixing: projected light.
(Paul Brierley)

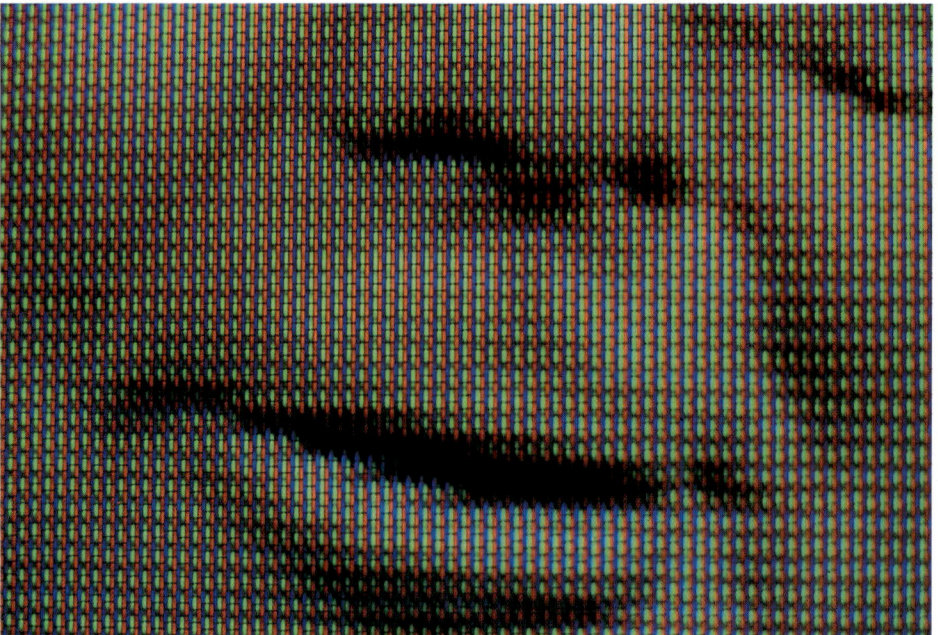

Additive colour mixing: colour television.
(Peter Shaw)

236

Subtractive colour mixing: the four-colour process used in printing.
(Peter Shaw)

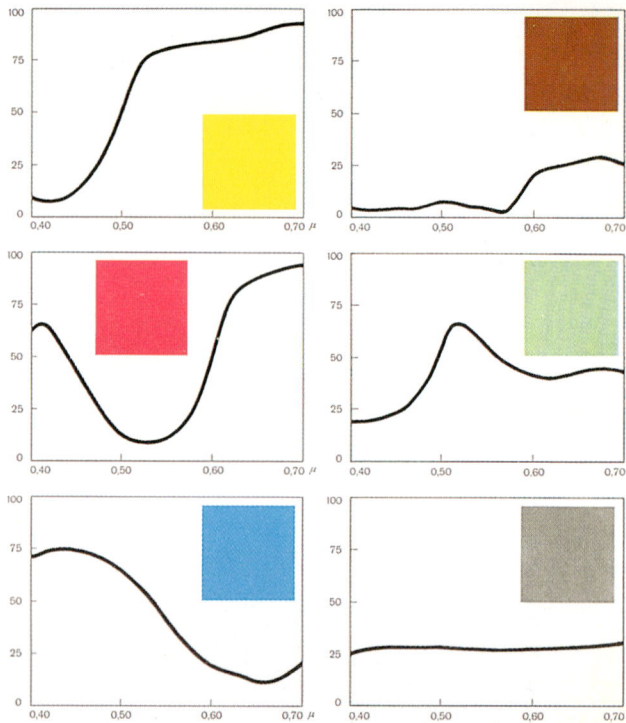

Spectral intensity distribution of various colours.

The C.I.E. chromaticity diagram.

Munsell colour tree.
(Munsell Color, 2441 N. Calvert S., Baltimore, MD 21218)

PMS colour booklet used for colour specification in the printing industry.
(Peter Shaw)

Printed triacetate fabric treated with optical brightener, photographed under white and in fluorescent light.
(*Bayer Farben Revue* No. 18, 1970, by permission of the editors. *Bayer Farben Revue* is published by Bayer AG, Leverkusen, West Germany.)

Influence of different processing oils on dyestuff absorption after the grey goods had been set at 160°C.
(*Bayer Farben Revue* No. 18, 1970, by permission of the editors. *Bayer Farben Revue* is published by Bayer AG, Leverkusen, West Germany.)

CLEAN COLOUR

The colour of greasy wool is not always a good guide to the colour of wool after scouring.

Clean colour measurement is related to both whiteness after processing and hence end use potential. Clean colour is reported as 2 numbers. For example:—

12/88

Degree of **YELLOWNESS**

— Normal range 8 (white) — 16 (yellow)

Degree of **BRIGHTNESS**

— Normal range 90 (bright) — 75 (dull)

white   yellow   bright   dull

Natural colours of cleaned wool. Although not yet accepted by the Australian Standards Association, a grading of this type is a useful predictor of the results to be expected from dyeing.
(Australian Wool Corporation)

Cotton occurs naturally in a range of colours, from pure white to the deep tan of this Guatemalan huipil.
(Courtesy University of Pennsylvania Museum; photograph Charles Dorkins. Reproduced from Frances Schaill Goodman, *The Embroidery of Mexico and Guatemala*. Charles Scribner's Sons, New York, 1976.)

241

A few of the colours obtainable with natural dyes. Dyes used are indigo, fustic, madder, butternut, walnut, maple, sumac, and logwood, singly and in combination; the threads and cloth are linen.
(Courtesy Pocumtuck Valley Memorial Association; photograph E. Irving Blomstrann. In Margery Burnham Howe, *Deerfield Embroidery*. Pocumtuck Valley Memorial Association, Deerfield Massachusetts, 1986.)

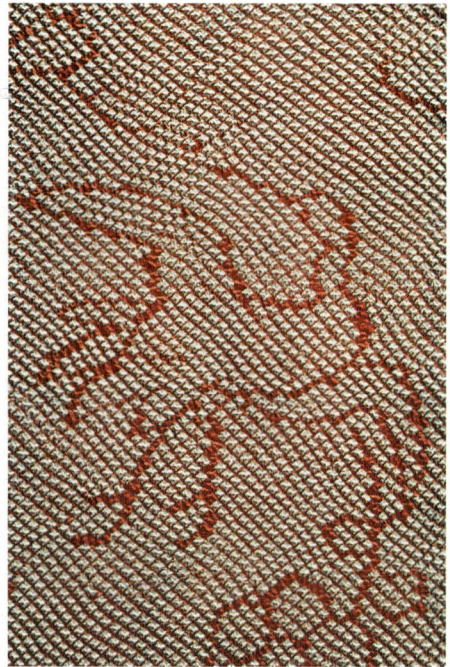

Tie-dyeing a silk kimono, Japan. As with all craft techniques, designs of extraordinary intricacy can be achieved by skilled workers.
(© Cary Wolinsky, Stock, Boston)

A rare velvet ikat from Central Asia. In ikat designs, yarns are dyed in patterns before weaving: sometimes just the warp or just the weft; occasionally both (the Indian patola is a double ikat). Velvet ikat demands even greater skill from the dyer and weaver, as allowance must be made for the pile.
(Courtesy A. and L. Copeland; photograph Dale Mann/Retrospect)

ONION

	wool	wool	linen	cotton	
					none
					alum
					chrome
					iron
					tin
					tannic acid

RHODODENDRON

	wool	wool	linen	cotton	
					none
					chrome
					copper sulfate
					iron
					tin
					tannic acid

The effect of different mordants on two natural dyes.
(David Bailey)

25 °C/200 min.
with soda ash   25 °C   40 °C   60 °C   80 °C

The effect of dyebath temperature on efficiency of dyeing with reactive dyes on cotton. (*Bayer Farben Revue* Special Edition No. 12, Levafix Dyestuffs, by permission of the editors. *Bayer Farben Revue* is published by Bayer AG, Leverkusen, West Germany.)

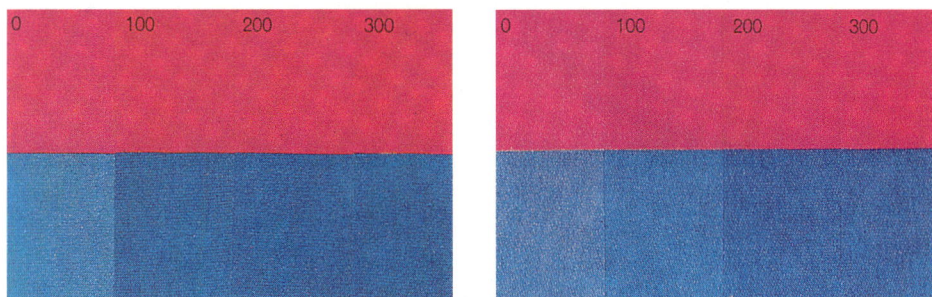

The effect of different salt concentrations (g/L) on uptake of reactive dyes by mercerized cotton. In both these tests, the dye was padded onto the fabric, which was then padded with an alkali salt fixer. The samples on the left were allowed to dry between the two padding operations.
(*Bayer Farben Revue* Special Edition No. 12, Levafix Dyestuffs, by permission of the editors. *Bayer Farben Revue* is published by Bayer AG, Leverkusen, West Germany.)

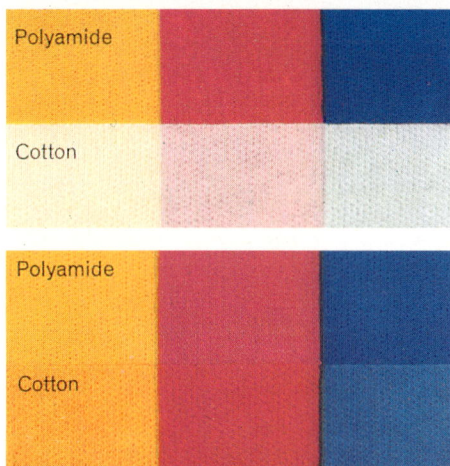

One bath dyeing of polyamide/cotton using reactive dyes. In the acid phase (top), the polyamide dyes strongly, but the cotton takes up very little dye. In the following alkaline phase (lower), the cotton is dyed to a similar depth of shade.
(*Bayer Farben Revue* No. 20, 1971, by permission of the editors. *Bayer Farben Revue* is published by Bayer AG, Leverkusen, West Germany.)

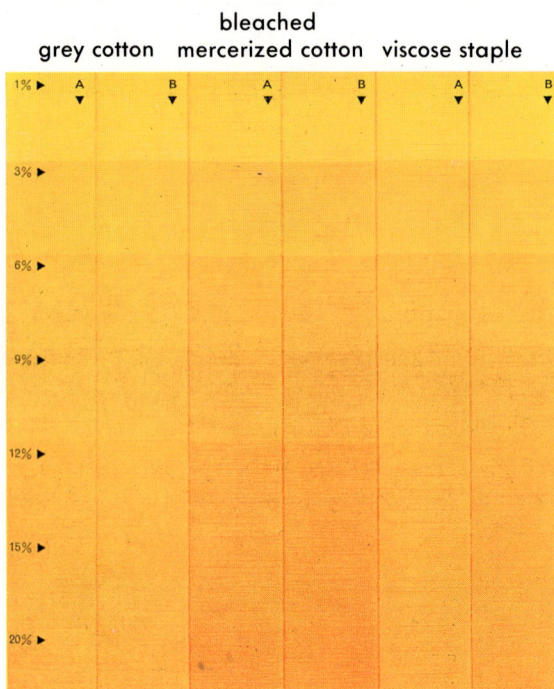

Effect of time on uptake of reactive dye. Percentage figures indicate dyebath concentration. A: treated with salt for 30 minutes and dyed with soda for 90 minutes at 50°C. B: treated with salt for 60 minutes and dyed with soda for 180 minutes at 50°C. Note the much greater dye uptake by mercerized cotton.
(*Bayer Farben Revue* Special Edition No. 12, Levafix Dyestuffs, by permission of the editors. *Bayer Farben Revue* is published by Bayer AG, Leverkusen, West Germany.)

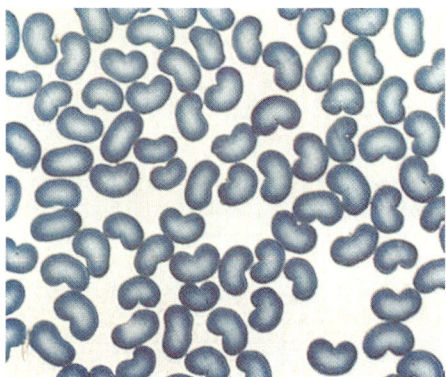

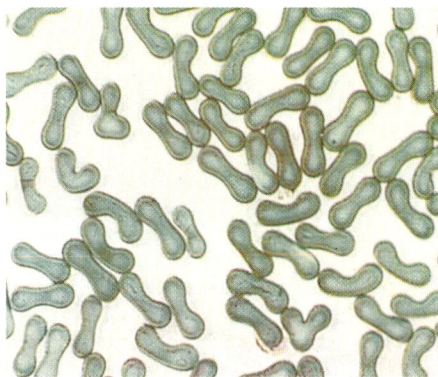

Poor dye penetration of Leacryl (left) and Dralon acrylic fibres (right), caused by excessive reduction in the steaming time.
(*Bayer Farben Revue* No. 15, 1969, by permission of the editors. *Bayer Farben Revue* is published by Bayer AG, Leverkusen, West Germany.)

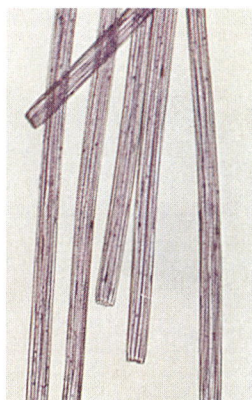

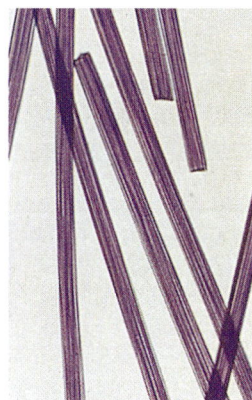

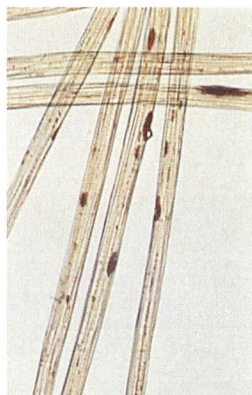

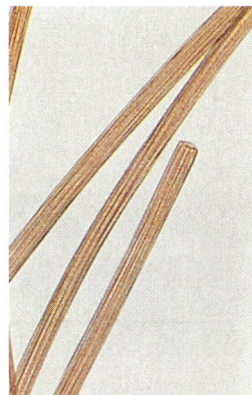

Effect of degree of dispersion of pigments on the colour yield in dope dyeing. The paler and darker dyeings have the same concentration of the pigment, but in the darker dyeings the pigments are more finely dispersed through the viscose.
(*Bayer Farben Revue* No. 15, 1969, by permission of the editors. *Bayer Farben Revue* is published by Bayer AG, Leverkusen, West Germany.)

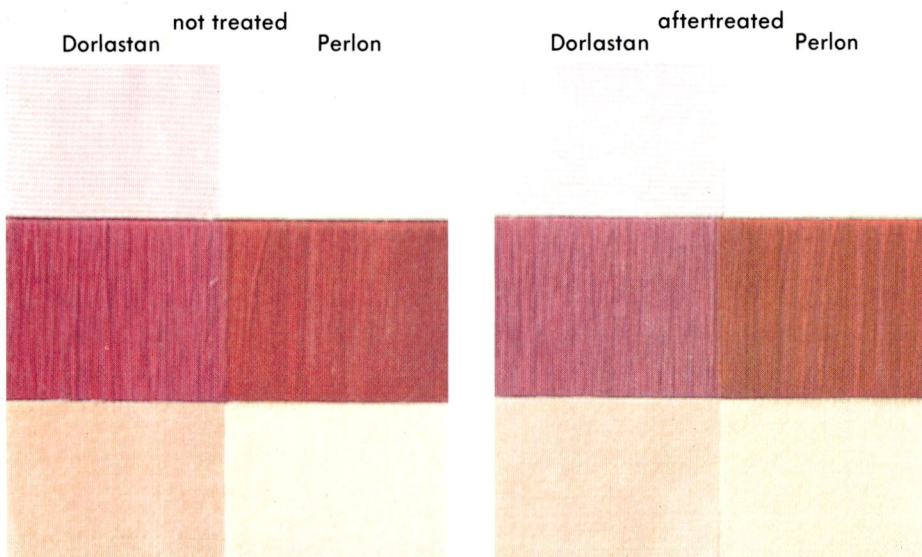

not treated

Dorlastan   Perlon

aftertreated

Dorlastan   Perlon

Many elastomerics are heat-sensitive, so heat cannot be used to fix dyes. The photographs compare the fastness to perspiration of Dorlastan polyurethane filaments and Perlon polyamide filaments, with and without aftertreatment with a binding agent. For the test, the printed filament yarns were sandwiched between knitted Perlon polyamide (top) and woven wool (lower).
(*Bayer Farben Revue* No. 12, 1967, by permission of the editors. *Bayer Farben Revue* is published by Bayer AG, Leverkusen, West Germany.)

wet   with 30 g/kg Perlit   dry   wet   without Perlit finish   dry

Finishes must be compatible with the dye and dyeing procedures used. These tests show the effect of the Perlit water-repellent finish on the rubbing fastness of a pigment.
(*Bayer Farben Revue* No. 24, 1975, by permission of the editors. *Bayer Farben Revue* is published by Bayer AG, Leverkusen, West Germany.)

Embroidered panel (Fire pond II)
Cream silk ground with gold kid applique,
embroidered with shades of brown, green,
white, and gold thread
Designed by Kathleen White
Scottish 1973
Purchased by National Gallery of Victoria, 1974
(National Gallery of Victoria)

Modern production techniques make a wide range of
effect yarns and finishes available to the textile artist.

Textile sample book
Block and roller printed barege, grenadine,
and merino fabrics in a variety of paisley floral patterns
Compiled by J. Claude Freres and Co., Paris
French 1856
Presented to the National Gallery of Victoria
by the Australian Wool Corporation, 1982
(National Gallery of Victoria)

Manufacturers must be able to communicate with their potential customers.

Getting the dye into the fibre

For all fibres, the basic process of dyeing is the same. The fibre is *immersed in a liquid* which contains the dye, and the dye molecules diffuse into the *amorphous* regions of the fibre.

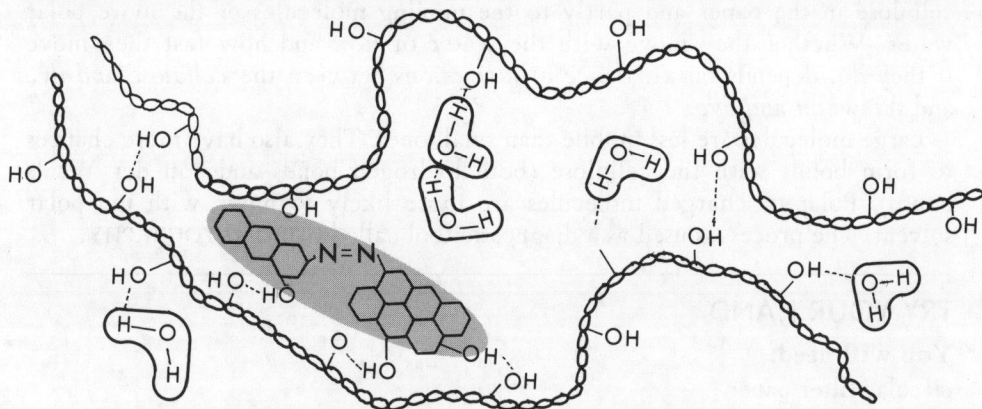

Dye molecule anchored in an amorphous region of a cellulosic fibre.

The cheapest carrier liquid is water, and so it is used wherever possible. Non-polar dyes used for non-polar fibres (disperse dyes) may be dissolved in some other solvent, or they may be applied as an *emulsion,* a suspension of tiny droplets of the solvent in water.

Solubility

With enough added polar groups, even large molecules can become soluble in water. Sulfonic acid (usually as the sodium salt, $-SO_3Na$) is the most commonly used solubilizing group for dyes.

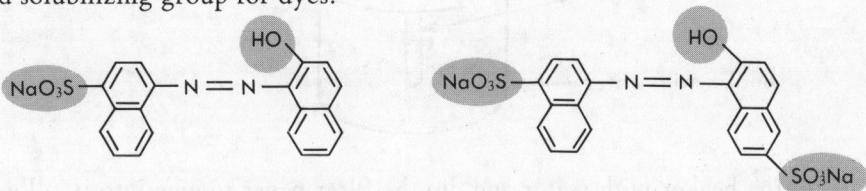

C.I. Acid Red 88 (Naphthalene Red JS) and C.I. Acid Red 13 (Naphthalene Red EAS). The dye molecule with the extra sulfonate group would be more soluble in water. The two dyes have the same chromophore (the azo-coupled naphthalene system), and auxochrome (the hydroxyl group), so have very similar spectra. (See p. 263.)

When choosing – or creating – dyes, it is often useful if they are water soluble; but it is also important to have a balance between the attraction the dye has for the molecules of the fibre, and the attraction it has for the molecules of water. If a dye is very soluble in water, it can be difficult to make it stay inside the fibres. As soon as· the dyed fabric is placed in water, the soluble dye molecules leave the fibres and colour the water. The dye is then described as having poor wash fastness.

Affinity of dye for fibre

Different dye molecules have different affinities for different materials. If a strip of filter paper is placed in water, the water will creep up along the paper strip by capillary action. If there are dye molecules on the paper, as the water flows past them they will be attracted partly to the stationary molecules of cellulose in the paper and partly to the moving molecules of the more polar water. Whether they move with the water or not, and how fast they move if they do, depends on a *balance of attractions* between the *cellulose and dye* and the *water and dye*.

Large molecules are less mobile than small ones. They also have more chances to form bonds with the cellulose (both hydrogen bonds and van der Waals bonds). Polar or charged molecules are more likely to move with the polar solvent. The process is used as a diagnostic tool called CHROMATOGRAPHY.

TRY YOUR HAND

You will need:
circular filter paper
scissors
dropper
beaker
small amount of Shirlastain A

What you do:
Place a drop of the stain at the centre of the filter paper and allow it to dry.
Cut a tongue on the filter paper and bend it as shown.

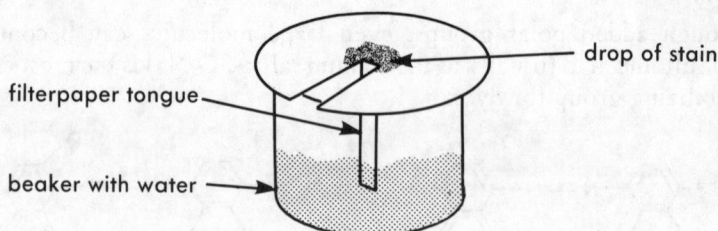

Half fill the beaker with water and dip the filter paper tongue into it, allowing the other part of the filter paper circle to rest on the rim of the beaker.
Leave for about 30 minutes, observe, then leave it for another hour.

1　What colour was the stain originally?
2　How many different dyes are used in Shirlastain A?
3　Which dye moves fastest? What do you think are the properties of this dye?
4　Which dye moves slowest? What do you think are the reasons for its slow motion?
5　Which dye would you expect to stain cellulosic fibres?
　　Check your ideas against the results of the staining tests in Chapter 6.
6　Why do you think that none of the dyes used in Shirlastain A is able to stain polyester or acrylics?

7 What type of dye would you expect to have an affinity for polyester? Would that dye be soluble in water? Why?

8 Would you expect one of the dyes used in Shirlastain A to give washfast dyeing on cotton fabrics? Which one? Why?

The dynamics of sorption

The dye molecules need to leave the solvent and *diffuse* into the amorphous regions of the fibre.

This process of diffusion takes some time. The time depends on the interactions between the dye, the fibre, and the liquid medium. Diffusion occurs through repeated sequences of *adsorption* and *desorption* of the dye on both external and internal fibre surfaces. Ideally, in a successful dyeing, all the dye from the dyebath is absorbed into the fibre.

At the begining of the dyeing process the dyebath is full of coloured molecules jostling against one another. Eventually one of them is knocked against the fibre surface. The fibre has some attractive forces for this molecule – polar sites for hydrogen bonding or charged sites for ionic bonding, or just masses of carbons for van der Waals bonding. In any case, the dye molecule is hooked – it is ADSORBED onto the surface of the fibre.

Dye in solution Dye adsorbed on fibre surface Dye absorbed into fibre

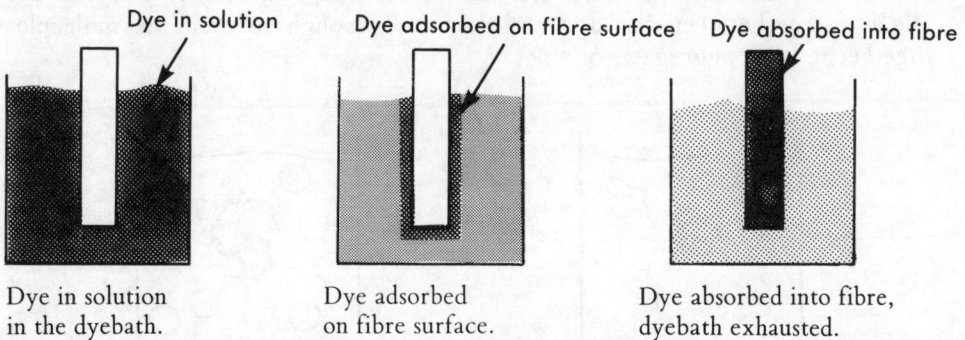

Dye in solution
in the dyebath.

Dye adsorbed
on fibre surface.

Dye absorbed into fibre,
dyebath exhausted.

Unless dye is anchored within the fibre, this is a reversible process.

If the forces between dye and fibre are very strong, then the dye molecule will stay put. If the forces are weak, then a few molecules of water – or another dye molecule – may knock the adsorbed dye out of position.

Very strong attractive forces between dye and fibre can create *streaky, unlevel dyeings* unless the dye liquor is circulated exceptionally evenly to all parts of the fabric to be dyed. If the dye molecules can adsorb and desorb, there is a greater chance that they will – at the end of the dyeing – be evenly distributed throughout the fabric. Such an effect is termed LEVEL DYEING.

If the dye molecules are *charged* ions – as indeed all water-soluble dyes are – then it is possible to *force them out of solution* by adding *salt* to the dyebath. The water molecules are attracted more strongly to the charged ions from the salt than they are to the bulky dye molecules. This makes it harder for the dye molecules to desorb from the surface of the fibre. In this way, the addition of salt helps to exhaust all the dye from the dyebath onto the fibre.

TRY YOUR HAND

You will need:

a solution of salt-sensitive dye, such as Direct Orange 26
2 large test-tubes
salt

What you do:

Place 10 mL of the dye solution in each of the test-tubes.

Add some salt to one of the test-tubes, shake the tube, and observe.

The untreated solution is dark but clear. What happened after the addition of salt?

Did the solution go cloudy?

What happened after an hour?

Did all the dye molecules form into large flocks and leave a clear solution?

Different dye molecules have different sensitivities to salt.

As the water molecules are now more attracted to the salt ions than to the dye molecules, the attraction of dye for water is no longer strong enough to overcome the attraction of dye molecule for dye molecule. The van der Waals forces between the dye molecules are enough to hold the molecules together in large aggregates or flocs.

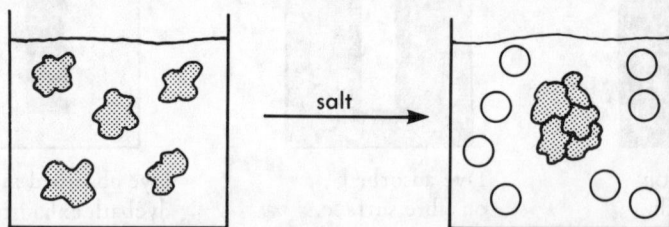

Dye in solution: the attraction of the water molecules for the dye is strong enough to overcome the attraction of the dye molecules for each other. When salt is added, the polarity of the water changes – it becomes more polar – and the attraction of the water molecules for each other is stronger than their attraction for the dye. The dye is therefore forced out of solution and forms large flocs.

TRY YOUR HAND

Use the solutions from the previous experiment. Place a strip of filter paper in each solution. Leave for 5 minutes, then remove. Which do you think will give a better dyeing – with salt or without? Why?

Now place the filter paper in the untreated solution, leave for 3 minutes then add salt. Leave another 2 minutes. Compare the three dyed filter paper strips.

Which is the lightest? The darkest? Why?

Would you recommend adding salt to the dyebath before the fabric is added? When is the best time to add salt? Why?

Penetrating the fibre

Diffusion is the process by which the dye molecules move from the surface of the fibre and are distributed throughout the fibre. The dyeing sites – polar, non-polar or ionic – available on the outside surface of the fibre are also available inside it, since the fibre is made up of the same molecules throughout.

Inside the fibre the sites which can be reached by dye molecules lie within *amorphous regions*. Amorphous regions have enough voids – spaces between fibre molecules – to accommodate dye molecules.

Dye molecules which are adsorbed on the outside of the fibre may break away (desorb) because of vibration due to heat or because of being bumped by another molecule. The inside of the fibre is empty of dye molecules – the dyebath and the outside of the fibre are full of them. This CONCENTRATION GRADIENT creates a push for the loose dye molecule to move further into the fibre rather than out into the dyebath.

Higher temperatures create more movement and more vibration among the molecules. This movement allows the molecules of the fibre to move apart – it creates temporary *voids* within the fibre. This increased amorphousness helps the penetration of the dye into the fibre. On the other hand, higher temperatures also increase the rate of bond breakage between dye and fibre molecules. This allows dye molecules to desorb, and, having greater vibrational energy, to move completely away from the fibre.

The *increased space* in the swollen fibre and the *increased rate of desorption* are finely balanced against one another to allow a certain proportion of the dye molecules to be absorbed into the fibre. The following graph shows the effect of increased temperature on the absorption of two different dyes by cotton fibres.

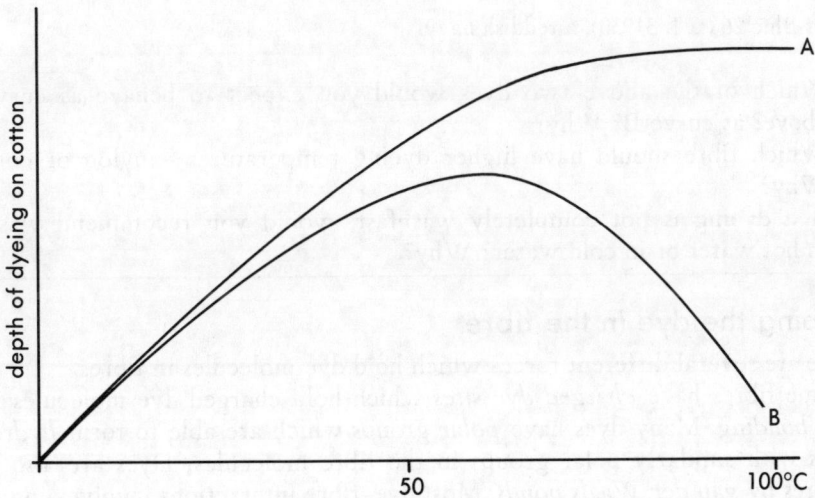

The effect of temperature on depth of dyeing on cotton with two dyes, A and B.

TRY YOUR HAND

1 Why do you think these two dyes react differently to increasing dyeing temperatures?
2 Which of the two do you think has stronger bonds with the cellulose of cotton?

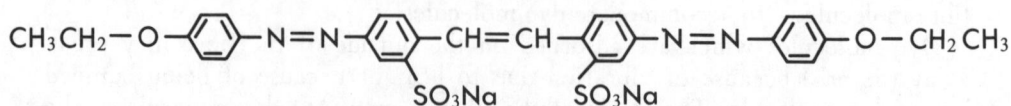

$$CH_3CH_2-O-\!\!\bigcirc\!\!-N=N-\!\!\bigcirc\!\!-CH=CH-\!\!\bigcirc\!\!-N=N-\!\!\bigcirc\!\!-O-CH_2CH_3$$
$$SO_3Na \qquad SO_3Na$$

Direct Yellow 12.

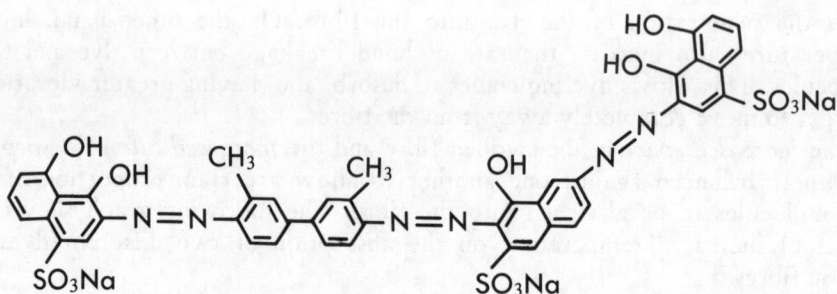

Direct Blue 26 (C.I. 31930), a reddish navy.

3 Which of the above two dyes would you expect to behave as curve A above? as curve B? Why?
4 Which fibre should have higher dyeing temperatures – nylon or cotton? Why?
5 If a dyeing is not completely washfast, would you recommend washing in hot water or in cold water? Why?

Keeping the dye in the fibre

There are several different forces which hold dye molecules in fibres.

Some fibres have *charged dye sites* which hold charged dye molecules with *ionic bonding*. Many dyes have *polar groups* which are able to form *hydrogen bonds* with similarly polar groups in the fibre molecules. Dyes are also held in fibres by *van der Waals bonds*. Most dye–fibre interactions involve a number of the above bonding systems.

Small molecules of dye move into the fibre easily – they may also move out quite readily. A number of different dyeing systems introduce small molecules which join up to make the large final dye molecule in the fibre itself. Such large molecules are then trapped in the fibre. Examples are azoic dyes and mordant or metallized dyes.

Occasionally, the solubility of the dye in water is changed after absorption into the fibre – as in vat dyes.

Other, more soluble dyes can be made to align themselves parallel to the molecules of the fibre, and so are able, through more intimate contact, to form more and stronger bonds with the fibre. Such an effect is achieved during *soaping* of vat dyes and azoic dyes.

Types of dye

Direct dyes

The dyes

DIRECT DYES are *long, flat, ribbon-like molecules* – some small, some larger. A typical direct dye is C.I. Direct Blue 81 (C.I. 34215), which was first synthesized in 1907.

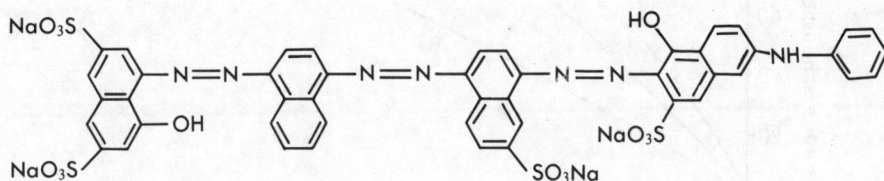

C.I. Direct Blue 81 (C.I. 34215).

Many of the dyes found in natural colouring matter (onion skins, lichens, flowers) are direct dyes. All direct dyes have enough charged groups ($-SO_3^-$: sulfonic acid) to carry the hydrophobic parts of the large molecule into solution.

All have special affinity for *cellulosic fibres,* because they can align effectively with the ribbon-like shapes of the cellulose molecules. In the course of dyeing, layers of dye molecules are built up over one another to form an aggregate group. This is less easily desorbed than a single molecule during later wet treatments.

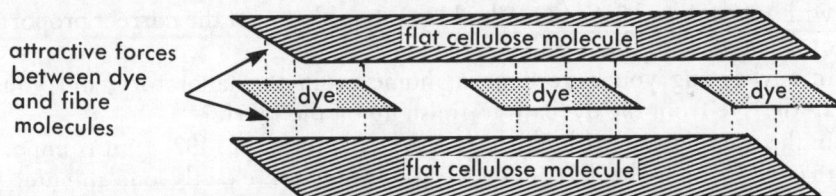

The flat dye molecules are able to align with the flat cellulose molecules, forming many van der Waals bonds along their length which hold them in place.

Direct dyes fall into various categories, depending on their size and shape. Some are able to move around the dyebath rapidly and to adsorb and desorb easily from the fibre. Such dyes distribute themselves evenly around the fabric to be dyed, and so are called *level* dyeing dyes.

Some dyes are strongly attached as soon as they are adsorbed onto the fibre. Because of the strength of their bonding, they will not desorb and distribute themselves around the fabric during dyeing. Therefore, unless precautions are taken, the dyeing will be streaky or *unlevel*.

Effect of time

TRY YOUR HAND

1   Which of the two dyes A and B would you expect to give a more level dyeing?

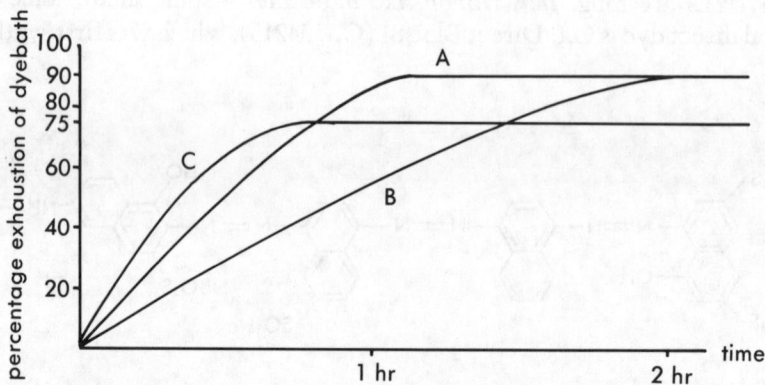

Exhaustion of dyebath against time, for three dyes, A, B, and C.

2   Which would be streaky? Why?
3   Is all of the dye from the dyebath exhausted – adsorbed – onto the fibres?
4   Why do both the A and B dyeing curves level out?
5   What is the difference between the behaviour of dyes A and B and the behaviour of dye C?
6   What could be a reason for dye C not exhausting onto the fibre as completely as A and B?
7   Would you expect to be able to observe the results of such difference? How?
8   You are a dyer and calculate that the correct shade of green is best obtained with 1% of A and 2% of C. Would you be able to get the correct proportions of A and C onto the fibre? Why?
9   In any dyeing you have done at home or in the laboratory, did you get all the dye from the dyebath to finish up on the fabric?
10 In the staining test with Shirlastain A (Chapter 6, pp. 192–3), it is important that all samples be left in the dye solution for exactly one minute. Can you give reasons from the graph above as to why such timing could be important for reproducible results?

The effect of salt

When salt is dissolved in water, its positive and negative ions separate and are independently mobile. In other words, a glass of water containing a spoonful of salt has many Na^+ and Cl^- ions moving about in it.

The effect of salt on the *solubility* of dyes has already been discussed. When direct dyes are used on cellulosic fibres, salt has an additional important effect.

Cellulosic fibres develop a natural negative charge (ZETA POTENTIAL) when they are wetted. This charge is not very great, but it is enough to repel negatively charged ions. The dye molecules are negatively charged in water, because their solubilizing sulfonate groups ($-SO_3^-Na^+$) separate into $-SO_3^-$ and Na^+ in water. The more sulfonate groups the dye has, the more soluble it is in water, but the more it is repelled by the negatively charged fibre surface.

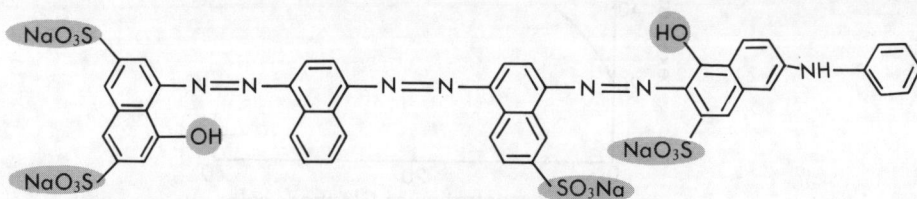

C.I. Direct Blue 81 (C.I.34215).

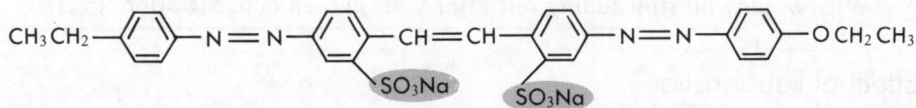

Direct Yellow 12.

For example, Direct Yellow 12 has only two $-SO_3^-$ groups, and can dye cotton quite effectively in water. Direct Blue 81, however, has four $-SO_3^-$ groups, and stains cellulose only lightly until salt is added to the dyebath. The many small Na^+ ions from the salt are attracted to the negative fibre surface and *neutralize* it. This allows the large dye molecule to approach. Once the

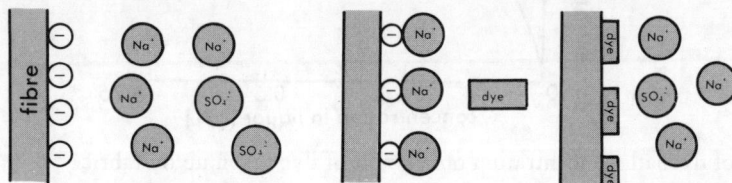

The small, highly mobile sodium ions from added salt are attracted to the negatively charged fibre surface. They thus neutralize the charge, allowing the large, negatively charged dye molecules to approach.

259

dye molecule is at the fibre surface, the van der Waals forces and polar attractions are strong enough to hold it there, and the dye molecule can be bonded successfully to the fibre molecules.

In industry, Glauber's salt, sodium sulfate (Na_2SO_4), is usually used in preference to common salt, NaCl. Glauber's salt delivers two sodium ions, and is also less corrosive to metals than is common salt.

TRY YOUR HAND

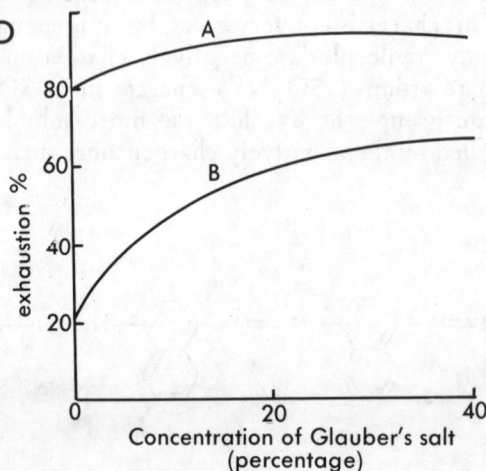

Effect of salt concentration on dyebath exhaustion for two dyes, A and B.

1 Which of the two dyes would you expect to have more sulfonic acid groups?
2 When would you say that enough salt has been added?
3 Why would you stop adding salt after your chosen concentration level?

Effect of liquor ratio

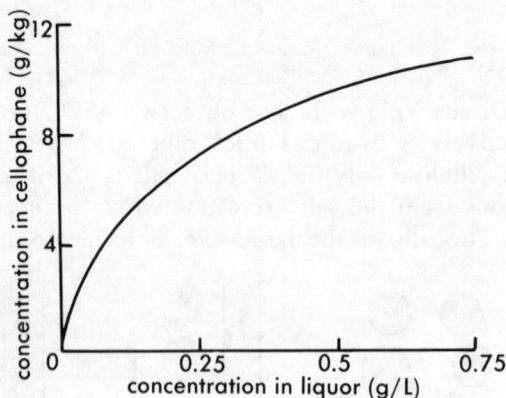

Effect of dyebath concentration on amount of dye taken up by fabric.

The above graph shows that if there are more molecules of dye in the dyebath initially, then more will – at the end of the dyeing – find their way into the fabric.

260

TRY YOUR HAND

You will need:

2.5 L beaker
250 mL beaker
concentrated solution of direct dye
2 identical samples of cotton fabric, each 5 g, labelled
2 test-tubes
heat source
stirrers

What you do:

Place equal amounts of the concentrated dye in each beaker, and fill the beakers with boiling water. Mark the level.

Add a wet fabric sample to each beaker.

Bring the beakers to the boil, and keep them stirred and simmering for 20 minutes. Keep the liquid level topped up to the mark.

After the 20 minutes, remove the samples.

Take 10 mL of the dye liquor remaining in the small beaker, and dilute it to 100 mL (i.e., ten times). Pour some of this diluted liquor into a test-tube, put some of the liquor from the large beaker into another test-tube, undiluted, and compare.

1　Which sample was dyed a darker shade?
2　Which dyebath had more dye left in it?
3　Can you give reasons for your findings?
4　Was one sample more uniformly coloured than the other? Why?
5　How important is stirring during dyeing?
6　Describe two ways in which the evenness of contact between fabric and dyebath can be improved.

Tie dyeing

Unlevel dyeing can be exploited to create decorative effects.

TRY YOUR HAND

You will need:

white cotton material that has not had a special finish applied to it: well-washed fabric, such as an old white pillowcase, is ideal
commercial hot-water dye, such as RIT or Dylon, or direct dye from a craft supplier
salt

What you do:

Fold and tie the fabric, using clothes pegs or rubber bands, as shown.

clip with clothes peg

tie with rubber bands

If dye is in powder form, mix it to a paste with cold water, and then add hot water to dissolve. Make up to full volume.

Wet the pillowcases before adding them to the dyebath.

Stir the fabric in the dyebath for 10 minutes.

First dissolve your salt in water separately. Add half the recommended amount of salt while stirring.

Stir gently for a further 10 minutes at the recommended temperature or at the boil.

Add second half of the salt solution.

Stir for further 20 minutes.

Cool, then rinse until only clear water comes out of the dyed fabric.

1 Did these instructions differ from those on the commercial packet? Can you give reasons for the differences?

2 Discuss what each of the instructions above achieves, and why they need to be followed in sequence.

Wash fastness

Direct dyes have an affinity for cellulosic fibres, but they are also soluble in water. Therefore, many direct dyes are only *moderately washfast*.

TRY YOUR HAND

You will need:

2 pieces of fabric dyed with direct dyes, each 5 cm × 3 cm (use the pieces from the previous exercises)

piece of white wool fabric
piece of white cotton fabric
beaker
soap solution
stirrer
source of heat
iron

What you do:

Keep one sample of dyed fabric as a reference.

Sandwich the other piece between the cotton and wool, as shown.

Place the fabric sandwich in the soap solution, and heat and stir.

Keep near boiling for 5 minutes.

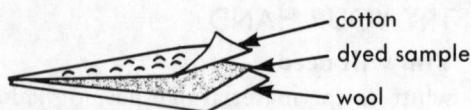
cotton
dyed sample
wool

Rinse thoroughly, blot dry between paper towels. Unpick the stitching and iron the sample, the cotton fabric, and the wool fabric until dry.

1 Compare the boiled sample with the reference sample. How much fading has occurred?

2 How much dye has been transferred to the wool fabric? to the white cotton fabric?

Acid dyes
The dyes

ACID DYE molecules are generally smaller than direct dye molecules. They are usually not such flat, ribbon-like molecules, and so they have less affinity for cellulosic fibres.

Like the direct dyes, acid dyes are coloured because of their system of conjugated alternating double and single bonds, and they are water soluble because they contain *sulfonic acid* groups.

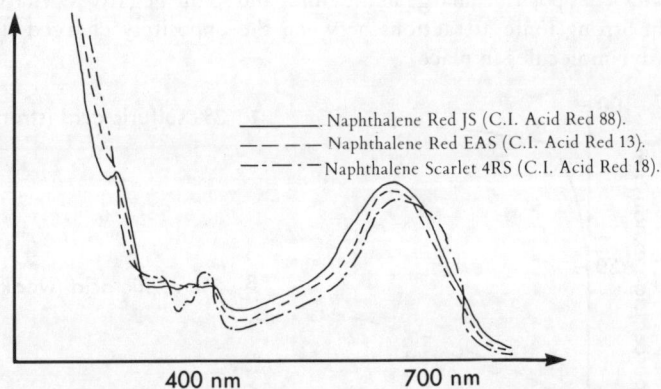

———— Naphthalene Red JS (C.I. Acid Red 88).
— — — — Naphthalene Red EAS (C.I. Acid Red 13).
— · — · — Naphthalene Scarlet 4RS (C.I. Acid Red 18).

The absorption spectra of these related dyes show the effect on the spectra of increasing the number of sulfonic acid groups.

Because these dyes are the sodium salts of sulfonic acids, they separate into ions in solution. The dye molecules carry one or more negative charges. If the number of sulfonic acid groups on a particular dye molecule is increased, the new dye molecules will be more soluble, and have a slightly different colour.

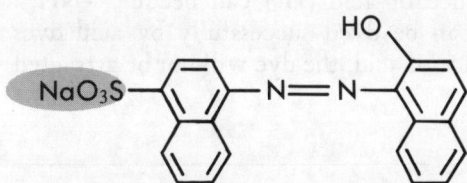

Naphthalene Red JS (C.I. Acid Red 88).

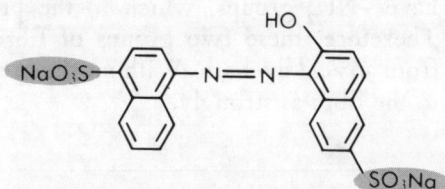

Naphthalene Red EAS (C.I. Acid Red 13).

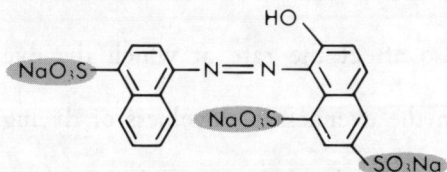

Naphthalene Scarlet 4RS (C.I. Acid Red 18).

263

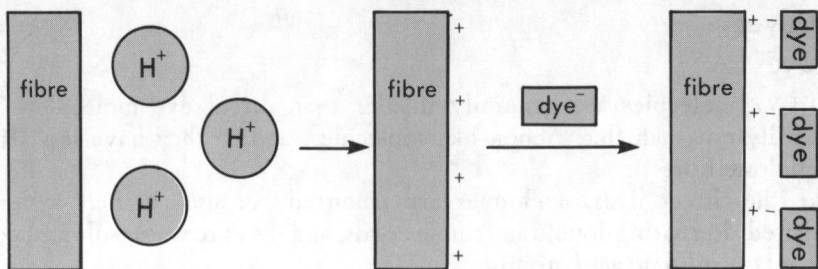

Hydrogen ions from the added acid are attracted to the basic groups in the fibre molecule. This creates a net positive charge in the fibre, and so the negatively charged dye molecules approach. Strong ionic attractions between the oppositely charged fibre and dye then hold the dye molecules in place.

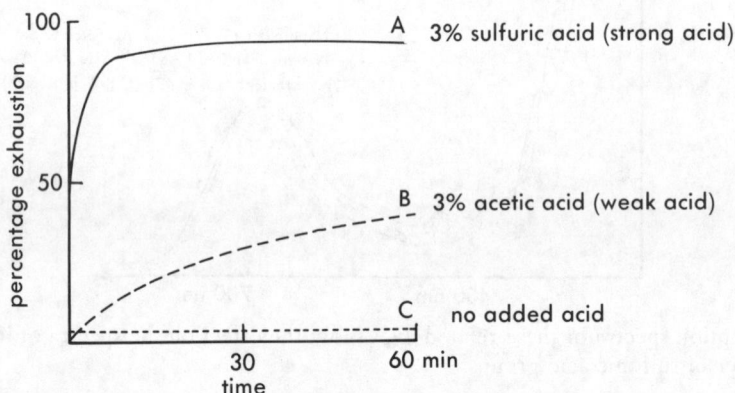

Exhaustion of dyebath against time for three different concentrations of acid in the dyebath. (This particular example is Solway Blue B (C.I. Acid Blue 43).)

Acid dyes are applied only to those fibres which have internal sites that can accept a *positive charge* in acid solutions. Both *protein fibres* and *polyamides* have $-NH_2$ groups, which in the presence of acid (H^+) can become $-NH_3^+$. Therefore, these two groups of fibres can be dyed successfully by acid dyes from an acid dyebath. Without the presence of acid, the dye will not be attracted to the fibre as strongly.

TRY YOUR HAND

1 Why was there no dye adsorbed without acid in the dyebath?
2 Why is there more dye adsorbed by the wool when there is more acid in the dyebath?
3 Does the amount of acid present also affect the rate at which the dye goes onto the fabric?
4 Would you expect any differences in the evenness or levelness of dyeing between curves A and B? Why?
5 Suggest a way in which you could promote level dyeing with acid dyes. Try your suggestion using test-tubes of dye solution and small fabric samples.

6 Would you expect all acid dyes to react in exactly the same way to the presence of acid?

7 What factors could influence the rate at which an acid dye is exhausted onto the fibre?

The effect of salt

It is preferable to have most of the dye exhausted from the dyebath onto the fibre. After all, no dyer wants to throw money down the drain! From the above graphs the *high exhaustion* of *acid solutions* can readily been seen. These conditions however also tend to produce *streaky, unlevel dyeings*.

In order to slow down the rate at which the *negatively charged dye molecules* are attracted to the *positive dye sites* on the wool, other negatively charged molecules are added to the solution. These negative ions – usually SO_4^{2-} from Glaubers salt (Na_2SO_4) – are small and therefore mobile. They reach the positive dye sites on the fibre before the slower moving, larger dye molecules. Since now there are fewer positive charges on the fibre, adsorption of the dye is slowed down. With slower adsorption, there is more time for the dye molecules to be evenly distributed around the fabric. The large, slow-moving dye molecule can knock a small sulfate ion from the salt away from the dye site, but is less easily dislodged in turn.

After the dye molecule is adsorbed, it is held in place with van der Waals bonds as well as by ionic attractions. Therefore, it will not be dislodged by other molecules. Since the charge on the dye molecule has been *cancelled* by its adsorption to the opposite charged site on the fibre, the dye loses its affinity to water. This cancellation of charge is also very important for holding acid dyes *inside* the fibre. The result is both *level* and *fast* dyeing.

Wash fastness

The final wash fastness of the dyeing depends on both the ionic and the van der Waals forces available to bond the dye to the fibre.

TRY YOUR HAND

1 Which of the dyes below would you expect to have better wash fastness? Why?

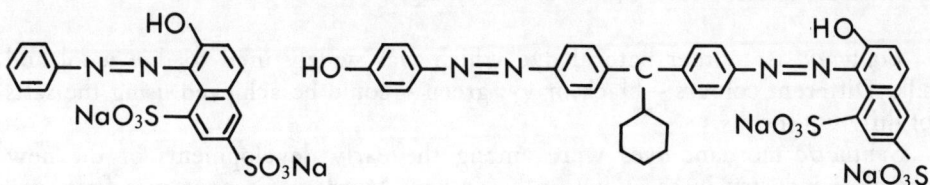

A: Acid Orange 10 (C.I. 16230).   B: Acid Orange 33 (C.I. 24780).

2   Soap solutions are alkaline. What happens if an acid-dyed wool or silk garment is washed at high temperatures in soapy water?

3   What kind of detergent is preferable for washing acid-dyed garments? .

4   You have washed a white cotton pillowcase with a silk piece dyed with acid dyes. The pillowcase was stained in the wash. Which of the two dyes A and B could be more easily washed out from the cotton (other than by stripping or bleaching)? Why?

5   Would you expect any acid dye to have affinity for cotton? Which ones? Why?

Those dyes which can form strong bonds with polymers – particularly with wool – are called ACID MILLING DYES. They can be used successfully on woollens which are *milled* during finishing. The milling or fulling process consists of pounding the woollen fabric in warm soapy water, in order to consolidate the fabric surface. Only large planar (flat) acid dyes are suitable for such treatment, as their shape allows many van der Waals bonds to form.

Mordant dyes

After-treatments

In the days before synthetic dyes were invented, it was found that if certain metallic salts were added to wool, the dyeings became very washfast. The metal salts were called MORDANTS from the French word *mordre,* 'to bite'. It was thought that these salts helped the dyes to bite onto the wool and so to hold fast during washing. *Madder* was a red mordant dye of great importance – the uniforms of British soldiers were dyed with it!

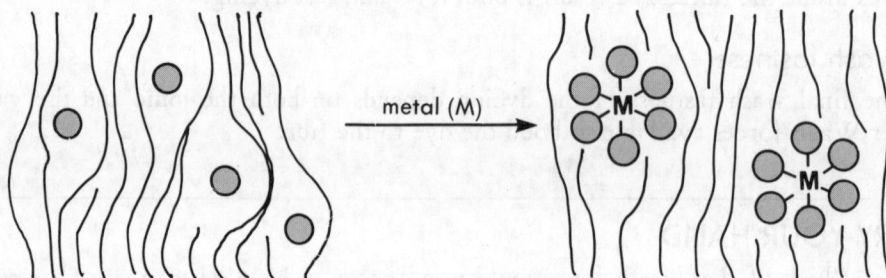

The small dye molecules pass freely in and out of the fibre. However, if certain metals are added, the dye molecules will bond to the metal to form a much larger molecule. If this occurs inside the fibre, the metal complex formed will be trapped inside the amorphous regions, resulting in a washfast dyeing.

Logwood is another natural dye which was widely used to dye wool and silk. Different colours – black, navy, green – could be achieved using the salts of different metals.

Synthetic mordant dyes were among the early developments of the new chemical industry in the nineteenth century. Mordant Green 4 was first synthesized in 1875. It is a *brownish olive* when treated with *chromium,* and a *yellowish green* when treated with *iron.*

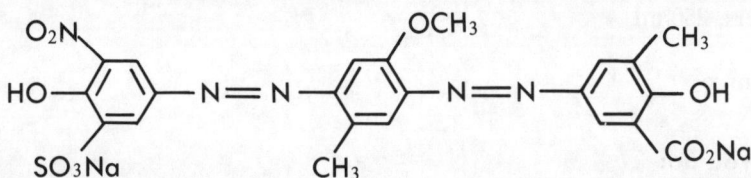

Mordant Red 80 (C.I. 26565), developed in 1899, is orange-brown and soluble in water. When wool dyed with Mordant Red 80 is treated with chromium acetate, a washfast dull red dyeing is achieved.

On its own, the small dye molecule does not have much attraction for the molecules of wool. When two dye molecules are *joined* by a co-ordinating metal atom, the increase in size makes the dye molecule large enough to be trapped in the fibre.

Premetallized dyes

Some dyes are prepared by manufacturers in such a way that the metal ions are already included in the dye molecule. Such dyes were first produced in 1915, and refined in 1955. An example is Acid Violet 78. This dye has a chromium atom which is able to accept spare electrons from the *carboxyl group* of wool protein.

This binds the dye directly to the fibre. The result is more washfast than normal mordant dyes, which merely aggregate within the fibre.

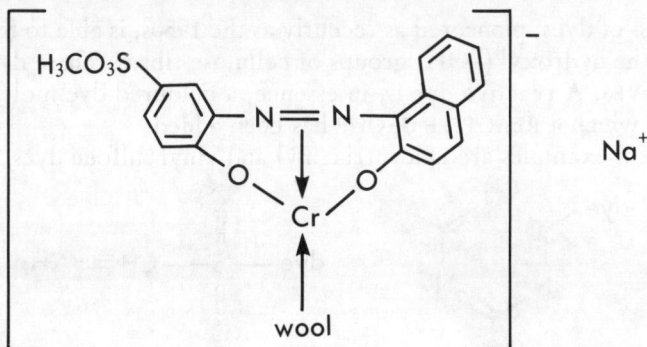

Acid Violet 78 (Irgalan Brown Violet DL). Because the chromium atom bonds directly to the wool, this dye has excellent wash fastness.

Dyeing with natural dyes

TRY YOUR HAND

You will need:

some onion skins, berries, flowers, or other source of natural dye, boiled to extract the colouring matter
white wool yarn – enough to use later to knit a scarf or other article
4 different mordants – ferrous sulfate (iron), potassium bichromate (chromium), copper sulfate (copper), and alum (aluminium) are all available from chemists

4 beakers, 250 mL
pot
heat source
stirrer

What you do:

Make a solution of each mordant by placing 2 heaped teaspoonfuls in a beaker and adding water to dissolve.

Wind the wool into a hank, divide it into four parts, and wet it thoroughly.

Boil each segment of the hank in a different mordant for 5 minutes.

Label each section with a tag.

Now dye the mordanted wool by heating it in the dye solution for 20 minutes.

Rinse, squeeze, dry.

1   Were all segments of the hank dyed the same colour?
2   What were the effects of the different mordants?
3   Choose the mordant which pleased you the most, and experiment with it as an after-treatment. First dye some fresh wool for 20 minutes, then add a solution of the mordant to the dyebath. Heat for 5 minutes. Does this give the same result as the pre-treated wool?
4   Test the wash fastness of the two methods of linking the dye molecules with the metal atoms. Which is more effective? more convenient?

Reactive dyes

The dyes

Another class of dyes, pioneered as recently as the 1950s, is able to form *covalent bonds* with the hydroxyl (–OH) groups of cellulose fibres. These dyes are called REACTIVE DYES. A reactive dye is, in essence, a coloured dye molecule, soluble in water, to which a REACTIVE GROUP has been added.

Two typical examples are chlorotriazinyl and vinyl sulfone dyes.

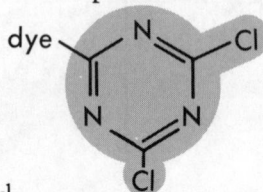

Chlorotriazinyl.

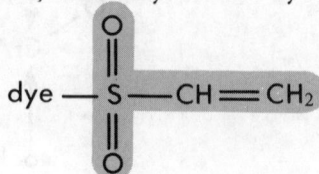

Vinyl sulfone.

The reactions with the –OH group of cellulose are:

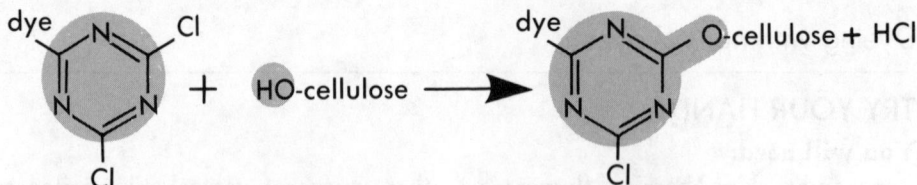

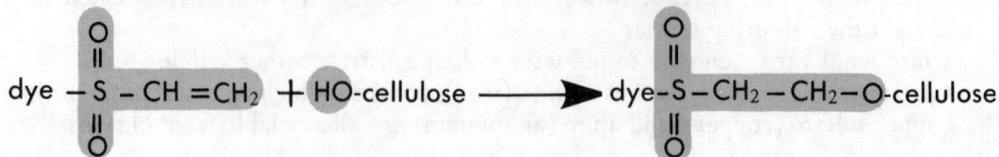

268

Depending on the reactivity of the dye, it may be applied under hot or cold conditions.

Cold-water reactive dyes can be used successfully with a variety of craft dyeing methods. Although they are expensive, they are also favoured by commercial dyers because of the high wash fastness of the results, and the energy savings from the low-temperature dyeing. They can also be applied with a pad-steam process, or by foam dying, or used in printing pastes.

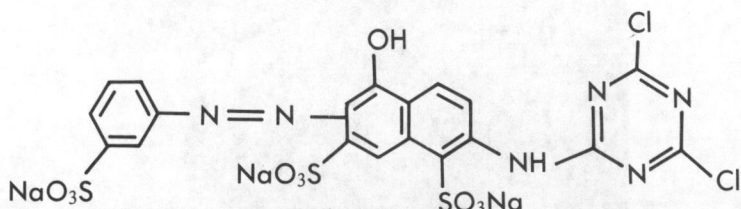

C.I. Reactive Red 17965, a dull bluish red developed in 1966. It dyes nylon in acid conditions and cotton in alkaline conditions, and so is suitable for blends of these fibres.

Batik

TRY YOUR HAND

This project will involve you in batik making. A RESIST pattern is applied with melted bees or paraffin wax to a white fabric. (Caution: do not overheat the wax when you are melting it.) The areas left unprotected are dyed with a pale shade of a reactive dye. Where the pale shade is needed in the design, more wax can be applied after the first dyeing has dried. The second waxing is then followed by a second dyeing. Sometimes a third waxing is applied to protect the second, darker colour, and then black or other dark contrast is

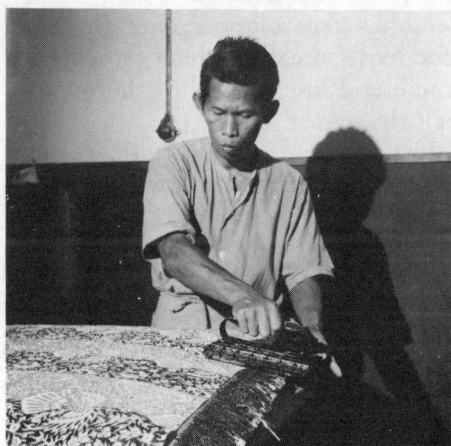

Traditional batik manufacture, Java. The wax may be applied by a metal stamp (called a cap), or carefully poured on the cloth from a canting tool.
(Photographic Archives of the Royal Tropical Institute, Amsterdam)

269

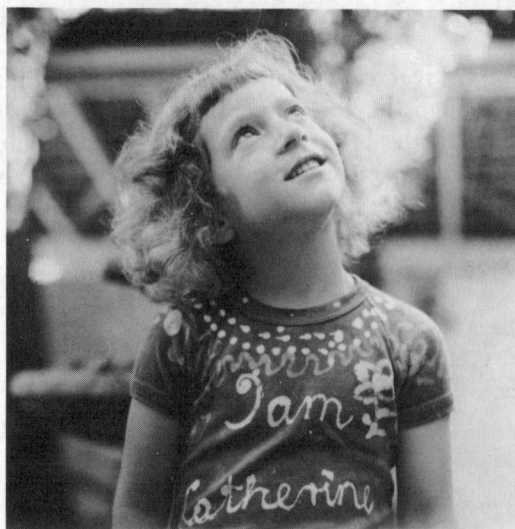

(Anne Fritz)

used for a finish. The cooled wax may be cracked to achieve fine lines throughout the design. (This effect is frowned upon by traditional Indonesian artists.)

You can make a simple T-shirt with your name on it. Choose a fabric with a high content of cotton or nylon. Polyester will not dye with batik dyes. Use a canting or a brush to apply the wax to the dry cloth, then dye with cold water dye.

In order to produce effective resistance to the dye, you must make sure the wax is hot enough to 'wet' the fabric and not just sit on the surface. The cold-water dyeings should be carried out in the following way:

First mix the dye into a paste and then dissolve in water. Make up to a dyebath with about 25:1 liquor ratio.

Wet the fabric and place in the dyebath. Stir for 10 minutes.

Add salt to the dyebath gradually, preferably from solution, over the next 10 minutes, while stirring. Now add the fixative – usually sodium bicarbonate or soda ash – an alkali. Stir for a further 20 minutes. Rinse your fabric thoroughly, and dry. Take care during drying not to melt the wax!

When you have finished dyeing your T-shirt to the desired pattern, remove the wax by boiling the shirt in water with a little detergent. It will probably need to be boiled several times to remove all the wax.

1   Why do you think that salt had to be added to the dyebath even though the dye is reactive?
2   Why was the salt added gradually?
3   Reactive dyes are coloured dye molecules with a reactive group added on. This reactive group can be readily broken off the dye; hydrolysis is speeded up by high temperatures or acid or alkaline conditions.

dye-NH-reactive group $\longrightarrow$ dye-NH$_2$ + HO-reactive group

Why was the fixative not added early during the dyeing process?

4   What would happen if you added the dye, the salt and the fixative to the dyebath all at once, and only then added the fabric?
5   Check the instructions on a packet of commercial cold-water dye. Why do you think they printed the instructions the way they did? Is your method the same? Which would be more effective? Why?
6   Why is reactive dyeing particularly suited for batik work? What other craft techniques would you use it for?

Disperse dyes

The dyes

DISPERSE DYES are coloured molecules which do not have strongly polar groups. Therefore they are only *sparingly soluble* in water.

Since they are *hydrophobic,* van der Waals forces are important in their bonding to fibres. They are compatible with the *hydrophobic fibres,* such as nylons, polyesters, acetates, and acrylics. Because they are not soluble in water, they are reasonably *washfast.*

In order for disperse dyeings to be carried out in aqueous dyebaths, the hydrophobic dyes are *dispersed* into a fine suspension with the help of detergents.

Typical disperse dyes are:

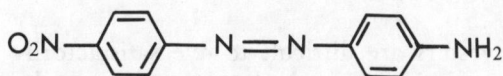

C.I. Disperse Orange 3

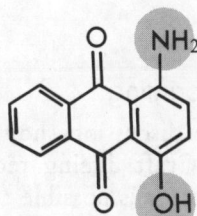

C.I. Disperse Red 15

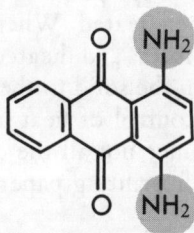

C.I. Disperse Violet 1

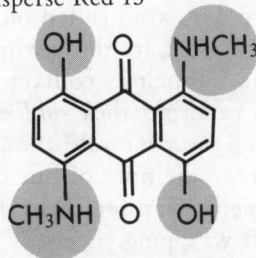

C.I. Disperse Blue 26

Note that the red, violet, and blue have the same chromophore (an anthroquinone), and differ only in their auxochromes.

Carriers

Disperse dyes have a natural affinity for hydrophobic fibres. They are readily attracted to cellulose acetates, polyamides, polyesters, polyolefins, acrylics. Cellulose acetates and polyamides are often dyed with disperse dyes. However, the molecular structure of polyesters and polyolefins is so densely packed and crystalline that these fibres cannot be successfully coloured under normal dyeing conditions.

It is necessary to *create space* for the dye molecules between the polymer chains. This can be done by:

(i) Heating the dyebath and fibres to about 130°C, under pressure. At this *high temperature,* the molecules of polyester move and vibrate sufficiently to create voids within the fibres which are large enough to accommodate the dye molecules.

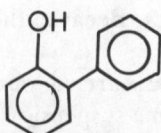

(ii) Adding *swelling agents* to the dyebaths. These are *small molecules* which can *move freely* among the chains, pushing them apart and creating spaces which become large enough for the dye molecules to enter. Such CARRIER molecules (diphenyl, benzoic acid, o-phenylphenol) allow dyeing to be more effectively carried out, faster, and at lower temperature.

Disadvantages of using carriers include their high cost and unpleasant smell. They are, however, remarkably effective.

Some commonly used carriers.

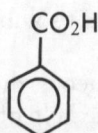

benzoic acid   o-phenylphenol   diphenyl

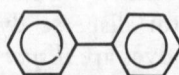

Transfer printing

The above discussion shows why polyesters are difficult to dye satisfactorily at home. Craft dyeing techniques are designed for hydrophilic fibres only. Nevertheless it is possible to gain some firsthand experience with disperse dyes through transfer printing.

Disperse dyes are printed on paper with the usual paper printing techniques. The dyes selected are those which vaporize easily when heated. When the paper is brought into close contact with hydrophobic fibres and heated, the dye molecules vaporize (they *sublimate* – go directly from the solid to the gas phase) and one by one penetrate the fabric. With careful control of heat and timing, accurate reproductions of fine designs are possible. Since not all the dye printed on the paper is transferred to the fabric, used transfer printing papers are often sold as gift wrapping paper.

TRY YOUR HAND

You will need:

5 pieces of 20 cm × 20 cm heat transfer paper. Do not choose the kind with a rubberized design

5 samples of white fabric of different fibre contents: nylon, acetate, polyester, polyester/cotton blend, cotton; each 20 cm × 20 cm

a clothes press if possible, large enough to press all five samples simultaneously to ensure that treatments are identical for each (If the clothes press is not available, an iron, held stationary for a carefully timed period, could be used.)

What you do:

Place all five fabric samples on the press, and smooth them flat.

Cover each with a piece of transfer paper, pattern side down.

Set the heat on dry, at cotton setting, and press for 30 seconds.

Remove the papers and compare the results on the different fabrics.

1   Did all the samples accept the same amount of colour?

2   Which was most coloured? Which least?

3   Can you explain why each sample accepted the disperse dyes to the extent that it did?

4   Identify the two properties of fibres which together help in allowing the penetration of disperse dyes.

5   It is possible to print much finer and more detailed patterns onto paper – with the photogravure process – than onto fabric. Often, transfer prints are characterized by *fineness of detail*. Also, transfer printing is restricted to the surface of the fabric – the dyes do not penetrate into fibres that are not in contact with the printed paper.

Take a careful look at printed fabrics in a store, and try to identify those that were coloured by heat transfer printing.

TRY YOUR HAND

You can make your own transfers by using disperse dyes painted on typing or other good quality close-grained paper. Use a fountain pen for fine lines. Remember the pattern will be transferred as its mirror image.

Polyester/cotton blends work well.

If you transfer your pattern to a fabric which is already dyed a pastel colour, there will be some distortion in the colour of the transfer. This may not matter if you choose your colours carefully.

Vat dyes

The dyes

VAT DYES are *large, flat, insoluble molecules.* Because of their insolubility and large size they are *extremely washfast.* Because of their shape they have an *affinity for cellulosic fibres.* They are, however, rather more expensive than direct dyes.

The characteristic property of these insoluble dyes is that they can be made soluble by *reduction.* That is, they can be made to undergo a chemical reaction in the absence of air which changes them into a water-soluble form. This soluble LEUCO (white, or colourless) form is also a flat or linear molecule.

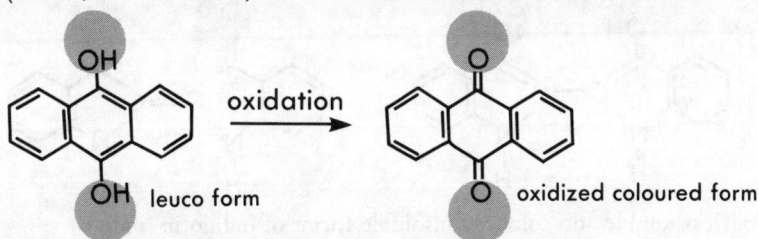

Anthraquinone: many vat dyes have an anthraquinone skeleton.

A woman in Narang, West Flores, Indonesia, dyeing cotton yarns with indigo.
(Mattiebelle Gittinger)

The soluble leuco form is applied to cotton from an alkaline reductive bath. When the alkali is neutralized with acetic acid, the oxygen in the air *oxidizes* the dye back to its insoluble form.

Indigo is such a vat dye. Traditional dyeing methods used fermentation to reduce the dye to its leuco form. Chemical reducing agents, such as sodium hydrosulfite (hydros), are used in industry.

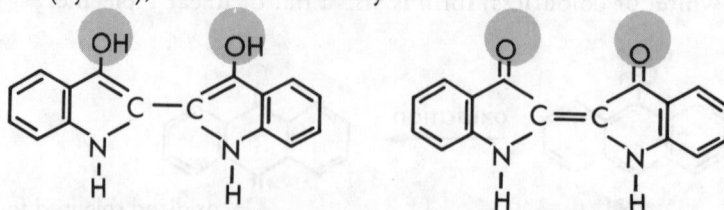

The colourless soluble and coloured insoluble forms of indigotin (indigo).

274

When the fabric is removed from the dyebath, some dye molecules are inside the fibres, and some are on the fibre surfaces. The oxidizing action of the air converts all of these dye molecules to their insoluble coloured form. The molecules inside the fibres are quite washfast, but those on the fabric surface can be rubbed off quite easily. *Soaping* – washing in near-boiling soapy water – is used to remove the excess dye molecules from the fabric surface. It also promotes the aggregation and orientation of the dye molecules within the fibres.

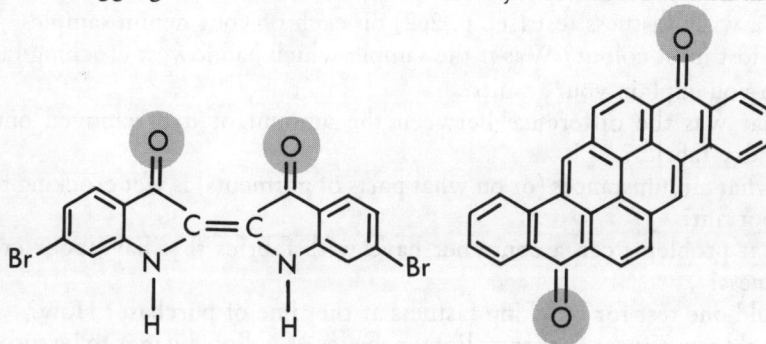

Tyrian Purple and C.I. Vat Orange 9 (Indanthrene Golden Orange G): vat dyes.

Light fastness

Vat dyes, when oxidized, are quite insoluble and therefore are extremely washfast. This, however, has no relationship to light fastness. Some vat dyes are sensitive to ultraviolet light, while other vat dyes have excellent light fastness.

Indigo is used for dyeing denim. The light fastness experiment in Chapter 7 explores this dye's sensitivity to light.

Crocking fastness

When insoluble dye molecules are adsorbed on the surface of fibres, they may not wash off readily, but they can be removed by rubbing the fabric against another abrading surface.

TRY YOUR HAND

You will need:
a variety of denim fabrics, labelled
twice as many pieces of white cotton fabric, 10 cm × 10 cm, also labelled
a cork

What you do:
Wrap white cotton fabric No. 1 over a cork as shown, and rub it firmly ten times across denim fabric No. 1.
Now wet another piece of white fabric and rub it across denim fabric No. 1 ten times, the same way.

Repeat for denim fabric No. 2, with dry and wet cottons Nos 2, and so on for all denim samples.

Display your results.

1 Do all the denims have the same crocking fastness?

2 Do you think the dye which was rubbed onto the white fabric will come out in the wash? Why?

3 Does crocking fastness correlate with wash fastness?
   Do a wash fastness test (see p. 262) on each of your denim samples. Which one lost most colour? Was it the sample which had lowest crocking fastness?
   Can you explain your results?

4 What was the difference between the amount of dye removed onto wet and dry fabrics?

5 In what circumstances (or on what parts of garments) is wet crocking fastness important?

6 What problems can a consumer have with fabrics that have poor crocking fastness?

7 Could one test for crocking fastness at the time of purchase? How?

8 Would you expect darker or lighter shades of indigo dyeings to be susceptible to poor crocking?

9 Can you give reasons for the light patches on these jeans?

Azoic dyes

AZOIC DYES are dyes which are formed *inside* fibres.

First a colourless component – a NAPHTHOL – is dyed onto the fibre. These small molecules are easily absorbed into hydrophilic fibres. The DIAZONIUM COUPLING component is then added. These two components combine inside the fibres to give the brightly coloured final dye. Different coupling components yield different colours with the same base.

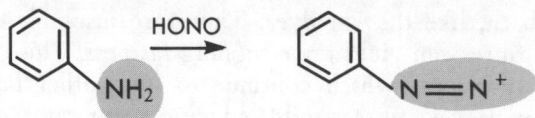

Formation of diazo ion from a diazonium coupling component.

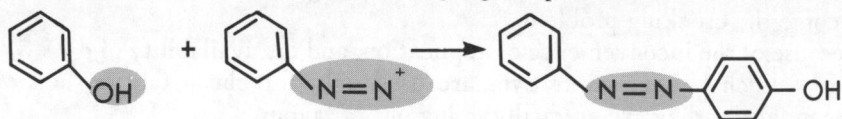

A colourless aromatic molecule (typically a naphthol, phenol or amine) is coupled with the diazo ion to form a coloured compound.

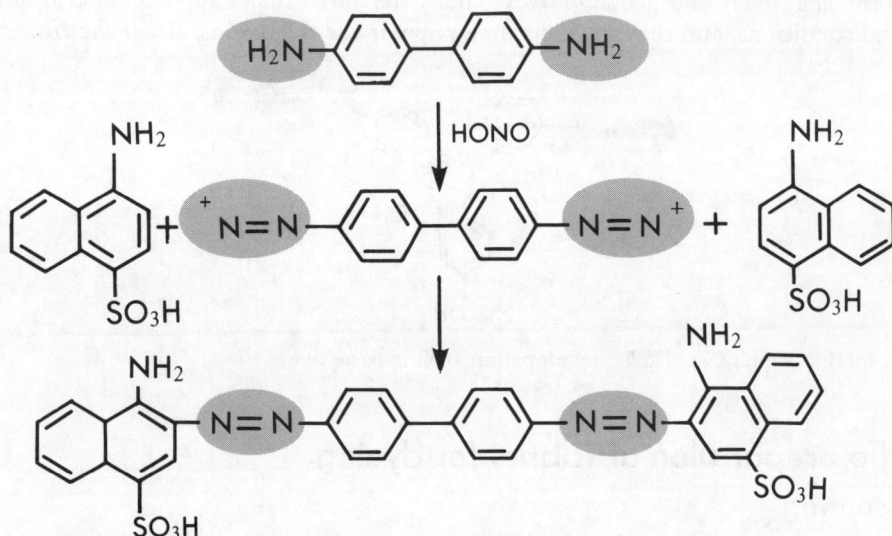

Formation of Congo Red, an azoic dye which is red in alkali and blue in acid.

Azoic dyeing takes place under cold conditions, to minimize other unwanted chemical reactions. Because the colourless components need to be uniformly distributed through the fabric, azoic dyeings can have *problems with levelness*. The dyeings are generally *bright* – especially reds and oranges – and wash-fast. Care must be taken to wash off any dyes formed on the fibre surface to prevent poor rubbing fastness. Azoic dyes are used on cottons and silks – particularly in batik dyeing. They are gradually being restricted to specialist dyeing applications.

Sulfur dyes

The range of SULFUR DYES is limited to dark, dull colours. Since these are *insoluble* molecules, the dyed fabrics have *excellent wash fastness*, but they have rather *poor light fastness*. Because of their *low cost*, they are still a source of dyeing navy blue and black shades on cellulosic fibres. Sulfur dyes are applied

277

from a reducing bath, like the vat dyes. Therefore insoluble oxidized dye on the surface of the fibres can yield *poor rubbing fastness*. This can be especially annoying in black garments which continue to stain other fabrics even after many washes. A quick wash with a mild reducing agent can remove the surplus dye without bleaching the fabric. Sulfur dyeings are usually *soaped* to help overcome the crocking problem.

Because of the inconvenience of application, and the availability of dyes which perform much better, sulfur dyes are used only for cheap fabrics, and even in cheap fabrics they are gradually going out of favour.

Basic dyes (cationic dyes)

CATIONIC DYES carry a *positive charge,* and so are named *basic* dyes in contrast to the *negatively* charged *acid dyes.* Today they are usually applied under mildly *acid* conditions, and they have become popular for the dyeing of *acrylic fibres.*

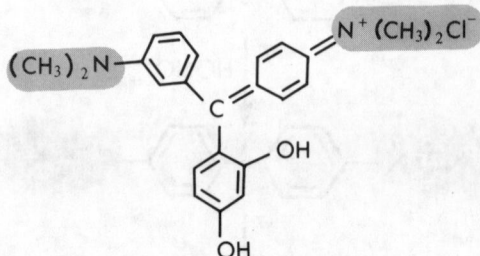

Resorcine Violet C.I. 43520, developed in 1883: a basic dye.

The preparation of fabrics for dyeing

Scouring

Oils are often used on fibres as lubricants, to help spinning efficiency. *Starches* and *waxes* are applied to warp yarns to protect them during weaving. These additives must be removed before the dyeing process, or the dyes may not penetrate the fabric evenly.

Most synthetic fabrics only need a light scouring – washing – process, which may be carried out in the dyeing machine just before the actual dyeing process. Others – cotton fabrics in particular – need extensive scouring, desizing, boiling and bleaching to remove impurities.

Scouring of *cottons* can be carried out in a *kier*. This is a large vat into which up to 4 tonne of fabric is packed in rope form. Water, detergent, alkali and bleach are then added, and the kier is allowed to boil (often pressurized to maintain 115°C) for 4 to 10 hours. Today scouring is more often done using high temperature processing or as a continuous process using large *J boxes.*

Wool is scoured in the fibre form, using non-ionic detergents to remove the natural fats and soil from the fibre. Therefore, before dyeing, only a mild washing and wetting out treatment is necessary to remove combing and spinning oils. Woollens with a heavy oil content may be scoured with alkali; this converts the fats to soap.

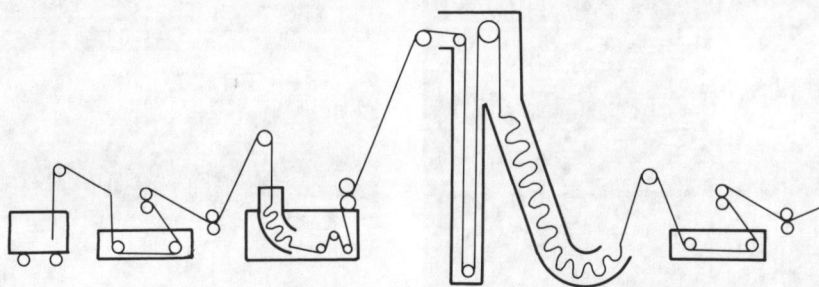

The J-box (so-called because of its shape) allows fabrics to be scoured at high temperatures in the absence of air. This is important as, under the harsh conditions, reaction with atmospheric oxygen could cause some breakdown in the fibre molecules. Continuous lengths of fabric can be scoured in this apparatus.

Bleaching

Cottons

The natural colour of the fabric needs to be removed before dye can be applied. Cotton fabrics may be bleached by either *chlorine* or *peroxide* bleaches.

The peroxide bleaches are hydrogen peroxide (H_2O_2), and sodium perborate, which is easier to handle and which liberates hydrogen peroxide in water. These are mild bleaches and safe to use on most fabrics.

Hypochlorite and sodium chlorite are also used as commercial bleaches. Hypochlorite causes greater degradation of cellulosic fibres than does chlorite, and to work effectively needs pretreatments to remove all oily impurities. Sodium chlorite and the peroxide bleaches offer the manufacturer savings in labour and energy, and are gentler on the fabrics.

Liquid bleach, which is usually a solution of sodium hypochlorite, and bleaching powder, which contains mainly calcium hypochlorite, are harsh bleaches. Care is needed in their home use, as they may destroy the resin finishes on cellulosic fabrics. Perborate, as a gentler action bleach, is recommended.

Wool

The mildest bleach used for whitening wool fabrics is hydrogen peroxide. Sodium perborate is also safe if not too alkaline, but chlorine bleaches must not be used under any circumstances, as they destroy the fibre.

The application of colour

Which stage of manufacture?

Textile materials can be coloured in different ways.

A white *fabric* may be printed with different pigments or dyed a solid colour. *Yarns* coloured by a dyeing process may be woven or knitted to yield a coloured or patterned fabric. The *fibres* from which the yarn is made may be dyed before spinning. Production is easier to control the later in the process the colouring is performed.

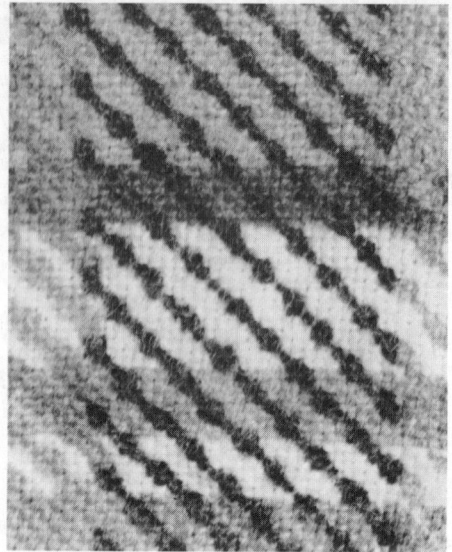

Twill weave with yarn-dyed yarns, and a plain weave printed to imitate twill. Which method of colouring would be cheaper, and why?

Where possible, colouring is performed late in production for another reason. Fashion colours change often, and fashion trends are often planned as much as two years in advance of the release of the new style. Colour is an important selling point, but not all colours promoted will be equally successful in a given selling season. Sometimes planners may misjudge the market. For this reason, it is less risky for the manufacturer to make the fabric first, and to apply the colour later. A manufacturer with a stock of white fabric can dye or print it as soon as retailers report which colours will sell best. This approach reduces the risk of a warehouse full of unsold stock.

Fibre dyeing

Some special colour effects cannot be dyed or printed on a fabric. For many fabrics – tartans, checks, brocades, tapestries, jacquard knits, inlaid effects – differently dyed *yarns* are needed. Tweeds and other textured fabrics are made from yarns spun from mixtures of differently coloured *fibres*. For such styles, staple fibres have to be separately dyed *before* the spinning process.

Dope dyeing

Synthetic fibres can be coloured before extrusion. This is mostly done for melt-spun fibres, particularly those which will be used in great quantities in one colour. An example is artificial turf, which is made from green polypropylene. The green colour is an insoluble pigment which is carefully mixed into the melted polypropylene before extrusion.

A *white* pigment is mixed into most synthetic polymers to make the fibres less transparent. The small grains of pigment (usually titanium dioxide) scatter the light and make the fibre less lustrous. This white pigment is called a DELUSTRANT.

280

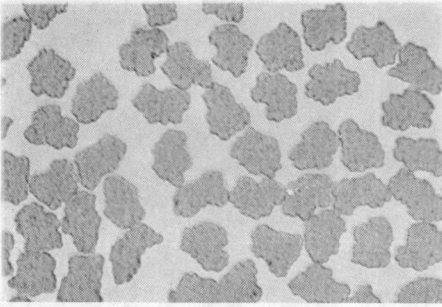

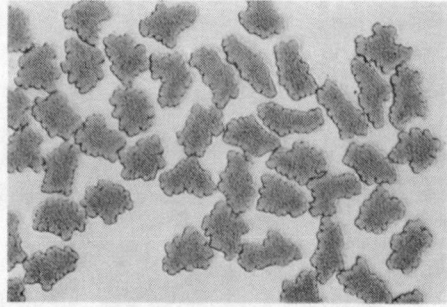

Dope–dyed viscose fibres (cross–section)

Stock dyeing

For tweeds and other fancy effects, the fibres have to be dyed before spinning. Different colours of fibres are then mixed in carefully measured proportions and processed into yarn.

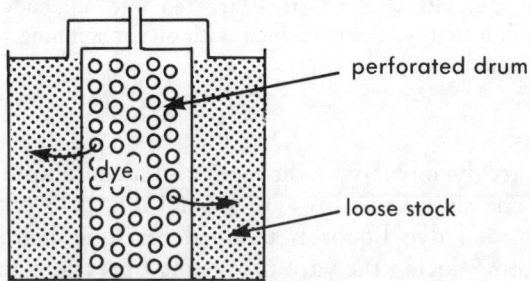

Stock dyeing: the dye is pumped through a perforated drum around which the stock is loosely packed. Pressures higher than atmospheric can be used.

Stock dyeing.
(Australian Wool Corporation)

281

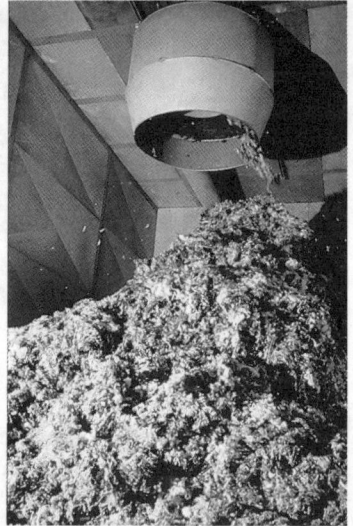

Carefully measured amounts of dyed stock are fed into a blender, where they are tumbled and mixed thoroughly, and combined with oil for spinning.
(Australian Wool Corporation)

Top dyeing

Sometimes fibres are dyed half-way through yarn manufacture – as carded sliver or combed tops. The slivers of fibres are wound into balls on perforated cores, placed on a dyer, and dye liquor is pumped through them. Any unevenness in dyeing is overcome during the yarn processing operations that follow.

Sometimes *wool tops* can be printed. If black stripes are printed on the top slivers (a process called MELANGE PRINTING), then each fibre will be coloured partly black and partly white. When the slivers are doubled and drawn, the

Unloading dyed wool tops.
(Australian Wool Corporation)

Melange printed wool sliver.
(C.S.I.R.O. Division of Textile Industry)

printed patterns get mixed up and the resulting yarn will be a subtle shade of grey. Such printed tops are often used for grey suitings and school uniforms. You can tell by using a magnifying glass whether the fibres used in a worsted have been melange printed.

Yarn dyeing

There are many fabrics which are made from different coloured yarns.

Yarns can be dyed in hank form, or loosely wound on special perforated cores. The dye liquor must circulate evenly through the fibres in the yarn for

(Pegg Whiteley Group, Leicester, England)

(Vald Henriksen A/S, Copenhagen, Denmark)

Yarn is packed so the dye will circulate evenly, the type of package depending on the fibre, dye, and yarn. The dye is pumped through the perforated core passing through the centre of the cheese or yarn package. Yarn dyeing machines may be designed to take either vertical or horizontal yarn carriers.

283

A top-loading yarn-dyeing machine, capable of taking a variety of material carriers and of operating above atmospheric pressure (i.e., at temperatures above 100°C). (Pegg Whiteley Group, Leicester, England)

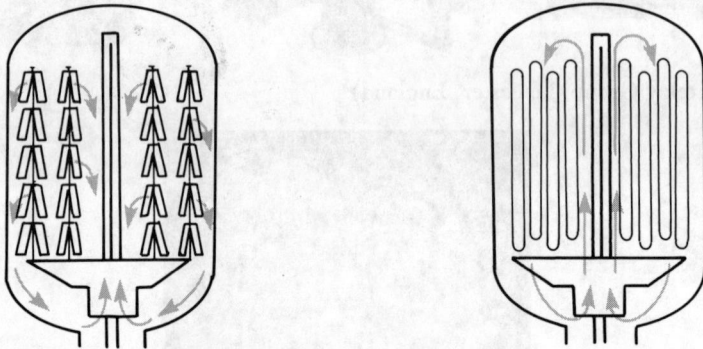

Yarn dyeing in packages (left), and hanks (right). The choice of methods and the form of package will depend on the type of yarn being dyed. (Vald Henriksen A/S, Denmark)

even dyeing. The yarns are wound loosely into the special CHEESE form for dyeing, and after dyeing are wound into a different package ready for cloth production. The extra handling operations add to the cost of fabric manufacture. Cheeses can be rewound more rapidly and cheaply than can hanks, and allow better liquor flow, and so are the preferred package.

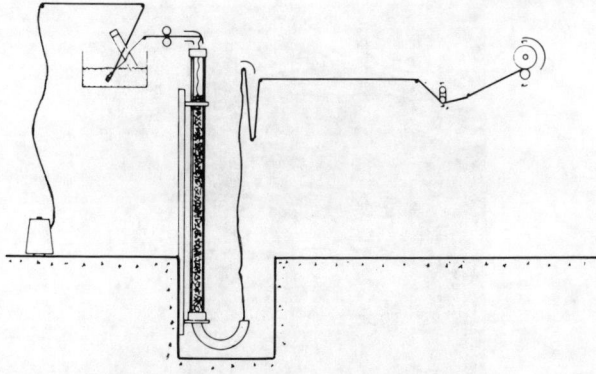

C.S.I.R.O. continuous yarn dyeing plant. Dyeing yarn in continuous length eliminates the time-consuming and therefore costly winding and rewinding required by conventional equipment.
(C.S.I.R.O. Division of Textile Industry)

Fabric dyeing

Many different kinds of machines have been developed for dyeing fabrics.

Winch

A WINCH has a large vat full of dye liquor. Ropes of fabric are sewn into a continuous circle. The winch drum pulls many such ropes from the dyebath at the one time and drops them back on the other side.

This system is well suited to dyeing soft cottons and fluffy woollens. However, it is not suitable for tightly constructed, stiff, or smooth fabrics, as it may crease or mark them permanently.

Winch dyeing takes a lot of water, which must be heated. It therefore is wasteful of energy and chemicals. It is usually an atmospheric process – it can dye only at temperatures below 100°C.

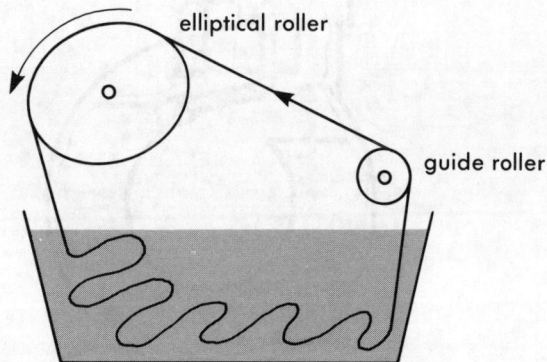

Winch dyer (schematic): an atmospheric pressure dyer requiring a large, heated dyebath.

This modern winch dyer has a low liquor ratio and can be designed to blow air into the fabric to eliminate crease marks. When in operation it is fully sealed, so air-sensitive vat dyes can be used economically. Carriers are essential if disperse dyes are to be applied in this atmospheric pressure machine.
(Bruckner Machinery Limited)

Jet dyer

An improved machine for rope dyeing is the high-temperature JET DYER. The jet dyer moves the fabric together with the dye liquor; this reduces the strain on the fabric. As it is fully enclosed, the jet dyer can be pressurized and heated up to 150°C for processing. It also uses less water – has a lower liquor ratio – than the conventional winch, and is therefore more economical of energy, water and chemicals.

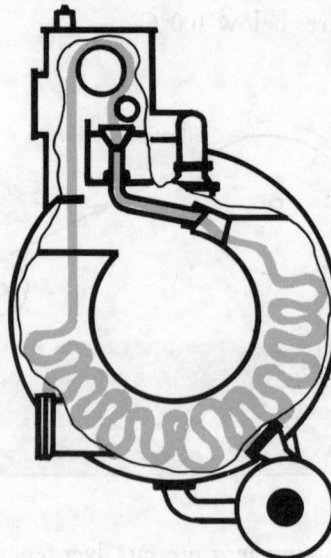

Jet dyer (schematic).

The fully enclosed jet dyer allows ropes of fabric to be dyed under pressure at temperatures above 100°C, and is economical of chemicals, water, and energy.
(© Skjern Foto, courtesy Vald Henriksen A/S, Copenhagen, Denmark)

Beam dyer

Some fabrics cannot be dyed in the rope form – they are too easily creased or stretched. Open-width dyeing is then more suitable. This can take many different forms.

For BEAM DYEING, the fabric is carefully wound on a hollow, perforated beam. The winding has to be both even and loose, to allow the dye to penetrate the fabric uniformly. The liquor is then pumped through the fabric. The beam dyer can be used for both atmospheric and high-temperature dyeing.

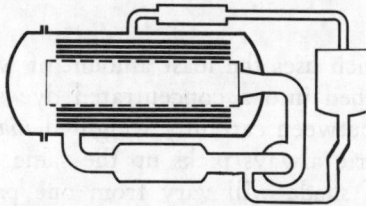

The beam dyer allows less robust fabrics to be dyed in open width and under pressure.

In the beam dyer, dye is pumped through fabric wrapped in open width around a perforated beam.
(Bayer)

Jig

The JIG DYER does not immerse the whole fabric into a dyebath. Rather, the fabric is passed back and forth through a concentrated solution of dye until most of the dye has been taken up into the fibres. This method leaves the fabric surface undisturbed, and is often used for dyeing fine worsted fabrics.

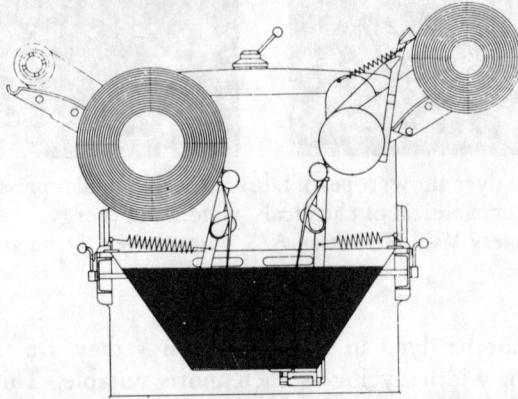

The jig dyer allows fabric from a roll to pass in open width through a dyebath. It oscillates the fabric to ensure even dye take-up. Operation is at atmospheric pressure (a disadvantage) but the jig dyer can take greater lengths of fabric than can jet, beam, or winch dyers.
(Vald Henriksen A/S, Denmark)

Padding

The dyeing method which uses the least amount of water is called PADDING. Here, the fabric is dipped into a concentrated dye solution; the excess dye liquor is squeezed out between carefully weighted *mangles*. It is important to make sure that the fabric always picks up the same amount of dye liquor – otherwise the depth of shade will vary from one part of the fabric to the other.

Unlike the previous dyeing methods, this is a *continuous* system. The dye liquor can be topped up to maintain constant concentration so the fabric always picks up the same amount of dye for an even shade.

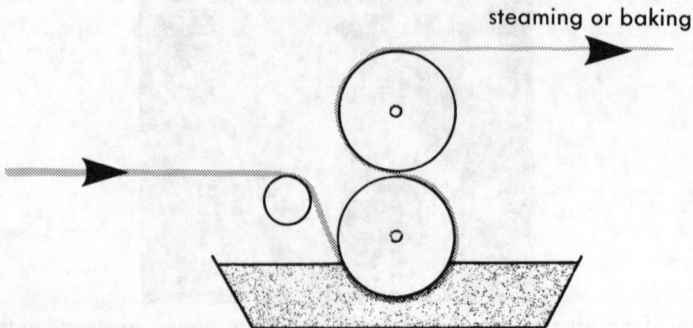

steaming or baking

Padding and foam dyeing both allow continuous lengths of fabric to be dyed, and are economical of dyestuffs.

After padding, the fabric may be *steamed* to allow the dye molecules to penetrate to the inside of the fibres. If cold-water reactive dyes are used, the padded batch may be left to stand at room temperature for several hours to allow the dye to penetrate the fibres. The excess dye is then washed off. This *cold pad batch* system saves on energy.

Such a system is not suitable for polyester and other fibres which need high-temperature dyeing. These thermoplastic fibres are dried after padding, then baked for five minutes in an oven at 210°C. After baking, the excess dye is washed off. This is called the *pad-thermosol* system of continuous dyeing.

Foam dyeing

Wet processing of textiles is expensive. The more water used, the more energy is needed for heating it. The wetter fabrics are, the more energy is needed to remove the water from them.

If a lot of air is bubbled through a concentrated solution of dye containing a *foaming agent* (detergent), then the FOAM formed can be spread smoothly and evenly on the surface of the fabric. When the foam breaks, the fabric is wetted, and the dye molecules enter the fibres. This way a minimum amount of water is used with maximum efficiency.

Foam application is now gaining popularity for many different finishing processes, as new efficient foaming agents enter the market.

TRY YOUR HAND

You will need:
foam hair dye (from chemist)
2 identical pieces of wool fabric, weighed
spatula
balance

What you do:
Shake the bottle well, and apply an even layer of foam to one of the pieces of wool fabric. Spread the foam with a spatula to break up the bubbles.
Examine how the dye penetrates the fibres.
Weigh the foamed fabric; wet the other fabric piece and weigh that.
1 How would you control the amount of dye taken up by the fabric in foam dyeing?
2 Compare the weights of the foamed and the wet fabric. Which would take less energy to dry? Why?
3 Why is a *wool* fabric suggested for this exercise?

Garment dyeing

Hosiery

The most common garment dyeing procedure is in *hosiery*. Stockings and pantyhose are knitted white, and are dyed after their construction is completed. Most garment dyeing takes place in a dry-cleaner type tumbling machine. The fragile hose are protected within a mesh laundry-bag. The less expensive lines of pantyhose are produced in limited colour ranges, and are often re-dyed from unsuccessful lighter dyeings.

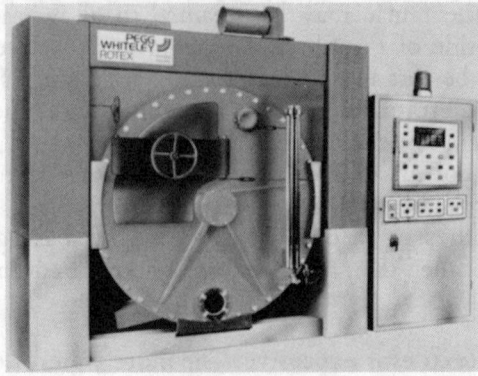

A hosiery-dyeing machine.
(Pegg Whiteley Group, Leicester, England)

TRY YOUR HAND

Choose a brand of cheap pantyhose in a particular colourway. Buy samples over a period of time – say over six months – or buy samples from a large department store with a rapid turnover of stock, and from a small corner store with much slower turnover.

When you have a range of samples, compare the colours.

Are all the colours the same? How wide a tolerance does the manufacturer allow? Why does the variation arise?

Printing

Block printing

Block printing uses engraved surfaces to apply dye to the fabric.

TRY YOUR HAND

You will need:
some large potatoes
deep biscuit cutters
printing ink
white or pale fabric
glass slab and rubber roller

What you do:
Cut 3 cm thick slices from your potatoes and block out shapes with the cutters. You can also do some extra carving with a knife.

Spread some printing paste on the glass slab with the roller, and when the roller is smoothly coated with ink, roll it onto the dry potato surface.

Press the design firmly onto the fabric.

Allow to dry, and iron to set.

Experiment with other vegetables – onions, cabbage, cross-cut celery – for variations in design.

In all these printing methods the dye is picked up by the raised surface of the design, and transferred directly to the fabric. The raised components of the carpet printer are made of a spongy, rubber material for more efficient dye pick-up – the deep carpet pile needs more dye to colour it than does a flat fabric surface.

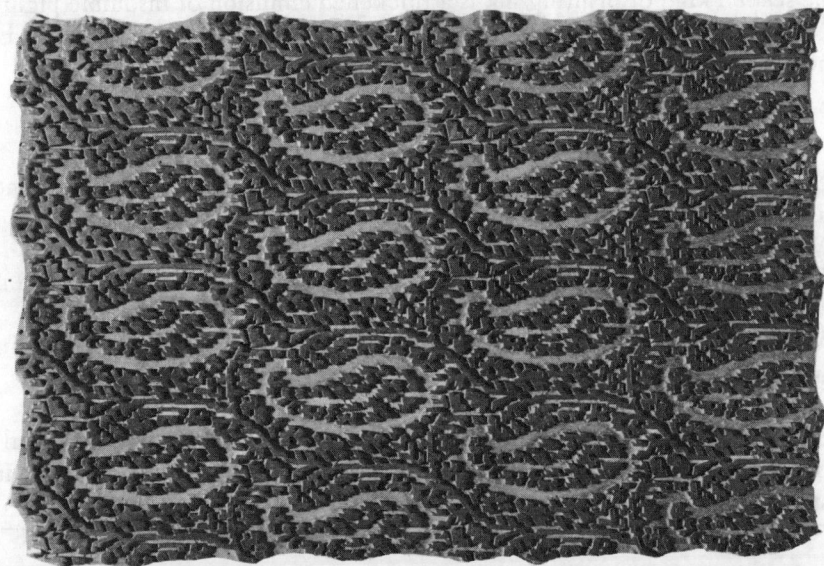

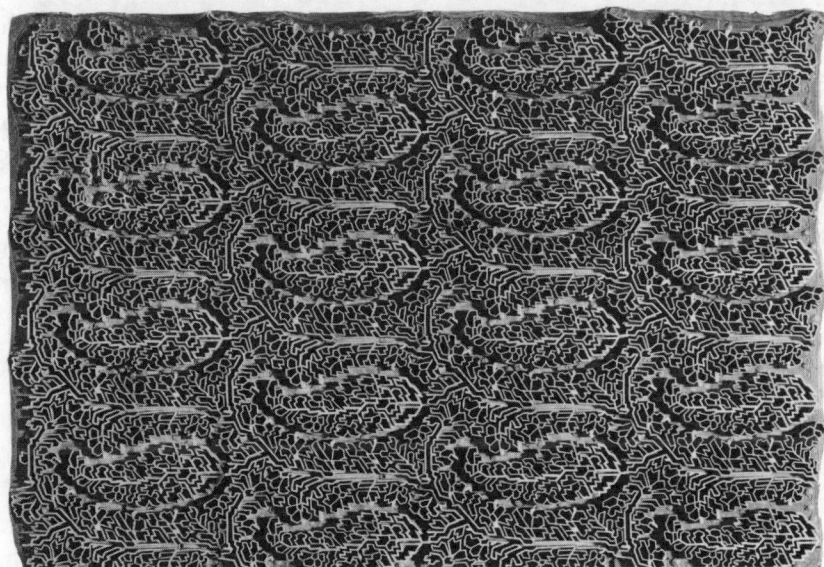

Hand printing blocks, a pair
Carved wood block and outline block of repeat metal sections of paisley pattern
Made by David Evans and Co. of Crayford, Kent
English, nineteenth century
Presented to the National Gallery of Victoria by Bruno and Christina Pichler
(National Gallery of Victoria)

Printing inks

There are two basic kinds of printing inks. One is a thickened solution of dye – hydrophilic dyes for hydrophilic fibres, disperse dyes for hydrophobic fibres. After printing, the fabric is steamed to allow the dye molecules to penetrate deep into the fibres.

The second kind of printing ink is a thickened emulsion of insoluble pigments. The thickener in this case is a resin – often thermoplastic – which holds the pigments in place on the fabric surface after drying and heating.

TRY YOUR HAND

Take a close look at your printed samples. Examine them by touch and by looking under the microscope.

1   What kind of printing ink did you use – pigment or dye?
2   Discuss some differences between the expected behaviours of the two types of printing inks.
3   What types of dye molecules would you expect to be included in printing inks designed for cotton fabrics? for polyester fabrics? for wool?
4   Could you use any one printing paste for all fabrics?
5   Suggest a way of making printing paste at home. Which commercial dye or pigment would you start with? What would you use for thickening – starch, glue, gelatine, or some other thickener? Test your ideas.

Roller printing

Roller printing uses metal rollers engraved with the design. The roller is dipped into the printing paste, all the raised parts are scraped clean, and the grooves carry the dye to the fabric.

The size of the design repeat is determined by the diameter of the rollers. Large rollers are far more expensive to buy and to prepare than small rollers – hence roller printing tends towards *small designs*.

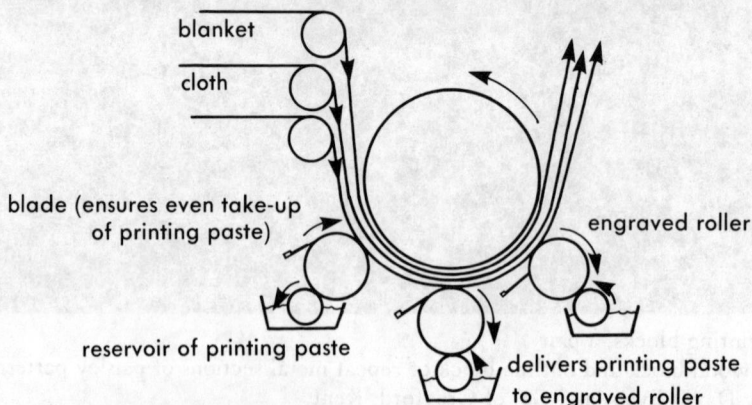

Roller printers can be equipped to handle as many as fourteen different colours, allowing for great complexity of design.

The first roller printing machine.
(*Bayer Farben Revue* No. 3, 1962,
by permission of the editors.
Bayer Farben Revue is published by
Bayer AG, Leverkusen, West Germany.)

Designs with a small repeat, such as this,
are more economical to print than are
designs with a large repeat.

The cost of roller preparation makes roller printing unsuitable for short runs
– 5000 metres of the one design is often used as a minimum figure. Therefore,
cheaper fabrics are generally printed by this method.

In the printed design, each colour application needs a separate roller. Printers
are usually equipped to cope with up to fourteen different colours.

A fabric designer at work.
(Gebrüder Sulzer Aktiengesellschaft, Winterthur, Switzerland)

TRY YOUR HAND

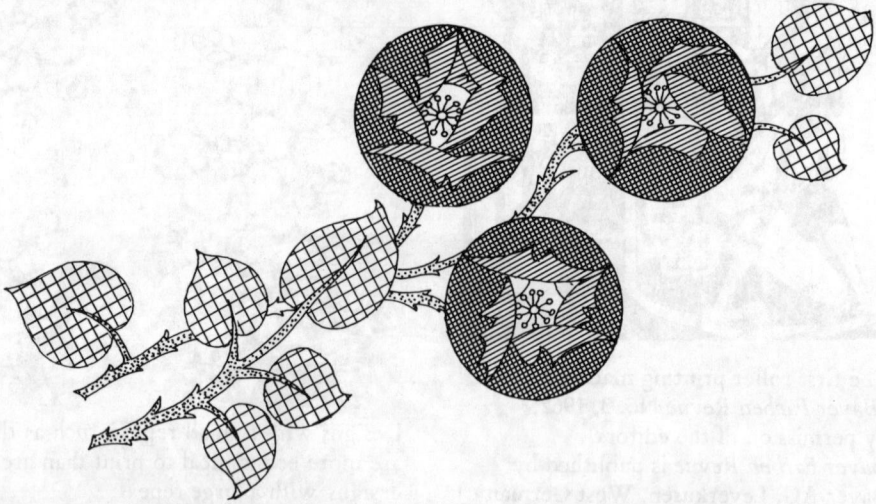

With tracing paper, copy those parts of the drawing which are coloured with dots. Make sure you mark the corners accurately. On another drawing, copy only the slashed parts. Continue until you have a separate drawing for each of the 'colours' used in the design. You have now carried out the colour separation process.

1 Why was it important to mark the corners of the design?
2 How many rollers will your design need? What diameter will they be?
3 Decide which colour to use for each colour separation. Colour your design accordingly.
4 Now change the colours used. A careful choice will create a totally new effect – a different colourway for your design.

Screen printing

The origins of this process were in stencil techniques, in which dye was pressed onto the fabric around a stencil made of paper. Later, instead of the delicate stencil being tied together with hair fibres, it was fixed to a screen made of fine silk.

Today's 'silk screens' are made of polyester filaments, and the *resist design* is developed photographically using an emulsion which hardens when exposed to light.

Each colour of the design needs a separate screen; the screens however are less expensive to produce than engraved metal rollers. *Large designs* are possible, and *shorter runs* are also economically justifiable. Although the process is *slow,* silk-screen printed fabrics have developed a reputation for *exclusive designs* on luxury or high-class fabrics. This is probably because of the high cost of traditional FLAT SCREEN printing methods.

The ROTARY SCREEN PRINTER uses less space on the factory floor, and is also able to print faster, though it still uses the same basic principles as the flat screen printer.

TRY YOUR HAND

Find examples of artists' work which uses silk-screen printing. Try galleries or craft outlets before you try books: it is always best to see the work for yourself. If you can, find out why the artists have employed this technique.

A nineteenth century Japanese stencil design. The Japanese stencils, made of mulberry paper held together by silk filaments, achieved extraordinary detail and delicacy. (Andrew W. Tuer, *Japanese Stencil Designs*. Dover Pictorial Archives Series, New York, 1967.)

Flat screen printer.
(Fritz Buser Limited)

Rotary screen printing. This technique is extremely efficient, and has largely replaced roller printing.
(Gebrüder Sulzer Aktiengesellschaft, Winterthur, Switzerland)

Transfer printing

In this recent technique, hydrophobic dyes are printed on shiny paper with ordinary paper printing technology. This can produce excellent detail in the design. The dyes used are all able to *sublime* (pass from the solid to the vapour form) at relatively low temperatures.

When a hydrophobic fabric is held in close contact with the printed paper and heated, the dyes are transferred from the paper to the fabric. Since the dye vaporizes as single molecules, it penetrates the hydrophobic fibres easily to produce a clear, fast print.

Check the experiments suggested on the 'disperse dyes' section of this chapter (p. 272) to investigate this process further.

Transfer printing does not need any water. It is relatively fast, and can be carried out by less skilled operators than are needed for other printing methods. (Success mostly depends on the skill of the paper printer.) Paper is cheaper than fabric, so if a design or a colourway is not a commercial success, only the paper needs to be thrown away. The fabric, left white until the last possible moment, can be used for other designs.

These factors have contributed to the success of heat transfer printing in the field of synthetic fabric colouring.

Wet heat transfer processing developed for wool and cotton/polyester blends has been less successful because with wet processing the energy savings are not significant.

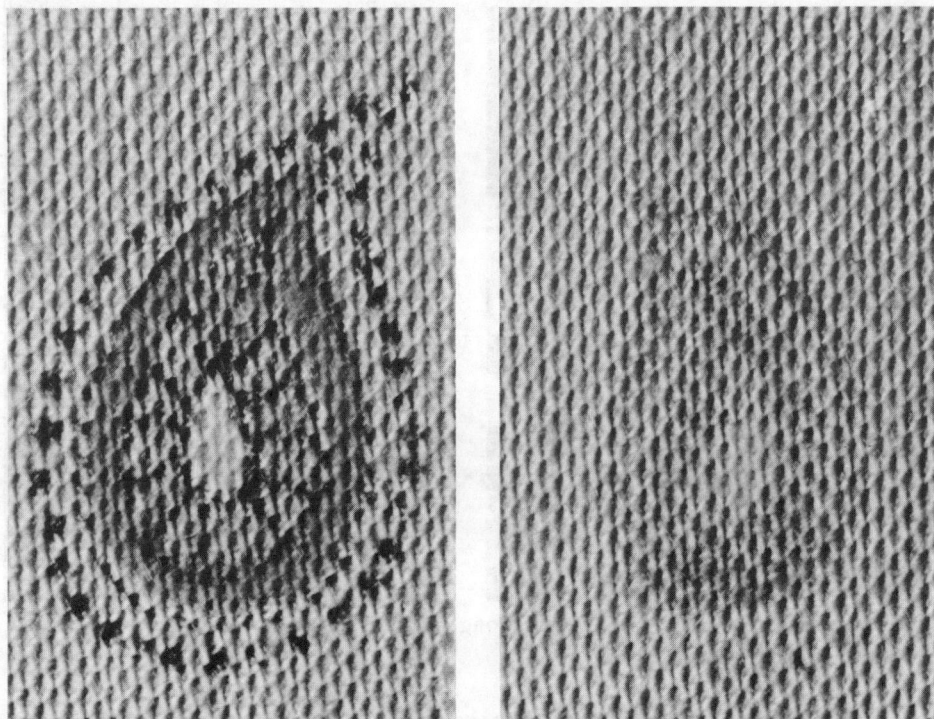

Transfer print on spun polyester (right and wrong sides). The fabric had been dyed to a uniform background colour before printing.

Resist printing

Batik is an example of traditional resist printing. The design is printed on the fabric in wax or other dye-resistant paste, and the fabric is then dyed. After dyeing the resist is removed, leaving white areas of design.

Resist prints (such as genuine, rather than silk-screened imitation, batik designs) can be readily identified by examination of the wrong sides of the fabric. In silk-screen prints, the coloured portions are less clear on the reverse than on the face. Since the resist print was *dipped* in dye solution, the colour should be equally strong on both sides, though the white resist may appear a little fuzzy on the reverse.

Printing techniques used on fine fabrics – voile, lawn – are harder to identify than the printing techniques used on thick fabrics, because of the ease with which the printing paste penetrates the fabric.

Discharge printing

Sometimes when a small, light-coloured design is needed on a dark background, a *discharge paste* is printed on dyed fabric. The chemicals in the paste destroy the dye molecules, and create white – or almost white – areas where they were applied. Other colours can then be printed onto the discharged parts of the design.

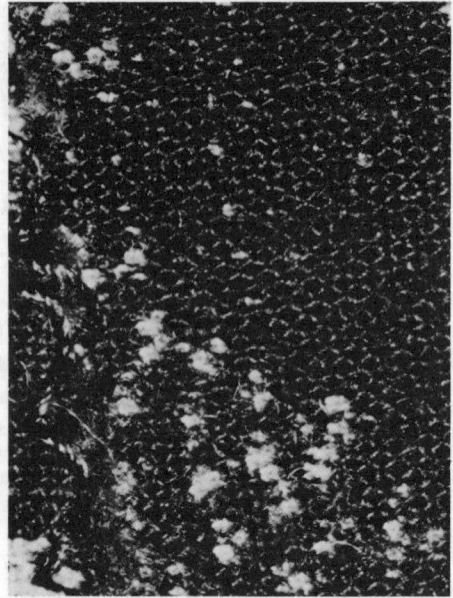

Discharge print on cotton (right and wrong sides). The small white flowers are much less distinct on the wrong side.

Discharge prints can be identified because the white is weaker on the reverse than on the face of the fabric, and also because it is rarely a pure white. The dark background colour is a solid shade throughout discharge prints, as in resist-printed fabrics.

TRY YOUR HAND

Go into a fabric shop and try to identify the different techniques that were used to colour the fabrics.

1 What is the percentage of plain dyed fabrics out of the total?
2 The percentage of yarn-dyed fabrics?
3 The percentage of fibre-dyed fabrics?
4 The percentage of prints?
5 Would you expect these proportions to hold true through different seasons? From one year to the next?
6 What are the variables that determine the methods used for colouring fabrics?
7 How many different printing techniques can you identify among the fabrics you surveyed?

Blends

Cross dyeing

Different types of fibres need different kinds of dyes. So when two types of fibres are blended in a fabric – as often happens – it is necessary to carry out two dyeing processes, one for each fibre.

For example, intimate blends of polyester and cotton are often used (p. 39)

If direct, vat, or reactive dyes are used to dye the cotton fibres, the polyester will remain uncoloured. The blend will have a soft 'heather' effect. While this may be attractive, it is not always desirable. For a uniform, smooth, solid colour, the polyester fibres need to be dyed with disperse dyes to the same shade as the cotton.

Which process to do first?

Cotton fibres can be dyed at low temperatures. At high temperatures, direct dyes may leave the cotton fibres, reactive dyes may decompose. Therefore dyeing of the polyester fibres, which is carried out under pressure at 130°C, is performed first.

After thorough rinsing and cleaning to remove excess disperse dyeing reagents, the cotton is dyed at lower temperatures, with whichever cotton dyeing process has been selected as most suitable. Colour matching is checked visually during the process, to ensure that the final colour effect is up to requirements.

For blends of wool and polyester, the polyester is again dyed first.

Union dyeing
Different fibres

Whereas cross dyeing deals separately with the components of a blend, union dyeing colours the two types of fibres simultaneously, but in different ways.

A non-commercial example of this is the use of Shirlastain A for fibre identification. When different fibres are placed together in the mixture of dyes, each absorbs different colours from the mix. (Do the exercises in fibre identification, Chapter 6, pp. 192–3, if you have not already done them.)

Deep-dye nylons

It is possible to change the affinity of nylon to different dyes by adding various chemical side-groups to the polymer.

For example, Fibremakers make three varieties of Nylon 6,6. The standard nylon is called Nylon 100. *Nylon 110* has extra $-NH_2$ groups and therefore can develop more positive charged sites for acid dyeing. It is called a *deep-dyeing* nylon. If a fabric woven in a check design from white Nylon 110 and Nylon 100 yarns is dyed, the check pattern will appear as lighter and darker shades of the dye colour.

Nylon 120 has acid groups added; therefore it can accept positively charged (basic) dyes.

All nylons can be dyed with disperse dyes, but that needs a separate dyeing process, i.e. cross dyeing. Disperse dyes will dye all the different nylons equally effectively.

TRY YOUR HAND

Design a tartan pattern using the three nylons mentioned above. You are given a red acid dye, a yellow basic dye and a blue disperse dye.

First, discuss how you would carry out the dyeings – in what order, at what temperatures, with which addition. Then colour in your design, showing the expected effect at the end of each dyeing process.

Survey different blends of fibres available to the consumer.
Could you dye any of them with union dyeing techniques?
What dyes would you use?
Could you use acid and basic dyes together in the one dyebath?
Name some blends that cannot be coloured in the one dyebath – ones that need cross-dyeing technology.

Cost considerations

In order to make the factory economically viable, the dyehouse manager must pay attention to many different factors.

Cost of reagents

Dyes and chemicals cost money: a dye formulation that is effective but cheap will be preferred to a more expensive alternative. Look back to the colour matching computer in Chapter 7 (p. 222) for the types of options a dyehouse manager must consider.

Chemical additives other than dyes also cost a great deal of money. Their concentration is calculated according to the amount of water used for the dyeing process.

TRY YOUR HAND

You have to dye 100 kg of cotton with a reactive dye. You need 80 g/L salt to push the dye onto the fibres. You also need 20 g/L soda ash as an alkaline fixer to complete the reaction between the dye and the fibre.
Calculate how much of each reagent you need if the dyeing machine chosen for the operation has a liquor ratio of (i) 15:1 (i.e. 15 litres water for each kilogram of fabric) in a jet dyer, (ii) 25:1 for a winch dyer.

The cost of energy

Liquor ratio

If high temperatures are used for dyeing, and if large quantities of dye liquor need to be heated, then much energy is used in the process.

TRY YOUR HAND

4.2 kilojoule is needed to raise the temperature of 1 litre of water by 1°C.
You have 100 kg of fabric to dye, and a choice of two processes. Your *jet dyer* has a liquor ratio of 15:1, and uses direct dyes that need to be heated from room temperature to 90°C. Your *winch dyer* has a liquor ratio of 25:1, and uses reactive dyes at 45°C.
1   Calculate the energy requirements of each process.
2   Which process would you choose if energy costs were the only consideration?

3   Reactive dyes are more expensive than direct dyes. Does this affect your
     choice?

Energy source

The cost of energy is also determined by the type of energy source used –
electricity, gas, oil or coal. The energy is used to convert water into steam
in the boilerhouse. This steam is then carried through pipes to the different
dye vats. The more steam allowed to flow through the copper coils in the
vats, the faster the dye liquor is heated.

Heat exchangers

Rather than waste the energy from the hot dyebath after dyeing is finished,
the hot liquor is piped through a heat exchanger, and is used to warm the
clean water waiting to be used for the next batch of dyeing.

Automation

Automatic control of the heating and cooling times used in a dyeing cycle
can eliminate human error. Microprocessors control the valves which fill the
machine to the required water level. They are programmed to control the rate
of heating, and to keep the dyebath at the correct temperature for the right
length of time for the fibre/formula combination. The drain valves are automatically opened at the end of the dyeing cycle, and then they are automatically
closed while pumps refill the vessel for rinsing and after-treatments. Dyes and
other chemicals can also be added automatically, at pre-programmed times.
This system can cut down on dyeing times, and so save energy used for dyebath
heating.

Drying processes

It takes a lot of energy to evaporate water from a wet fabric. So the more
water is squeezed out of the fabric – between mangles or in a centrifuge –
before drying, the less costly is the drying process.

There are different ways of heating the fabric in the dryer: hot drum, hot
air, infrared radiation. Even radio frequency and microwaves are now being
developed for greater efficiency in drying fabrics.

Water pollution

The cost of water

The bodies responsible for the supply and purification of water in a community
charge higher water rates for industries than for households. This is because
of all the *waste* that factories discharge into the sewage system. Industrial water
can be polluted by oils, acids, alkalis, metal salts (which may be poisonous),
detergent, dyes and other chemicals, and by dirt and heat.

Naturally, the less water used by an industry, the less costly are the water
rates and anti-pollution measures. New processes, such as foam application of
dyes and finishes, are being developed to minimize the amount of water and
energy consumed during processing.

Pollution control measures

Some countries have such strict laws regulating what may and may not be poured into rivers that industries must invest a great deal of money into ensuring that pollutants are removed from the waste water.

Pollution has the potential to cause much misery and to cause serious damage to the environment. Environment protection agencies must strike a balance between control of pollution and the cost of control measures. If industries cannot afford the cost of implementing pollution control regulations, they may close down or move off shore.

In order to minimize pollution, biodegradable detergents can be used. Any excess acidity or alkalinity needs to be neutralized.

Sludge and solid contamination are allowed to settle out, and are collected separately. Disinfected sludge can be dried and used as a fertilizer, though this is warranted only where large quantities of organic matter are removed from fibres, as in a large wool-scouring plant.

The *temperature* of the dumped water is important. The water must not overheat the sewage pipes through which it flows. Hot water can interfere with the delicate biological processes involved in digesting the impurities in the water. Therefore waste water is allowed to cool before discharge into the sewage system.

Sewage treatment includes sedimentation and filtration to remove solid contaminants, and aeration to allow aerobic bacteria to digest biodegradable matter before the waste is emptied into our waterways.

TRY YOUR HAND

Read the following editorial from *Australasian Textiles*.

New Victorian EPA bill – on balance, a more sensible approach

The textile industry has suffered more than many of its overseas competitors, as far as environmental legislation is concerned. For most plants, considerable expenditure has been involved in meeting in some cases very tight regulations. For example, Bradmill's Kotara plant (now National Textiles) had to install filters throughout its humidity control system for externally discharged air – at a cost of $120,000. Several dyehouses in Melbourne and elsewhere have had to spend similar amounts and more to deal with their air discharges from stenters. In some cases, environmental regulations have led to plant re-location – at considerable cost – for example Tennyson Textiles' move to Dandenong and the transfer of the Bradmill dyehouse at Newtown to Rutherford.

The fact that high cost may be involved in satisfying Victorian EPA regulations has not been allowed – until recently – to be considered by the EPA as a factor in deciding its requirements for a particular plant. In 1977 a decision of the High Court of Australia precluded the EPA from taking account of economic considerations. This resulted in unnecessarily harsh requirements on industry.

Now this has been overturned in a new bill recently enacted by the Victorian Government. The bill included a clause recognising the need for consideration

If industry is polluting our sewerage system we'll get to the bottom of it.

Industry is often one of the biggest culprits behind beach and river pollution problems.

So we inspect factories and plants throughout the Sydney metropolitan area to make sure they comply with the regulations about discharges into the Sydney Sewer System. If we find they don't, they'll be up for significant fines. In extreme cases, we can require that all discharges to the sewer be suspended.

Last year, for example, we detected and prevented 1,090 violations.

And our inspection programmes are being stepped up all the time.

So if you think no one will notice the sump oil your truck drivers dump into the sewer, or the inadequate treatment of waste water at your factories, or that occasional illegal discharge, think again. You could be making an expensive mistake.

To people who've got something to worry about, this may sound like a threat. They're right.

To people who enjoy Sydney's rivers and beaches, it'll probably sound more like a reassurance that they are going to become increasingly pollution free.

If you want to know more about the regulations we impose on industry (or want to report any violations you suspect) just ring us on 436 0359.

THE WATER BOARD
METROPOLITAN WATER SEWERAGE AND DRAINAGE BOARD, SYDNEY.

(Metropolitan Water Sewerage and Drainage Board, Sydney)

to be given to economic as well as social factors in environmental decision making.

This change – and several others – has been made largely as a result of pressure from the Victorian Chamber of Manufactures, and the Chamber should be congratulated for its efforts.

Other improvements which should have significant benefits for industry, the EPA and the community in general include:

- A rationalisation of licence requirements which entails the elimination of licences for smaller, less offensive discharges.
- The development of a new Works Approval System to apply a 'best practical means' approach and precede the issue of a licence, which would then be an operational matter.
- New procedures to process applications within four weeks if there are not objections and eight weeks if there are objections.
- A reduction in the time-span for third-party appeals to be lodged.
- The review of State Environment Protection Policies to introduce greater flexibility to consider individual cases, having regard to local environment matters.
- Increased emphasis on assistance and negotiation with industry to resolve pollution problems, particularly in the noise-emission area where clear guidance will be provided at the planning and design stage.

The bill increases the power and role of the EPA in some respects, including the introduction of a new power to direct the installation and use of specific pollution-control equipment. However, the Victorian Government has given an assurance that this power will be used only in the event of a breakdown in negotiations with industry – and then only if the equipment is reasonably available and would be consistent with a works approval.

On the whole, the bill allows a more sensible approach to environment protection.

Australasian Textiles, 3/84

1 Why have EPA regulations been changed?
2 Find out from textiles industry magazines, local textile industries, unions representing textile workers, and other interested groups, their views on the current situation.
   a Do you consider that the regulations are accepted by all groups?
   b What are the arguments put forward by each group for or against maintaining the current situation?
   c Do you agree with any of these arguments? Why?
   d How do these various groups try to have the regulations changed in their favour? Do you think they should do this?
3 Are there any textiles industries operating in Australia which are favoured by strict regulations in another country? How is their environmental impact controlled in Australia? How does the Australian approach compare with the stricter regulations overseas?

Consumer questions answered

Q1 I have been knitting a cardigan and ran short of just one ball of wool. I bought an extra ball with the same colour name and number, but it shows up as an odd-coloured streak in my knitting. Why?

A You were probably not able to match the dyelot of the bulk of the wool when you bought the new ball. Although the same dyeing recipe was used for both lots (hence the same shade name), one batch may have been heated more slowly, or for a longer period, or to a slightly different temperature than the other. This created differences in the rate and extent of absorption of the dye molecules into the fibres. These differences in absorption then resulted in slight colour differences between the two batches. It is not commercially feasible to carry out adjustments to the dyebath composition until a perfect match is achieved. Therefore, the responsibility of choosing yarn from one dyelot rests with the consumer.

Q2 I have read somewhere that a drop of vinegar in the wash will stop woollens from losing their colour, and will give them a soft handle. Is this true?

A If you use a neutral synthetic detergent instead of soap then indeed you may add a *little* vinegar to your wash. The acid in vinegar creates positive sites on the wool, and helps to bind any acid dye molecules. Also, the wool is left very slightly acid, which gives a soft handle to the fabric. The use of cold instead of hot water also reduces the likelihood of dye loss.

Q3 I have washed a red T-shirt with other clothes and now some of the garments are stained pink. Others in the same wash-load were not affected. What can I do about this?

A Your T-shirt was probably dyed with a direct dye. During heating and agitation, and with the help of detergents, some of the dye molecules left the cotton fibres. From the water they could then be absorbed into other hydrophilic materials. Some molecules went back to the T-shirt, others to white garments. In cellulosic fibres, the dye molecules lined up with the cellulose polymer and were anchored with hydrogen bonds and van der Waals forces. They were less firmly attached to wool – if there was any in the wash – and not absorbed into polyester at all. A very light staining may have occurred in nylons. The few dye molecules absorbed by nylon would be only loosely bound, and can be removed easily in later washes.

The removal of stains from cellulosic fibres is much harder, because of the many bonds which hold each dye molecule in place. Hot scouring may remove some of the dye molecules, and subsequent bleaching may destroy the colour of the remainder by oxidation. It must be remembered that harsh treatments readily damage rayons, and that even cottons may be damaged by strong chlorine bleaches.

Q4 I perspire heavily, and some of my shirts show stains under arms – these stains won't wash out. Is there anything I can do?

A The components of perspiration vary from person to person, but generally contain salt, water, fats and organic materials. The discoloration may be the result of some of the organic components of perspiration combining with metal components of the dye; the stain may be a salt ring; or the acid of fresh perspiration (or the alkali from stale perspiration) may decompose some of the dyes. Direct, metallic and disperse dyes on acetates are particularly sensitive to perspiration damage.

The dyer should select dyes with high perspiration fastness for sportswear and similar applications. The consumer can also guard against staining by the use of anti-perspirants and fabric guards.

Enzyme soakers and some nappy cleaners can be used with some success on this type of stain.

Q5 I have often had success with dyeing garments at home, but when I tried to dye my bedroom curtains, the dye just would not take. Why?

A Dye packs and solutions designed for home use are either mixtures of direct and acid dyes (hot-water dyes) or are reactive dyes (cold-water dyes).

These three classes of dyes are suitable for cellulosics, protein fibres, and nylon. Two common types of fibres which are hydrophobic, and have no affinity for water-soluble dyes, are polyesters and acrylics. Hence, items made from either polyester or acrylic fibres cannot be dyed successfully in the home.

Q6 I have spilled some fruit juice on a white cotton shirt – what can I do about it?

A Fruits contain naturally occuring dye molecules. These dyes penetrate into the fibre. Unless they are removed (e.g. by capillary action, as when salt is placed on the stain) while the stain is still fresh, the dye molecules will form hydrogen and other bonds with the polymer chains, and will be very difficult to remove. Careful bleaching is then the only remedy.

Hydrophobic fibres are not prone to such staining, though they do tend to hold on to grease stains most tenaciously.

Q7 I have a black velour track suit. Whenever I wear it, my hands become dirty very quickly. What can I do about this?

A It seems that in efforts to get the velour a very deep shade of black, a lot of dye was applied to the fabric – this is called over-dyeing. Since there were more dye molecules than dye sites, some of the black molecules are now easily rubbed off the fabric and onto your hand. As they may not be water-soluble molecules, it could be difficult to remove them by washing; still, a warm wash with plenty of detergent could improve matters slightly.

Q8 I dropped some of baby's bottle disinfectant on a pure wool jumper. At the time it was not noticeable, but the jumper has now developed a brown stain. How can I remove it?

A The disinfectant is actually a chlorine bleach – it oxidized the wool protein into dark-coloured products. The brown stain cannot be removed from the wool without causing too much damage to the fibres.

Q9 Why are some bleaches claimed to be safe for all fabrics?

A Sodium perborate is a powder which liberates hydrogen peroxide when dissolved in water. Hydrogen peroxide is a mild oxidizing agent which will act as a safe bleach for all kinds of fibres. It may not produce spectacular whitening effects, but it will not damage the fibres to which it is applied.

Q10 How can I tell whether I am buying a genuine batik or an imitation?

A A genuine batik is dyed – so the coloured parts of the design appear exactly the same shade on both sides of the fabric. Any white segments have been batik-resisted from one side only (except in extraordinarily fine pieces of hand-painted batik tulis). So the white portions of the design are usually clearer on the right side of the fabric.

Imitation batik is silk screened, and so the dye may not penetrate evenly to the wrong side of the fabric. The two sides of the fabric may appear slightly different shades. Sometimes, the clear lines of 'crackling' printed on the imitation batik can also give away its lack of authenticity.

9 Yarns

This chapter will help you, as a consumer, answer such questions as:
- What makes some synthetic fabrics more comfortable than others?
- Why do some men's suits develop a shiny 'seat' in use?
- What do the numbers on a reel of sewing cotton mean?
- What is the meaning of '8 ply' or '12 ply' in knitting wools?
- What is the difference between washable and non-washable crepe?
- How can one fibre – such as polyester – imitate such widely differing materials as wool, silk, linen and cotton?
- How do boxed pantyhose differ from variety store cheapies?
- How can a pure cotton surgical bandage be as stretchy and elastic as something made of rubber fibres?

Properties of yarns

Different types of yarns

Early in the history of textiles only a few different fibres were available for use. Nevertheless, from these few fibres many different fabrics were prepared. Often the differences between different fabrics and their properties lie not so much in the choice of fibre or the complexity of weave as in the characteristics of the yarns used.

TRY YOUR HAND

Collect samples of plain-weave polyester fabrics: chiffon, georgette, sailcloth, staple, linen-like: as many different weights and textures as you can find.

Pull a few yarns from each sample and examine them under the microscope at 40× magnification. You will find that some of these yarns are made from smooth or textured continuous *filaments;* others from short *staple* fibres. Some will have *no twist;* others will have very *high twist.* Some twisted yarns will be a *single strand;* others will have a number of *single yarns plied together.* Some yarns will be *thick;* others *thin.*

Tabulate fabric characteristics: *surface texture* (smooth, rough, hairy, crinkled), *handle* (harsh, smooth, soft), *weight and drape* (light, medium, heavy), and *performance requirements* (strength, abrasion resistance, comfort) against the yarn characteristics you have observed.

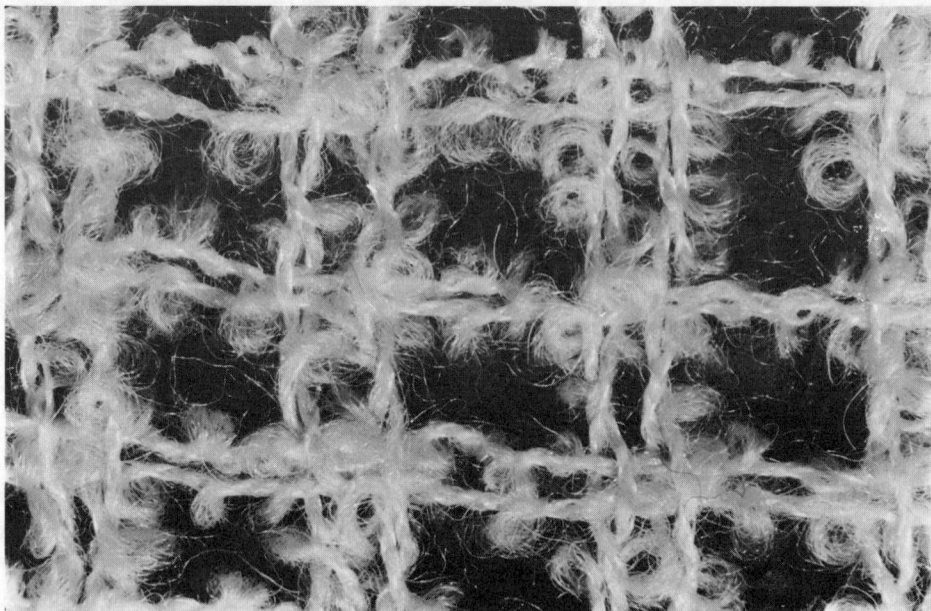

A three-ply bouclé acrylic yarn in a curtain fabric.

Do you find any correlation between the properties of the fabrics and the construction of the yarns?

TRY YOUR HAND

Go into a fabric shop with a magnifying glass and ask to see some wool fabrics. Try to evaluate the effects of twist, ply, hairiness and yarn thickness on the properties of the fabrics.

Yarn concepts

Yarn count

Early in the history of textile manufacture, different spinners – often geographically and culturally isolated from one another – measured the thickness of yarns in different ways.

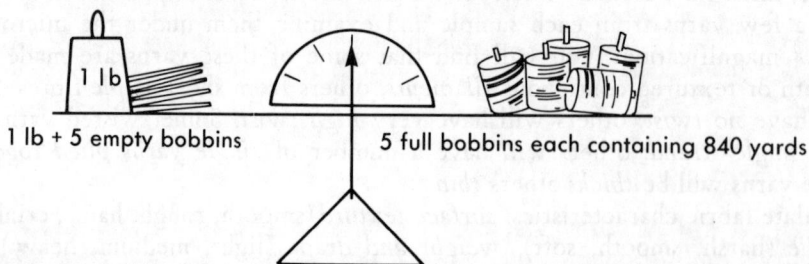

1 lb + 5 empty bobbins   5 full bobbins each containing 840 yards

A yarn count of 5.

All the bobbins on the spinning-jenny (the first mass-production spinner) filled up at the same time; usually the bobbins were changed after 840 yards of cotton were wound onto them. To estimate the thickness of the yarns, the spinner *counted* how many full bobbins were needed to balance a weight of one pound. The *higher* the COUNT (i.e. the more lengths of 840 yards needed to weigh 1 lb) the *finer* was the yarn.

TRY YOUR HAND

From a haberdashery shop buy two reels of cotton yarn – one which has 40s and one which has 60s on the label.
1   Which yarn is thicker?
2   What do you think the numbers mean?
3   Do both threads cost the same per metre? Why?
4   Are both sewing threads recommended by the salesperson for the same applications?
5   Can you find any other threads (e.g. embroidery) that have counts shown on their labels?

For a long time, in spite of their complexity, these systems were so well accepted in the trade that their use even spread to the measurement of fibre fineness.

The fineness of wool fibres was estimated – by people with experience – according to the count of the finest yarns that could be spun from them. A coarse wool was a 36s to 42s count; fine merino went from 72s to 80s counts. These 'quality numbers' are still in use, though often today fibre fineness is quoted in thousandths of a millimetre – microns – as the average diameter of the wool fibres in the staple.

In most yarn count systems, the *higher* the *count*, the *finer* the thread. In the silk denier system, now used for synthetic filaments as well, the *higher* the *denier*, the *thicker* the thread. Thus 30s cotton is twice the thickness of 60s; 30 denier silk is half as thick as 60 denier.

Systems of yarn count

System	Definition
1s cotton count	1 bobbin of 840 yards made from 1 lb fibre
1s linen count	1 bobbin of 300 yards made from 1 lb fibre
1s worsted count	1 bobbin of 560 yards made from 1 lb fibre
1s woollen count (Yorkshire system)	1 bobbin of 256 yards made from 1 lb fibre (There are at least 14 woollen count systems)
denier (silk, and, by extension, other filament fibres)	mass in gram of 9000 metres of yarn
tex	mass in gram of 1000 metres of yarn

TRY YOUR HAND

Hosiery is still identified by denier. Find examples of 15 denier, 30 denier, 40 denier, and 70 denier hose. Investigate the effect of denier on sheerness, strength and bulk.

Even though denier is not a fully metric system, its use has been widely accepted in textiles. The *strength* of yarns and of fibres is usually expressed in grams per denier. If, for instance, a 10 denier nylon thread can support 84 gram weight before breaking, its tensile strength is given as 8.4 g/den. *Tex* is a fully metric unit of count. (1 tex = 1 g/1000 m, 2 Tex = 2 g/1000 m.) Can you express the count of a 10 denier yarn in tex?

Torque

TORQUE is a rotary force. Untwisted yarns have no torque; yarns with high twist levels have high torque – a *tendency to untwist*.

Torque in a twisted yarn can be eliminated with suitable plying, so the twist of the strands around each other cancels the twist within each strand. If all the fibres in the yarn are parallel to the yarn axis, there is no tendency to untwist.

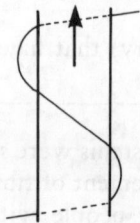

S (left-hand) twist.   Z (right-hand) twist.

Residual torque lends excellent drape properties to fabrics, but can create handling problems during manufacture. This is because a yarn will try to get rid of residual torque by snarling and curling and twisting up on itself. The *liveliness* may be used to produce interesting surface effects or stretch properties in fabrics.

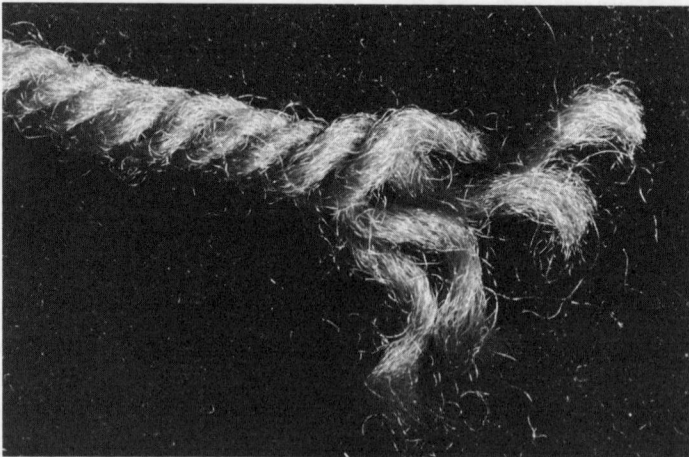

Knitting wool: Z twist in each ply and S twist in the yarn results in each fibre being aligned along the yarn axis. There is no residual torque, so the yarn is easy to handle and has no tendency to untwist.

TRY YOUR HAND

You will need:
10 strands of 4 ply knitting yarn, each 3 m long
a large bunch of keys on a hook, as a weight

What you do:
Hold one end of the bundle of yarns, while a friend holds the other end.
Twist the yarn bundle in a clockwise direction.
Can you feel the torque developing? When you feel that the strands cannot be twisted any further, hang the weight at the centre of the rope.
Bring the two ends together. When it is allowed to hang free the weight will rotate as the two halves of the twisted rope turn about each other to cancel the twist you inserted.
When the turning stops, unhook the weight.
1 Does the rope untwist itself?
2 Did you give the yarns an S or a Z twist?
3 What type of twist does the plied rope have?
4 What is the direction of the individual strands in the rope?

Hairiness

Yarns made from continuous *filaments* are *not hairy*. Those made from staple fibres have fibre ends protruding out of the fabric. The *hairiness* of yarns (and therefore of fabrics) depends on the *staple length* of the fibres used in it, on the level of *twist and plying,* and on fabric *finishing*.

TRY YOUR HAND

Take samples of different fabrics – such as woollen, worsted, denim, poplin, crepe, jersey – and drape each over the palm of your hand. Raise your palm to eye level, and check whether there are any fibres protruding from the surface. Can you relate the fabric hairiness you observe to the factors listed above?

Flexibility

The drape of fabrics can depend to a large extent on the softness or flexibility of its yarns.
    The *finer* the individual *fibres* that the yarn is made of, the *more flexible* is the yarn.
    Loose twist also helps yarn *softness*.
    Different methods of yarn manufacture produce yarns with different 'bending rigidity'. Methods which align fibres will produce stiffer yarns than methods which allow fibres to remain less orientated. For example, compare the results of the worsted and woollen spinning (p. 335). Woollen yarns are more pliable – flexible – than are worsted yarns.

Cover

The thickness of a yarn is not necessarily related to its count in tex. That is, the thickness depends on more than the actual amount of fibre in the yarn. Some yarns are loosely constructed, and so contain a lot of air: they will have a relatively low count. Others are tightly twisted and plied: that gives them a high count.

If a fabric is woven or knitted from *bulky* yarns, it will be *softer* and more *pliable* than if it is manufactured with the same number of yarns per centimetre or stitches per centimetre from a tightly constructed yarn of the same count. More importantly, *bulky* yarns have *better cover:* they create less transparent fabrics.

Texturing can be used to add bulk and improve cover in filament yarns.

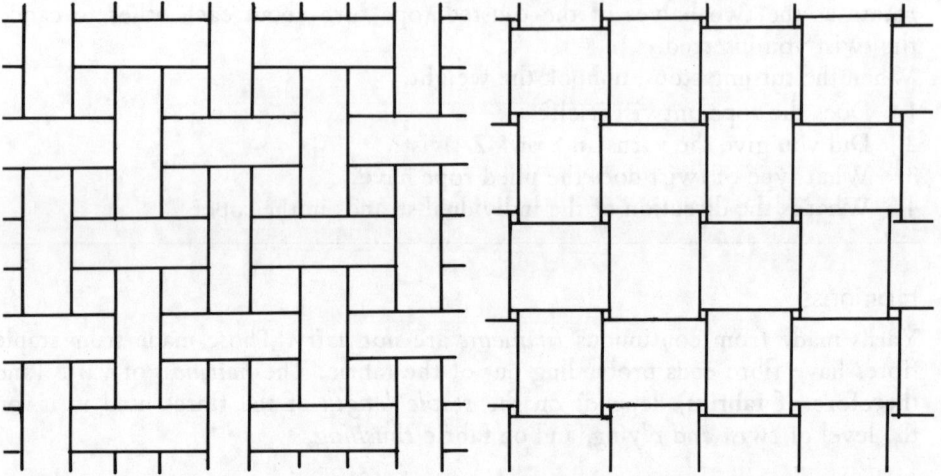

High and low cover with the same count (same number of threads per centimetre).

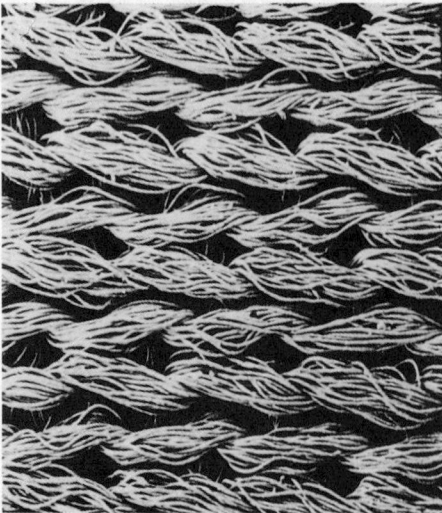

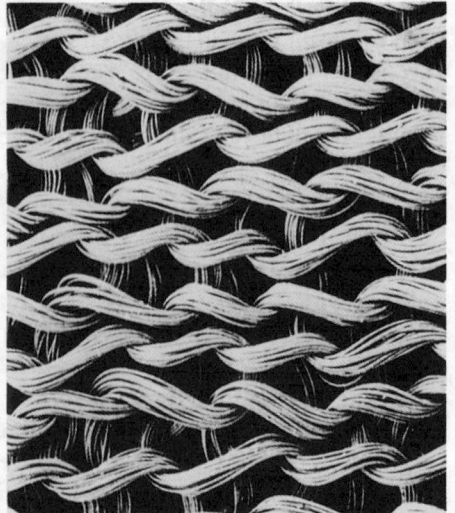

Weft-knit jerseys from bulked and flat yarns.
(David Bailey)

TRY YOUR HAND

You will need:
1 ball each of 10 different knitting yarns labelled as 8 ply
6 mm knitting needles
accurate balance
magnifying glass
metric graph paper

What you do:

To measure bulk:
Measure the diameter of each yarn in millimetres by placing the yarn in a relaxed state on metric graph paper.
Now stretch the yarn as taut as you can, and measure its new diameter.
Bulkiness, B can be found by:

$$B = \frac{\text{original diameter} - \text{taut diameter}}{\text{original diameter}}$$

The larger the value of B, the loftier or bulkier the yarn, that is, the more air is incorporated in its structure.
Because bulk is strictly a function of the *volume* of the yarn, a more precise formula would use the cross-sectional area:

$$B = \frac{(\text{original diameter})^2 - (\text{taut diameter})^2}{(\text{original diameter})^2}$$

To measure count:
Measure out exactly 10 metres of each yarn.
Find the mass in gram of each piece.
Tex, the metric measure of yarn count, is the mass in gram of 1000 metres of yarn. Therefore, the tex value of the yarns will be 100 times the mass in gram of the 10 metre samples.
The higher the tex value, the more fibre in the yarn.
Note: Different fibres have different densities, and so will give different tex values for the same yarn thickness. These differences, however, are very small, and in practice can be ignored.

To examine twist and ply structure:
Take samples of each yarn apart, and check: (i) the ply structure; (ii) twist levels (the number of turns per centimetre) in the yarn and in the plies; (iii) the direction of twist in the yarn and in each ply.
You will need to use the magnifying glass for some of these measurements.

2 ply twist, high twist   low twist   3 ply twist

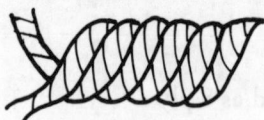

S twist in both ply and singles creates an unbalanced 'twist lively' yarn structure.

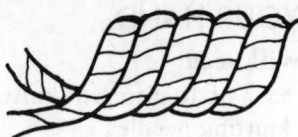

S twist in ply cancels Z in singles – balanced structure.

To examine the effects of bulk, count and structure:

With the 6 mm knitting needles, knit a sample 10 stitches × 10 rows of each yarn.

The same tension must be used for each sample: the same person should do all the knitting.

Measure the size of each knitted sample (length and width).

Weigh each sample, and calculate how much yarn was used in it:

$$\text{length of yarn used (metres)} = \frac{\text{(mass in gram)} \times 1000}{\text{count in tex}}$$

Examine your knitted sample. Is it loose or transparent or tight or heavy or fluffy or stretchy or soft or harsh or stiff or limp or any combination of these?

1   Tabulate your results. For each yarn name, list fibre type, ply structure, twist structure (S or Z), twist levels, tex count, bulk, area or size of knitted sample (mm² or mm × mm), mass of knitted sample, length of yarn in knitted sample, evaluation.

2   What effect does bulk have on the appeal of knitted samples?

3   What is the effect of twist levels on bulk?

4   Which yarn has the best cover? the worst? Is bulk a fair measure of cover? Why?

5   What effect does yarn count have on the stiffness of the knitted samples?

6   How does ply structure affect the surface smoothness? How does it affect bulk?

7   What does the mass of the samples relate to? Plot graphs of: tex against mass; tex against area; bulk against mass; bulk against area. A simple relation will give a straight line.

8   Does your table show any other relations between properties? For example, do you find that fibre type is related to bulk or mass of knitted sample, or that bulk is related to ply structure?

9   Does 'nominal ply' relate to actual yarn plying? Does it relate to yarn count? to yarn diameter? Why is it used by manufacturers?

10   How can you use the information in the tables if you plan to knit a jumper, and do not have a pattern designed especially for your chosen yarn?

11   Would all the yarns be suited for the same end-uses? For each sample, suggest an appropriate end-use, such as winter-weight jacket, shawl, knee rug. What criteria have you used in matching the samples to the end-use?

12   What criteria would you use in evaluating the appeal of each of these samples? Are these criteria applicable to every end-use, or would they change with end-use?

Quality

For most uses – excepting special fancy styles – yarns have to be even. Along their entire length they need uniform properties of strength, diameter, hairiness, colour, fibre composition, and twist levels. They must be free of knots and of tangled fibres (NEPS) and other imperfections.

Throughout the different stages of processing the yarn is checked to ensure final quality. The finished yarn is then checked for evenness of diameter and for evenness of strength – samples are statistically chosen from each manufactured batch to make sure that all the yarn produced conforms to specified quality levels. Even in fancy yarns, thickness is closely controlled.

The control processes include measurements of sliver weight, atmospheric humidity, number of neps and knots, strength at various intervals and the evenness of the final yarn. Evenness is in a way more important than maximum strength. A yarn with a high average strength but great variability may break more readily than a uniform yarn of lower average strength.

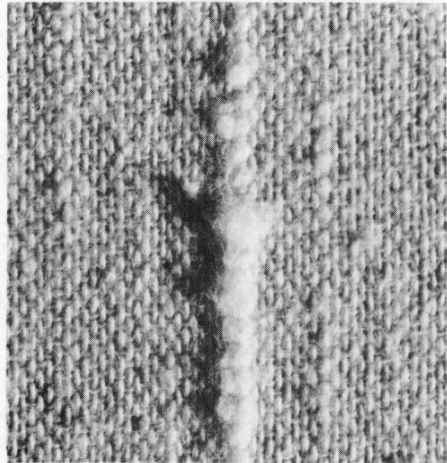

The importance of quality control. Thick spots on yarns may snag in the healds and break during weaving, or may cause unsightly faults in the fabric. Electronic detection devices locate thick and thin faults in yarns before winding.
(Zellweger Uster AG, Switzerland)

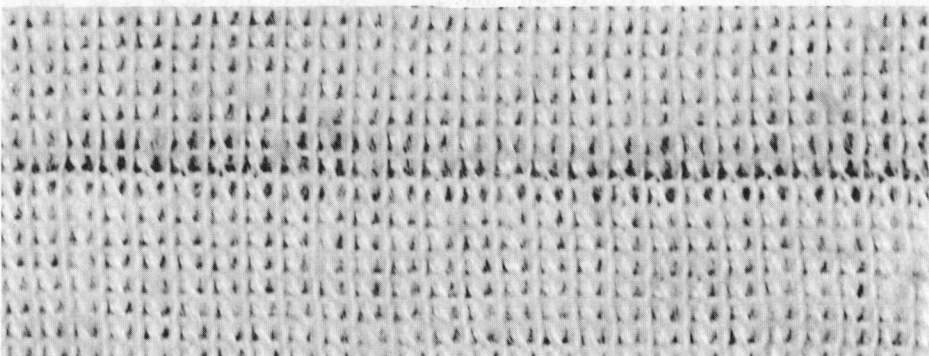

A thin yarn shows clearly as a fault in this knit.

TRY YOUR HAND

Select a variety of fabrics, such as poplin, interlock, jersey, lawn, percale, gingham.

Hold the fabrics against a light source, and check the evenness of the yarns in transmitted light.

1 Identify which fabrics have been manufactured from a filament yarn, and which from a staple yarn.
2 Which yarns were more even, filament or staple?
3 Did all staple yarns have the same level of evenness? Did all filament yarns?
4 Did you notice the unevenness of yarns on the fabric surface before you looked through the cloth?
5 Was any of the irregularity intentional?
6 How do you think irregularity of diameter affects the strength of yarns?

TRY YOUR HAND

You will need:

potato
knitting needle
kniting yarn
some unspun wool

What you do:

Make a potato-and-knitting-needle spindle as shown.

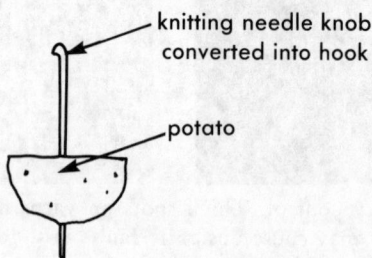

knitting needle knob
converted into hook

potato

Tie a length of knitting yarn to the knitting needle shaft of the spindle, and guide it around the potato weight and to the knob of the needle, as shown. This is the starter yarn.

Hold the loose wool fibres in one hand.

Join the starter yarn to some fibres between the thumb and forefinger of the other hand.

Hold tight, and twirl the spindle in the same direction as the twist in the starter yarn.

Next, with your free hand, grip the yarn and fibres close to the hand holding the wool. With the wool hand, slowly pull the bundle of fibres away, so as to DRAFT out a parallel strand of fibres.

As you feel the twist from the spindle climb up along the yarn to reach your fingers, move your thumb and forefinger up along the drafted strand. This allows the twist to run up, forming your new yarn.

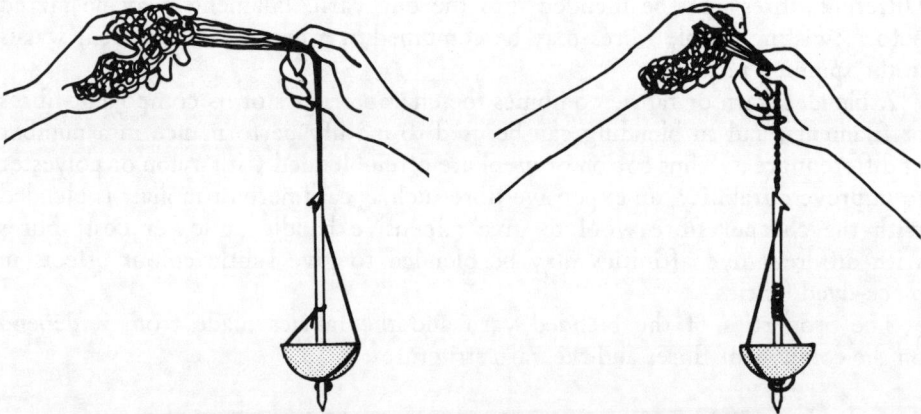

1 Is your new yarn even?
2 Which parts have the most twist, the thick or the thin?
3 On your sample, test your answer to Question 6 in the exercise above.
   Repeat your experiment several times to check your observations.

The importance of twist

Staple fibres are held in the yarn by means of *twist* (except in the case of twistless staple yarns in which they are glued together). The twist in the yarn pushes the fibres towards the centre of the yarn and towards each other. This creates a fibre-to-fibre friction which stops the fibres slipping past one another when the yarn is pulled.

In low twist yarns, breakage occurs because fibres slip past one another. In very high twist yarns, the *tension* created in the fibres by the twist can be so great that a slight pull on the yarn can bring the fibres to their breaking point. Therefore there is an optimum amount of twist for each staple yarn which creates maximum yarn strength.

TRY YOUR HAND

Take a thick, loose knitting yarn, and increase its twist. What happened to the diameter of the yarn? Hold the yarn at points 10 cm apart and try to break the yarn. How has increased twist affected the ability of fibres to slip over one another?

With a spinning wheel and some wool spin a fairly uneven yarn.

Watch the distribution of twist. Allow some slubs to form, then allow the yarn to overtwist in a narrow part. If you try hard enough, you can allow so much twist to accumulate in one part of the yarn that it will snap of its own accord!

Which are the weak spots in your yarn: the thick slubs or the narrow neckings?

Blends

Different fibres may be blended into the one yarn. Filaments may be mixed before twisting; staple fibres may be combined at a number of different stages in the spinning process.

A blended yarn or fabric combines the characteristics of its component fibres or filaments, and so blending can be used to modify performance in a number of different areas. Thus cotton or wool are often blended with nylon or polyester to improve durability; an expensive fibre such as cashmere or mohair is blended with the cheaper fibre wool to give expensive handle at lower cost; fibres with different dye affinities may be blended to give subtle colour effects in piece-dyed fabrics.

The properties of the blended yarn and the fabrics made from it depend on the component fibres and the yarn structure.

A programmable bale opener and a weighing hopper feeder. These allow precisely measured amounts of cotton fibre from a number of bales to be blended, thus ensuring uniformity of fibre quality.
(Trützschler GmbH and Co KG, Mönchengladbach, West Germany)

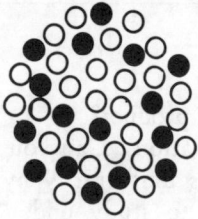

Intimate blend.   Non-uniform blend.   Core and sheath.

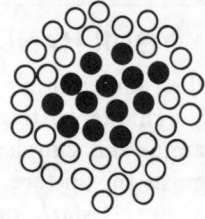

The different fibres in the yarns may be combined in a number of ways. If blending takes place early during processing, eg. during opening), the result is a more intimate and *uniform* blend. The coarser fibres will give the fabric its characteristic touch or handle, masking the effects of any fine (and costlier) fibres in the blend. If fibres of the same diameter but different stiffness (rigidity) are blended, the stiffer fibre, because it resists twisting, will remain on the outside of the yarn, and so will dictate the handle. The Selfil and Coverspun techniques (p. 348 and p. 349) and some other spinning processes are designed to give a *non-uniform* distribution of the component fibres. Thus each fibre in a 50/50 blend need not contribute equally to the final yarn or fabric properties. In practice, it is very difficult to predict exactly what the characteristics of a blend will be.

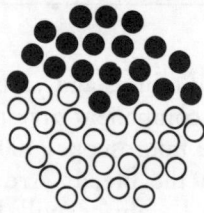

Viyella, a non-uniform blend of wool and cotton. The wool fibres, being coarser than cotton, end up on the outside of the yarns, giving a wool handle to a fabric which launders like cotton.
(Vivian Robinson)

TRY YOUR HAND

Collect some samples of blended fabrics and yarns.

What characteristics does each component give to the final product?

Examine the samples under the microscope (100× magnification). Can you distinguish the different fibres in the blend? Are the blends uniform? Why, or why not? (*Marl*, or cross-dyed, fabrics will be easiest to study under the microscope: many school uniform worsteds are such blends.)

Filament yarns

Twisted filaments

Although filaments (silk or synthetic) can be used in untwisted form (FLAT yarns) in fabric manufacture, *twist* may be added to a filament yarn: to make it more *compact,* to create *torque* (twist liveliness), or to give *texture* to a fabric. Twisted filament yarns are called THROWN yarns.

Filament and staple yarns can be given high levels of twist, and combined with other twisted yarns to give a variety of effects.

Textured filament yarns

There are many different ways of adding properties of texture, bulk and elasticity to filament yarns. Thermoplastic filaments are particularly suited to *texturing*.

Generally, the multifilament yarns are distorted by twisting or squashing in some way, and are heated. When allowed to cool, they keep their deformed state, and will spring back if stretched. Various degrees of residual stretch can be left in a yarn by a second heat treatment.

Some processes introduce mainly stretch, others mainly bulk, yet others mainly texture. Not all texturing processes need heat.

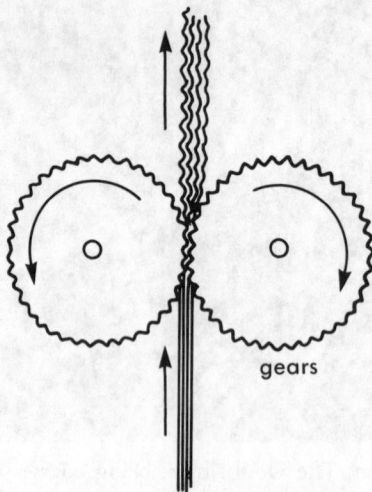

gears

Gear crimping is a simple method of texturing thermoplastic filaments.

Crimping machinery for texturing thermoplastic filament yarns. (Barmer Maschinenfabrik AG, Remscheid-Lennep, West Germany)

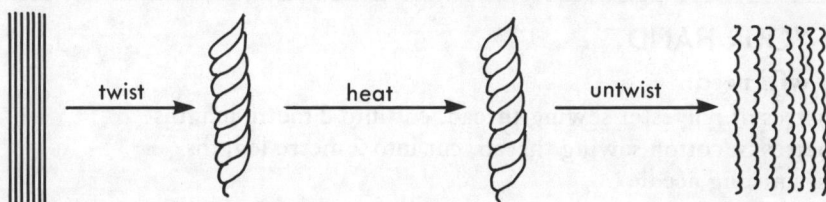

False twist texturing.

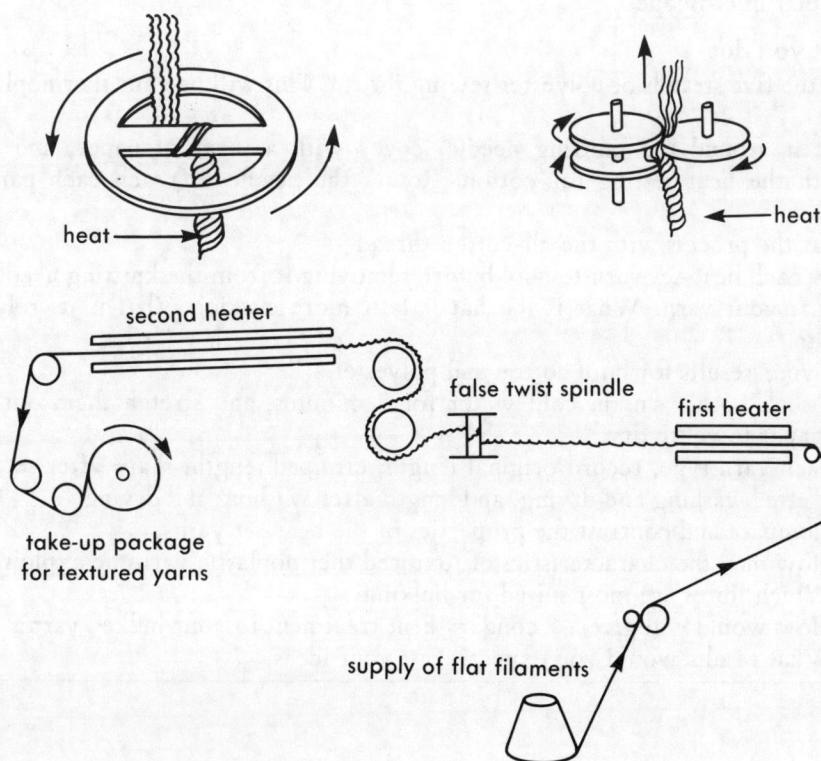

Methods of creating false twist.

False twist texturing

This is the most common method of filament texturing.

The multifilament strands are twisted, heated and untwisted. A second heating process – during which the yarn is kept at a controlled extension – sets in the required amount of bulk.

Without the second heating process, the yarn is bulky and stretchy. This is suitable for hosiery and sportswear, as in Helanca, used for tracksuiting, stretch pants, and swimwear. With the second heating process the bulk is retained while some of the extensibility is lost, as in Crimplene, which is suitable for outerwear applications.

These processes are very fast – a modern friction twister can produce 1200 m of yarn each minute.

TRY YOUR HAND

You will need:

10 metres of polyester sewing thread, cut into 2 metre lengths
10 metres of cotton sewing thread, cut into 2 metre lengths
metal knitting needle
paper
iron
ruler
binocular microscope

What you do:

Take the five strands of polyester sewing thread. This will be your thermoplastic yarn.

Wind it around the knitting needle, cover with a sheet of paper, and iron it with the heat setting for cotton. Rotate the needle to reach each part of the yarn.

Repeat the process with the all-cotton thread.

Allow each heat-set yarn to cool before removing it from the knitting needle.

Look at your yarn. What is its shape? Is it more elastic? What is its relaxed length?

Note your results for both cotton and polyester.

Now soak both yarns in cold water for 5 minutes, and stretch them out flat on a paper towel to dry.

For each yarn type, record original length, crimped length, shape after setting, shape after washing and drying, and length after washing and drying.

1   Compare and contrast the properties of the heat-set yarns.
2   How may the characteristics of textured thermoplastic yarns be exploited?
3   Which fibres are most suited for bulking?
4   How would you give a secondary heat treatment to your bulked yarn?
5   What results would you expect? Test your ideas.

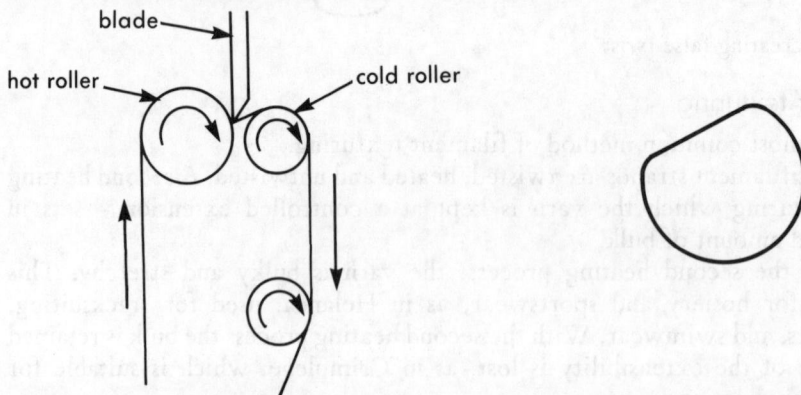

Edge crimping produces a filament slightly flattened on one side, with a tendency to coil.

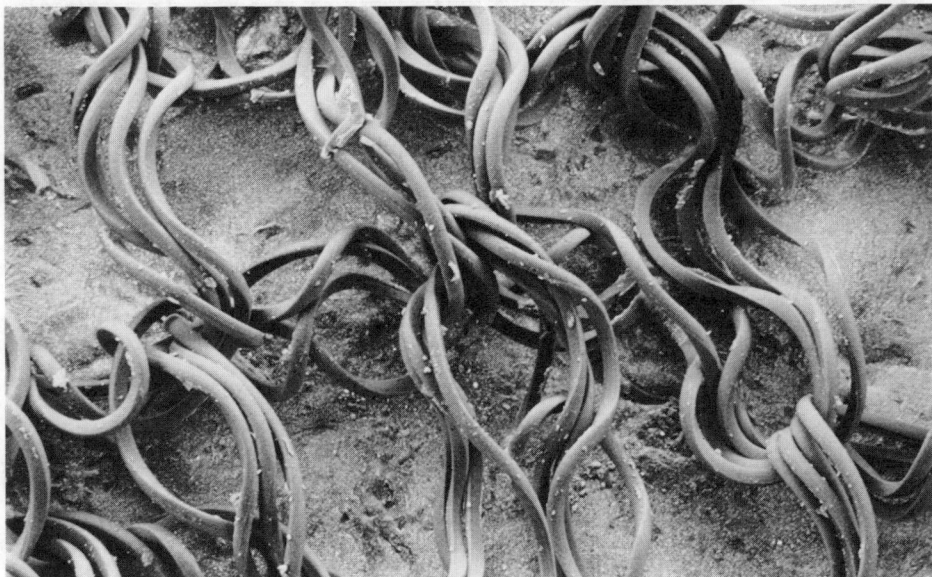

Edge-crimped nylon in women's stockings.
(Vivian Robinson)

Edge crimping

In this process, the hot filaments are crimped or curled in their passage over the edge of a blade. During this sharp turn, the outer edge of the filament is stretched, and the inner edge is compressed. This makes one side of the filament longer than the other, and so gives the filament a permanent helical shape.

The helical filaments are very elastic, and are mainly used for hosiery.

Knit-deknit

This heat-setting process is used for creating fancy bouclé effects from continuous filament yarns. The final effect depends on the gauges of the original and final knitting machines.

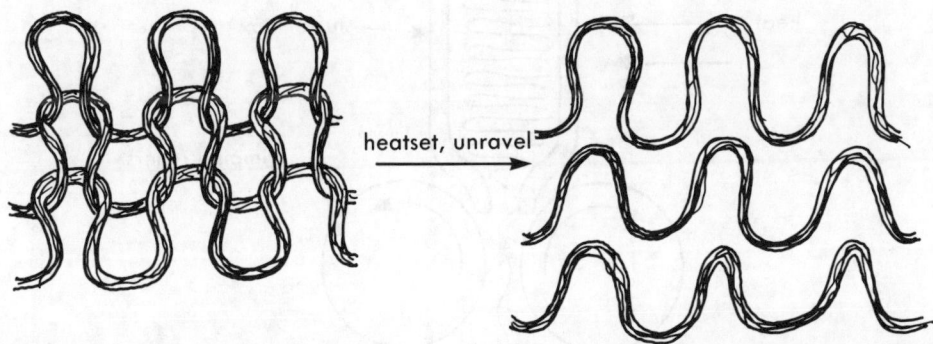

heatset, unravel

Knit-deknit produces a kinky yarn.

TRY YOUR HAND

You will need:

ball 3 ply nylon knitting yarn
pairs of knitting needles, sizes 2 mm, 3 mm, and 4 mm
iron

What you do:

Knit up the ball of yarn, using 3 mm needles and plain stitch.
Press the fabric under a pressing cloth, at cotton setting.
Allow the fabric to cool, and then unravel it.
Examine the shape of the yarn.
Now knit the yarn into samples, one with the same needles as the original fabric, one with the finer needles, and one with the thicker needles. Use stocking stitch.

1   Compare and contrast the effect achieved by the textured yarn in each case, and compare with the original fabric.
2   What is the effect of the texture on the extensibility and elasticity of the fabric?

Stuffer box

When thermoplastic multifilament yarns are squashed into a heated container, they are heat set into a bulky, zig-zag shape. This type of yarn is not as extensible as the false-twist or edge-crimp yarns, but has good bulk with a rather harsh handle.

Such yarns are mainly used in upholstery, carpeting and outerwear.

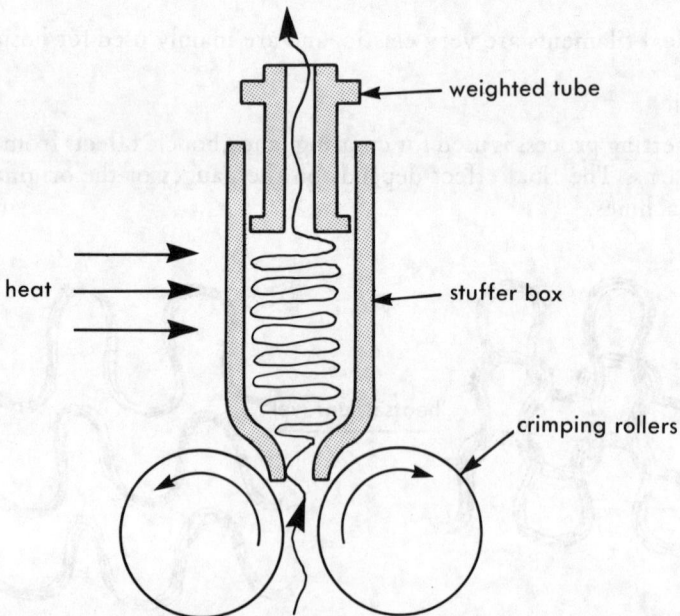

Stuffer box.

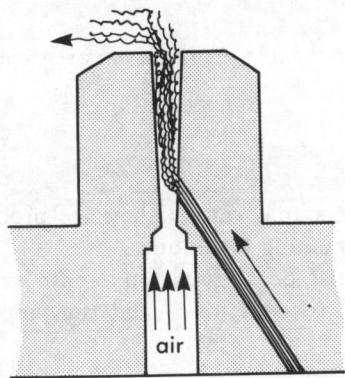

Air-jet texturing produces a looped textured yarn without creating extra bulk or increasing elasticity.

Air-jet texturing

Not all methods of filament texturing require thermoplastic yarns. Some create texture without the use of heat. The Taslan method of yarn texturing blows loops of filaments out of a loosely twisted bundle. This gives the yarn a less slippery texture, but no extra extensibility and only a little extra bulk.

Taslanized yarns are used as *non-slip* synthetic *sewing threads*, for woven goods with a 'natural' handle, and for upholstery fabrics.

Fasciated yarns

The word 'fascis' means a small bundle of sticks (these were the symbols of power that Roman senators carried). When fine parallel filaments are exposed to sharp bursts of air at right angles to them, some of the outer filaments break and their ends wrap around the other filaments to form a 'tied' bundle. These fasciated yarns have a particularly *linen-like* handle when woven into cloth.

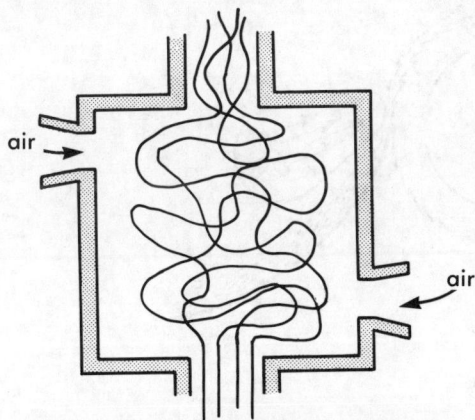

Fasciated yarns are produced when a strong air stream causes some filaments to break and wrap around others in the yarn.

High bulk yarns

Acrylic fibres cannot be heat set. Instead, their *shrinkage* properties are exploited to create bulk (p. 153).

Yarns from staple fibres

Seventy per cent of all yarns used today are staple yarns. They are produced on machinery developed from ancient hand-spinning techniques.

For short fibres to form a yarn, they must be aligned, and must overlap – usually a yarn is at least 15–30 fibres in cross-section – and they must be *twisted* to hold the fibres in place.

Fibre preparation

Much of the energy consumed in the spinning of staple yarns is expended in the preparatory processes of cleaning, blending, and aligning the fibres. These processes thus contribute greatly to the cost of the final yarn. In addition, the final character of the yarn may be largely determined by the nature of the preparatory processes the fibres go through before spinning.

Cleaning: cotton

All natural fibres need to be cleaned before processing.

Cotton *seeds* are removed by a machine called a GIN. This has a knife-like comb to catch the seeds and permit the passage of the cotton fibres. The cotton fibres are then tumbled about in a fast-moving air stream to loosen dust, twigs and other debris, which is collected at sieve-like screens.

The cleaned cotton fibres (LINT) are then baled for transport; the seeds are sold for oil.

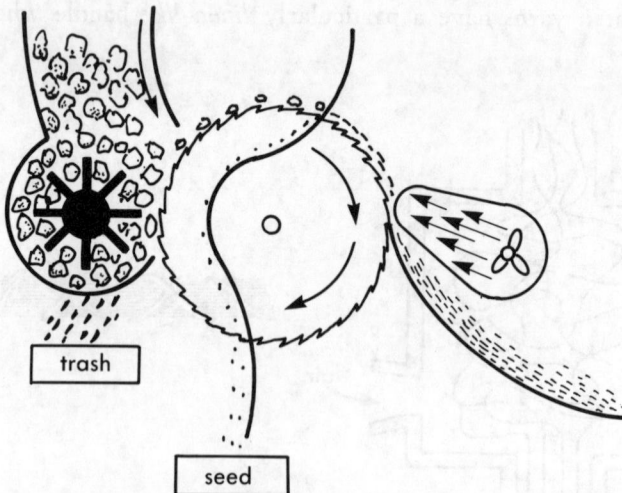

trash

seed

The cotton gin.

When the bales of cotton reach the mill, they are OPENED. The opener takes a quantity of cotton fibre from a number of bales in turn to eliminate the effects of any variation in quality between bales. The same blending effect may be achieved by feeding the contents of each bale into a hopper and passing them together into the next stage of the spinning process.

Automatic opening plant with bale opening machines.
(Rieter Machine Works Limited, CH-8406 Winterthur, Switzerland)

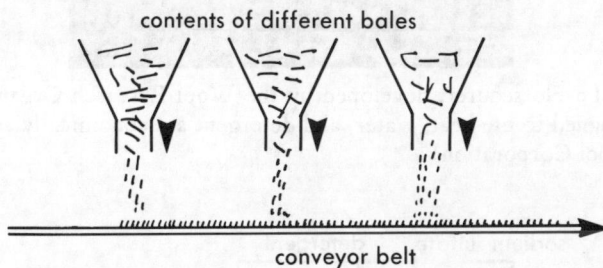

contents of different bales

conveyor belt

Blending of cotton.

Cleaning: wool

Wool is full of suint and grease, which is removed by SCOURING. Solvent scouring has been tried, but the most popular method of cleaning wool remains passing the wool through successive bowls of water and detergent. The scoured wool is then rinsed in clean water, squeezed, and dried in hot air.

327

Scouring wool to remove
dirt, suint, and grease.
(Australian Wool Corporation)

Washing the scoured wool.
No detergent or grease must remain.
(Australian Wool Corporation)

Scouring is costly, in water and energy. Industrial processes aim to save both water and energy. One approach is to use the scouring liquor on a counter-flow principle, so the grease and dirt are gradually concentrated. Water savings of up to 75 per cent can be achieved. The liquor is then centrifuged to separate the dirt, which settles as sediment and may be used as fertilizer, and the wool fat or *lanolin,* which is a valuable raw material for the cosmetics industry.

Because of potential effluent problems, many Japanese buyers circumvent their country's strict anti-pollution laws and scour their wool in Australia, before exporting it to Japan for further processing.

The WRONZ Lo-Flo scourer, developed by the Wool Research Organization of New Zealand, is designed to use heat, water, and detergent as economically as possible. (Australian Wool Corporation)

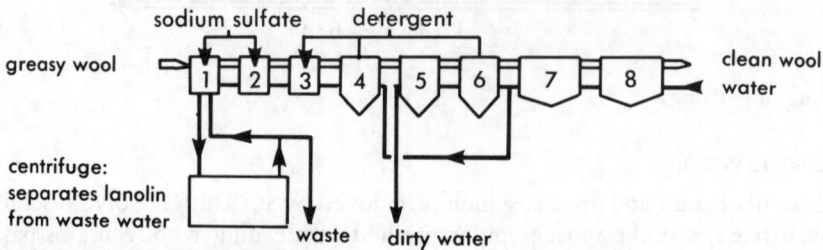

The Wronz Lo-Flo scourer (schematic).
(C.S.I.R.O. Division of Textile Industry)

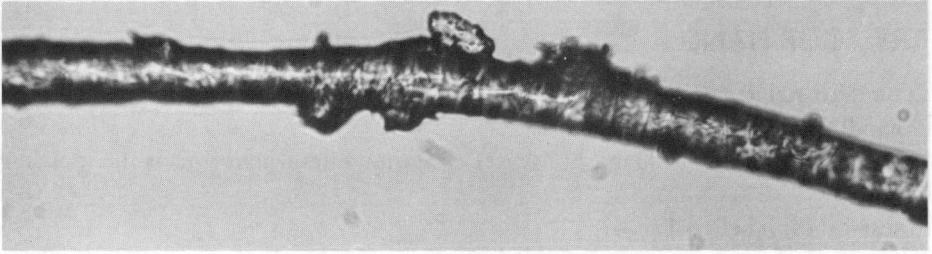

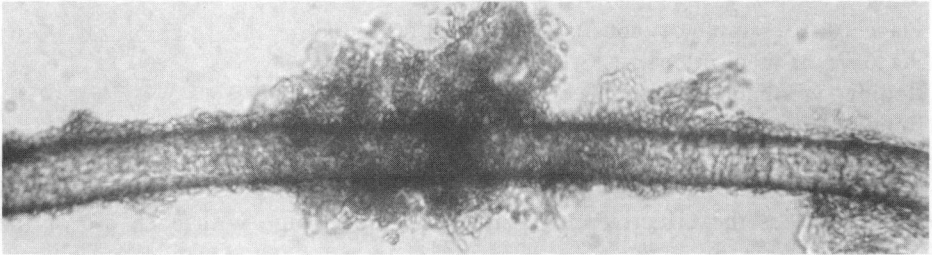

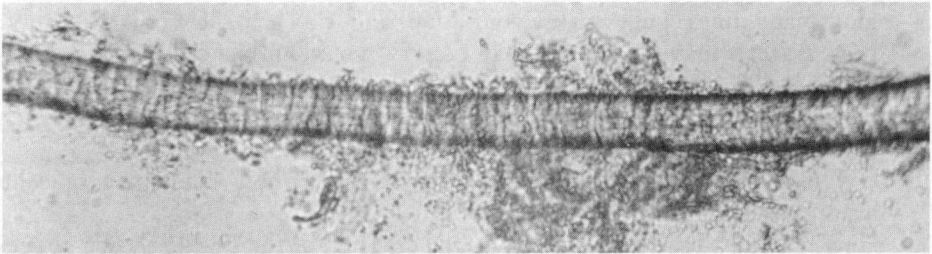

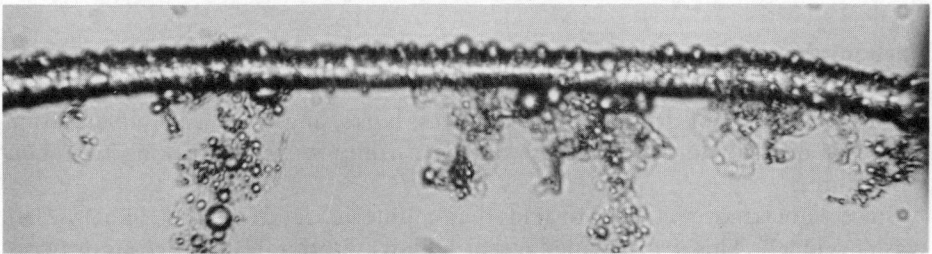

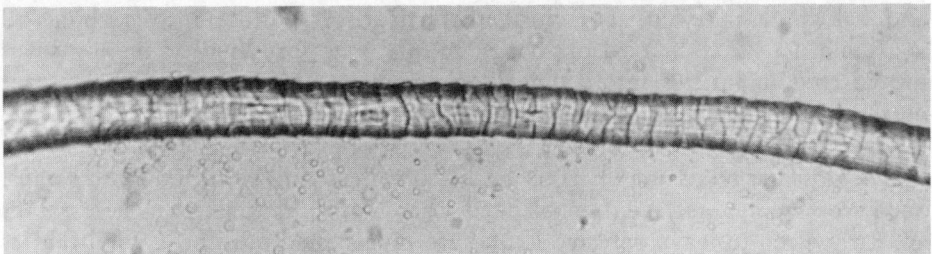

Scouring of wool. Detergent molecules attach themselves to the grease, breaking it up and carrying it into solution, leaving the clean wool fibre.
(C.S.I.R.O. Division of Textile Industry)

329

TRY YOUR HAND

You will need:
5 handfuls (5 g each) of raw wool
a 250 mL beaker with 200 mL hot water containing detergent (this is the scouring liquor)
6 test-tubes, labelled 1 to 6

What you do:
Place a 1 mL sample of scouring liquor in a test-tube.
Place 5 g of wool in beaker. Stir gently a few times.
Remove wool and squeeze. Keep for reference.
Repeat these three steps for each new greasy wool sample.

1   How does the colour and content of the scouring liquor change with each use?
2   How does the effectiveness of the scouring change with each use of the liquor?
3   Does the dirty water clean the wool at all?
4   How many times could you use your liquor and still clean the wool noticeably?
5   If you are a manufacturer, what factors would influence your standards of acceptable cleaning?
6   If you wished to recover lanolin (wool fat) from the scouring liquor, which batch would you use?
7   If you had to heat the water used for scouring, how could you save on heating costs?
8   How do your observations relate to the diagram of the Lo-Flo shown above?

Carbonizing

Some *poorer qualities* of fleece – such as from sheep that grazed dry or scrubby pastures – are full of dried twigs and leaves, burrs, and other vegetable matter. It is very difficult to remove these during carding without breaking too many wool fibres.

Since wool is more resistant to acids than cellulosic materials are, CARBONIZING was developed. This is performed either half-way through the scouring process, or at its end. The scoured wool is dipped in dilute sulfuric acid, quickly dried, and baked in an oven. This process turns the cellulose into carbon; all vegetable matter is charred or 'burnt' by the acid. Crushing or flexing turns the charred impurities into a fine black powder which is easily shaken out during further processing. The carbonized wool is then neutralized to remove traces of acid.

Old wool garments and blankets are often collected to be torn up into scraps. Such wool scraps may contain blended viscose and other fibres. These scraps are carbonized to leave only wool, and are reprocessed into those wool fabrics and garments which do not bear the 'pure new wool' label. Whilst carbonizing does not destroy protein fibres, it certainly does not help to soften their handle!

TRY YOUR HAND

Find samples of wool garments, some labelled 'pure new wool', others labelled '100% wool'.

1   Which is more likely to have been made from carbonized wool?
2   Compare the handle and softness of the differently labelled fabrics. How does the handle relate to the labels they carry?

Staple yarns from filaments

When the filament output of many spinnerets is collected into a large bundle, the resultant mass is called a TOW. Tow can contain up to tens of thousands of filaments.

To produce *staple fibres* the untwisted tow is cut or torn into shorter lengths on specialized machinery. The length of the staple can be adjusted to that of either wool or cotton: the synthetic fibres can thus be processed on standard staple spinning systems to produce yarns which are similar in character to various natural fibre yarns.

Direct spinning

With this method, filament tow is converted into yarn in one direct operation. Light tow – only about 4000 denier – is used. It is broken into staple lengths by tension created between two pairs of tightly gripping nip rollers moving at different speeds. There is some randomness in the exact points where the filaments break, so the fibres vary in length, just as natural staples do. The broken fibres are supported on a slanted rubber apron and carried to the drafting rollers of a ring spinning system.

This method is one of the fastest for staple production, but it cannot be used to process blends. It is rarely used, however, despite the speed of production, as the machinery is expensive and yarn quality is a problem: weak spots may occur where broken fibre ends are clustered.

Fibre alignment

Carding

After the cleaning process the fibres point every which way. CARDING allows the tufts of fibres to be organized into a web in which a large percentage of the fibres point in the one direction, which will eventually be the direction of yarn formation.

Cotton fibres are air-blown through ducts in carefully weighed quantities. The fibres are carried around a cylinder coated with card wire, against which a series of wire-coated flats is moved. Although the tips of the wire teeth do not touch, the fibres are pulled against each other and are partly untangled, forming a web. During the process, burrs and other vegetable matter are removed as TRASH.

Wool, being a longer and weaker fibre than cotton, needs much gentler carding, and so the fibres are aligned through a greater number of less vigorous steps.

Wool carding machinery.
(Gebrüder Sulzer Aktiengesellschaft, Winterthur, Switzerland)

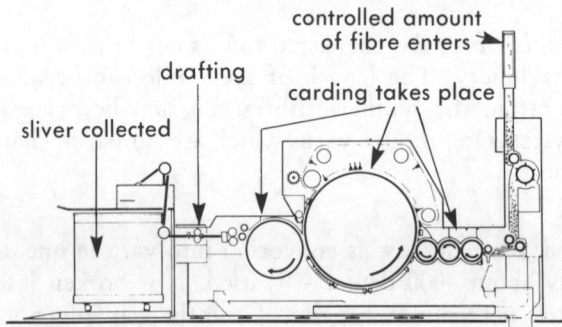

Cotton card (schematic). The fibre enters on the right. The fibres are gently aligned as they pass a series of wire teeth; the carded web is then drafted to form a thick sliver, which is coiled into the drum on the left.
(Schubert and Salzer, Ingolstadt, West Germany)

Accurate control of the quantity of fibre fed to the cards ensures a uniform SLIVER goes on to the next process, and goes a long way towards ensuring uniformity in the final product. For hairy yarns such as *woollens* or *carded cotton*, the fibres can proceed from this point to the spinning frame.

On the card, many fibres are processed together. A card web typically has enough fibres laid side by side for more than 100 yarns. Therefore there is a need to reduce the carded sliver to smaller dimensions before spinning.

Wool carding.
(Australian Wool Corporation)

The fibres are aligned, but the sliver has no strength

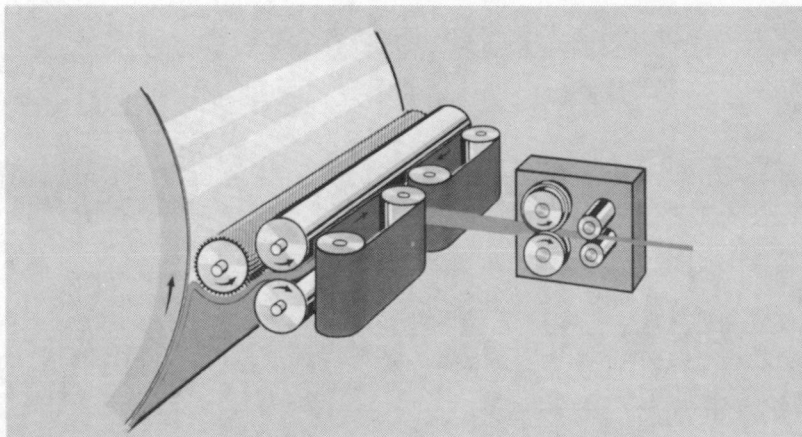

Apron drafting system designed to produce level (even thickness) sliver.
(Schubert and Salzer, Ingolstadt, West Germany)

At the end of the card, the web may be collected to form a sliver for further processing. For WOOLLEN yarns, the production sequence takes a short cut at this point – instead of forming one sliver, the web is divided into narrow strips by tapes. The strips of web are rubbed between oscillating rubber aprons to form thin TAPE-CONDENSED slivers, which proceed directly to the spinning frames. Because the fibres are only partially aligned in the web, woollen yarns have a characteristically fuzzy appearance.

The carded web is converted into sliver in preparation for spinning.
(*Textiles*, a periodical published by Shirley Institute, Manchester, U.K.)

The sliver (here, cotton) is coiled into large drums which will supply the spinning frames.
(Trützschler GmbH and Co KG, Mönchengladbach, West Germany)

Drawing

For worsteds (fine suitings), semi-worsteds (outerwear, carpet yarns), and the finest cottons, further processing is necessary after carding.

To ensure uniform thickness, 6–8 slivers are placed side by side and *drafted* out to the original thickness of one sliver. This process aligns the fibres by pulling them against each other, and also helps to cancel out thick and thin spots. For *cotton,* a DRAWFRAME with drafting rollers is used. The wool sliver is thicker than a cotton sliver, and it also contains a greater range of fibre *lengths.* For successful drafting, a system of pins called a GILLBOX is used to control the fibres, and to help straighten and align them.

Evenness may be further ensured by the use of an *autoleveller* device, which changes the speeds of the drafting rollers to compensate for thick or thin spots.

Drawing and gilling may each be repeated a number of times to ensure uniform slivers and aligned fibres.

Blending

Staple fibres may be blended during opening or carding, or by repeated drawing.

Combing

Staples – whether natural or artificial – contain a distribution of different lengths of fibres. The shortest of these fibres may be unsuitable for spinning fine yarns; they cannot turn through many twists, and their ends stick out of the yarn in many places.

For carded cottons and semi-worsted yarns, drawing or gilling is the end of the preparatory process. But for WORSTEDS and for *combed cottons,* all the short fibres must be removed from the slivers, and the fibres must be as parallel as possible. This is achieved by COMBING.

The combing process removes the short fibres as NOIL, and leaves the long fibres as the TOP. Noil will also contain neps, knots, and vegetable trash.

For all stages in the handling of staple fibres, machinery is designed so the longer fibres are not broken, and the shorter fibres are kept controlled. The upright (French) comb for wool and the cotton comb both work on the same principle, but the machine for the longer wool fibres is much larger in all its dimensions than the cotton comb.

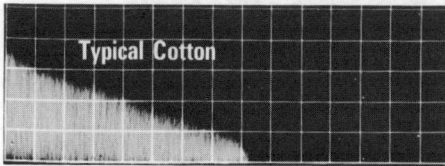

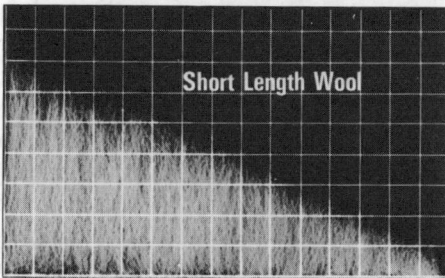

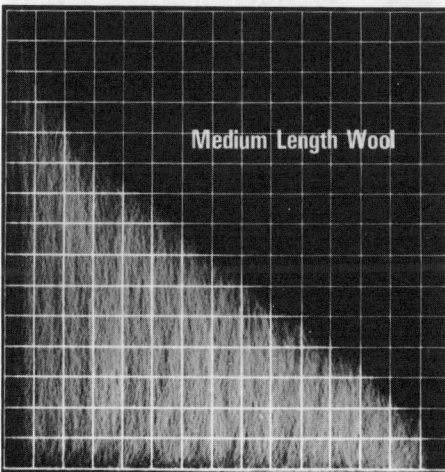

Typical fibre length distributions for cotton and fine and medium wool. (C.S.I.R.O. Divison of Textile Industry)

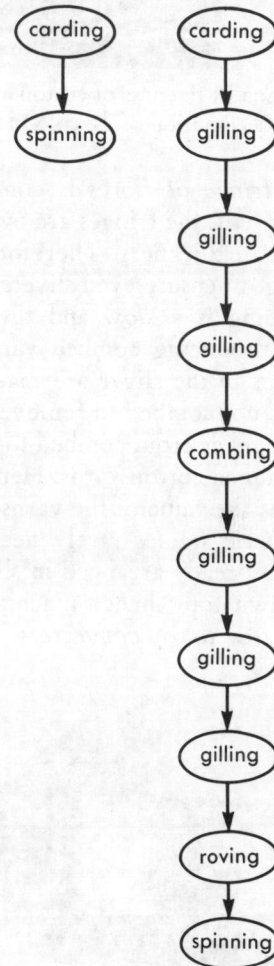

Production stages of woollen and worsted yarns.

335

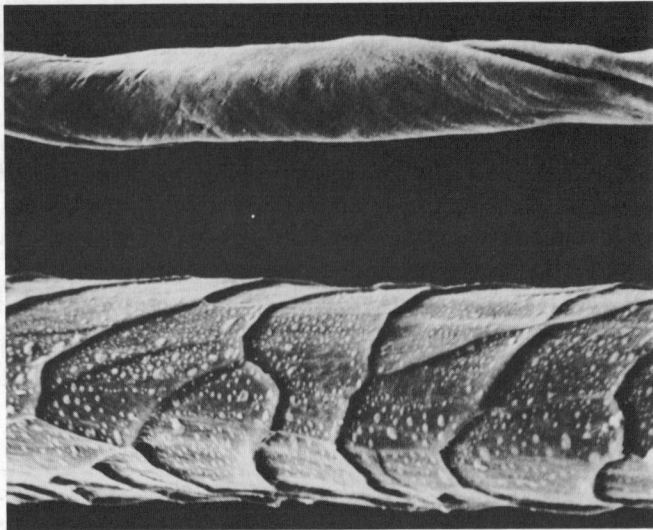

Comparison of fineness of cotton and fine wool.
(C.S.I.R.O. Division of Textile Industry)

One *fringe of fibres* is combed out at a time – first the front half, then the back half; the fringes are overlapped to form the *top,* which is not altogether uniform in thickness. Therefore, a number of drawing processes must follow combing, to ensure even slivers for spinning.

Combing is a *slow* and therefore *expensive* process. Because of the high cost of producing combed yarns, a compromise is sometimes made, whereby the fibres in the sliver are made parallel by repeated drawing and gilling, but they are not combed to remove the short fibres.

Yarns made from combed tops are smoother, finer, and stronger than woollen or uncombed cotton yarns. Hence, despite the cost of the slow process, combing increases the value of the yarns.

Synthetic staples rarely need to be combed. If they are produced as *tow* they are already arranged in sliver form, with all the characteristics of combed and drawn tops: hence machines producing tow from synthetic filaments are termed *tow to top converters.*

Tow to top converters allow synthetic filament tow to be converted directly into staple fibre slivers.

336

Roving – the final step before spinning

The production of ROVINGS follows gilling or drawing, and is basically a reduction in the diameter of the sliver. In order to hold the fine group of fibres together, a small amount of twist is inserted. For cotton the *speedframe,* and for wool the *roving frame,* carry out this function.

The sliver is drafted out between pairs of rollers turning at successively higher speeds. Final fibre control is achieved with an apron draft system. At the point where the largest drafting ratio occurs, a rubber apron presses the fibres together. This prevents the bundles of fibres slipping past one another, controls floating fibres, and allows even drafting to take place. The twist is inserted by a hollow flyer.

The thick sliver in the drums is drafted to produce the finer rovings, which have a small amount of twist inserted to give them strength.

Drafting the rovings on a cotton speedframe.
(Zinser Textilmaschinen GmbH, Elbersbach, West Germany)

roving

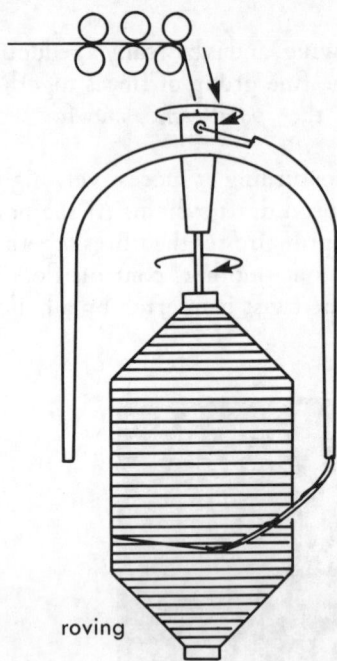

The hollow flyer.

The hollow flyer which inserts the small amount of twist necessary to hold the rovings together. (Zinser Textilmaschinen GmbH, Ebersbach, West Germany)

A cotton speedframe. (Zinser Textilmaschinen GmbH, Ebersbach, West Germany)

Some of the most recent cotton spinning processes use the slivers directly from the card for making yarn, thus eliminating the need for drawing, combing and the making of rovings. However, as the character of such yarns differs from those produced by traditional methods, these preparatory processes have not yet been superseded in the interests of economy.

Spinning of staple yarns

Mule spinning

In traditional hand spinning, the fibres are drafted to the correct thickness, and then the twist from a rotating spindle, which also serves as the carrier of the twisted yarn, is allowed to run up to the point of drafting. This is the method you used with the potato spindles (p. 316).

This process is imitated in MULE SPINNING. Mule spinning is slow, and therefore expensive. It is a stop-start process: while the fibres are being drafted, they are not being spun, and while they are being spun, they are not being drafted. In addition, the machinery required occupies a large floor area – and that also is expensive – and it is difficult to maintain. For these reasons, although mule spinning yields excellent fine yarns from short fibres, it is becoming increasingly rare.

Ring spinning

In spite of many other innovations, the most widely used spinning system uses the RING SPINNING FRAME. Ninety per cent of all staple yarns are ring spun.

The roving (or, for greater evenness, two rovings) is reduced to the required thickness by an apron system of drafting. A *rotating spindle* inserts one turn of twist into the yarn for each revolution it makes. The yarn then passes under the hook of a TRAVELLER and is wound onto the *rotating bobbin*. The traveller, which is pulled around a ring-shaped path by the yarn, controls the speed at which the spun yarn is wound onto the package.

Hand spinning in Hungary.

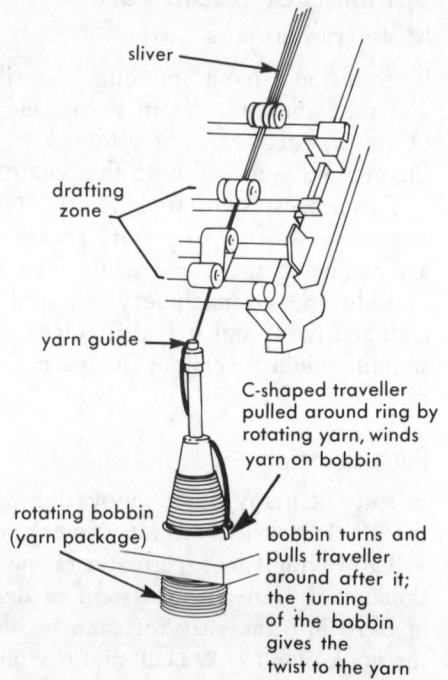
The mechanism of ring spinning.

High draft worsted ring spinning. Worsted yarns require the wool fibres to be kept precisely aligned, and here drafting is achieved by two sets of rollers instead of the traditional apron system.
(Officine Gaudino, Cossato, Italy)

Every time a bobbin is filled it must be taken off the spindle (DOFFED) and replaced by a fresh bobbin. This limits the *length* of the spun yarn in the package, and the *speed* of production. Doffing and PIECING-UP (knotting) are automatic on only the most modern ring spinners, and so most ring spinning is still quite labour-intensive.

Ring spinning frame. Twist is inserted as the yarn is wound on the spinning bobbin.

Wool spinning frames are larger than cotton frames, to handle the longer fibres. Frames used to produce *woollen* yarns often have special ratchets at the top of each spindle which vibrate the array, allowing a *weak, soft, fluffy* yarn to be spun at *low yarn tension.*

Ring spinning is a mechanized extrapolation from hand spinning, and so the machinery is simple in concept and maintenance is relatively straightforward. It produces fine quality yarns; better than other spinning systems. For these

Automatic doffing on a ring spinning frame. When the sensor registers that the bobbins are full, the full bobbins are automatically lifted off onto a conveyor belt and empty tubes are dropped onto the spindles.
(Officine Gaudino, Cossato, Italy)

Changing bobbins (doffing) on a cotton speedframe.
(Zinser Textilmaschinen GmbH, Ebersbach, West Germany)

With ring-spinning machines, supporting the rings on a mount of soft elastomeric washers from the support rail has considerably reduced noise levels in the spinning room. (C.S.I.R.O. Division of Textile Industry)

reasons, ring spinning is likely to keep its position in the market, in both Western countries and developing countries, for a long time yet.

Inventions such as rubber supports for the rings have helped to reduce the *noise* in the spinning room. Rotating rings reduce the *friction* between the traveller and the ring, allowing greater spinning speeds. Precision engineering of other components has generally improved efficiency.

There are, however, problems with ring spinning. The ring spinning spindle carries a whole package of spun yarn, and must rotate this load once for each turn of twist it inserts on the new yarn. This is very *wasteful of energy*. This energy cost, and the *limit on the size* of the rotating package, limit the speed and output of the whole process. Ring-spun yarn is produced at rates of 10–20 metres of yarn per minute per spindle. Other methods of spinning have been able to increase the speed of yarn production to 500 metres per minute and more.

All new spinning systems attempt to achieve higher efficiency by eliminating the rotating package.

Electronically controlled open-end rotor spinner. (Rieter Machine Works Limited, CH-8406 Winterthur, Switzerland)

This open-end rotor spinning frame for cotton includes sliver precleaning (trash removal). (Savio, Pordenone, Italy)

342

Open-end spinning

Open-end rotor spinning was the first commercially feasible process which eliminated the rotating yarn package.

The open-end spinner uses *card sliver* (usually drawn once or twice), and thereby *eliminates many of the preparatory steps* needed for ring spinning. A BEATER separates the individual fibres in the card sliver. These are thrown in a steady stream into the inside groove of a high-speed ROTOR – rather like candy floss. As the rotor turns, it twists the fibres around each other, and they are removed in yarn form through the centre. The finished yarn is wound on a *large, stationary* package.

The rotor used for open-end spinning must have a diameter close to the length of the fibres: longer fibres therefore need larger rotors. Larger rotors use more energy when they are rotated, and so are more expensive to operate. Open-end spinning is therefore most cost-effective for *short* fibres. It is the cheapest method of producing *coarse cotton yarns,* such as those used in denim, canvas, and sheeting.

Every modification to a spinning system changes the characteristics of the yarn. Open-end spun yarns do not have the same characteristics as ring-spun yarns and so, although they can be produced five times faster, they need to find market acceptance. They have captured a small part of the market and, as fabric manufacturers come to appreciate their properties and possibilities, they can be expected to become more popular.

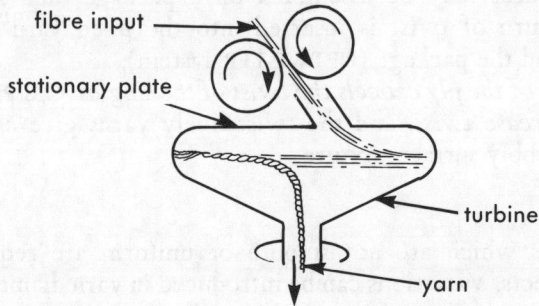

Open-end rotor spinning.

Comparison of ring-spun and open-end spun cotton yarns

Open-end rotor spun	Ring-spun
weaker	stronger
structure is a core of 80 per cent of the fibres, with an outer layer of wrapped and trapped fibres	uniform twisted structure
	yarn 'blooms' when cut: suitable for velvets
fewer faults	softer
more regular	more suited for polyester blends
much more abrasion resistant	
more durable	
does not squash easily	
more twist-lively in blends	

Winding

The yarn packages produced by the spinning frame are usually not appropriate for subsequent stages of production. The yarn is therefore rewound, for dyeing, plying, weaving, knitting.

During winding, automatic detectors check the yarn for slubs and weak spots. Any fault is cut out and the yarn rejoined by splicing or knotting.

Plying and fancy yarns

Plying

The yarn leaving the ring spinning frame is a SINGLES YARN: it consists of a single twisted strand. It is a fine yarn with unbalanced torque.

Yarn in this form may be woven or knitted, after treatment such as steaming, heat setting, or coating with size to reduce handling difficulties caused by the unbalanced torque. The greater the twist, the more lively the yarn, and the greater the handling difficulties.

For many other applications, such as ropes, hand knitting and craft yarns, sewing threads, many carpet yarns, and some fabrics, the singles yarn is PLIED: two or more singles yarns are twisted around each other.

In general, where a *large* arrangement of fibres is to be held securely together, individual thin strands are combined: plying is thus usual in *thick* yarns. It may also be used to make a yarn more *even,* to provide *strength,* and for *torque-free* behaviour.

The singles yarns may be assembled on a package, and simply pulled off together. One turn of twist is inserted into the plied yarn for each turn of the singles around the package (UPTWISTER system).

Usually, *twist of the ply cancels the twist of the singles.* CREPE yarns, however, are plied to *increase twist,* and the twist-lively yarns give crepe fabrics their characteristic bubbly surface texture.

Fancy effects

Sometimes yarns which are not smooth or uniform are required for special fabric design effects. Variations can be introduced in yarn diameter, composition, and the twist and colour of singles used in a ply.

Slub yarns

Thick-and-thin yarns are produced by irregular drafting of the roving during spinning. This can be achieved by a stick-and-slip effect when the rear drafting rollers are weighted periodically. Care must be taken so the SLUBS (thick patches) do not form at precisely regular intervals, but that each metre of yarn has approximately the same number of slubs.

Tweed yarns

These have small quantities of coloured neps and small tufts of different fibres sprinkled onto the carded web just prior to tape condensing (woollen system). Often the base web which forms the yarn is made from an intimate blend of differently coloured fibres. The exact quantities of each stock-dyed fibre component must be carefully measured into the feed hopper of the card, and there the fibres must be tumbled to achieve a uniform mix.

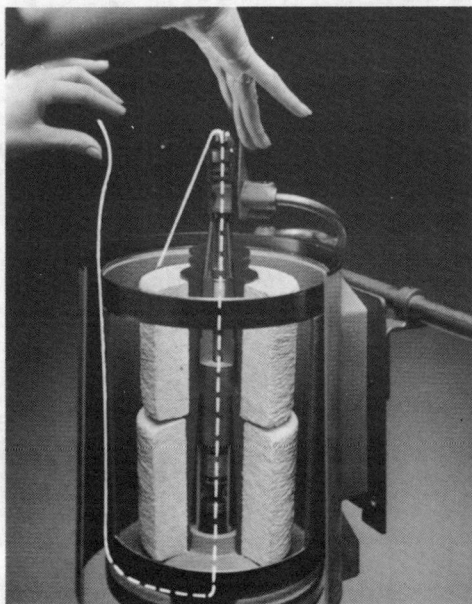

Plying. This modern two-for-one twister is equipped with automatic pneumatic threading. (Savio, Pordenone, Italy)

Subtle colour effects can be achieved in this way (e.g. Berber carpet yarns, Harris tweed, 'crafty-look' knitting yarns).

Loop yarns

A variety of bouclé, loop and snarl effects can be achieved during plying. If one of the plied strands is delivered through a separate pair of rollers which rotates faster than the delivery rollers for the other strands, then that singles component will form loops (or even snarls), which are held in place as the other strands twist about them.

Recent developments in spinning

Improvement of the *working environment* has concentrated on the elimination of dust and fly (airborne fibres). This is achieved by vacuum suction on machines such as cards and by blow-down and sweep-up devices. This not only improves the comfort of the workers, but also assures better final yarn quality.

At the spinning and winding stages, *automatic knotters* move along the spinning frames and tie or splice any broken ends. The full bobbins are automatically replaced by fresh ones, and are carried to the winding stations. One operator can now supervise over 300 spindles. With automation, the output of each worker has increased.

Many other processes previously performed manually are now handled automatically, and many mills have integrated computer materials handling systems. Spinning, which once was extremely *labour* intensive, has now become *capital* intensive.

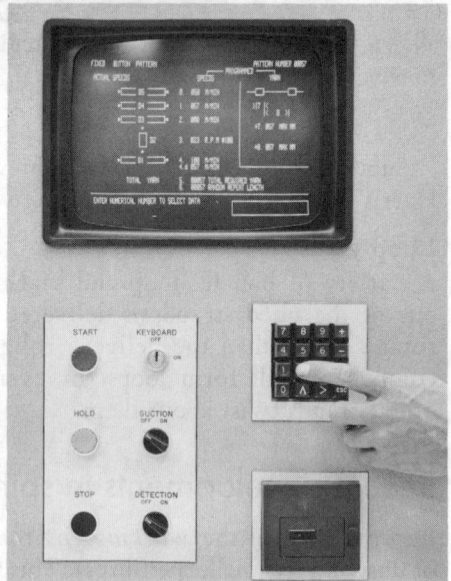

The electronic fancy yarn machine can be programmed to produce a variety of fancy yarns, including bouclé and slub. It can work direct from sliver or roving, and can handle staple lengths of up to 220 mm.
(Gemmill and Dunsmore Limited, Preston, England)

New spinning techniques

Ring spinning machinery has been successfully modified to produce a variety of combination yarns. Core yarns and fancy yarns, combinations of staple and filament, and elastomeric covered yarns are all adaptations of ring-spun staples.

Whenever a spinning system is modified, even slightly, the resultant yarn has a new set of properties. These new properties must be explored and exploited, and, above all, accepted by fabric manufacturers.

On the whole, private consumers have very little influence on developments in yarn manufacture: it is the fabric manufacturer who uses most of the yarn produced.

Core yarns

Core yarns are spun on conventional ring spinning frames, with the drafted staple fibres twisting themselves around a filament core introduced just after the drafting zone.

Elastic core yarns – such as hat elastic – were the first application. The rubber core was covered (under tension) by a cotton layer to improve wearer comfort.

Woolfil

Woolfil uses a similar principle, to combine a polyester or nylon core with a wool cover to produce fine and strong worsted-style yarns. Worsted yarns need long, and therefore expensive, fibres. The Woolfil system can use much shorter fibres, and so is cheaper. Because the staple fibres are wrapped around the strong filament core, the yarn has an all-wool handle.

With modern equipment, such as this worsted ring spinning frame, one operator can supervise many spindles.
(Gebrüder Sulzer Aktiengesellschaft, Winterthur, Switzerland)

347

Selfil

The Selfil system, developed by the C.S.I.R.O., wraps strong filaments around a drafted strand of staple fibres, such as wool. The wrapping is achieved by inserting an alternating false twist by means of friction discs. Two filaments are produced out of phase, unlike the self-twist system, and so the yarn structure is both strong and stable, and does not need any further plying or twisting. The yarns are suitable for fine wool-type knits and for woven goods, and are produced at around 300 metres per minute.

Selfil is no longer manufactured because of a decline in the knitting market at the time of its launch, and also because the Selfil yarn was not much cheaper than conventional yarns.

Selfil.
(C.S.I.R.O. Division of Textile Industry)

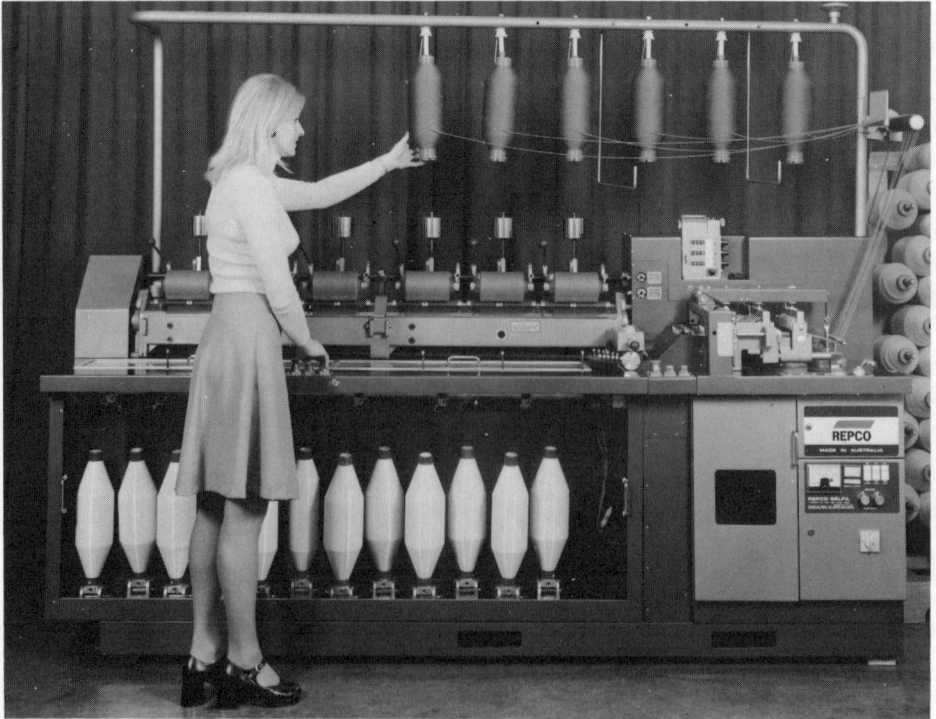

The Selfil system wraps staple fibres such as wool with a strong filament.
(C.S.I.R.O. Division of Textile Industry)

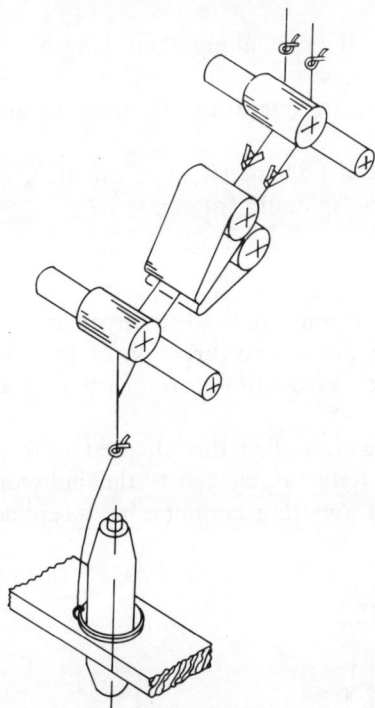

Sirospun creates a two-ply yarn by drafting two rovings side by side. (C.S.I.R.O. Division of Textile Industry)

The self-twist spinner produces two-ply worsted-style yarns in a single spinning operation. (C.S.I.R.O. Division of Textile Industry)

Coverspun

Coverspun also wraps a staple core with a strong filament. Instead of using a ring spinner, the method uses the uptwister principle to twist the filaments around the staple core. The yarns produced are well suited for knitted fabrics.

Sirospun

Sirospun is a modified ring spinning system which drafts two wool rovings side by side, and twists them together to give a yarn resembling a two-ply worsted.

Self-twist spinning

Self-twist spinning was developed to produce two-ply worsted style yarns from longer fibres.

Two drafted roving are passed between a pair of oscillating rollers. S and Z twist, alternating with the oscillations of the rollers, is introduced into both rovings. When portions of the rovings with *similar* twist are allowed to touch, they rotate around each other to *cancel their torque* with the *opposing ply twist*. The resulting two-ply yarn has stable *S and Z ply regions alternating* as the twist in the rovings alternated, separated by *twistless zones* where the direction changed.

Self-twist yarns are highly extensible and elastic, with regularly spaced *weak spots* in the twistless zones. These weak spots mean that only long fibres are suited for self-twist spinning, as fibre–fibre friction along their length is all that holds the yarn together in the twistless zones.

Extra twist is added to the yarns in a separate operation, in order to make them perform more like conventional yarns.

Advantages of self-twist spinning are *speed* (200 metres per minute), *low capital cost,* and the *small area occupied* by the spinning frames.

Other spinning methods

In Dr Ernest Fehrer's *DREF System,* rotating drums are used to introduce twist by friction. Fibres are introduced at right angles to the direction of the yarn, and high twist is inserted to form a compact yarn suitable for carpeting and furnishing applications.

Staple fibres can be *electrostatically* charged, and in this charged state are introduced into a rotating electromagnetic field which twists the individual fibres around each other. This system is still awaiting commercial acceptance because of the very high voltages required.

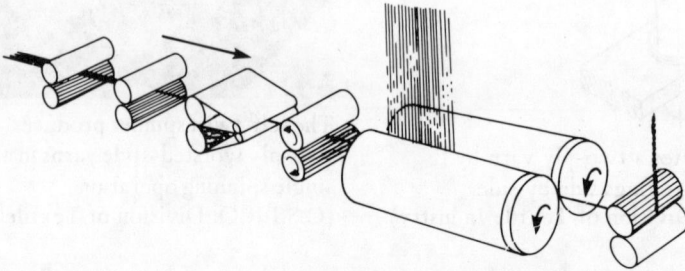

The DREF system introduces the wrapping fibres at right angles to the core fibres.

Another method where twist is inserted without rotating the yarn package is *air vortex* spinning. This method can be modifed for both short and long fibres. The drafted fibres are introduced into a chamber where two jets of air, injected at right angles to the fibres, create a vortex. The vortex twists the fibres around each other, and, in most applications, around a core filament which is incorporated to give the yarn strength.

The *Murata spinner* uses two air jet nozzles to insert false twist. The sliver passes through two successive air jets which insert opposing false twist. Cotton/polyester blends are most suited for this new process, which produces a very fine fasciated yarn. The system is very strongly marketed by the Murata Corporation, and may become a commercially important technique.

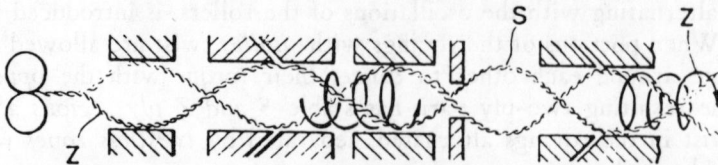

Murata air vortex spinning.
(C.S.I.R.O. Division of Textile Industry)

Yarns without twist

Twistless staple yarns

Glued yarns

These yarns can only be used in *tightly woven* fabrics, where the interlacings of the weave hold the fibres in place.

The finely drafted roving is passed through quick-drying glue, heated, and wound onto a package. After weaving, hot wet processing removes the glue, and leaves the unadulterated, untwisted, flat smooth yarns in the fabric.

The fabrics have *low air permeability* because of the tight structure, and are thin, because of the flatness of the untwisted yarns. The special properties of such fabrics limit the application of untwisted yarns.

Bobtex

Bobtex yarns were developed in Canada. Staple fibres are stuck on a molten filament to form an outer sheath. Later developments use adhesive over a drawn rigid polymer core.

Despite many years of experimentation, these yarns are not yet a commercial success.

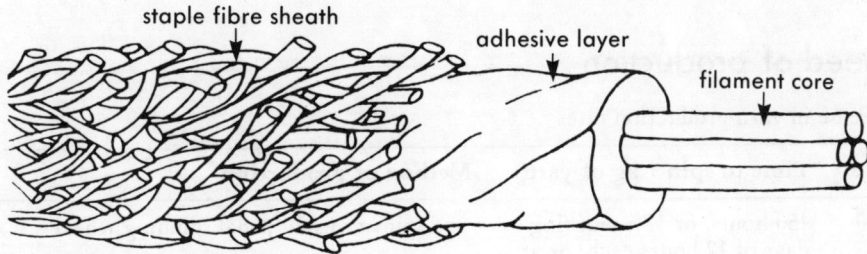

Bobtex yarn.
(C.S.I.R.O. Division of Textile Industry)

Neither filament nor staple

Slit film yarns

Slit film yarns are a speciality group. They are decorative, often metallic. A fine layer of aluminium is deposited on a film of polyester, and is then protected by a further film of polyester on top. The silvery film sandwich may be dyed different colours before it is split into fine strips to make metallic (e.g. Lurex) threads. Because of their thin, rectangular cross-section, slit film yarns are rather sharp and harsh to touch, and they are often twisted with other, softer components to make the final yarn.

Fibrillated yarns

Yarns can be made from polypropylene film by *fibrillation*. An embossed polypropylene film is first stretched lengthwise and then sideways, to create a web of interconnected fibrils. The film is then divided by slitting it into

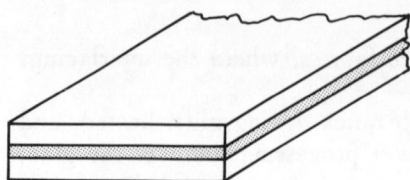

Cross-section of a slit-film yarn.   Embossed film for fibrillation.

lengthwise strips. The strips are then twisted into yarns of the required thickness. Such fibrillated yarns are mostly used for cordage and as twines. Other applications include pile yarns for artificial turf and carpets, and raffia.

Some fibrillated yarns are not given a twist. These flat yarns (tape yarns) are woven into fabrics which can replace jute sacking and hessian, and are also increasingly used as a base cloth for the manufacture of carpets.

There is some overlap between the uses of slit film and fibrillated yarn.

Speed of production

Increase in yarn production rates

Date	Time to spin 1 kg of yarn	Method of production
1750	150 hours, or 12½ working days of 12 hours each, or at least two 6-day weeks	one spinster using pedal spinning wheel
1770	60 hours or five 12-hour days or nearly one 6-day week	one spinster using one spinning-jenny with eight spindles
1810	10 hours or nearly one working day	one spinning mule with several spindles
1830	4 hours	improved mule spinning, water-powered, with many spindles
1870	2 hours	improved, steam-powered, flyer spinning with many spindles
1880	50 minutes	steam-powered ring spinning with many spindles and spindle speeds of 2000 revolutions per minute (r.p.m.)
1940	20 minutes	since 1900s, electrically-powered ring spinning, with many spindles and spindle speeds of 7000 r.p.m.
1970	30 seconds	ring spinning with many spindles and spindle speeds of 15 000 r.p.m.
1977	10 seconds	many open-end spinning heads, with rotor speeds of nearly 100 000 r.p.m.

The efforts to produce faster and more efficient spinning have led to a bewildering selection of machinery for the manufacturer. Each machine has its own advantages and disadvantages, and is suited to a particular range of raw materials. The products of different techniques perform slightly differently when they are woven or knitted into fabric. Each machine requires its own trained operators and maintenance technicians.

The amount of yarn manufactured in one 45-hour working week in the 1940s was produced within one 8-hour shift in the 1970s. The yarn produced in the 1970s is superior in quality to the yarn of the 1940s.

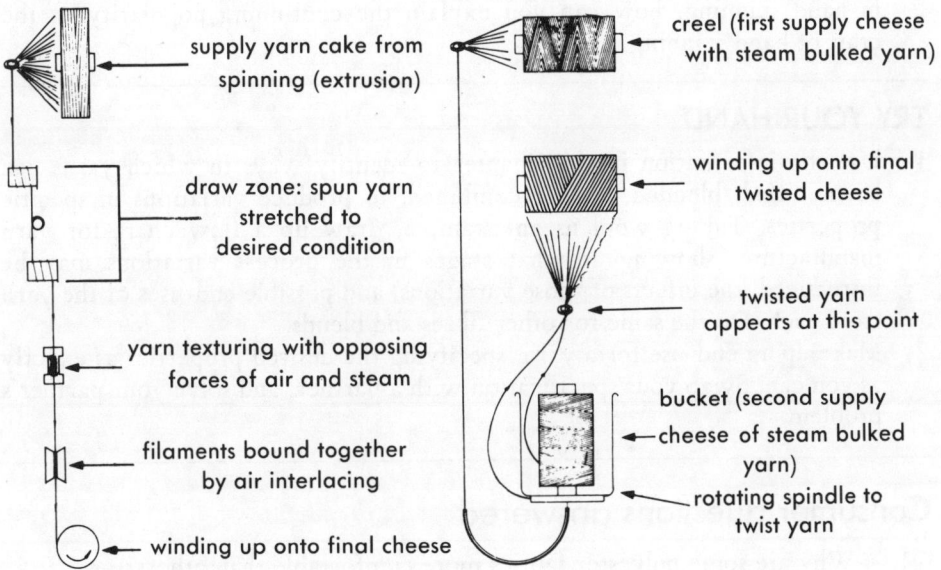

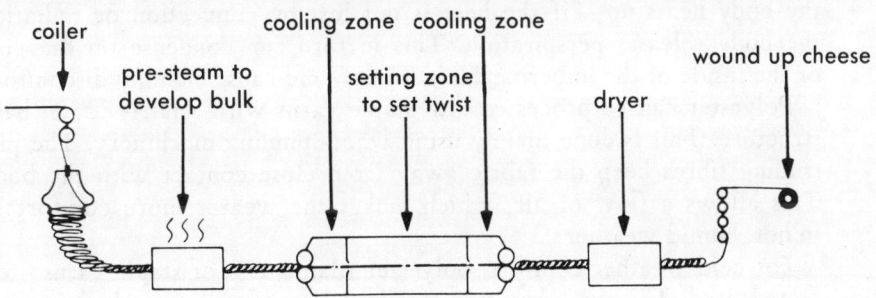

supply yarn cake from spinning (extrusion)

draw zone: spun yarn stretched to desired condition

yarn texturing with opposing forces of air and steam

filaments bound together by air interlacing

winding up onto final cheese

creel (first supply cheese with steam bulked yarn)

winding up onto final twisted cheese

twisted yarn appears at this point

bucket (second supply cheese of steam bulked yarn)

rotating spindle to twist yarn

coiler

pre-steam to develop bulk

cooling zone

cooling zone

setting zone to set twist

dryer

wound up cheese

The production of high quality nylon carpet yarns. Steam bulking produces yarns suitable for level loop carpet and tufting. Twisted and bulked yarns made by twisting together two or three twisted textured threadlines, then steam bulking and setting, are suited to resilient cut pile carpets.
(Fibremakers)

TRY YOUR HAND

1   By what factor has production increased between 1770, 1870, and 1970?
2   List possible reasons for the increased number of turns per minute that can be achieved by the more modern equipment.
3   Why do you think the quality of yarns has improved with the increased production rates?
4   Why could a spinster with an eight-spindle spinning-jenny not spin eight times as fast as a spinster with a single spinning wheel?
5   Given the efficiency of mechanized spinning and the amount of effort involved in hand spinning, how can you explain the continuing popularity of the craft of hand spinning?

TRY YOUR HAND

1   Use the information in this chapter to identify ways in which yarns can be modified, blended, and/or combined, to produce variations in specific properties. Taking wool as an example, draw up a flow chart for yarn manufacture, showing at what stages in the process variations may be introduced, the effects of those variations, and possible end-uses of the yarn from each. Do the same for other fibres and blends.
2   Make up an end-use for a yarn, specifying the desired properties as exactly as you can. Swap your specification with a partner, and solve your partner's problem.

Consumer questions answered

Q1   Why are some polyester fabrics more comfortable than others?
A   Closely structured polyester filament fabrics can be quite uncomfortable, because the hydrophobic filaments lie close to one another and allow neither air nor water vapour to pass through. With any physical exertion, the body heats up – if the heat is not lost by convection or radiation, the body releases perspiration. This in turn can condense on the body or the inside of the impermeable garment, and cause clammy discomfort.

Polyester can be processed into staple yarns with a fairly loose, hairy structure. This is done mainly using wool-spinning machinery. The protruding fibres keep the fabric away from close contact with the body. This allows a flow of air, which makes the wearer more comfortable in hot, humid weather.

For cold weather comfort, polyester filaments – or staple yarns – can be textured. This way they trap small pockets of still air, which increase the insulation value of the garment.

Q2   Most wool fabrics have a soft, fuzzy surface. Why are pure wool men's suitings so smooth?
A   Worsted suitings are made from fine smooth yarns. The wool chosen for them is fine, with a small crimp. A combing process removes all the short fibres that would later stick out of the yarns. The fibres are

laid parallel to each other by gilling and spun into firm yarn. The plying twist helps to hold fibre ends safely tucked away from the surface.

Fabrics made from such wool yarns have the characteristic of showing the weave patterns clearly. The secret of making such yarns was long held by the Flemish spinners and weavers who settled in Worstead in England in the fourteenth century.

Q3 Why do some worsteds develop a shine in use – such as on the seats of pants?

A In worsted yarns the wool fibres are held parallel to one another. The dull look of the new fabric is due to very slight variations in the yarn surface. During use, the fine yarns get pressed into the fabric, and the fibres are lined up even more exactly. This way a smooth, shiny surface is produced.

Steaming and vigorous brushing can alleviate the shine temporarily, but the original hairless-but-dull texture will never be recovered.

Twill and other long-float fabric structures (Chapter 10) are more prone to shine than are plain weaves. *Woollens* never become shiny – they are made from yarns in which the fibres point in many different directions.

Q4 Can I predict by looking at a fabric whether it will crease or not?

A Creasing and a lack of resilience are the result of many different factors which combine in a fabric. Fibres differ in their abilities to recover from bending or creasing. For a given fibre, yarn and fabric structure can also play an important role.

Generally, loosely spun yarns and loosely constructed fabrics give a better chance for fibres to move out of the way when the stress of creasing is applied. So, the fibres are less likely to be stretched past their elastic limit, and are more able to pull the garment back into its original, creaseless state.

Tightly spun yarns and tightly constructed fabrics, conversely, are more prone to creasing.

Q5 What is the difference between washable and non-washable crepes?

A Washable crepe is made from crimped hydrophobic filaments. The crepe effect is due to the irregular interlacings of the warp yarns in the fabric structure. These fibres are not affected by water.

Non-washable crepe also has an irregular crepe weave, but the yarns have a more complex structure. Hydrophilic fibres – wool, silk, or viscose – are twisted tightly around each other into a crepe yarn structure. When the fabric is wetted, the fibres swell. So each swollen fibre needs to be wound around other, thicker fibres. The new 'wet' path is longer, so the fibres move, bringing in some of their length to accomplish the larger diameter turn. This makes each yarn shrink in length and swell in diameter. When the moisture evaporates, it is possible – though sometimes difficult – to reset the fabric to its original dimensions by ironing. Ironing may give a shine to the otherwise dull surface of the crepe fabric.

Q6 How can one fibre – such as polyester – take on the character of such widely differing materials as wool, silk, linen or cotton?

A   The versatility of synthetic fibres arises from variations in techniques of yarn manufacture. The filaments are produced in diameters that resemble the fibre they are to imitate. They are then cut to a suitable staple length or, to imitate silk, left as filaments. Wool-length staple synthetics are processed on wool carding machinery to imitate wool; short staple fibres are worked on cotton drawframes and spinners to produce cotton-type yarns.

Q7   Is there any difference between the yarns used for expensive boxed pantyhose and the variety store cheapies?

A   Yes. Although the total thickness of both multifilament yarns is 15 denier, the more expensive hose are made of six finer filaments, the cheaper hose of four coarser filaments. This gives the expensive garment softness of handle and a sheer appearance, without unduly weakening it. Unless the hose have been knitted by a special non-run technique, however, they are just as easily ruined by snags and sharp toenails.

     After dyeing, the more expensive hose are boarded and heat set. The heat setting reduces the extensibility of the yarns, so the hose are less likely to bag around knees and wrinkle around ankles, but they are also less able to stretch around thighs and seat.

     The reason for the extra cost is that the production of finer filaments requires more careful techniques, finer spinnerets, and tighter quality control at all stages of manufacture. Even small variations in thickness (perhaps due to uneven pumping during melt spinning) can cause large variations in the fine filaments, which then need to be rejected as seconds. The extra handling and inspections at the boarding stage also contribute to the cost.

Q8   What do '8 ply', '5 ply' mean in knitting wools?

A   In technical textile terminology 'ply' indicates the number of single strands of yarn that are twisted together. In practical terms however – in the specialized field of hand knitting yarns – 'ply' has come to indicate the thickness of the knitting yarns. Ply is not an accurate measure of yarn count. It cannot be relied on to produce the same size stitches on the same needles from all different kinds of same ply yarns (see p. 314). At best, ply is an indication of effective thickness, rather than a measure of it.

Q9   Is yarn twist important in determining the final look of a fabric?

A   Yes. Often the whole character of a fabric depends on yarn twist alone. Both elephant crepe and chiffon are made of plain woven, loosely set, highly twisted yarns.

     In elephant crepe, all the weft threads are twisted in the same direction. The twist of the weft displaces the warp yarns in a rhythmic way – this gives rise to large 'hill and valley' effects, which run in the warp direction along the fabric.

Q10 Most polyester/cotton blends contain either 35/65 or 65/35 per cent of the components. Why?

A   This is mainly a marketing angle: consumers have learnt the kind of properties to expect from each of these popular blends. The 35/65 poly/

Pure cotton crepe bandages owe their elasticity to their yarn structure.
(Vivian Robinson)

cotton is more absorbent, the 65/35 poly/cotton is more resilient and drip-dry.

Other blends of other fibres – often as many as five or six combined together – are also being introduced on the market. With such complications it is becoming increasingly difficult for the consumer to predict the final properties of the textile.

Q11 How can elastic bandages be made of pure cotton when cotton is not an elastic fibre?

A The elasticity of such bandages is achieved by giving very high twist and the same direction of plying twist to the warp yarns of the bandage. The twisting makes the yarns curl into a helical shape. On stretching, the helix extends; on release, it pulls back.

Q12 What is raffia?

A Raffia was once made from a fibre obtained from the leaves of the raffia palm. Modern raffia is a fibrillated film of viscose rayon, lightly twisted into twine.

10 Fabric manufacture

This chapter will help you answer questions such as:
- How can firehoses be woven in a tube form without any seams?
- How can pure cotton fabrics be waterproof?
- How are synthetic furs produced?
- How can the most complicated patterns (birdseye, houndstooth, Prince of Wales check) be produced by the simplest weaves?
- How modern are the most modern looms?
- What is 'real' crepe?
- Why is velour an expensive fabric?
- How do fabrics get their names?
- Why do some knits stretch out of shape and others not?
- Why do interlinings keep their shape?
- What is the difference between stretch terry cloth and terry towelling?
- How can I tell warp from weft in woven fabrics?
- Why is it important to cut fabric 'on the grain'?

Textile fabrics are manufactured with many different properties. Some are strong, some supple, some rigid, some easily stretched; some are used for their attractive appearance, others for industrial purposes.

Most of the fabrics produced in the world are either woven or knitted. Nonwovens, such as felts and bark cloth, have an ancient history, but their large-scale production is very recent.

The way a fabric is constructed to a large extent determines its properties. Generally, knitted fabrics are extensible and flexible; wovens are strong and less easily stretched; nonwovens are relatively stiff fabrics. In all these different constructions, if the yarns or fibres are pushed tight against one another, the fabric becomes more rigid. If there is plenty of space between the yarns or fibres, the fabric becomes softer and more pliable.

The consumer making choices about textiles is mainly concerned with fabrics for fashion, clothing, and household use, yet a large proportion of the textiles produced are used for industrial purposes, or as invisible textiles in the linings of shoes, in upholstery, and in carpets. In all these uses, an understanding of how textiles are produced can give consumers an idea of the limitations – in price and properties – of the fabrics they buy.

There is a wide range of skills involved in making different fabrics, in getting the best performance out of each loom or machine. Often specific skills for making particular fabrics were developed in different towns. The fabrics became known by the name of the town where they were produced, and where the secrets of their manufacture were jealously guarded.

TRY YOUR HAND

1   These places – countries, cities, regions, rivers – are associated with particular fabric names. What are these fabrics?
    Calcutta, Paisley, Kashmir, Worstead, Shangtung, Madras, Jersey, Milan, Damascus, Bedford (United States), Manchester, Cheviot, Venice, Morocco, Saxony, Tweed, Turkey, Brussels
2   What are the origins of these fabric names?
    muslin, ottoman, pongee, surah, hopsack, cheesecloth, ticking, sailcloth, sackcloth, honeycomb, evenweave

Weaving

Interlacing the yarns

The interlacing of twigs or reeds for making baskets was possibly the earliest form of weaving, followed by interlacing strips of skin or furry hide. As long as 30 000 years ago, Stone Age workers wove fabrics from linen threads and fashioned them into garments: these were found preserved in the mud of the Swiss lakes and discovered in recent archaeological diggings.

Weaving with a backstrap loom, Todo, West Flores, Indonesia.
(Mattiebelle Gittinger)

The earliest looms used threads hung from a crossbar and weighted with stones. The WEFT was threaded through these taut WARP threads. A second crossbar, added as a weight at the bottom, allowed better control of the warp tension. Such looms have been invented and used by many cultures throughout the ages. The Navajo Indians used it in the vertical form; the Chinese and various South American cultures developed it into horizontal and backstrap looms.

TRY YOUR HAND

If you haven't tried weaving before, you can try needle-weaving on a simple frame loom. Prepare the frame as illustrated and tie the string tightly to form your warp. Use a thick tapestry needle to thread carpet wool or other craft yarns across the warp threads to produce an imaginative design.

Some quite complex work is done on simple looms. Investigate the possibilities further through craft books and local craft associations.

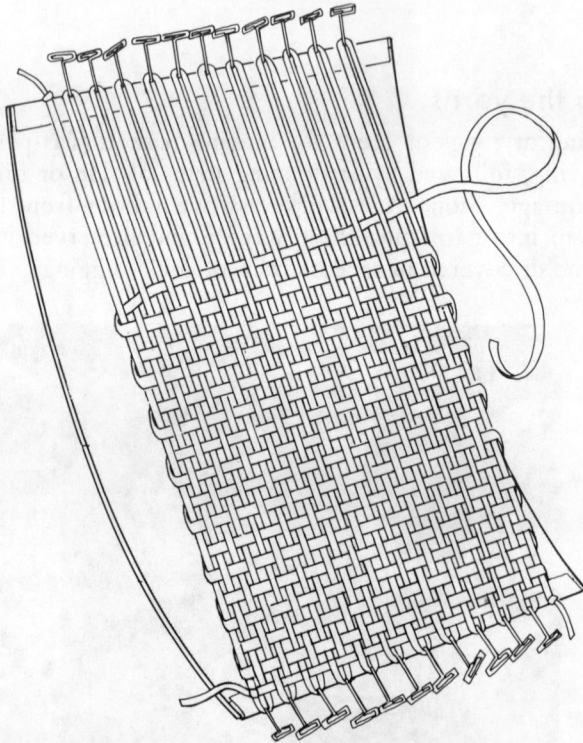

Forming the shed

The threading of the weft over and under each warp yarn is a slow and painstaking task. Weavers realized the advantage of lifting more than one warp thread at a time, and inserting the weft between the raised and lower warp threads. The space between these threads is called the SHED.

A simple loom has two frames which hold wires with eyelets. These are called HEALDS or HEDDLES. The warp yarns pass through these eyelets: odd warp yarns in the first frame, even warp yarns in the second frame. When the first frame is lifted, all the odd warp yarns are raised with it, and the shed is formed between the even and odd threads. When the first frame is down and the second frame is up, the opposite shed is formed. Such faster weaving using shed formation was first developed on horizontal looms.

If the warp threads are lifted in special patterns, a patterned cloth results. Fragments of silk cloth found on Yin period bronzes prove that the Chinese were able to weave intricate damask patterns as long as 3500 years ago.

Shed open to receive the weft.
(Grob and Co. Ltd, CH-8810 Horgen, Switzerland)

Threading the warp through the healds. Although automatic warp threaders may be used, threading the warp by hand or machine remains a time-consuming and therefore expensive process. Longer runs are thus more economical than short runs, as the setting-up costs are the same for each.
(Australian Wool Corporation)

TRY YOUR HAND

Many of the features of this Thai Hilltribe loom are found in the most modern looms. Look at the illustration of the modern loom and see if you can find the corresponding parts in its early version.

(Anna Janca)

(Weefautomaten Picañol N. V., Ieper, Belgium)

Passing the weft

The shed is formed in one movement by raising the heald frame (SHAFT). Once this is done, the speed of weaving is governed by how quickly the weft can be inserted and the shed changed again. So the rate of cloth production depends on how quickly the weft thread can pass from one end of the cloth to the other.

Until the invention of Kay's flying shuttle in 1738, all weft yarn was carried across the shed by hand-thrown shuttles. In John Kay's loom, the weaver pulled a string, which made a hammer move along a track and hit the shuttle, which then 'flew' across the shed and carried the weft yarn with it. This speeded up the rate of cloth production by a factor of eight.

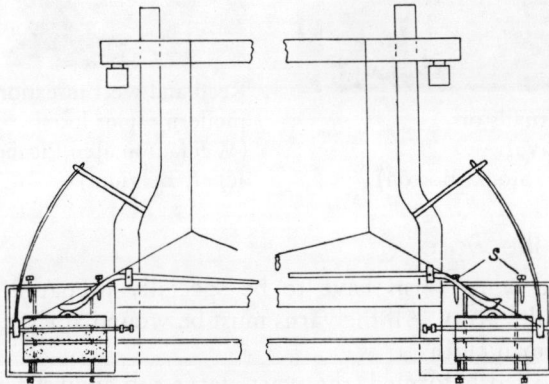

Fly shuttle device with double shuttle boxes.
(Ulla Cyrus, *Manual of Swedish Hand Weaving*. Charles T. Branford Company, Boston, 1956)

Other parts of the loom

In addition to opening the shed and passing the weft through it, a loom must also have the ability to press or BEAT UP the newly-laid weft yarn against the cloth. Once this is done, all the warp yarns must be moved forward or LET OFF the beam by exactly the same amount for each weft thread inserted. In this way, the next weft can be beaten up to the same point as the previous one and the distance between the successive PICKS (wefts) in the cloth is exactly the same.

In early hand-looms the beating up was done with a comb; Japanese silk weavers use their serrated fingernails; in modern looms the warp threads pass between the comb-like teeth of the REED which moves forward at the end of each pick to beat up the weft.

The length of *warp let-off* must be balanced by *cloth take-up* to keep the warp yarns at the constant tension a uniform fabric requires. The length of the finished cloth will be *slightly less* than the length of the warp on the beam. There is some loss at each end of the fabric, and some of the length of the warp yarns is taken up in their path around the weft yarns. The thicker the weft, the more warp is taken up in this way.

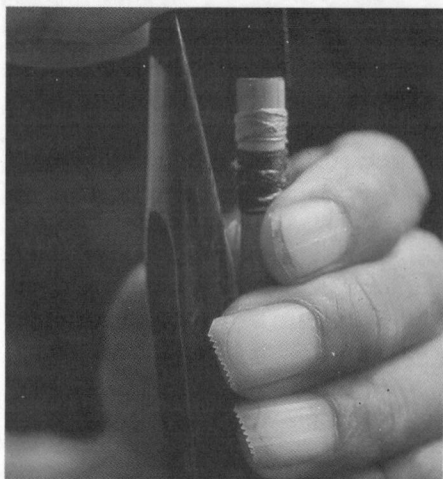

The serrated fingernails of
a Japanese silk weaver.
(© Cary Wolinsky, Stock, Boston)

Reed and weft insertion on a
modern rapier loom.
(Weefautomaten Picañol N. V.,
Ieper, Belgium)

Preparation of the warp

In all looms, the warp yarns have to be specially prepared so they form a
uniform and parallel array. All the yarns must be wound at exactly equal tension
to produce a uniform cloth.

Each time the shed is formed, the warp yarns rub against one another. Such
rubbing can make the warp yarns fuzzy and weak, and affect the appearance
of the fabric. In order to protect the warp threads, the yarns are SIZED –
coated with starch or other protective substance. This makes them smooth and
fairly stiff: the freshly woven GREY or GREIGE cloth has a coarse, stiff handle.
During finishing the starch or size is removed, and the true handle and drape
of the fabric is developed.

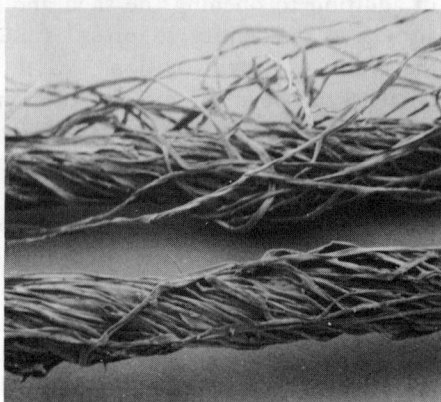

Winding the warp yarn on the
warp beam. For successful weaving,
the warp yarns must be wound
parallel and at equal tension.
(Australian Wool Corporation)

Sized and unsized yarns.
(*Textiles*, a periodical published by
Shirley Institute, Manchester, U.K.)

Patterning

The pattern of a woven cloth depends on the way in which the warp and weft are interlaced, and on the colour of the yarns used. At each cross-over point of the weave, one yarn is up (and visible), the other yarn is down (and hidden).

A skilled hand-weaver can produce extremely complex patterns on a simple loom. In industry, however, economics and speed of production outweigh creative ingenuity, and simple looms are limited to simple patterns.

In simple looms, many warps threads are tied into a frame or harness. When the frame moves up or down, all the warp threads in it move together. The more frames a loom has, the greater is the possibility for varying the pattern. A dobby loom has up to 40 frames, so a design that stretches over 40 warp threads can be woven on it. A simpler cam or tappet type loom has only 8–16 frames, and is limited to weaving plain, twill or satin weaves. Some complex Jacquard looms are able to move each warp thread up or down independently. In such a loom, a nearly infinite variety of woven patterns is possible.

Naturally a more complex loom can do simple weaves as well.

Tappet loom

In a TAPPET loom, the heddles are raised and lowered by the action of cams on levers. These cams imitate the action of the old-fashioned foot-pedal looms.

Most high-speed looms which are used to produce simple fabrics in large quantities use tappets to move the frames up or down. Although in theory two frames are enough for a plain weave, in practice 8–16-shaft looms tend to be used, as then the load is distributed more evenly across the loom.

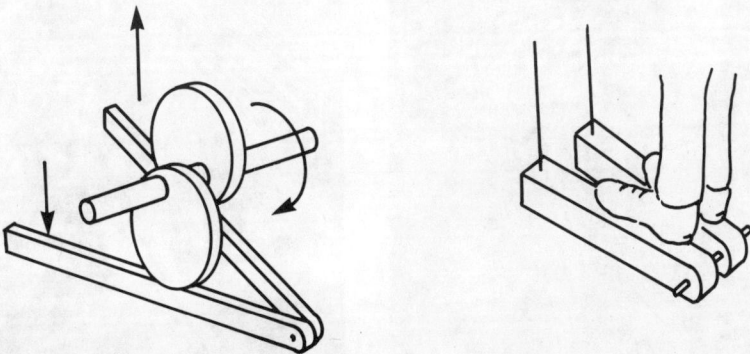

Tappets and cams imitate the action of the earlier treadles.

Dobby loom

Complex patterns need complicated cam shapes, and are more easily woven on a DOBBY LOOM. Fabrics typically woven on a dobby loom are complex twills, satins, crepe, piqué, honeycomb, and double cloth.

In the dobby system, each frame is attached to a lifting device which receives instructions from a punched paper roll. The instructions to lift or not to lift

A plain weave crepe de chine. This could have been woven on a tappet loom. (Vivian Robinson)

Twill weave acetate foulard, which could have been woven on a tappet loom, but probably was produced on a dobby loom. (Vivian Robinson)

The complex moss crepe weave in this nylon and acetate crepe would have required a dobby loom. (Vivian Robinson)

A peg-operated dobby mechanism on a craft loom. (Swedish Handcrafts)

are coded by the holes in the paper – rather like a pianola roll, or the paper tape used on early computers.

The dobby mechanism can control up to 40 frames at a time. This means that a pattern is repeated across the fabric after every 40 warp yarns.

TRY YOUR HAND
Find some examples of twill or satin or other fancy-weave fabrics and examine them under a magnifying glass. How many warp threads are there for each pattern repeat?

Jacquard loom
The Jacquard loom is capable of producing the most elaborate colour and texture effects in woven fabrics.

Each warp thread can be controlled individually. It is threaded through a wire which, instead of being mounted in a frame, is tied by a string to a lifting rod. The lifting rod in turn is connected to a control box, where it is governed by instructions on a punched card. The modern Jacquard loom is a direct descendant of the loom developed by Joseph-Marie Jacquard in the early nineteenth century; its eighteenth century forerunner was the first use of punched cards for data storage.

Jacquard looms.
(Gebrüder Sulzer Aktiengesellschaft, Winterthur, Switzerland)

This small Jacquard panel would have required a different pattern of lifting the warp threads for each pass of the weft.
(Dale Mann/Retrospect)

A Jacquard loom set up to weave damask tablecloths may have only two threads per lifting rod; other looms may have more. Seven threads attached to a lifting rod will repeat the pattern seven times across the width of the loom.

The effect of coloured yarns

Interesting patterns can often be created using a simple weave and contrasting colour of yarn. Pinstripe, houndstooth, Prince of Wales checks, birdseye, and tartans, are examples of traditional coloured yarn effects.

Extremely elaborate colour effects can be obtained with a Jacquard loom, such as in coloured tapestries and brocades.

TRY YOUR HAND

In the diagram on the left a twill weave has been shown. The dots in the design indicate that a warp yarn passes over a weft and hides it. A space in the square indicates that the weft is over the warp. If all the warp yarns were black and all the weft yarns white, the twill would show clearly (centre). If, however, the warp yarns were two white and two black, then the pattern would change as shown on the right.

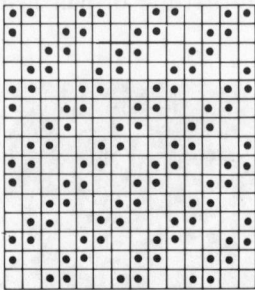

Twill weave.

Twill weave with warp and weft different colours.

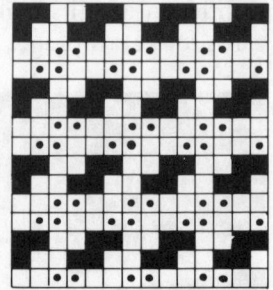

Twill weave with warp yarns two black, two white.

Copy the original twill weave, and try out what would happen if both warp and weft were coloured 2 black, 2 white. What if they were 1 black 1 white? 3 black 1 white? Any other combination?

Can you have colour effects with plain weave? See if you can create a pinstripe or a step or a houndstooth pattern from a plain weave.

Texture and pile

Fabrics produced on the tappet, dobby, or Jacquard looms described so far may have quite complex textures, as in satins and crepes, but they are flat. A *raised* texture can be produced by the introduction of an extra set of yarns. These form the pile or loops or other texture effect, and can be added as either warp or weft.

Texture, pile or loop *warp* yarns are carried on their own separate warp beam, which allows a greater length of pile warp than of ground warp to be let off for each pick. Such warp effect fabrics include velvets, woven carpets, towelling, fur fabrics, clipspot, and woven seersucker.

Velvet, velveteen and corduroy

As described above, the pile yarn of VELVET is carried on a separate *warp* beam, which allows a greater length of pile warp than ground warp to be let off for each pick. With each pick, a wire is inserted between the pile warp yarns and the ground warp: this holds the pile in raised loops. As the wire is withdrawn a few picks later, it cuts the tops of the loops to give the pile. *Depth* of pile depends only on the amount of pile yarn let off for each pick. It is independent of *density,* that is, depth of pile is not affected by how closely spaced the special effect warp yarns are.

VELVETEENS and CORDUROY are woven with the special effect *weft* yarn forming FLOATS on the surface of the fabric, in rows for corduroy, staggered for velveteens. These floats are then cut to form the pile. A deep pile can therefore be made only from long floats; long floats can only be made by the special effect yarn passing over many threads in the ground fabric. Therefore deep pile velveteens and corduroys can be produced only at the expense of density.

Velvets and velveteens can be made with identical pile depth and density, but usually the weft-pile velveteen has a shorter pile.

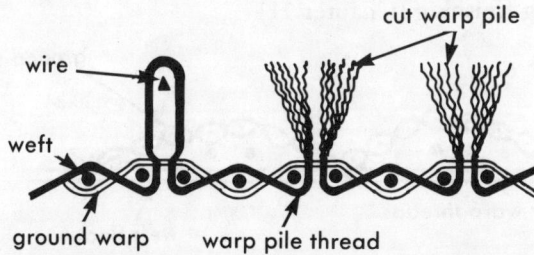

The manufacture of velvet (wire method). Special warp yarns form the pile.
(*Bayer Farben Revue* No. 17, 1969, by permission of the editors. *Bayer Farben Revue* is published by Bayer AG, Leverkusen, West Germany.)

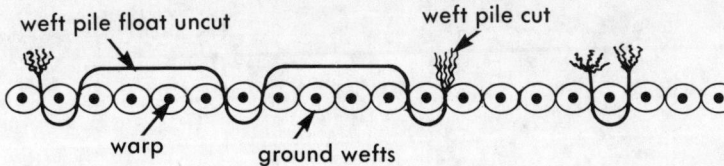

Velveteen and corduroy. Long weft floats are cut to form the pile.

TRY YOUR HAND

Buy a sample of each velvet, velveteen, pinwale cord, stretch velour, and any other cut-pile fabrics you can find. Compare the fabrics for drape and handle. For what uses is each fabric most suited? (type of garment, furnishing fabric) Are the fabrics interchangeable in these uses?

If all samples were bought at the same shop, their relative prices can be assumed to reflect their relative cost to the retailer. How do the prices of the different pile fabrics compare? How would you explain the differences in price between them?

Piqué

A certain proportion of the length of a weft yarn is used up as the weft interlaces with the warp yarns. The fewer warp yarns interlaced, the shorter the weft across the fabric.

If a fabric is woven with regular, long floating wefts and a ground fabric with a plain or twill weave, the floating wefts, which follow a shorter path, will pull the fabric surface into hills and furrows. The textured fabric produced in this way is called PIQUÉ.

The cord effect of piqué can be accentuated by the presence of STUFFER YARNS, thick warp yarns which lie between the floats and the plain-weave face of the fabric and which serve as padding.

The sculptured effect of piqué shows only when the tensions of weaving are released during finishing (Chapter 11).

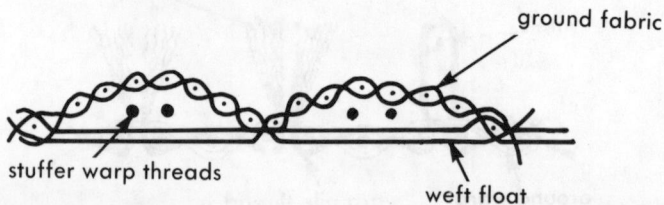

In piqué, weft floats pull the fabric into hills and valleys to give a sculptured effect.

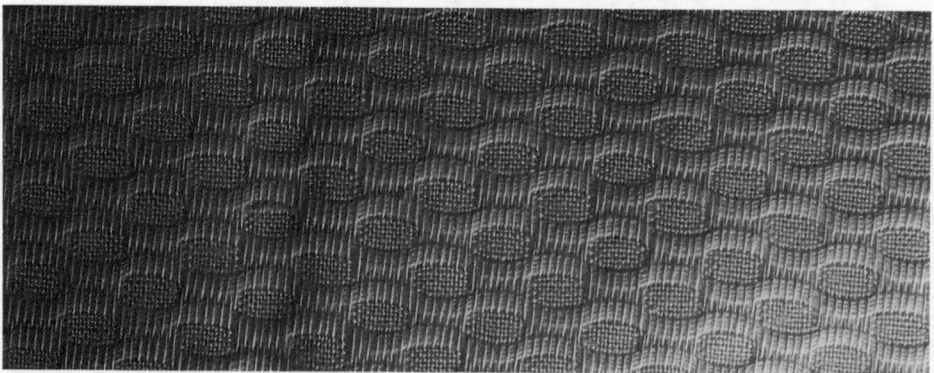

Piqué.
(David Bailey)

370

Towelling

TERRY TOWELLING is a warp-pile fabric. Thick pile yarns are housed on a separate warp beam under slack tension. When the weft is beaten up, the pile warp yarns buckle into loops; the taut ground warp yarns form a firm, strong foundation fabric.

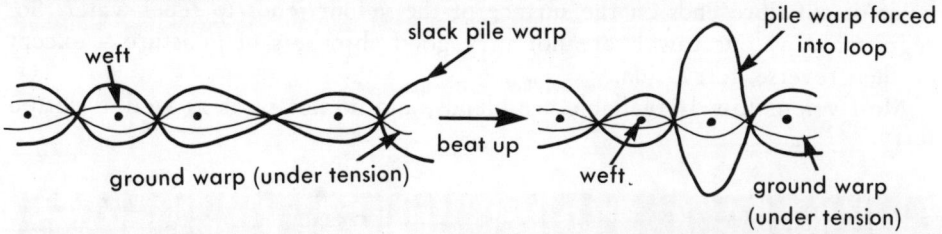

Three-pick terry. When the cloth is beaten up the slack warp pile forms loops.

A Jacquard mechanism can be used on individual pile warp yarns to create a patterned raised pile. Reverse-colour patterns are also possible.

VELOUR TOWELLING is made by cutting off the tops of the loops on one side of the terry structure. There are a number of factors that make such a product expensive. First of all, the loops must be long, so that an even pile can be left after cutting. As much as 30 per cent of the pile can be lost in

Terry loom weaving multiple widths.
(Gebrüder Sulzer Aktiengesellschaft, Winterthur, Switzerland)

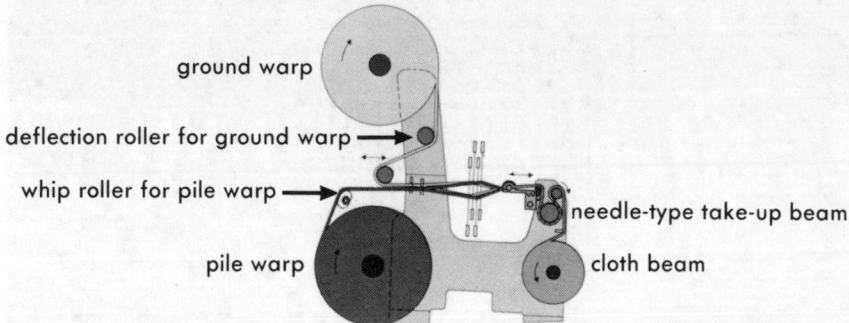

Terry loom (schematic)
(Gebrüder Sulzer Aktiengesellschaft, Winterthur, Switzerland)

this way, and pile yarn is expensive. Secondly, the base fabric needs to be closely woven so the cut pile yarns do not readily fall out in use. Even so, only one side of the terry fabric can be shorn to velour. Thick pile yarns are needed for an attractive, uniform velour effect. Velour towels are therefore generally of high quality. Although such towels are pleasant to touch, the uniform layer of cut fibre ends on the surface of the velour tends to repel water. So, regrettably, velour towels are not very good absorbers of moisture – except on their reverse, terry, side.

Most velour towels available are blends, and do not wear as well as cotton terry.

Three-pick terry. Two wefts inserted at some distance from the fell (already woven and beaten up cloth).
(David Bailey)

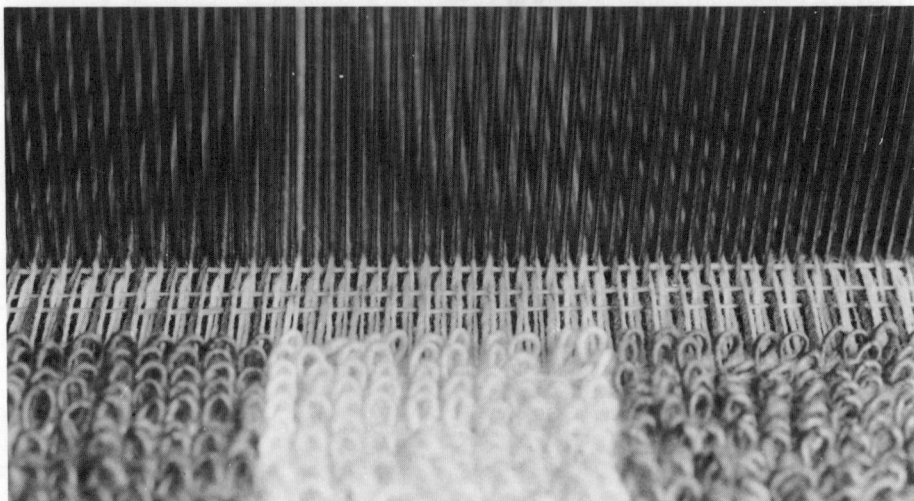

The third weft (pick) is inserted.
(David Bailey)

The three wefts are beaten up together. The slack pile warp buckles to form the loops of the pile, while the taut ground warps form a firm ground fabric. (David Bailey)

Three-pick terry fabric.
(*Textiles*, a periodical published by Shirley Institute, Manchester, England)

TRY YOUR HAND

Investigate the differences between high quality and low quality hand towels or towelling nappies. (How will you decide what is and is not a high quality towel?)

Take a sample of each, and compare and contrast number of loops per 10 cm (both warp and weft directions); and number of yarns per cm in both warp and weft.

Cut a 10 cm length from each towel, and withdraw a pile yarn. Measure the stretched length of this yarn, and compare both the *length* and the *thickness* of the threads removed from each towel.

1   Can you tell a good quality towel from an inferior one by just looking at it?
2   Was price an accurate indicator of quality?
3   What constitutes 'good value' when buying towels?

Double cloth

Some heavy fabrics, such as blanketing or upholstery, have a different woven pattern on each face. Such fabrics are called DOUBLE CLOTHS. Each face is made from a different set of warp, and sometimes weft, threads.

Two layers of fabric are woven simultaneously on the one loom, with an occasional warp or weft passing from one layer to the other to hold the double cloth together. Genuine double cloths are very warm because of the air trapped between the two layers.

Firehoses are made in this way. They cannot be seamed, as a seam would be a line of weakness in the hose and could rupture with the pressure of water. Instead, the hoses are woven as a double cloth, joined only at the edges where the weft passes from one layer to the other: a tubular fabric off a normal loom.

Double cloth: picking pattern for a typical structure.

Double cloth (face).
(David Bailey)

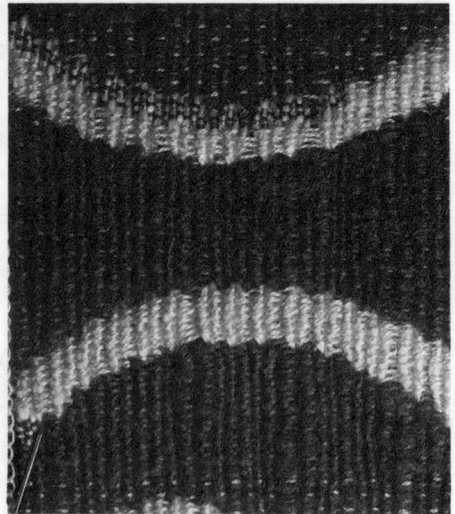

Double cloth (reverse).
(David Bailey)

Velvets may be woven as double cloths. The warp pile yarn passes from one layer to the other, and is cut later to separate the two layers and form the pile.

Double cloths may be imitated by glueing two fabrics together, but the qualities of the adhesive may adversely affect the properties of the cloth.

374

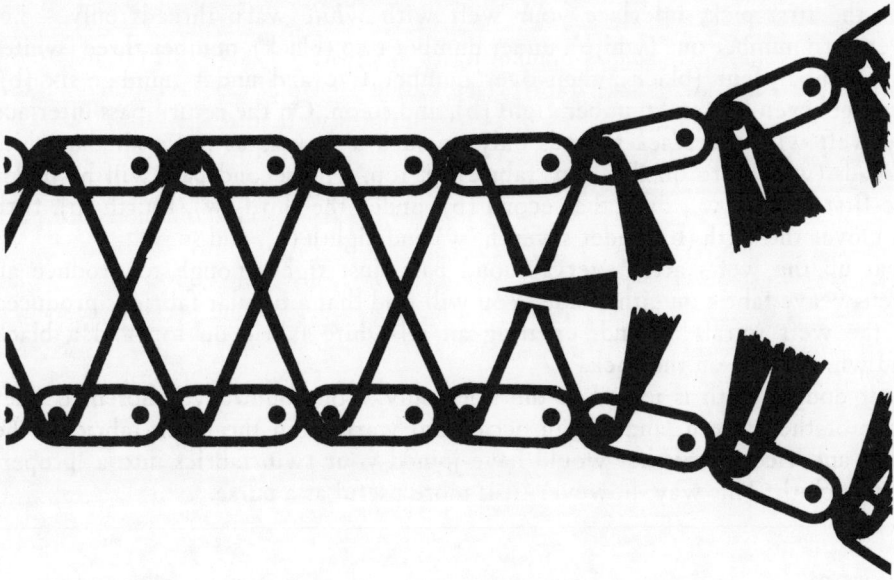

Manufacture of velvet by weaving it as a double cloth. This method uses less pile yarn than does the wire method, and so is more economical.

(*Bayer Farben Revue* No. 17, 1969, by permission of the editors. *Bayer Farben Revue* is published by Bayer AG, Leverkusen, West Germany.)

TRY YOUR HAND

You can make a double cloth on a simple frame loom, and use it as a purse or shoulder bag.

Set your warp up as alternate threads of black and white, closely spaced. For your weft, use only white yarns. Carpet yarn or 12 ply knitting wool will give quick, effective results.

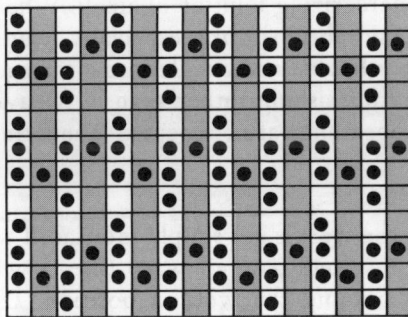

Picking pattern for the double cloth purse. The dots indicate that warp passes over the weft.

For the first pick, interlace your weft with *white* warp threads only – i.e. over yarn number one (white), under number two (black), number three (white) and number four (black); then over number five and under number six (b), number seven (w) and number eight (b), and so on. On the return pass interlace the weft with the *black* threads only, in such a way as to leave all the white threads (which are on the 'top' fabric) on top. The second pass will be under the first thread (w), over the second (b), under the third (w), fourth (b), fifth (w), over the sixth (b), under seventh (w) and eighth (b), and so on.

Beat up the wefts after every second pass, just tight enough to produce an even-weave fabric on either side. You will find that a tubular fabric is produced as the weft spirals around, creating an all-white fabric on top and a black and white check on the back.

Your double cloth is joined at the sides only. You could have chosen to alter some of the pattern, and lift an occasional warp from the black fabric to the front and vice versa: that would have joined your twin fabrics into a 'proper' double cloth. This way, however, it is more useful as a purse.

Clipspot

Swiss dotted or clipspot effects can be achieved by having only selected warp threads on the extra warp beam. These are woven into the ground fabric by a dobby or Jacquard mechanism. Between the 'spots' of design are long floats of warp threads which are cut off during finishing. To do away with this rather expensive procedure, imitation dotted fabrics can be produced by bonding FLOCK (short fibres) to a base fabric in a spot pattern.

Although modern adhesives produce fabrics with satisfactory wear properties, the imitation is inferior visually to the woven product.

Seersucker

Woven SEERSUCKER needs a loom with two beams. A slack beam supplies the warp threads to the bubbly parts of the fabric; a tight beam holds the warp yarns that form the stretched, narrow strips.

It is possible to make imitation seersucker by printing plain woven cotton fabric in stripes with a mercerizing solution of alkali. Where the alkali is applied, it swells the fibres and the fabric shrinks, especially lengthwise. This shrinkage makes the unprinted areas bubble and buckle into a seersucker effect.

Seersucker produced by finishing rather than by weaving is less satisfactory in keeping its original appearance after repeated laundering and ironing.

Leno weaves

In LENO woven fabrics, the weft is not simply interlaced with the warp. Rather, after each weft yarn is inserted, pairs of warp threads working together twist about each other, thus holding the weft in position. Leno weaves allow no slippage of yarns, even in loosely constructed fabrics, and can create interesting lacy effects.

In order to allow the warp threads to twist about one another, they are threaded through leno heddles (DOUP needles) which make the yarns work

Woven seersucker.
(David Bailey)

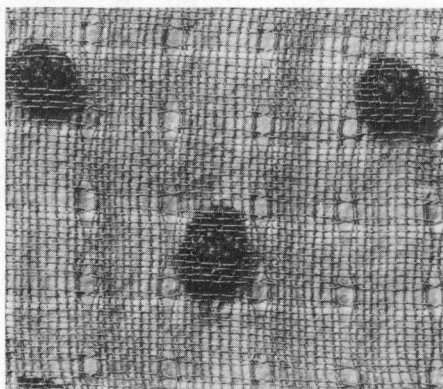

A Swiss dotted fabric with additional
leno patterning (reverse side).

in pairs. As the doup needle rises, it pushes the warp to the side. On the next
pick, the warp yarn is pushed in the opposite direction.

Because the twisted warp does not slip over the weft, many modern looms
used leno attachments at the sides to create strong selvedges. This is especially
useful for rapier and jet looms, and looms which weave a number of fabric
widths at the one time, as the cut ends of their wefts need to be secured
(p. 380–84).

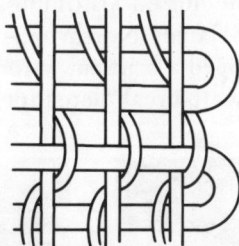

Leno selvedge.

Speed and efficiency

The speed at which woven fabric can be produced depends on the speed at
which the weft yarns can be inserted: the more weft inserted in a given time,
the more fabric produced. For this reason, if a fabric is not BALANCED (that
is, does not have equal COUNTS (numbers of yarns per centimetre) in the warp
and weft directions), it is made with fewer weft than warp threads.

Improvements to traditional looms

John Kay's invention of the *flying shuttle* (1738), which replaced the hand-
thrown shuttle, was a revolutionary step which speeded up production eightfold.

The first *powered* loom was built by Edmund Cartwright in 1774.

At the introduction of the steam-powered loom, organized mobs of hand-
weavers (Luddites) stormed factories and smashed the power looms which

threatened their livelihood. In 1790, an English mill equipped with 400 power looms was burned to the ground. Workers found that skills developed in long apprenticeships and years of employment were now obsolete, and there was massive dislocation of the labour force. However, the increased rate of cloth production lowered the unit cost, and the market actually increased. More people were buying more cloth, and this in turn created employment.

Shuttle looms

In a traditional loom the weft is carried across the shed in a shuttle. As this is the slowest and least efficient step in weaving, it is the area where greatest improvements can be made.

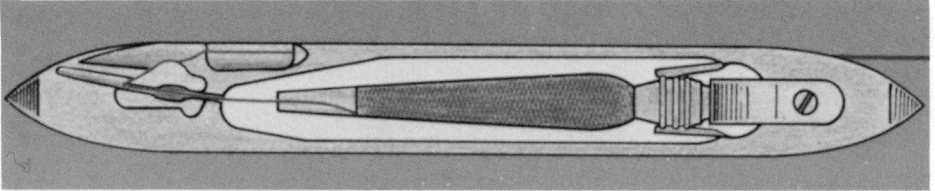

Modern shuttle (schematic), showing the pirn or bobbin.
(Weefautomaten Picañol N. V., Ieper, Belgium)

The shuttle contains a small bobbin called a PIRN. In early looms, each time the pirn emptied of weft yarn the weaver had to stop the loom and place a full pirn in the shuttle. If he missed his timing, an empty shuttle left unsightly faults in the fabric. In 1895 James Northrop – an English mechanic working in the United States – developed a brilliant automatic loom. His patented design included features such as a weft-break detection device (to stop the loom when

The shuttle box on a modern loom permits precise positioning and braking of the shuttle.
(Weefautomaten Picañol N. V., Ieper, Belgium)

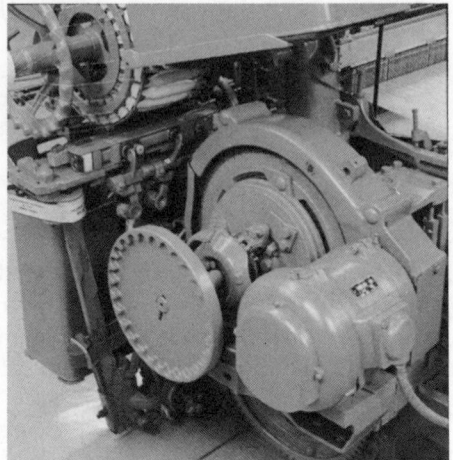

Fresh pirns ready to be dropped automatically into the shuttle, with minimum disruption to the weaving.
(Weefautomaten Picañol N. V., Ieper, Belgium)

Warp-stop motion. Should a warp yarn break, a rider falls and makes electrical contact with a bar below, stopping the loom action.
(David Bailey)

the weft ran out), an automatic pirn changer, and an effective warp stop motion. Gradually one weaver was able to look after more than one loom at a time, and productivity increased further.

In spite of its success, the fly shuttle is basically cumbersome. It is heavy, and it needs a lot of energy to be hit hard enough to fly across the shed and deposit one light thread of weft. If hit out of control, it is dangerous. It is also very noisy: *deafness* has been a traditional burden to weavers ever since the days of the industrial revolution.

Today, legislation restricts the maximum noise level allowed in a weaving mill. This, together with energy and efficiency considerations, has given impetus to the development of the *shuttleless* looms.

Shuttleless looms

The projectile loom

In the PROJECTILE LOOM – such as that produced by Sulzer – a *single length* of weft is held by a small gripper which is then hit across the loom. Instead of small packages of yarn being carried back and forth across the loom, a large cone of yarn supplies the weft from the side of the loom. Picking is from one side only. A number of grippers are used, one carrying the weft while the others return to the starting position.

The gripper is a much smaller projectile than a flying shuttle, so it needs much less energy to cross the loom. It travels faster, and can go further. Since much of the production time is taken up by accelerating and decelerating the weft carrier, a wider loom, which needs fewer stops and starts, can be used efficiently. Often, a projectile loom is used to *weave several widths of cloth simultaneously,* the light projectile carrying a single weft thread across all widths in the one action.

Whereas the yarn from a shuttle is unwound as it is laid across the shed, the gripper of a projectile loom drags a weft past all the warp yarns before it is laid down. This creates some strain on the yarn, so projectile looms are *not suited for weaving delicate or fragile weft yarns.*

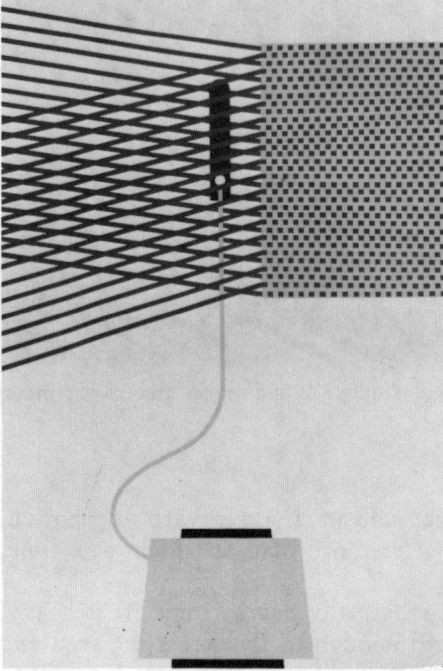

Action of the projectile loom.
(Gebrüder Sulzer Aktiengesellschaft, Winterthur, Switzerland)

Four-colour projectile loom with a maximum working width of 3930 mm. Note the large yarn packages on the side of the loom feeding into the projectile mechanism. This particular machine has an electronic dobby.
(Gebrüder Sulzer Aktiengesellschaft, Winterthur, Switzerland)

The small, light projectile allows very rapid weft insertion. It is particularly suited to weaving multiple widths, and can handle heavy and bulky yarns.
(Gebrüder Sulzer Aktiengesellschaft, Winterthur, Switzerland)

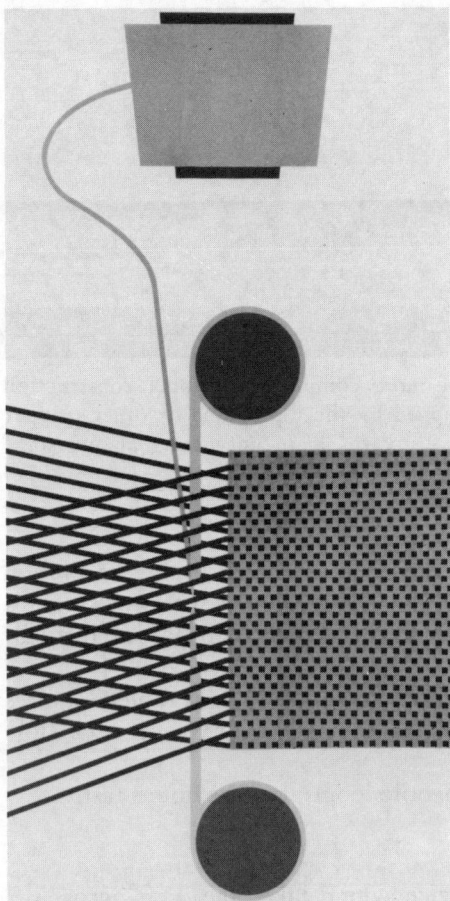

Rapier loom using a flexible rapier. This form of rapier is economical of factory floorspace. Note the yarn packages on the right, and the dobby mechanism on the left. (Weefautomaten Picañol N. V., Ieper, Belgium)

Action of the rapier loom. (Gebrüder Sulzer Aktiengesellschaft, Winterthur, Switzerland)

Flexible rapier loom with electronic dobby and microprocesser. (Weefautomaten Picañol N. V., Ieper, Belgium)

Rapier looms

In the RAPIER LOOM, a gripper mounted on a rapier carries the weft half-way across the shed, and transfers it to another rapier-mounted gripper which takes it the remaining distance. The first rapier is back in position by the time the pick is completed.

Transfer of the weft yarn from the right-hand gripper to the left-hand gripper, which carries the yarn to the left selvedge. Transfer occurs without physical contact between the grippers. (Weefautomaten Picañol N. V., Ieper, Belgium)

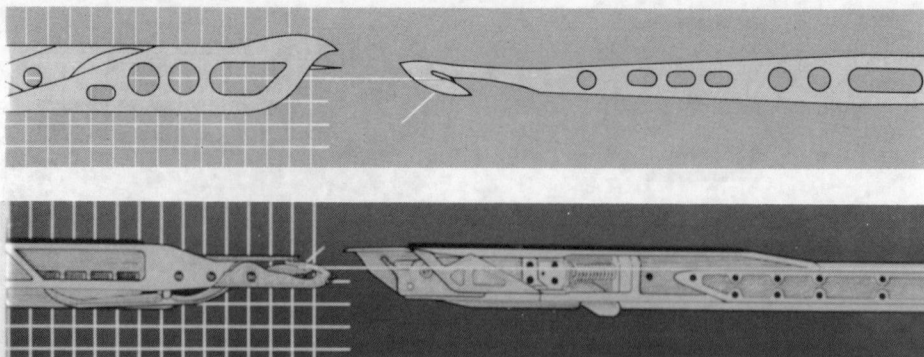

Two forms of weft transfer (schematic). The more complicated gripper construction is designed so that the top of the weft is gripped by the nip of one or other gripper at all times. This prevents highly twisted weft yarns from untwisting at any stage during the insertion process.
(Weefautomaten Picañol N. V., Ieper, Belgium)

Because the gripper is guided by the rapier, not in free flight, its motion is smooth and controlled. It places much less strain on the weft than does a projectile loom, and so rapier looms are capable of weaving delicate and difficult wefts.

The rapier itself may be rigid, flexible, or telescopic. A rigid rapier requires a loom twice the width of the cloth, and factory floor-space is expensive. Flexible rapiers are the most common.

Rapier looms are even quieter than projectile looms, but are not as fast.

Jet looms

To reduce further the weight of the device which takes the weft across the loom, blasts of air and jets of water have been developed as successful weft carriers. Both airjet and waterjet looms are quiet and efficient.

AIRJET LOOMS are mainly used to weave staple yarns. A jet of air has less carrying power than a jet of water, and so generally narrower fabrics are woven on airjet looms. WATERJET LOOMS are best suited for weaving hydrophobic

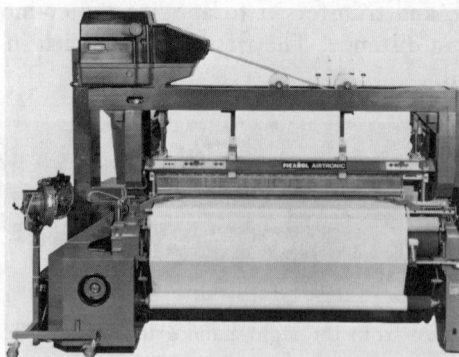

A two-colour airjet dobby loom.
(Weefautomaten Picañol N. V., Ieper, Belgium)

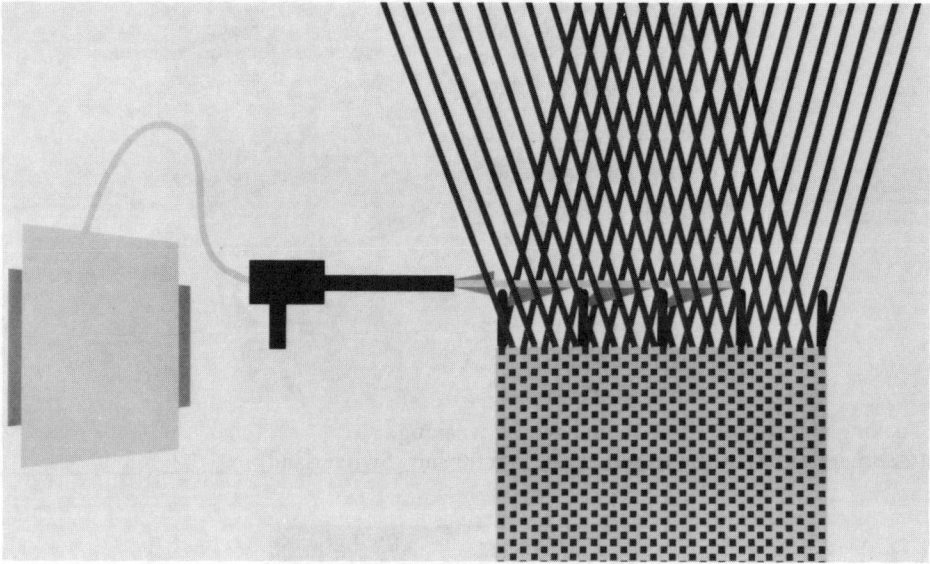

Action of the airjet loom.
(Gebrüder Sulzer Aktiengesellschaft, Winterthur, Switzerland)

filament yarns, though the most recent models can also handle polyester/cotton blends as well. Wet fabric cannot be stored in a roll, so the latest waterjet looms are equipped with efficient drying units. These units use vacuum suction and heat to remove the water from the fabric.

In addition to their speed (three times the speed of shuttle weaving) and quiet operation, jet looms occupy relatively little space. This allows them to be placed close to each other and so fewer weavers are needed to supervise the looms.

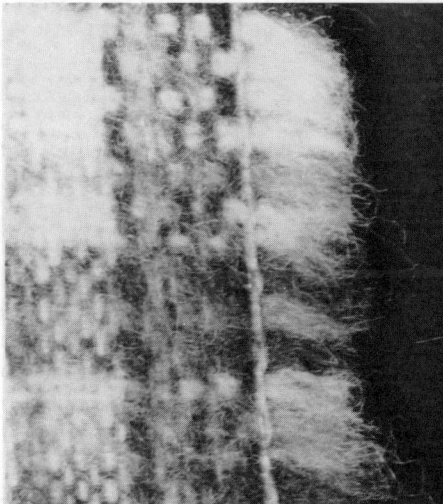

Selvedge of a woollen fabric
woven on a shuttleless loom.

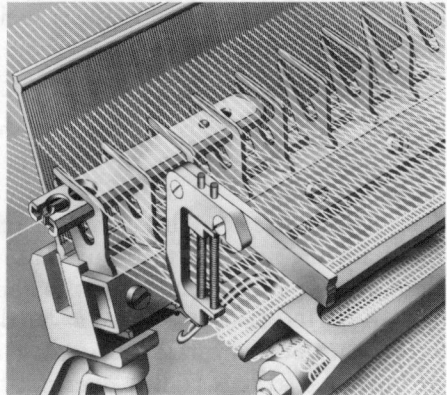

Selvedges can be neatened by tucking the cut weft ends back into the fabric with the next weft insertion (schematic). (Gebrüder Sulzer Aktiengesellschaft, Winterthur, Switzerland)

Tucking unit for selvedges in multiwidth weaving.
(Gebrüder Sulzer Aktiengesellschaft, Winterthur, Switzerland)

Another method of neatening selvedges. The weft is inserted through a few extra warp yarns at the fabric edge, and this narrow strip is trimmed off.
(Weefautomaten Picañol N. V., Ieper, Belgium)

Multished looms

All the above systems require the shed of warp yarns to be open while the weft passes from one end to the other. The latest idea in weaving divides the loom into sections. The heddles in each section can move independently. As soon as the weft has been laid in one section, that shed can change and be ready for the next weft. In this way, up to 6 weft yarns can be inserted during the time taken for the first weft to complete its passage across the loom.

MULTISHED LOOMS require a shuttle, because a carrier could not pull a weft across the loom after the first shed changed. A shuttle allows the thread to be unwound as it is laid: it does not have to be dragged through. The shuttles used are small, carrying only enough thread for one weft pass, and are replenished after each passage.

The multiple weft insertion of the multiphase loom makes it the fastest method of cloth production, but it is suited only to high speed production of plain weaves.

Weaving production rates

Date	Working hours needed to weave 100 m of plain fabric	Working conditions	Technology involved
1750	400 hours (5½ weeks)	72 hour working week	hand loom worked at home by one or two weavers
1790	100 hours (1½ weeks)	72 hour working week	John Kay's flying shuttle – cottage weaving
1810	100 hours (1½ weeks)	72 hour working week	power looms start to be used; factory system begins
1840	14 hours	72 hour working week	power looms take over, especially in United States
1900	9 hours	72 hour working week	Northrop automatic loom
1950	50 minutes	48 hour working week	one weaver supervising 24 automatic looms
1970	25 minutes	40 hour working week	one weaver supervising 20 projectile (Sulzer) looms
1980	10 minutes	35 hour working week	multished looms weaving 3 widths of fabric at a time

TRY YOUR HAND

Look at the above table.

1   How much fabric could a weaver produce during a week's work in 1750? in 1980?
2   Assume that weavers were as well paid in 1750 as they are now, relatively speaking.
    How many metres of cloth did a weaver weave in a week in 1750?
    How many metres did a week's work by one weaver produce in 1980?
    Would you expect the price of fabric to have risen or dropped since 1750? Why? By how much?
    How much fabric could each weaver afford to buy then? Now?
3   How did the working hours of the weavers change over the years? What are the differences in the skills required for a weaver then and now?
4   Some developing countries, such as India, export hand-woven goods to the West. Find some hand-woven Indian cotton goods. How do they compare in price to similar machine-woven goods? Does the difference in price reflect the difference in working hours involved in their production? What does this suggest about the role of such cottage industry – does it aid development, or is it sweated labour?

Hand weaving Harris Tweed in the
Hebrides to supply an exclusive market.
(Harris Tweed Association)

Harris Tweed is able to survive by trading
on its reputation as an exclusive textile.
(Harris Tweed Association)

A modern weaving mill (here, equipped with flexible rapier looms). Very few staff
are needed for supervision.
(Gebrüder Sulzer Aktiengesellschaft, Winterthur, Switzerland)

Narrow fabrics and braids

Some narrow fabrics may be knitted, but most ribbons, tapes and edgings are woven on special narrow fabric looms. These looms can weave a number of ribbons or tapes side by side. Each narrow fabric has its own rapier or shuttle, but they all share frames, reed, and patterning mechanism.

BRAIDING involves interlacing warp threads without the use of weft. This allows all the yarn to be at an angle to the direction of the braid, and helps in making the structure flexible. In Japan, braid-making has reached the intricacy and stature of an art form.

Shoelaces, and decorative ricrac braids, are the most common braids.

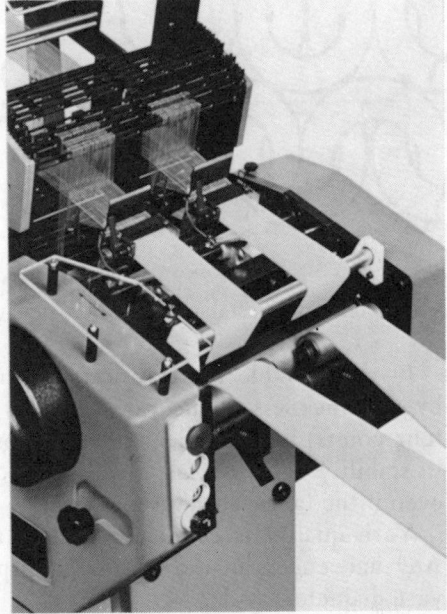

Narrow fabric high-speed needle loom.
(Bonas Machine Company Ltd, Sunderland, England)

Jacquard-weave narrow fabric, face and reverse.

Knitting

Properties of knitted fabrics

Knitted fabrics are made of interconnected loops. When these rounded loops are stretched they straighten and extend with the pulling force. The more space in the fabric, the more room there is for movement, hence the greater the extension.

Stable (left) and unstable (right) knit structures.

In many stable knit fabrics, the distance between the loops is determined by the thickness of the yarns. In this way, each knitted loop is supported (at four points) by the other knitted loops around it. This support gives the fabric its stability. *Thermoplastic yarns* can, of course, be heat-set for greater stability, even if the fabric structure is open.

Yarn quality is even more important for knitting than it is for weaving. Any unevenness in the yarn is multiplied fourfold, because each loop is four yarn diameters wide.

Some knit structures hold the yarns at greater tension than do others. This means that the different types of knitted fabric vary in extensibility. In general, rib or double jersey is more extensible than plain jersey, which in turn is more extensible than tuck-stitched fabric. Warp-knit tricot is the least extensible.

TRY YOUR HAND

Using 4.5 mm knitting needles, and 8 ply wool, knit a plain knit structure of 15 stitches by 15 rows. Repeat with the same needles with 3 ply wool, and with 12 ply wool. Examine the extensibility, softness and resilience of each structure.

Can you explain the behaviour of your samples in terms of the diagram above? How would you go about picking the right needle for hand knitting yarn if you did not know its ply and had no knitting instructions?

How could you estimate the length of yarn needed to knit one row?

Mechanization of knitting

The mechanization of knitting started in 1589 when the Reverend William Lee of Cambridge invented the stocking frame. His invention was completely original and without precedent. Queen Elizabeth refused to grant him a patent because of lobbying by hand knitters who were worried that this new machine might take away their livelihood. After all, it did knit six times as fast as a diligent hand worker! Lee took his invention to France, where it was accepted.

Until the 1950s, knitting was much less productive than weaving. In 1900, when 10 hours on a loom was sufficient to weave 100 metres of cloth, a knitter using a hand-operated flat knitting machine had to work for 300 hours to knit 100 metres of cloth.

The circular knitting machine was the turning point in knitting productivity. Equipped with up to 108 yarn feeds, such a machine can now produce the same 100 m in less than 3 hours. Today, on the average, warp and weft knitting is three times as fast as weaving (averaged over different looms). Therefore, knitted fabrics are increasing in their availability and popularity. Because of lower cost, excellent design range and comfort properties, and rapid production, they are gradually replacing woven cloth in many applications.

Specialization of machinery

The size of the loops in machine-knitted fabrics is determined by how close the knitting needles are set. Naturally, the closer the needles, the finer and smaller they have to be. Fine gauge knitting machines are suited only for knitting fine yarns, and coarse gauge machines can make stable fabrics from thick yarns only. Looms are far more flexible than knitting machines in the range of fabrics they can produce.

There are other ways, also, in which knitting machines are specialized to make a particular type of product. For example, single-jersey machines cannot knit rib fabric, and double-jersey machines cannot do warp knitting. A machine has to be specially designed to knit fully-fashioned garments, and it cannot be converted to make – say – rib scarves or continuous yardage. A sock-maker will not knit pantyhose and vice versa. Nevertheless, the variation in design and scope of knitted fabrics is enormous.

In knitting, as in weaving, there are few limits to variations in colour and pattern. Multicolour yarns are used in Jacquard knitting machines for a nearly infinite range of patterns. Textures based on pile and on lace effects are possible. Stretch and other 'effect' yarns can be included to widen design possibilities further. In addition to surface design, it is possible to shape the outline of complete garments, or even to knit three dimensional constructions such as gloves and hats.

Weft knitting

WEFT KNITTING is so called because the yarn travels across the fabric, like the weft in weaving.

389

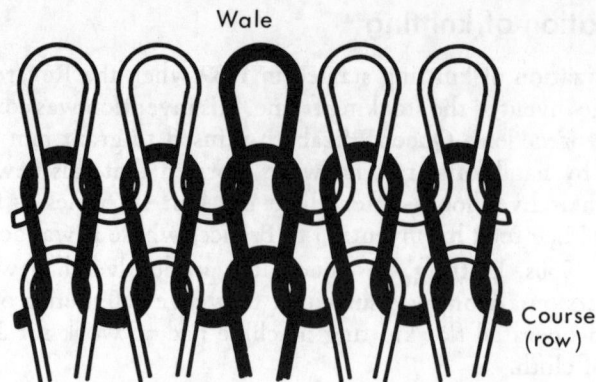

Wale

Course (row)

Single-knit jersey

The original knitting frame was the earliest form of a FLAT-BED KNITTING MACHINE. It had a row of needles which moved up and down in turn when a cam passed over them. A yarn feed attached to the cam fed the yarn to each needle when it was at its highest raised position. The cam and yarn feed were passed back and forth across the needles, forming one row of loops on each passage. Since all the needles of such a machine face one way, all the loops produced also face in one direction. The fabric has a distinct face and back.

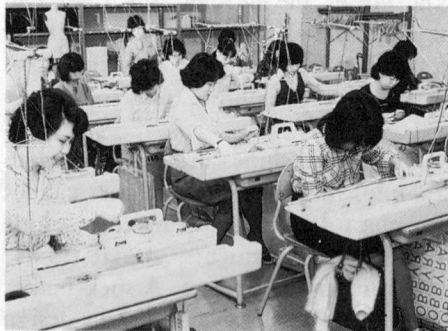

Room of flat-bed knitting machines. Most home knitting machines are of this general type.
(Australian Wool Corporation)

Weft knitting machine with needles raised to accept the yarn from the yarn guide.
(David Bailey)

TRY YOUR HAND

This exercise illustrates the process of weft knitting.

Five students play the role of the needles, and one, holding a ball of yarn, the role of the yarn guide.

The needles stand in a straight line, shoulder to shoulder, right hand extended in front.

The yarn guide ties a loose loop around the right wrist of the first needle, and then makes a loose loop around the right wrist of each needle in turn.

The left hands of the needles act as tensioning devices, to hold the loops tight. When the yarn guide makes a return trip, she or he feeds new yarn to the needles. The new yarn is pulled through the old loop, forming new loop. The left hands still hold down the old loops, keeping the tension even.

After a few passes there should be a large, stretchy, net-like fabric. This is your giant weft-knit jersey.

The circular knitting machine

In flat-bed knitting, only one yarn can be fed across the needle bed at a time. This means that at any one time only one needle is raised. A few are on the way up or down and all the rest are idle, waiting for the yarn feed to reach them. The abrupt stop-and-start movement of the cam and yarn feed at the two ends of the needle bed adds to the inefficiency of the flat-bed knitting machine. The CIRCULAR KNITTING MACHINE is an improvement in both speed and efficiency.

In this machine the needles – and the fabric held by them – are stationary. The yarn package and cam and feed rotate around the circle. Some circular knitting machines work the other way: the needles, fabric and cloth take-up roller rotate together, and the yarn packages, cams and yarn guides are stationary.

Fabrics produced on a circular knitting machine are *tubular*. Machines of various diameters are used for different purposes. Narrow diameter circles are used to make socks or legwarmers, medium diameter machines knit singlets without a side seam, small, medium or large. Any finer adjustments to the sizing are done during *finishing*. If the fabric is not properly set during finishing, then the garment will shrink or expand, relaxing to the size that it originally had on the knitting machine (p. 453). Fabrics made on large diameter knitting machines are usually slit and processed as flat yardgoods.

The smooth motion of the circular knitting machine means that it can work at quite high speeds, but, as in the flat-bed knitting frame, most of the needles are idle most of the time. This inefficiency is overcome by the use of *multiple yarn feeds*.

The MULTIFEED KNITTER feeds a new yarn to each needle as soon as it has finished its last stitch. In this way, many rows can be knitted in one revolution. Each needle needs time to complete its motion, so it is not possible to have a separate yarn feed for each needle. However, with refined needle shapes and cam paths, a modern multifeed knitter can have as many as 108 yarn feeds around a 30 cm diameter circular machine: it knits 108 rows for each revolution. Such a machine can produce a metre of fabric in 20 minutes.

TRY YOUR HAND

This exercise illustrates the multifeed circular knitting frame.

Six to eight students stand shoulder-to-shoulder in a circle, as the needles.

Three students, each with a different coloured ball of yarn, are the yarn feeds.

With just one yarn feed, make several rows of weft-knit jersey, rotating the 'frame' anti-clockwise. How does the fabric compare with the flat-bed jersey made earlier?

Now, space the other yarn feeds around the circle, and knit with yarn from all three yarn feeds.

What is the effect of the multiple yarn feeds on speed of production?

How is the structure of the jersey affected by the multiple yarn feeds?

Double knit

DOUBLE KNIT is the knit equivalent of double cloth: a double thickness knitted in the one action.

Double-knit fabrics are produced on machines where two sets of needles face each other at right-angles. On a *flat-bed machine* the needles are arranged in an upside-down V. The needles on one bed face the spaces between the needles on the other bed; the yarn is fed between them. Such flat-bed machines are used to manufacture scarves, and gloves and other three-dimensional objects. In a circular machine, the right-angle between the needles is created by a dial-and-cylinder arrangement.

Some knitting machines – such as those used to make hosiery – use a double cylinder. The needles here have a hook on both ends. For a double knit structure, every second needle is transferred to the top cylinder. For a single knit, all the needles work on the lower cylinder.

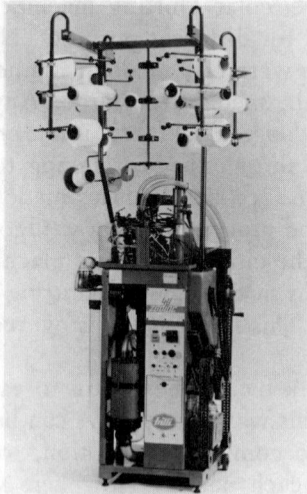

Pantyhose knitting machine (Zodiac 4 Ultra/S). This particular machine can knit the elasticized waistband, plain or micromesh, and double welt.
(Edizioni Scientifiche e Techniche Mondadori).

Dial-and-cylinder arrangement of a pantyhose knitter.

Both sets of needles prepared to accept the yarn from multiple yarn feeds on a circular double knitting machine.

Interlock

Interlock is a form of double knit. It differs, however, in that on one pass of the yarn guide every second needle is missed out. These needles are used on the next pass of the yarn. The yarns of the first and second feeds cross over one another between loops – this makes it more difficult for a 'run' to form if a stitch is dropped.

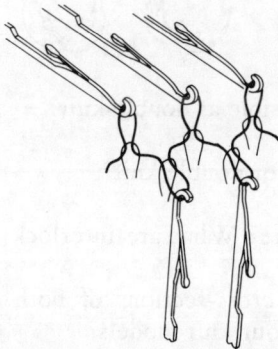

Rib (one row).

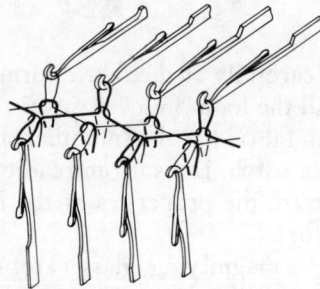

Interlock.

TRY YOUR HAND

Set up two lines of student 'needles' facing each other, about 50 cm apart. Cast on loops as shown in the diagram for double knit. Knit four rows; check the structure closely.

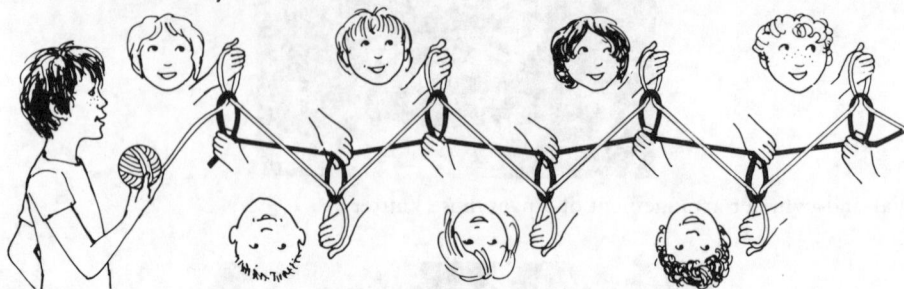

Double knit.

Which way do the loops face? Why?

Move the needles sideways to stretch the fabric. Which is more extensible, single jersey or double jersey?

How does the zig-zag arrangement of yarns affect the properties of the fabric?

Drop one stitch. Does a ladder form?

Unravel the knit, and line up the needles once again. This time, follow the knit diagram of interlock jersey. Use two different colours of weft, one for the odd, the other for the even rows.

Interlock.

Look carefully at the fabric formed. Is it as extensible as double knit?

Are all the loops in level rows?

Which fabric has the smoother surface, interlock or double knit?

Drop a stitch. Does it run readily?

What are the properties of the interlock structure? What are interlock fabrics used for?

With a magnifying glass carefully look at the cross-sections of both double jersey and rib fabrics. How do they compare to your knit models?

Can rib knitting be done on a single jersey knitting machine?

Patterning

Patterning is introduced into the knitted fabric by the use of coloured yarns and by varying the pattern of stitch formation: holding on to stitches without knitting them, transferring loops from one needle to another, and so on.

The smallest unit in a knit (the loop) is larger for the same size of yarn than the smallest unit in a woven structure (the interlacing). Because of this, knitted designs cannot be quite as finely detailed as woven designs. Nevertheless, Jacquard-style 6-colour designs and three-dimensional or lacy textures can be produced very effectively.

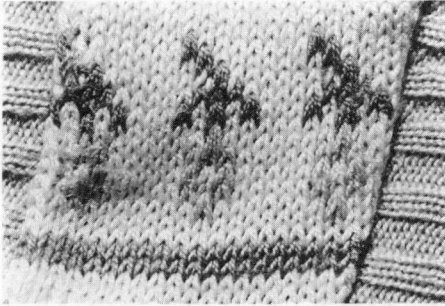

Jacquard patterned weft knit, face and reverse.

Colour

The yarn feeder is attached to the cam, and presents the yarn to the needles at precisely the right moment as the needle descends from its highest position. If there is more than one yarn feeder for each cam, then a choice can be made as to which colour of yarn is to be guided into the hooks of the needles.

The yarn *not* fed into the hook of the needle will not form a loop, and floats on the back of the knit. Such floats are typical of Fair Isle knits, which are made on a single-bed knitting machine.

Multiple yarnfeeds for colour and pattern.
(Wildt Mellor Bromley Ltd, Leicester, England)

If a double bed of needles is used, the floats can be taken up into loops on the back of the fabric. Such designs are typical of Jacquard double-knit jerseys.

In some modern knitting machines, the instructions can be given to the needles and yarn feeders electronically. The machines are connected to a computer console, and are able to switch over to new designs in a matter of minutes.

Texture

Texture is introduced into knit fabrics by stretching selected loops out of position.

The simplest way to do this is what is known as a slipstitch in hand knitting: instead of pulling a new loop through, the old loop is not knitted, but is retained for the next row where it will be knitted together with the new loop. This is known as a TUCK STITCH.

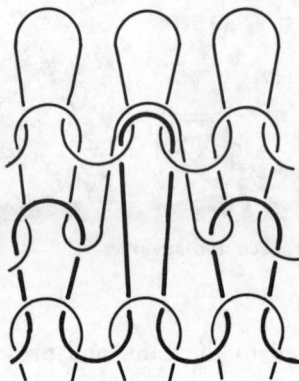

Tuck stitch.

The retained loop needs to stretch twice as far as the other loops, and so tucking increases the tension in the yarn. If every second row in a double knit is tucked, a very firm fabric (known as CARDIGAN STITCH) is formed. This fabric structure is widely used for knit suits and other outerwear.

Shaping

When a flat piece of fabric is cut into the shape of a garment, some of it is wasted as offcuts. In a knit, special precautions must be taken so the cut stitches do not run. These two problems mean there are advantages in producing a fully finished garment panel with sealed edges. FULLY-FASHIONED panels can be knitted on a straight-bed single jersey knitting machine.

In hand knitting, narrowing is achieved by knitting two stitches together as one. In a knitting machine, however, each needle forms its own line of stitches. In order to narrow the fabric, a stitch must be transferred from one needle to another. This necessitates a special mechanism. A fine needle like a loop holder (TRANSFER POINT) descends on the knitting needle in the 'up' position, moves down with the needle through the old loop, and thereby allows the loop to be transferred to it. The transfer point then moves up and to the side, and lowers the captured loop onto the new needle.

A fully-fashioned garment shows characteristic bunching marks plus slanted stitches near the edge of the panel. Usually single-bed machines are used for

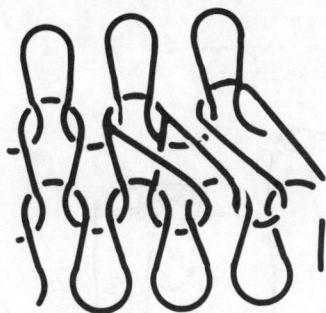
Stitch transfer to narrow the fabric.

Shaping in a fully-fashioned garment.

fully fashioning. Single-bed machines, however, cannot knit rib basques, so such ribbed segments are knitted separately on a double-bed machine and transferred to the garment knitter at the start of each new panel.

Hosiery

Socks

The first knitting machine produced hose that had to be seamed together at the back. Such *fully-fashioned* hosiery is still made on straight-bed machines, but all other hose, including socks, is made on circular knitting machines. The diameter of these machines is the diameter of the garments produced. The gauge – i.e. number of needles to the inch – relates to the thickness of the yarns used and to the final density or sheerness of the fabric.

TRY YOUR HAND

You will need a fully-fashioned sock and cardboard tube about 10 cm in diameter. Unpick the seam on the top front of the sock, and slip the sock onto the tube. You will find that the heel sections and the toe sections form two identical pockets that stand out from the side of the tube.

This is the form in which the sock had been knitted on a circular machine. Where the pockets were being formed, the stitches were held on the needles at the edges, gradually narrowing to a triangle shape, then the held stitches were picked up as the fabric was widened for the lower half of the pocket. When the heel section was completed, the foot was produced by circular knitting, then another pocket for the toe. The sock was then divided from the following one by a separating thread made of alginate or PVA. During dyeing or finishing the separating thread dissolved in the water and the socks separated. Each sock was then completed by machine stitching the lower opening. This is the seam on the top of the toe which you have unpicked.

Pantyhose

There are a number of different ways in which pantyhose can be produced. The most common method involves knitting two tubes with reinforced tops, seaming the toes, then cutting the reinforced sections and seaming to make the finished garment. This however involves much handling and sewing after

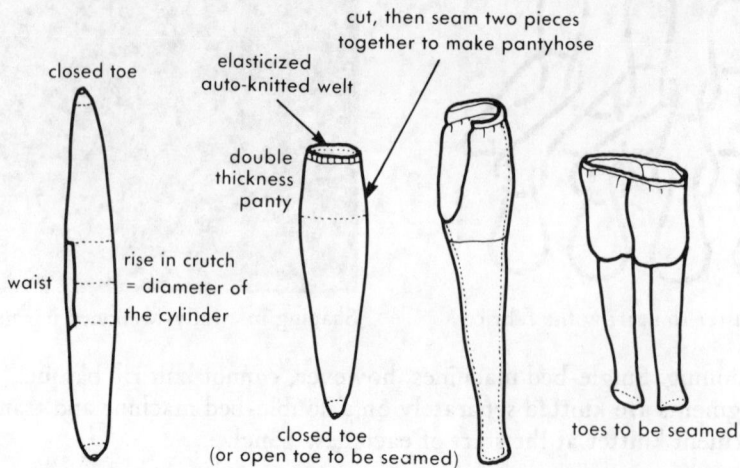

closed toe

elasticized auto-knitted welt

cut, then seam two pieces together to make pantyhose

double thickness panty

waist

rise in crutch = diameter of the cylinder

closed toe (or open toe to be seamed)

toes to be seamed

Methods of manufacturing pantyhose.

the knitting operation. Many automated devices have been developed to improve the speed and efficiency of hosiery production.

TRY YOUR HAND

Collect as many different brands and styles of pantyhose as you can. Make note of packaging ('image') and price.

Examine the hose for method of construction. Is there any correlation between the cost and method of manufacture? Between cost and 'image'? Cost and packaging?

Knitted pile fabrics

Stretch terry

This fabric has become popular for many applications, from car seat covers to track suits. It is a single-knit loop-pile fabric, in which the pile is formed over specially shaped SINKERS inserted between the knitting needles.

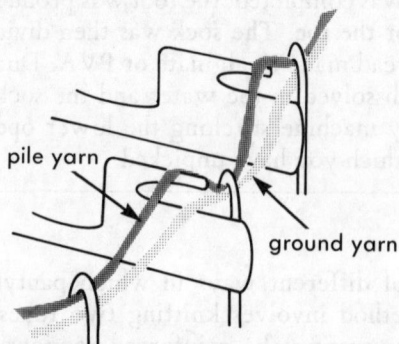

pile yarn

ground yarn

Manufacture of stretch terry.

398

Luxury knitted velour semi-dull polyester upholstery in the Mitsubishi Magna Elite. (Mitsubishi Australia)

Knitted velour

Knitted velours are used for both upholstery and luxurious sportswear. Basically they are loop pile knits with long loops. During finishing the loops are sheared, losing as much as 30 per cent from their length. Therefore, because of the cost of yarn wastage, a good quality knit velour is rather more expensive than its terry counterpart.

Laid-in yarns

It is possible to include yarns in a knitted structure without making them form loops. Such laid-in yarns do not need to be as fine, smooth or flexible as other knitting yarns. The yarn to be laid in is fed between the existing loop and the new yarn feed, so as to miss the hook of the needle.

If thick ground yarns are used, they may cover the laid-in yarns. The elastic yarns in surgical hose and in the tops of socks are used in this way. At other times, the laid-in yarns may be special effect yarns for improving either the appearance or the insulation properties of the fabric.

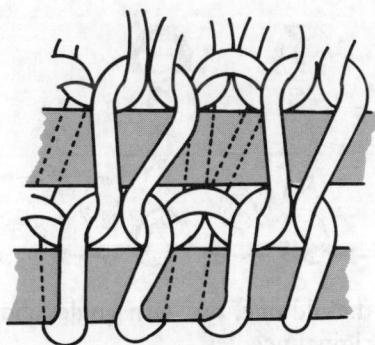

Laid-in yarns in weft knitting.

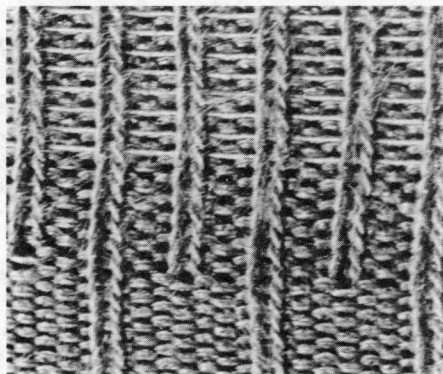

Laid-in elastic yarns at the top of a sock.

Brushed knit

A favourite for windcheaters, this fabric is also made as a single-knit jersey. A thick yarn is allowed to float on the back of the fabric, being 'laid in' to the stitches at intervals but not taking part in loop formation. In the unfinished state this fabric snags easily, but when the fibres are matted by brushing, a firm, stable fabric results (see p. 444, Chapter 11).

Warp knitting

Warp-knit tricot

The term TRICOT is generally used to indicate warp knit fabrics, while the word jersey is used for weft knits. Whereas in weft knitting a single yarn goes past each needle in succession *across* the fabric, in a *warp*-knitting machine *each needle has its own yarn* which goes the *length* of the fabric.

If each needle performed the loop-forming cycle time after time with its own yarn, the result would be a large number of unconnected crocheted chains. If, however, the yarn, after forming a loop on one needle, is passed on to its neighbour, then the crocheted chains are joined and a fabric results.

TRY YOUR HAND

For a role-play representation of warp knitting, six 'needles' should be faced with six 'yarn guides'.

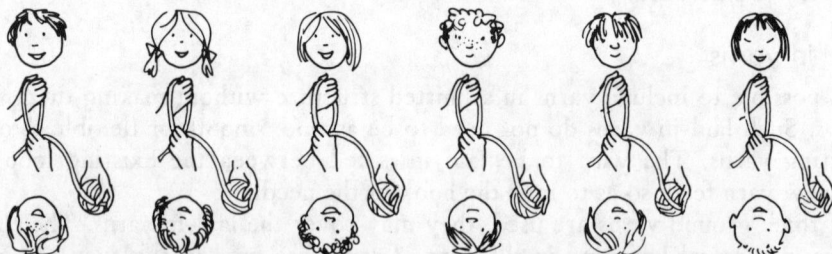

After casting on one loop on each needle, each yarn guide should move one step to the right, then feed yarn to the needle now in front of her.

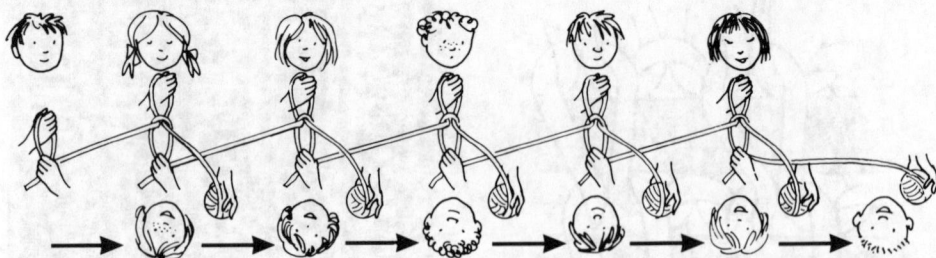

After the new loop has been pulled through the old one, the yarn guides should step back to the left, and feed yarn to their original needles.

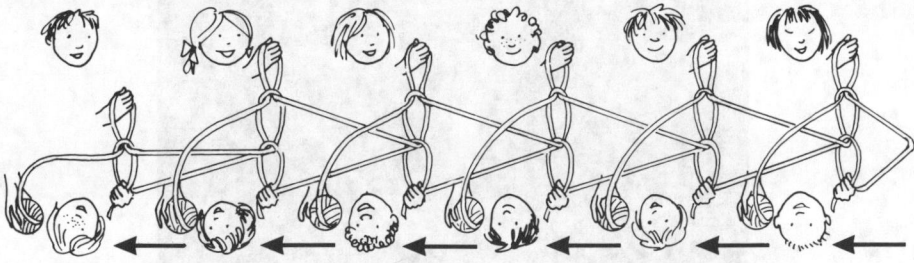

Continue to the right for row number 3, to the left for row 4, and so on.
1 Were any needles idle at any time?
2 Is this a faster or a slower process than weft knitting?
3 What are the differences between the two fabric types?
You can create a variety of textures and effects by having more than one yarn guide for each needle. While guide A goes from needle 1 to needle 2, guide B can go from needle 1 to needle 4, and so on. Such structures, involving more yarns, are more stable. They also resist runs if a loop is dropped or broken.

The machine

In warp knitting all the needles work together: a full row of knitting is completed in the time it takes a needle to complete its up-and-down cycle. Therefore, warp knitting is very productive. The use of trouble-free, easy-slip filament yarns further improves knitting efficiency.

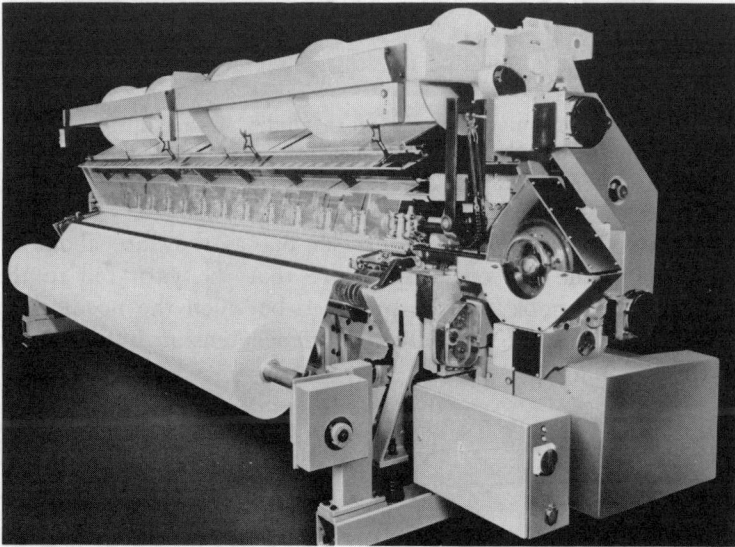

Warp-knitting machine.
(Karl Mayer Textilmaschinenfabrik GmbH, Obertshausen, Germany)

Warp-knitting machines usually use fine filament yarn and fine-gauge bearded needles. Since the bearded needles do not have an automatically closing latch,

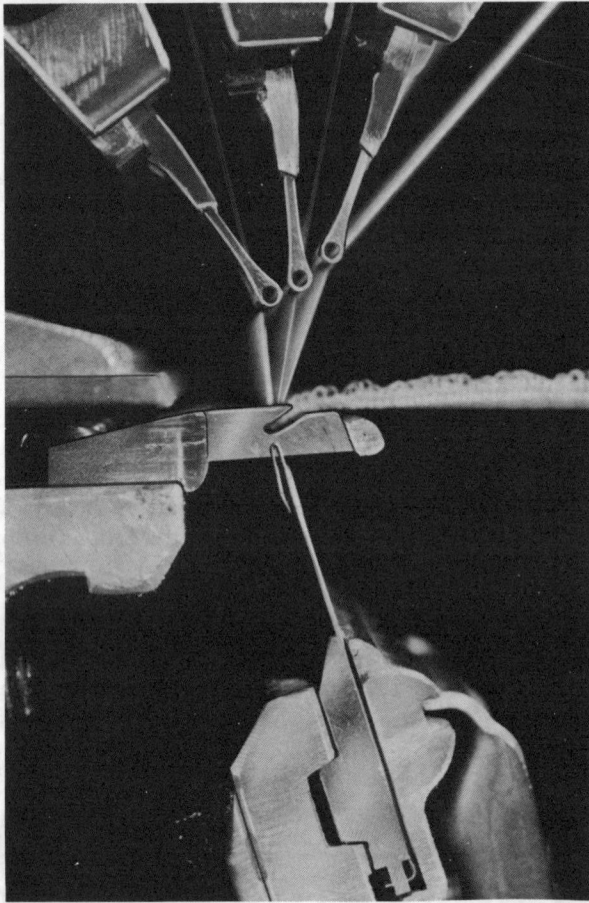

Parts of the warp-knitting machine. The yarn guides at the top feed the yarn to the needle below, while the nosed sinker in the centre forms the pile yarn into loops. (Karl Mayer Textilmaschinenfabrik GmbH, Obertshausen, Germany)

their hooks need to be closed by a presser bar as they are about to pull the new loop through the old one. To ensure that the yarn fed to the needles is not missed by the hook, sinkers are used between the needles. The fabric produced and the existing loop are held under tension until the moment of new loop formation.

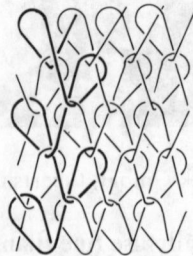

Path of one yarn in warp knitting.

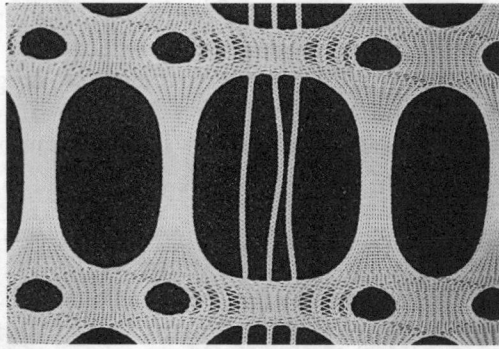

Warp-knit curtaining.
(David Bailey)

Colour and texture

Open, lacy effects can be achieved in warp knitting by having more needles than yarns, and no overlapping yarns. On the other hand, fairly rigid knits that can imitate woven fabrics can be made by having a greater number of overlapped yarns, using more than one yarn feed for each needle.

Extra coloured yarns from extra warp beams may be introduced for a surface effect.

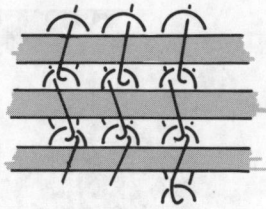

Laid-in yarns in warp knitting.

Warp-knit fabrics can be recognized by the small vertical loops visible on the face of the fabric and a zigzag line of overlaps on the back. These zigzags appear as slightly larger lines of loops going across the back of the fabric, at right angles to the wales on the face.

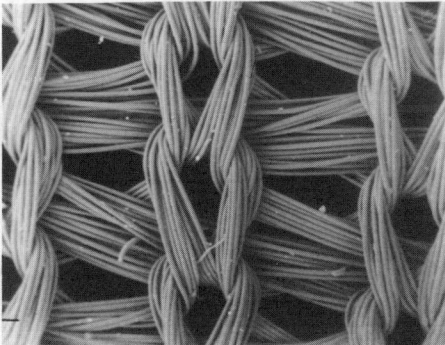

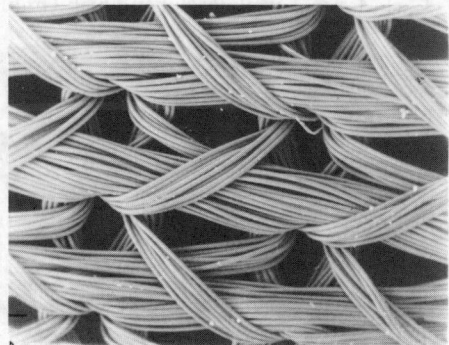

Face (left) and reverse side (right) of warp-knit tricot.
(Vivian Robinson)

403

Underwear made from synthetic filaments is often warp-knitted. Brushed nylon, so often used for warm nightwear, is a warp-knit fabric made from fine filament yarns. During finishing, the filaments crossing over at the back of the fabric are raised with rotary wire brushes to create the fluffy surface. Similarly constructed but heavier polyester warp knits can be used for brushed sportswear or leisurewear.

Some of the huge range of fabrics that may be manufactured by warp knitting. (Karl Mayer Textilmaschinenfabrik GmbH, Obertshausen, Germany)

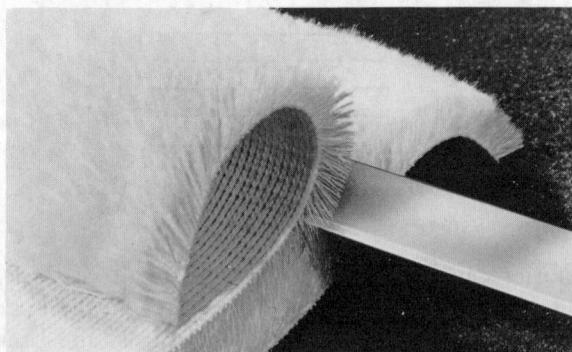

Manufacture of pile knits by slitting double warp knit. (Karl Mayer Textilmaschinenfabrik GmbH, Obertshausen, Germany)

Sliver knit printed to imitate fur. Instead of a pile yarn, slivers of synthetic fibre are inserted into the ground fabric. (*Textiles*, a periodical published by Shirley Institute, Manchester, U.K)

404

Raschel

RASCHEL MACHINES are warp-knitting machines which have one or two needle bars equipped with latch or compound needles. These are able to accept much thicker yarns than can the bearded needles of a warp knitter. Raschels use staple and fancy yarns to knit upholstery and curtaining fabrics, and laces.

For complex designs, in addition to multiple warp beams the Raschel machine is fitted with special creels to supply extra yarns.

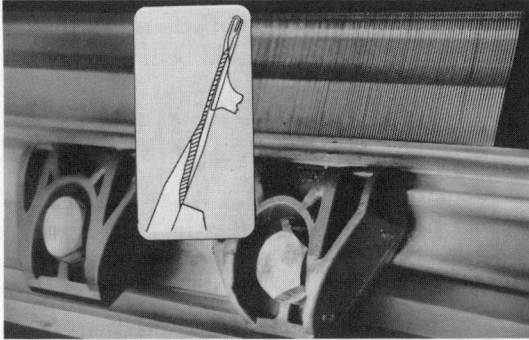

Compound needle as used in Raschel knitting machines.
(Karl Mayer Textilmaschinenfabrik GmbH, Obertshausen, Germany)

Jacquard Raschel machine.
(Karl Mayer Textilmaschinenfabrik GmbH, Obertshausen, Germany)

Raschel lace. Designs of enormous complexity may be achieved on these machines.
(Karl Mayer Textilmaschinenfabrik GmbH, Obertshausen, Germany)

Lace

Lace has a very long history: elaborate decorative networks have been found in 3000-year-old Egyptian tombs. It has been a fashion item in Europe since the seventh century, and reached its peak in the elaborate costumes of the sixteenth and seventeenth centuries. By this time, many centres in Europe were famous for regional styles and techniques of lace-making: Alençon and Chantilly in France, Brussels in Flanders, and many other cities gave their names to particular varieties of lace. In Venice in the fifteenth century, the locally-produced lace was such an important item of trade that Venetian residents were banned from wearing it, so there would be more for export.

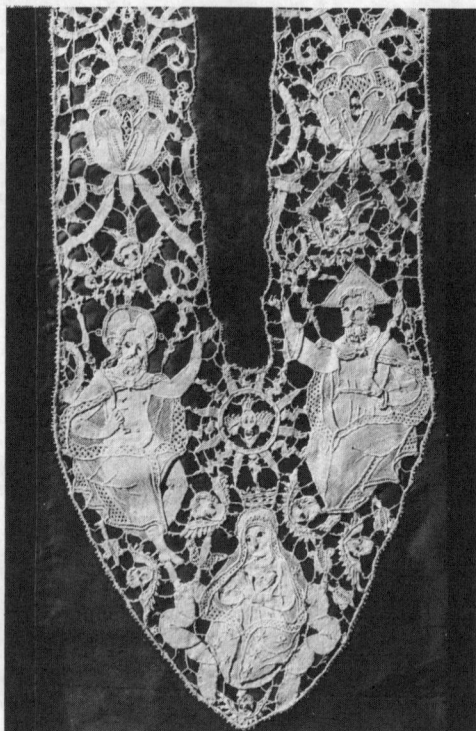

Ornament for the neck of an alb
Needlepoint lace (punto in aria)
Venetian school, sixteenth century
Felton Bequest 1964
(National Gallery of Victoria)

Two main techniques were used to make the lace. NEEDLEPOINT is made by laying out a fine framework of yarns to form a pattern, and filling in the design with fine buttonhole stitches. Venetian point is of this type. BOBBIN (PILLOW) LACE is a quite different technique which uses many threads at one time. The threads, held on bobbins, are twisted around each other to form a network pattern, which is held in shape by pins on a supporting pillow. Both these techniques can be dated to at least the fifteenth century, and may be older. Bobbin lace may be a direct descendant of the fishing net.

Torchon (bobbin) lace, about 1900. Simple designs such as this which could be made quickly were able to compete with the machine-made laces, whereas the more delicate and intricate laces could not.

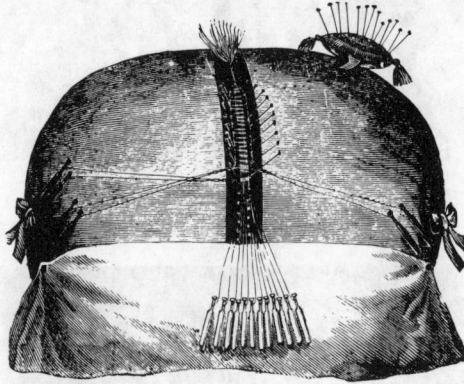

Dressed pillow with lace in progress.

The first machine-made lace dates from around 1770. This was a knitted net fabric made on a modified flat-bed stocking frame, inferior to the hand-made product, but cheap. Some traditional lace-workers were able to use this cheap, mass-produced net as a foundation for their designs, which were made separately and stitched on, but other regional designs were not able to be adapted. In 1809, John Heathcote in England invented a machine which produced a good imitation of some of the traditional bobbin foundations or grounds. This was often hand-embroidered to produce a patterned lace fabric.

A machine similar to Heathcote's was invented, also in 1809, by John Leavers. This was able to produce a completely machine-made bobbin-type lace, with or without ground, with only final details to be added to the lace by hand. By 1841, no hand work was necessary.

The modern Leavers lace machine has some 40 000 parts, almost all of them moving when the machine is operating. It takes an operator two to three weeks to thread, and produces continuous lace several metres wide which can later be slit into widths. In the Leavers machine the yarns are twisted around one another by some four thousand paper-thin bobbins which slide to and fro between shifting warp threads to create the pattern.

The Schiffli lacemaking machine is used to create guipure and other *re-embroidered* laces. These laces are worked on a woven background fabric. Sometimes this fabric is made of alginate threads, with cotton embroidery. When placed into water, the fabric which served as support during manufacture dissolves during wet finishing and leaves the delicate network of the stitched lace.

Some modern machine-made bobbin-style laces.
(David Bailey)

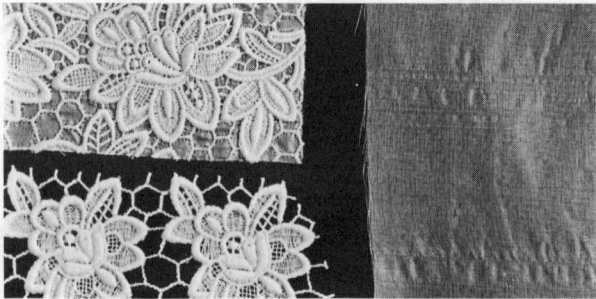

Guipure lace embroidered on water-soluble polyvinyl alcohol (Solvron) cloth.
(Courtesy Nitivy Corporation, Tokyo, Japan; photograph Dale Mann/Retrospect)

Adding to the ground fabric

Tufting

Candlewick chenille, imitation furs and other pile designs are easily achieved on a woven fabric base by TUFTING. This is a method of inserting tufts of yarn into the face of the fabric by the use of a needle punch operating from the back.

The most widespread application of this technique is in the manufacture of carpets: 80 per cent of carpets are tufted.

In SLIVER KNITTING, a pile of sliver rather than yarn is inserted in a knit fabric. Many fleecy car seat covers are manufactured in this way.

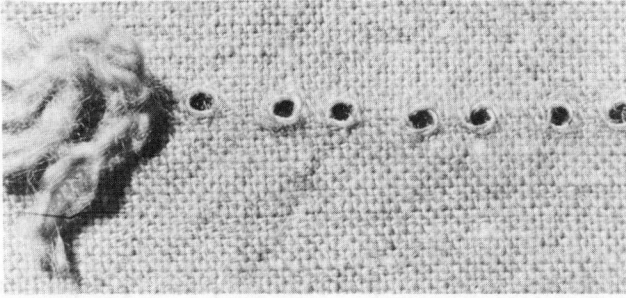

Chenille made by chain-stitching the pile yarn into the ground fabric.

Multiple-needle sewing machine for quilting. Quilting machines are required to stitch through padded or insulated materials up to 10 centimetres thick, and sew a wide range of fabrics, vinyl, and leather.
(Fales Machine Company, Walpole, Massachusetts, U.S.A.)

Quilting

QUILTING is the stitching together of two pieces of fabric with a layer of wadding or other filling between them. The lines of stitching create a pattern and divide the middle layer into stable pockets of good thermal properties. Quilting as a means of producing warm bedspreads became a popular community activity in eighteenth-century America. Today, quilting is not only used for bedding but also for furnishings, fashion, skiwear and leisurewear.

The quilting is carried out by multineedle sewing-machine stitchers. The main shortcoming of the quilted structure is that the stitching threads may abrade away and break during use.

A new type of quilting machine, the Pinsonic, is able to *weld* quilts made of thermoplastic fibres. The ultrasonic vibrations it generates melt the fibres along the design lines and produce a welded quilt. Such quilts are claimed to be more durable, since the stitching does not use breakable yarns. The process itself is seven times as fast as conventional quilting methods.

Nonwovens

Fabrics can be made from fibres as well as from yarns. In nonwoven (and non-knitted!) fabrics, the fibres may be held together in a number of different ways.

In some nonwovens – as in batts or wadding – the friction between the fibres is all that is available to resist the pulling force when the fabric is stretched. If the fabric needs greater stability or strength, additional cementing – stitching, adhesives or fibre-to-fibre welding – is needed.

The process of nonwoven fabric production depends on the properties required at the final application, and on the fibres that are used. For example, cheap and reasonably durable absorbent dustcloths can be manufactured from viscose staple. The fibres are carded into a web, which is then passed over strong jets of water which force the fibres apart and form a pattern of small holes. These holes can absorb dust or other things from the surface being wiped. The web is weak at this stage, so to make it withstand abrasion the fibres are bound together with glue. If the whole fabric is coated with glue it will take on the properties of the binder and become hydrophobic. A hydrophobic fabric is not suitable as a wet wiper, so a compromise is made: the glue (together with dye which makes it colourful) is printed on the web in a pattern. The result is a fabric which is both strong and absorbent.

TRY YOUR HAND

see safety note, Chapter 8, p. 232

You will need:
a roll of surgical cotton wool
PVA wood glue
container for mixing glue
few drops of food colouring
potato masher or biscuit cutter

What you do:
Mix 1 tablespoon of PVA glue with a cup of water and add the food colouring.
Cut the cotton wool sheet into 3 pieces, 10–20 cm long.
Keep one piece of cotton wool web intact.
With the potato masher or biscuit cutters, print the second piece with the glue to make a pattern.
Soak the third piece in the glue mix.
Dry the samples by first rolling them between two sheets of absorbent paper, then ironing lightly.
Compare the absorbency, strength, and flexibility of the three fabrics.
Design an experiment that will test these properties with reasonable accuracy, and tabulate your results to compare them.
Under the microscope, observe the structure of the three webs.
1   What is the effect of the adhesive on the properties and structure of the web?
2   How does this affect the performance of the final product?
3   In what other ways could you hold the fibres of the web together?

The great advantage in manufacturing nonwoven fabrics is the speed with which the final fabric can be produced. All yarn preparation steps are eliminated, and the fabric production itself is faster than conventional methods. To produce 500 000 metres of woven sheeting requires 2 months of yarn preparation, 3 months of weaving on 50 looms, and 1 month for finishing and inspection: a total of 6 months to produce and deliver the order. Most producers of nonwoven fabric can deliver the same quantity of sheeting within 2 months from receipt of the order. Not only are production rates higher for nonwovens, but the process is more automated, requiring less labour than even the most modern of conventional knitting or weaving systems. The nonwoven process is also efficient in its use of energy.

Relative rates of fabric production

Fabric manufacturing method	Typical rate of fabric production
weaving	1 m/min
knitting	2 m/min
nonwoven	100 m/min

Nonwoven fabrics can be engineered to give a wide variety of properties. Nevertheless their aesthetic properties (handle, drape, appearance) are such that they are not in direct competition with conventional fabrics in the outerwear market. Woven and knitted fabrics in this application will not be replaced by nonwovens in the near future. Currently, the main areas of growth in nonwovens are in geotextiles, medical and hospital uses, disposable products, and filters.

Tapa (bark) cloth, manufactured in many Pacific countries including Papua New Guinea, Tonga, and Fiji, is one of the oldest nonwoven textiles.

Web formation

A nonwoven fabric is basically a web of fibres held together in some way. The web may be made of either staple fibres or filaments, or from portions of a polymer film.

Webs from staple fibres

Carding is a time-honoured way of making a web from staple fibres. In a carded web the fibres are aligned more or less parallel to each other and to the direction in which the card produces the web. Such webs are stronger when pulled lengthwise than crosswise, because there is more friction between the fibres in a lengthwise direction.

Straight folder for producing a parallel-laid web from carded fibres.
(William Tatham Limited, Rochdale, England)

Carded webs are usually thin; they may be too thin for some nonwoven end-uses. To increase the final thickness, a number of webs can be layered. The layers may be kept parallel, or cross-laid. There is less directional variation in strength in a cross-laid web.

For many purposes it is important that the strength of a web should be the same in all directions. To achieve this, the fibres which make up the web must be orientated equally in all directions.

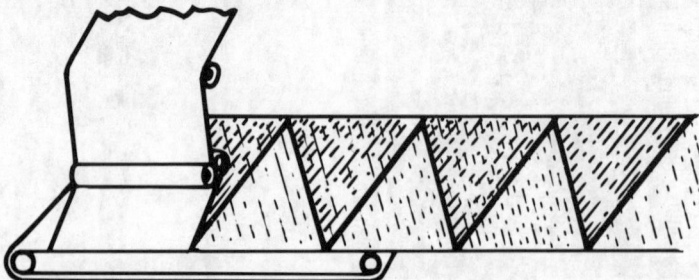

Formation of a cross-laid web. The properties of cross-laid webs do not vary with direction as much as do those of straight-laid webs.

The *Rando-Webber* creates such a randomly orientated web by blowing the fibres about in a stream of air and then sucking them onto the surface of a perforated drum to form a layer. This randomizing process produces a remarkably uniform web from staple fibres. DRY-LAID (carded or air-laid) webs account for three-quarters of nonwovens produced.

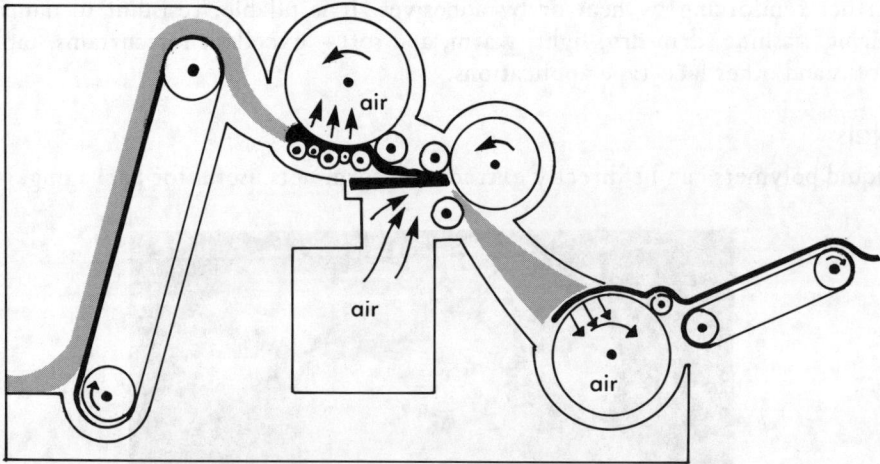

The Rando-Webber gives a randomly orientated web, with no directionality in its properties.

Other ways of forming webs from short staple fibres are wet laying, electrostatic web formation, and spraying.

WET LAYING is used in papermaking. The pulped fibres are mixed with water and then scooped into uniform layers on wire screens or on rotating, perforated drums. Short, pulped acrylic fibres are made into a wet-laid web from a salt solution. As the water evaporates, the salt chemically bonds the fibres into a strong, synthetic, waterproof paper.

In ELECTROSTATIC LAYING, fine fibres are given a static electric charge between the plates of a condenser, and are then allowed to fall on a moving belt to form a randomly orientated but uniform web.

Short *thermoplastic* fibres can be SPRAYED onto a belt to produce a random web. They are subsequently *fused* by the application of heat and pressure.

Webs from filament

It is possible to tangle *filaments* together to form a web. Such webs are much stronger than webs made from staple fibres.

Freshly extruded filaments are allowed to drop in curls and spirals onto a moving belt. The belt may contain patterns outlined in pins; if so, the filaments arrange themselves around the pins and form attractive, lace-like patterns. The thermoplastic filaments are welded to each other and form a strong fabric suitable for curtains, tablecloths and similar applications. Sometimes, the filaments are textured before web formation – this allows greater extensibility of the fabric in use.

Spun-laced webs

A new method of entangling fibres to create lace-like nonwoven fabrics uses fine, precisely controlled, jets of water. When the jets pass through the web of fibres, they form a small vortex at each point of contact. This creates sufficient fibre movement to entangle the fibres. The resultant fabric (full of carefully positioned holes where the jets of water passed through) does not need any further reinforcing by heat or by adhesives. It is pliable, resistant to damage during washing, drip-dry, light, warm, and soft – excellent for curtains, table-cloths and other lace-type applications.

Nets

Liquid polymers can be directly extruded to form nets useful for packaging.

Extruded net used for packaging.

Webs from films

One of the fastest methods of producing a web of some strength is to make small cuts in a *film*. The nets so formed are used as central support for laminated blankets and also as packaging.

TRY YOUR HAND

Cut slits in a piece of paper as shown:

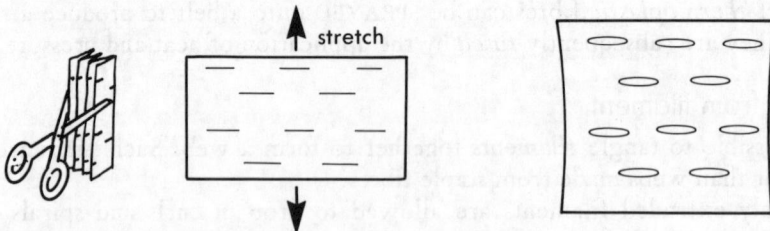

How did the cuts change the properties of the paper?

What textile properties would slitting add to a sheet of plastic?

In what applications would these be an improvement over the properties of unslit plastic film?

To what applications would the unslit film be better suited, and why?

Webs can also be made by *fibrillating* plastic films. The film is embossed and stretched lengthwise to orientate the crystalline areas. It is then stretched sideways to separate them into a network of interconnected fibrils. Such fibrillated webs (mostly of polypropylene) are used for yarn manufacture.

Bonding systems
Friction

All fibres are held in a yarn or a fabric by frictional forces. Some fibres are slippery, others have a better 'hold' on neighbouring fibres because of the frictional properties of their surfaces.

Wool and a few other hair fibres have greater friction when rubbed in one direction than in the other. These fibres have the unique property of forming *felts* when agitated in soapy water. This spontaneous matting of fibres is called TRUE FELTING. Fibres without the directional frictional properties of wool can also form felts, though they need more direct external intervention to tangle the fibres.

The closer the fibres are to each other in a felt – as in any other textile structure – the stronger is their hold on one another. So the denser the felt, the less easy it is to distort it.

True felts

TRY YOUR HAND

Obtain a selection of craft felts in a variety of thicknesses.
Cut 2 cm wide strips from each, and compare their flexibility.
Devise an experiment to compare their strengths by pulling until they break.
Identify the fibres in each felt by examining them under a microscope.
What conclusions can you draw about the properties of felt?
If stretched, will felt return to its original shape after the pulling force is removed?
What effect does this have on the uses of felt?
What other tests could you use to compile a comprehensive list of the properties of felts?

In felt-making, the carded webs of fibres are laid over one another to the desired thickness and then wetted with soapy water and agitated between weighted vibrating rubber aprons. The longer the vibration continues, the more compact the felt will become.

True felts are made from carded webs of wool, rabbit fur, beaver fur, or blends of these. The greater the proportion of fur, the better the felt. Rabbit or beaver fur are finer fibres than wool, with no crimp – therefore they make denser, firmer felts.

Needled felts

Fibres which have no directional frictional properties need to be mechanically entangled in order to form felts. This is done using barbed needles.

The products of needle felting are used for carpets, underfelts, upholstery and blankets, and in industrial applications.

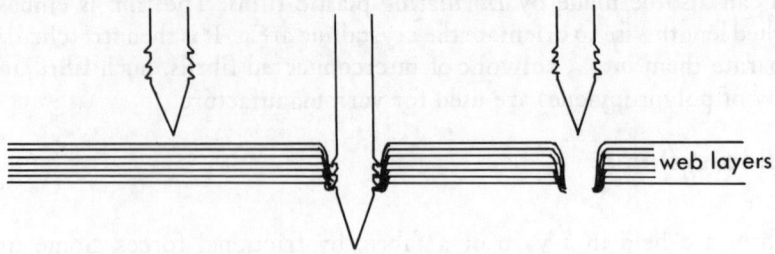

web layers

Needle punching. The barbed needle tangles together some fibres from each layer of the web.

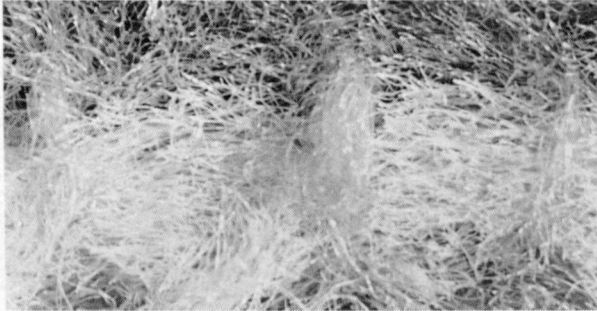

Cross-section of a needle-punched bonded fabric. The upper layer is dyed cuprammonium staple, the lower is modacrylic staple.
(*Bayer Farben Revue* No. 3, 1962, by permission of the editors. *Bayer Farben Revue* is published by Bayer AG, Leverkusen, West Germany.)

Adhesives

For most nonwoven applications, fibre-to-fibre friction does not provide enough strength. Adhesives can be effective in holding the fibres together. It is, however, important not to use too much adhesive or the natural properties of the fibres may be masked, giving the fabric the properties of the adhesive rather than of the fibres.

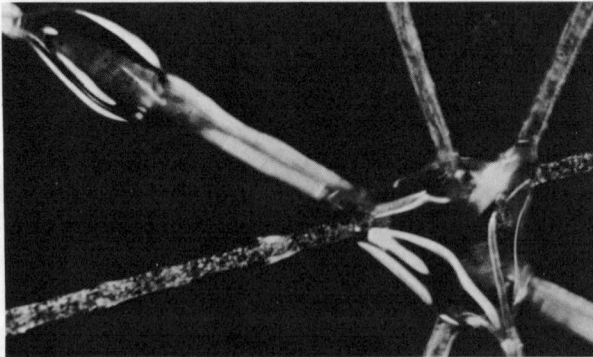

Adhesive bonding on a nonwoven.
(*Bayer Farben Revue* Special Issue No. 9, by permission of the editors. *Bayer Farben Revue* is published by Bayer AG, Leverkusen, West Germany.)

416

The adhesive can be applied to the web as a printed pattern or as a sprinkling of powder or a spray. For certain purposes (e.g. to produce a hydrophobic coating on a web of hydrophilic fibres, such as used for nappy liners) the whole web can be immersed in the adhesive. Some nonwovens may contain as much as 70 per cent adhesive.

The properties of an adhesive-bonded nonwoven depend to quite a large extent on the properties of the adhesive polymer used. Commonly used adhesives are polyvinyl acetate (PVA glue), polyacrylonitrile, polystyrene, PVC, and urea formaldehyde. Of these, PVC and polyacrylonitrile yield the softest fabrics.

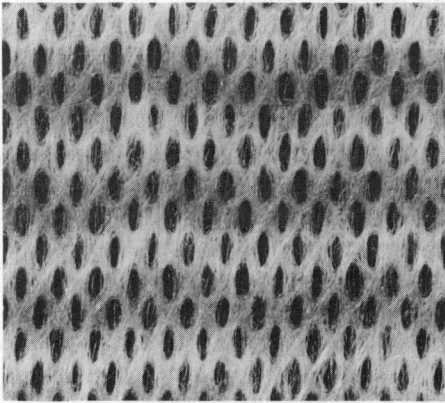

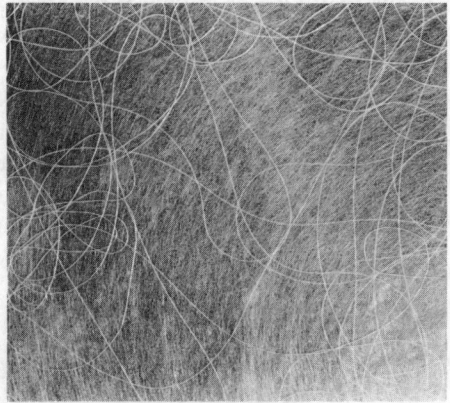

Adhesive bonding: a cellulosic wiper, with the adhesive applied in coloured stripes, and resin-impregnated fibreglass used to reinforce pipes.
(Courtesy Regina Glass Fibre, Ballarat, Victoria)

Heat bonding

When a web of thermoplastic fibres or filaments is heated, the fibres soften and melt into each other at their points of contact. Since the fibres cross over each other at many points, there are many rigid welding spots in the nonwoven fabric. This makes heat-bonded fabrics rather stiff and inextensible.

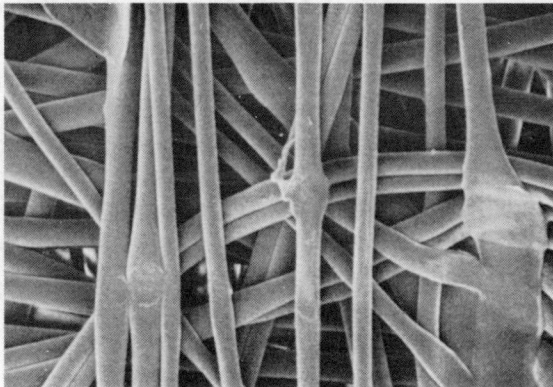

Melt-bonding at selected points to give extra stability to a spun-bonded polypropylene geotextile (Du Pont's Typar).
(Du Pont)

In order to produce a softer fabric without the use of adhesives, thermoplastic fibres with a low melting point are blended in the web with fibres of a higher melting point. On heating only the low melting-point fibres melt. This way, fewer weld points are formed, and the resultant fabric has more desirable textile properties. Such melt-welded fabrics are called MELDED fabrics.

Some polyamide filaments are produced with a special temperature-resistant core and a low melting-point sheath. These are specially designed to produce excellent melded fabrics.

Stitch bonding

Webs may also be given extra strength by stitching them through with yarns. Such structures are usually more flexible and less paper-like than heat-bonded or adhesive-bonded nonwovens.

Arachne (Czechoslovakia) and *Malipol* (East Germany) are two patented systems of stitch bonding. They are based on the principle of warp knitting, but with needles designed to stitch through webs of various thicknesses. A web of fibres, or a web of yarns, or a cheap fabric forms the base. Sharp-pointed needles pierce the base and loop binding yarns through it. In this way, relatively cheap but stable fabrics, with or without pile, can be produced.

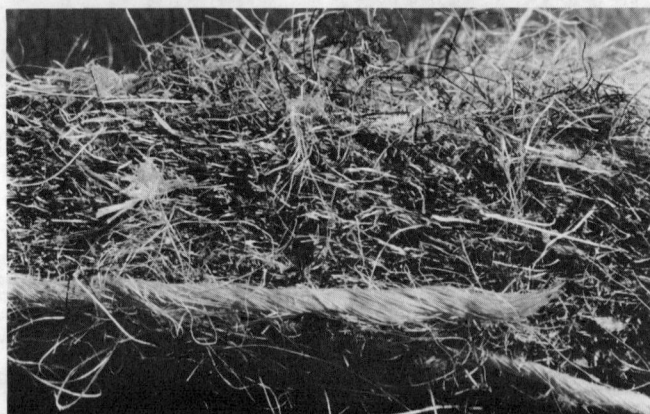

Carpet underfelts are often stitch-bonded, and may incorporate a layer of loosely woven hessian for extra strength.

Carpets

Carpets may be constructed by a large variety of processes. They may be woven, tufted, knitted, or nonwoven; they may be made from a number of different fibres and blends, and coloured by many different processes.

Traditional methods

Traditionally, carpets have been produced by weaving or by the still older techniques of knotting. With these methods it is possible to have a different coloured yarn for each tuft. This allows intricate, colourful patterns to be created. The finer the yarns used, and the closer they are packed, the more flowing and detailed the designs can be. The durability of such carpets depends mainly on the density of the fibres in the pile.

Hand-knotted carpets

The oldest known hand-knotted carpet – found during excavations in Mongolia – was made 2500 years ago. The same technique is used today, and even similar designs, in making oriental rugs. Here rows of knots are twisted around warp threads, and held in place by weft threads.

The finest 'city' rugs, such as those made in Nain or Isfahan in Iran, may have over 1½ million knots in each square metre (more than 150 knots to each square cm). Other carpets, often with more angular designs, are woven by nomadic tribes using thicker yarns and only 6–9 knots per square centimetre. These rugs are still extremely serviceable on account of the tough wool used in them. By comparison, the average tufted carpet could have a similar number of tufts, but using thinner, softer yarns. Hand-knotted rugs are generally expected to give over fifty years of service, compared to the much shorter life expectancy of woven or tufted carpets.

One of the simplest knots used in hand-knotted carpets.

Traditional rug-producing countries are Iran, Turkey, Afghanistan, the Central Asian republics of the U.S.S.R., and China. Each city or region developed characteristic designs and qualities of carpet over the centuries.

After the Second World War, some governments established carpet-weaving centres in non-traditional areas, such as Pakistan, India, Kashmir. These new industries are organized as a factory system, and cater for the needs of Western markets by adapting traditional designs from other places. While these carpets are also closely hand knotted and attractive, they are certainly not collectors' items and – because of the commercial quality of wool used in them – they are generally less hardy than the Iranian or Caucasian 'originals'. Some are even made of viscose rayon, and sold under the name of Art Silk or Queen Silk. While these rugs look beautifully silky when new, they will deteriorate within a few months of use.

Woven carpets

Woven carpets have established traditional reputations for durability. Indeed, their structure allows a very close jamming of yarns in the pile, which results in excellent serviceability. Nevertheless, the quality of a carpet depends not on the method of manufacture but on the effective pile density. It is possible to obtain woven carpets which are inferior to some of the top class tufteds.

Weaving is a much slower and more expensive method of carpet manufacture than tufting. Therefore, cost comparisons alone are not an accurate guide to the serviceability of carpets.

The three most important woven carpet systems are Axminster, Wilton, and Brussels. Axminster carpets always have level cut pile, which is formed from warp yarns supplied from creels behind the loom. A Jacquard mechanism selects the particular colour – from a choice of six or eight – that is needed for a given tuft point in the pattern. Spool Axminster uses spools of yarn corresponding to the design. These have an unlimited number of colours and no Jacquard mechanism is required. In Wilton carpets, those yarns not showing on the surface form a cushioned backing in the woven structure. Wilton and Brussels may have cut or loop tufts; Brussels carpets are rarely made today.

Extra or 'stuffer' warp and weft yarns may be added to woven carpet structures to increase dimensional stability.

Narrow Axminster carpet loom, with the Jacquard pattern cards above and the completed carpet below. The creel with its supply of warp yarns is behind the loom.
(David Bailey)

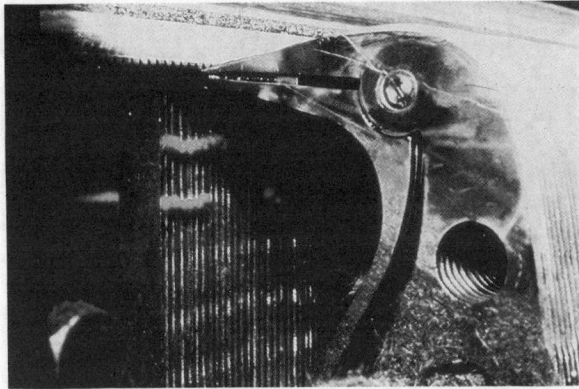

Axminster. The bird's head gripper draws the pile yarns through to just the right length. Behind the fabric and just in front of the yarn holders a sharp knife cuts off the required length of pile yarn. A comb then pushes the cut end around to form a U.
(David Bailey)

After the tufts are in position a rapier mechanism inserts the jute or polypropylene weft.
(David Bailey)

Tufting

About 80 per cent of all carpets available have been produced by tufting. Tufts of pile yarn are stitched into a backing fabric by a row of needles, and held there by latex adhesive.

The quality of the carpet is determined largely by how closely the fibres are packed in its surface. In a tufted carpet, this is regulated by a combination of yarn thickness, machine gauge (how closely the needles are spaced), and the length of the stitches.

The cost of a tufted carpet depends mainly on two factors: how much yarn is used in the carpet, and how expensive that yarn is. Fine, highly twisted yarns are more costly to produce than are loose, thick yarns.

Tufting needles inserting the tufts in woven polypropylene backing.
(David Bailey)

TRY YOUR HAND

A simple way to test for pile density is to press your fingers down into the carpet and try to move the pile from side to side.

Try this test on a number of different carpets.

Do you find that cut pile moves more or less easily than loop pile? deep pile than short pile?

Would you expect a carpet with a deep pile to last longer than a carpet with short pile? Why? Which would keep its new appearance for longer? Why?

Tufted carpets may be produced as looped or as cut tufts. Some carpets are produced with a combination of high and low loops or cut tufts. Some are made from Berber or tweed yarns which are blended from specially dyed fibres. Others are made from white yarns, and are dyed or printed with patterns after the tufting is completed.

All tufted carpets are finished with a coating of adhesive and a backing of hessian. The effectiveness of the latex glue in holding the tufts in place is proportional to the force needed to pull the backing away from the carpet.

Other carpets

Although tufting still holds a large share of the market, there are a number of different nonwoven methods of carpet manufacture that are successfully used today.

The Stoddard carpet

In Stoddard carpets, sheets of parallel yarns are pushed into the adhesive-coated surface of the backing fabric. The carpet is produced as a double cloth with the pile forming a velvet-type cut surface. The pile density can be very high, but the strength with which the tufts are held in the carpet depends on the effectiveness of the adhesive alone.

Knit carpets

Pile structures can be conveniently knitted – either loop or cut form – on Raschel machines. The speed of production is about the same as for tufting, but the fabric is far more flexible. If thermoplastic yarns are used to form the backing fabric, the carpet can be successfully moulded and heat set into shape. Knit carpets are often used for car interiors.

Blackburn rivet head tufting

This tufting machine inserts individual pre-cut tufts into a backing fabric, with the aid of a row of hollow needles. Each needle is connected to eight containers, each with a differently coloured yarn. Electronic controls can select any one of the eight colours for each tuft position in the design. This system has the ability to produce woven carpet patterns at the speed of tufting. Although a dense pile structure is possible, limits on serviceability are determined by the adhesive holding the tufts in place, by the backing fabric jammed full of tufts, and by the relatively loose twist of the yarn in the pile.

Nonwoven carpets

It is possible to make floor coverings directly from fibres, bypassing the costly yarn manufacturing processes. For such carpets to be serviceable, the fibres must be tightly entangled, and well anchored in a stable backing.

Needle punching of cross-laid carded webs of blended fibres is the usual method for producing underfelts and nonwoven carpet squares. The tangled fibres are held in place by a bituminous coating and a protective layer of nonwoven fabric as the backing.

This floor covering is able to keep its shape and size under various conditions and is well suited to be cut into tile shapes that fit together accurately. Carpet tiles have the advantage of design flexibility, and allow wear on the carpet to be distributed by rotation of tile positions around a room. Some carpet tile manufacturers have developed an excellent reputation for their products by using tough, hard-wearing fibres in their nonwoven blends. Other manufacturers produce less durable carpet tiles. On the whole quality still depends not on styling but on fibre type and fibre density.

Colouring and patterning
Millitron

Computer technology has produced a system for printing pile yarns just before they are tufted into the primary backing cloth. Accurate control of the size and position of each colour strip on each yarn, together with precise tufting tension, ensures that each loop or tuft in the carpet pile can have an individually determined colour. This produces woven-style carpet patterns at the speed of tufting. In addition, the designs can be changed without slowing production.

Carpet printing

Because of the influence of traditional carpet designs, machinery manufacturers have attempted in various ways to produce carpets which are fast to make but which look as though they have been woven. Some have been quite successful, some less so. As the faster processes have replaced woven carpets in the market-place, consumers have gradually come to accept different carpet appearances. The great Berber marketing excercise and the growing popularity of velvet and other plain designs have established the tufted carpet in its own right.

Printing the carpet after tufting is completed opens up a whole new world of flexibility in carpet designs. Colour is put on the carpet by printing, dripping, or by electronically controlled jets, to give soft, mottled, cloud-like designs that do not show soiling as quickly as do plain carpets.

The latest technology makes use of jets that deposit foamed dye on the carpet pile in carefully controlled geometric patterns. It is expected that, with refinement of patterning techniques, the most sophisticated designs can be made available for the general market.

Consumer questions answered

Q1 Why is it important to cut fabrics on the grain?

A The grain of a fabric is the direction in which the warp threads run; in knits it is the direction of the wales. During manufacture much tension is placed on fabrics in this direction. If the finishing is perfectly carried out, all the stress is removed from the warp yarns. In most cases, however, a small amount of tension is still left in the fabric, and during wear and laundering this tension is relaxed. The relaxation of tension causes shrinkage in the warp direction. This may be no more than 2 or 3 per cent; but if half a garment is cut in the warp direction and half in the weft direction, it will shrink unevenly, and look very odd indeed.

Q2 Is 'real' crepe always made of silk?

A No. The word *crepe* refers to a special kind of woven fabric with a matt, slightly pebbly, uneven surface. This effect is achieved by interlacing the warp and weft yarns as randomly as possible, but never allowing either to go over more than three threads at a time.

Crepe weaves are usually woven on looms with a dobby attachment. A *moss* crepe – which is made of wool – has a random design over more than 40 threads.

In order to increase the randomness of the crepe, and to help with the desired dull, powdery surface effect, crepe fabrics are usually – though not always – woven from *crepe yarns*. Crepe yarns are highly twisted, sometimes combined from a number of components. If made from hydro-philic fibres, crepe yarns shrink when wetted. Hydrophobic fibres make washable crepes.

Q3 Why are interlinings mostly made from nonwoven fabrics?

A In nonwoven fabrics the fibres are held together – whether by heat welding or adhesives – at many points. This gives most nonwovens excellent dimensional stability. They have uniform properties of stretch and drape in all directions, and do not fray.

Nonwovens can be produced in a variety of stiffnesses, and can be made extensible by a system of slits.

All these properties contribute to the suitability of nonwovens as hidden textiles to be used in conjunction with both woven and knitted fabrics of various weights.

Q4 What is the difference between stretch terry and terry towelling?

A Stretch terry is a weft knit fabric; its loops have been formed over sinkers between the knitting needles in the machine.

Terry towelling is woven on a special loom with two warp beams. The loops are formed by unrolling the pile yarns from the slack tension beam. The ground fabric is firm and not as extensible as a knit. Because of the different end uses, stretch terry is usually a finer fabric than terry towelling.

Q5 Why is velour an expensive fabric?

A Velour is the general name applied to long-pile fabrics, though today it mostly refers to cut-pile knits.

Such fabrics need thick pile yarns made from parallel, long fibres. The pile is formed as loops during knitting, and needs to be very long, because during shearing up to 30 per cent of its length is cut and wasted.

Although stretch velour and stretch terry are formed in basically the same way, the extra yarn and particular finishing required for a smooth, rich effect makes velour a costly fabric.

Q6 Why do some knits stretch out of shape and some not?

A Knits that have a lot of space between their yarns lose their shape much more readily than firm, densely knitted fabrics do. Tucking stitches, which further tighten the knitted loops, help the knit to keep its shape.

Q7 How can one tell warp from weft in a fabric?

A To find the grain of a fabric, the following may be useful guidelines:
(i) Selvedge runs in warp direction
(ii) Warp yarns are often finer and denser than weft yarns
(iii) Weft yarn is usually more 'fancy' – more likely to have slubs and possible weak spots than warp yarns
(iv) Fewer weft than warp yarns in many fabrics.

Q8 What is houndstooth check?

A Houndstooth is a type of two-tone colour design which is produced by using a twill weave and either a two-black two-white or four-black four-white order of colouring in both warp and weft.

Q9 What is so special about double weave cloths?

A For furnishings, a double weave fabric is likely to give better service because even if the yarns of the top fabric are under stress, the yarns of the backing cloth will tend to prevent the fabric from stretching out of shape. With such reinforcement, fancy fabrics which would otherwise be too weak for use as upholstery covers can be quite serviceable.

Also, a greater variety of patterns is possible: when some yarns are not required in the surface design, they are tucked away in the back cloth.

For blankets, the double layer of fabrics holds a layer of still air that has extra insulation value. Reversible coats and capes are also made very effectively from double weave fabrics which have a different design on each side.

Q10 How is velvet different from velveteen?

A   Velvet is a warp-pile fabric. The length of its pile can be varied independently of the structure of the ground cloth. Therefore it is possible to have velvets with pile which is both long and dense, and the backing fabric to be relatively loosely woven and therefore flexible. Velveteen is a weft-pile fabric. For it to have a dense pile the yarns of the ground cloth must be tightly woven together, and pile density is achieved at the expense of depth.

Therefore, velveteens are usually stiffer fabrics with a much poorer drape than velvets, and with a shorter pile.

Q11   How can I tell a good quality towel from a poor quality one?

A   In order for the towel to be serviceable, it needs to have long, evenly formed loops in its pile. The pile should not be pulled out easily, so it also needs to have a firm ground weave. The yarns of the pile should be fairly loosely twisted for greater absorbency; yet if they have too many protruding fibre ends they are likely to have been spun from short fibres which will readily be lost from the towel during washing. Loss of fibre not only creates troublesome lint, it also leads to the loss of absorbency and soon results in a threadbare towel. Some manufacturers try to hide poor quality pile yarns under a layer of size, but such cheap towels feel harsh to the touch and are not absorbent until the size is removed during repeated laundering.

Q12   Why do all woven fabrics have a selvedge, when knitted jerseys don't?

A   Selvedges are important during the finishing of woven fabrics. They serve as firm anchoring points for the grippers of the stenter which stretches and pulls the fabric to its final dimensions (p. 432).

Knitted fabrics are more extensible. Since less force is applied to them in finishing, and since their interlooped yarns are less likely to pull out of the fabric during finishing, most knits do not need special reinforcing at the edges. Some jerseys with a tendency to curl or fray have a strip of adhesive applied as a kind of false selvedge for processing.

Q13   A selvedge is such a neat way to finish a seam – why is it recommended that it be cut off before the pattern is laid out on the fabric?

A   The selvedge will certainly not fray when incorporated into a garment. It will, however, have different shrinkage properties from the rest of the fabric. This is because the number – and even the thickness – of yarns used in the selvedge is different from that in the cloth itself.

Q14   How can I match a correct knitting needle to my hand knitting yarn if I don't have any knitting instructions?

A   For a balanced knit structure, each loop created by the knitting needle needs to be just large enough to accommodate two yarns. Therefore the diameter of the needle needs to be twice that of the knitting yarn. Place the yarn on a measuring tape, and measure its diameter (in millimetres) under a slight tension. Double this number, and you have the metric number of the knitting needle needed.

Q15   If it is important to cut cloths on the grain of the fabric, how can one judge the grain correctly in a weft-knit jersey where the rows are skew?

A   The skewness of weft-knit jersey is due to the many feeds on the circular knitting machine. The tension during manufacture is in the direction of the wales, so the line of wales is the true grain of the cloth.

Q16   How is bias binding different from other edgings?

A   Bias binding is a plain woven, wide fabric which has been cut into strips on the bias, sewn into long lengths and the edges (which fray easily) folded. The bias cut means the binding is very elastic along its length.

    Most other edgings are either woven or braided narrow fabrics. Woven tape edgings are less easy to fit around curved seams than braids or bias binding are.

Q17   Why are interlock fabrics favoured for underwear?

A   Interlock fabrics have a smooth surface which is comfortable next to the skin, they are more extensible than single jersey fabrics, and they do not ladder as readily.

Q18   How are firehoses waterproof?

A   The simple answer is that they are not, and they are not meant to be. Firehoses are very tightly woven from a loosely twisted cotton. The cotton fibres swell when they are wet by the water in the hose, and this further closes the tight structure. However, some moisture seeps to the outside of the hose, and this keeps it cool and protects it from being burnt through, should sparks land on it. The hoses are woven as tubes, as a seam would place some strain on the fabric which could create a line of weakness and rupture under the pressure of the water.

Q19   I made a tailored suit of unbleached 'homespun' linen. The jacket hangs well, but the skirt has bagged around the seat. Why is this?

A   Linen fibres are highly crystalline, and therefore inelastic: they will not stretch. The problem lies not with the fibre, but with the fabric construction. Linen has a fairly stiff drape, and so to improve the 'hang' of linen fabrics, they are often woven loosely. This allows the yarns to move slightly, and makes the drape more graceful, but where the fabric is subject to tension the yarns can move to create the unsightly bagging. Linen garments are often treated with easy-care resins to overcome this problem; alternatively, they may be lightly starched after each wash.

Q20   Why is the cut surface of a plush towel less absorbent than the loop pile face?

A   The fibres of the cut pile stand up straight and form a smooth, level surface. This has no irregularities which can break the surface tension of a water droplet. On the loop side, fine capillaries are formed between the fibres in the loops, and they suck the water into the fabric.

Q21   Do all hand-woven carpets last longer than machine-made carpets?

A   No. Hand-woven carpets are usually made from heavy-duty wool, with a tight construction. If a machine-made carpet is woven or tufted as closely from the same materials, it can keep its appearance for just as long. Some hand-knotted carpets are made from less durable fibres, and so, like the average machine-made carpet, will not give the 50 or more years' service expected from traditional Oriental rugs.

Q22   Why is Thai silk so stiff and shiny?

A   Most silks are lustrous and soft. The fibre in Thai silk is the same as that used for any other silk fabric; the yarn and fabric structures however are different.

The yarns are not twisted. Instead, the filaments of silk are laid next to one another. These flat, ribbon-like yarns are woven very close to one another. The lack of space in the fabric gives a stiff, board-like effect, a little like the fabric used for sails of boats.

The flatness of the yarns contributes to the shiny effect.

Q23 Why are most work overalls made from gabardine or drill?

A Gabardine is a fine twill, and drill is a coarser twill fabric. Denim, another heavy-duty fabric, also has a twill construction.

Twill weaves have fewer interlacings than do plain weaves, and so allow the yarns to pack closer to one another. This produces a firm, heavy-duty fabric which is thicker and more resilient than a plain woven cloth.

Twill fabrics shed their wrinkles more readily because the lower number of interlacings allows the yarns more freedom of movement in the fabric structure. Plain weaves, because of their greater number of interlacings, have lower bias stretch, drape less gracefully, and while they are less easily distorted they do not regain their shape as easily. Graceful drape is not important in work clothes, but 'give' is important for the comfort of the wearer.

██ Finishing

When a fabric is removed from the loom or knitting machine, it has a stiff, rather rough or coarse handle from the size and lubricating oils used in manufacture. The yarns used in its construction will have been stretched, probably unevenly.

Finishing involves removing the size and oils, bleaching, relaxing tensions in the fabric, slitting tubular knits, and straightening wefts. It also involves all the after-treatments which modify the properties of the textile: all the processes which alter the appearance and performance of the fabric.

This chapter looks at the ways the woven or knitted fabric is modified to meet the needs of the consumer, and answers many questions that arise during the selection, care, and use of textiles:

- How can I tell if a sheet is of good quality?
- How can I make bonded blinds at home?
- How are fabrics protected against mildew?
- Why do some knit garments develop holes at the seams?
- Why does a showerproof coat lose its ability to repel water?
- Why do some singlets shrink drastically in length after washing?
- Can socks really 'eat' foot odour?

and many more.

Routine finishing procedures

Scouring and desizing

Yarns used for weaving are coated with *size*, a mixture of adhesive and lubricant that protects them from friction damage. Knitting yarns are lubricated with *oils*.

GREIGE fabrics straight off the loom or knitting machine have a harsh, unpleasant handle. They may also be soiled during manufacture. Desizing of woven goods, scouring, and bleaching are often the first steps in fabric finishing. They must be performed thoroughly: any size remaining will result in faulty dyeing of the fabric.

Scouring – a hot wash with detergents – will remove water-soluble dressings, such as most oils used on knit goods, and most dirt and soiling. Starch, a major component of most sizes, is not water-soluble, and must first be broken down by enzyme treatments.

Washing the fabric to remove size and oils.
(Australian Wool Corporation)

Both scouring and desizing are *wet treatments*. They are therefore *expensive:* energy is needed to dry the fabrics; and water and the pollution control measures for the waste are costly. Wet processes are always completed before the dry processes are performed, to minimize the energy costs.

Softening

The *flexing* of fabric during wet treatment not only helps to remove the size, but also contributes to improving the handle.

TRY YOUR HAND

Take a sheet of white paper and crumple it up thoroughly. Now flatten and iron it, then crumple and iron again.
How does the flexibility of the processed paper compare to that of the original sheet?
During crumpling, some of the bonds holding the fibres of the paper together were broken.
How does this relate to the mechanical softening of fabrics by flexing?
Why do tumble-dried towels feel softer than towels dried on a line in the sun?

Chemical softening includes the use of silicone lubricants or cationic agents added to the fabric just before it is dried. The silicone acts as a lubricant, allowing the fibres to slip past one another readily. The cationic agent gives each fibre a positive charge: the fibres then repel one another slightly and are more free to move. Both these treatments reduce the fabric's resistance to bending, observed as greater flexibility and softer drape.

Softeners added to the rinse during laundering perform the same function.

430

Removing accumulated tensions

During manufacture, fibres are stretched repeatedly: in drafting, when twist is inserted; when the shed is opened in weaving; when the yarn is fed to the needles in knitting. Warp yarns are generally subject to more tension than are weft yarns. Unless this tension is removed during finishing, the fabric will relax during laundering. Such relaxation can cause shrinkage of more than 30 per cent in some knit goods, and 15 per cent in woven goods.

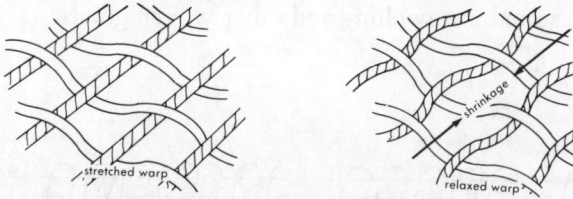

stretched warp

shrinkage

relaxed warp

Stretched and relaxed warp.

Skying fabric allows most creases to be removed and opens out the fabric for later stentering.
(Australian Wool Corporation)

Wet processing, such as scouring and dyeing, tends to relax these tensions, particularly in *hydrophilic* fibres. For *hydrophobic* fibres, *heat* treatments can aid relaxation.

Occasionally, woven fabrics are produced with a *skewed* or a *bowed* weft: these faults must be corrected before the fabric is dried and set into its final form, or uneven shrinkage will occur.

Skewed or bowed wefts are corrected by a STENTER. The stenter is a fabric dryer in which the fabric is hooked on gripper pins – tenterhooks – on each side. These pins are on a chain, and each side can be adjusted to move faster or slower to stretch the fabric warpwise or weftwise. The pin holes can often be seen in a selvedge: one of the main reasons for fabrics having a selvedge is so they may be gripped by the stenter without fraying. Knits and other fabrics with no selvedge may have glue applied to their edges so they can withstand the process. Some stenters have grippers with clips instead of pins, but these are less suited to towelling and other woven goods.

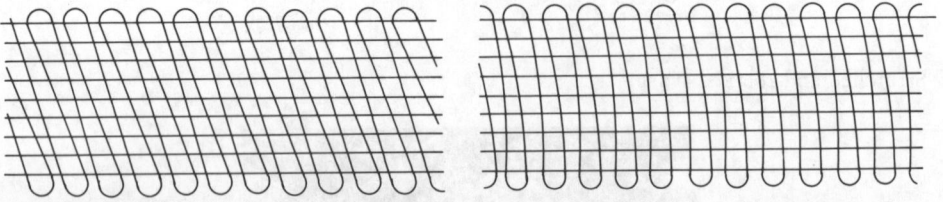

Skewed and bowed wefts.

Stentering.
(Australian Wool Corporation)

Before entering the stenter, the fabric is steamed and relaxed by overfeeding in the *warp* direction. Washing tests on the unfinished fabric indicate the extent to which it will relax under hot, wet conditions. The percentage shrinkage expected is then artificially induced by overfeeding under moist, hot conditions. The fabric is then dried in the relaxed state, and will not shrink or distort further during wear and care.

Warp overfeeding allows the fabric to relax to its expected shrunk length.
(Brückner Apparatebau GmbH, Erbach, West Germany)

If the weft is *skew,* it can be straightened by feeding one selvedge of the cloth faster than the other. *Bowed* wefts can be corrected by passing the fabric over specially designed curved rollers. On modern stenters automatic detection devices check for such faults, and initiate suitable corrective measures. Straight – 'on the grain' – weft is important for successful tailoring. In knitteds, the direction of *wales* should be taken as a guide to the grain, rather than the direction of the rows.

Body-sized tubular knits, destined for singlets and T-shirts and other garments, are usually not stentered. Instead, they are dried in a loop or drum dryer or a tumble-dryer. Wider tubular knits may be slit and processed in open width, through stenters.

Stentering and drying may not be sufficient to relax all the tensions in a fabric. Various other processes have been developed to remove any remaining stresses.

SANFORIZING is a method patented for woven *cotton* goods. It compresses the fabric between a thick blanket and a hot metal shoe. The new, compressed form can be further stabilized by cross-linking resins (SANFOR-SET.) *Rayons* cannot be stabilized by compressive shrinking because they swell and weaken so much on wetting that any setting process is cancelled by stretching. They can, however, be successfully treated with cross-linking resins.

Continuous pressing.
(Australian Wool Corporation)

Wool fabrics can be stabilized by steaming and cooling on perforated rollers – DECATIZING. LONDON SHRINKAGE is a labour-intensive method of wetting and drying fine worsted suitings between layers of cotton blankets. It achieves optimum relaxation, but is increasingly rare. Neither of these processes prevents *felting* shrinkage: resin treatments are needed as well (pp. 93–4).

Fabrics made from *thermoplastic* fibres do not need to be compressed, as they can be *heat set* in their final form. During use they will keep their set dimensions unless the heat setting temperature is exceeded – as sometimes happens during drying in the home laundry. *Acrylics* – which cannot be heat set – are particularly prone to heat shrinkage, especially if left in a dryer at high temperatures.

If, during treatment to prevent shrinkage, a fabric is *over*-compressed (the warp yarns end up *shorter* than they were before weaving), washing and tumble-drying will result in *drooping* of the fabric rather than shrinkage.

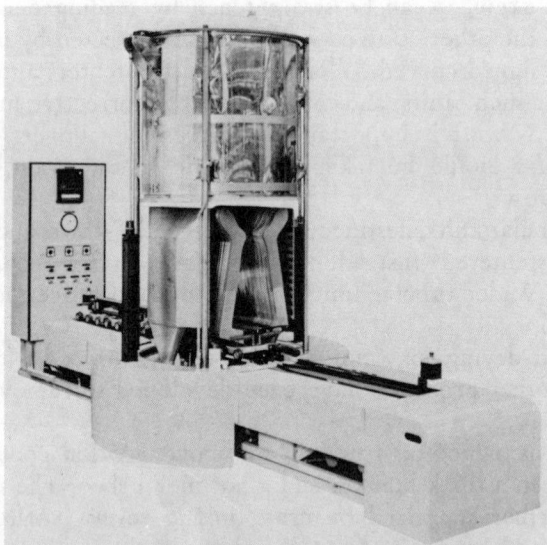

Hosiery and sock boarding machine.
(Pegg Whiteley Group, Leicester, England)

After finishing, the fabric is blocked, or folded into bolts, for sale.
(Australian Wool Corporation)

TRY YOUR HAND

Take a 10 cm × 10 cm piece of single-knit cotton jersey.

Seal the stitches at the top and bottom with PVA glue, and wet the fabric.

Stretch it lengthwise (in the direction of the wales) as far as it will go, and pin it to a board (a chopping board and sewing pins are ideal).

Allow it to dry; you may use a hairdryer.

Measure the stretched length and resultant width accurately.

Remove the fabric from the board, and wash in warm, soapy water, squeezing it thoroughly.

Lie the wet sample flat, without stretching, and measure it again.

Now, tumble-dry it, lie it flat, and measure it again.

Tabulate your results, quoting percentage change from the stretched measurements.

1   How do your results relate to the shrinkage of fabrics during use?
2   What percentage lengthwise shrinkage would make a pair of trousers unwearable?
3   How much can the sleeve of a jumper shrink before it is unserviceable?
4   What happened to the widthwise dimensions of the knitted fabric?
5   Woven fabrics sometimes shrink in the weft direction also. What percentage shrinkage can a shirt take before it becomes a full size too small?

Note: A 4 cm change in waist measurement corresponds to one size change.

Pile

Carpets, velvets, plush, velveteens, corduroy, and other cut-pile fabrics are usually treated by forcing *steam* through from the back to the pile face of the fabric. This allows the twist in the yarns to relax and unwind slightly, making each tuft wider and fluffier, creating a better surface cover. The effect is called 'bursting' or 'blooming' of the pile. Brushing with brushes rotating in opposite directions makes the pile stand upright.

Cropping machine for worsteds: no protruding fibres may be left in the fabric. (Australian Wool Corporation)

Singeing and cropping

For *clear printing* and a *smooth finish,* the surface of fabrics needs to be free of small protruding fibres. These fibres can be removed by *cropping* with a sharp rotary blade. A more effective method is that of *singeing* – exposing the surface of the fast-moving fabric to carefully controlled gas flames. This method is mostly used for *cotton* fabrics. The protruding fibre ends are burnt off. The heat makes the ends of *staple thermoplastic* fibres in blends shrink back into the body of the fabric, and this is thought to lessen the likelihood of pilling during wear.

Additions to the ground fabric

Filling and weighting

Some fabrics – particularly *woven cottons and linens* – are given extra body, weight and smoothness by the addition of *starches* in the final finishing operation. These starches fill the spaces between the yarns, and when pressed give a smooth, almost shiny finish to the fabric. After some wear and laundering, however, the *filler is worn off,* and the consumer is left with the original open-weave, loose-structured, lighter-weight fabric.

Silk fabrics are sometimes given added weight by allowing the fibres to absorb *tin salts,* which are not removed during washing. The weighting of silk *improves the drape* but *reduces the strength and abrasion resistance* of the fabric. (See pp. 108–9.)

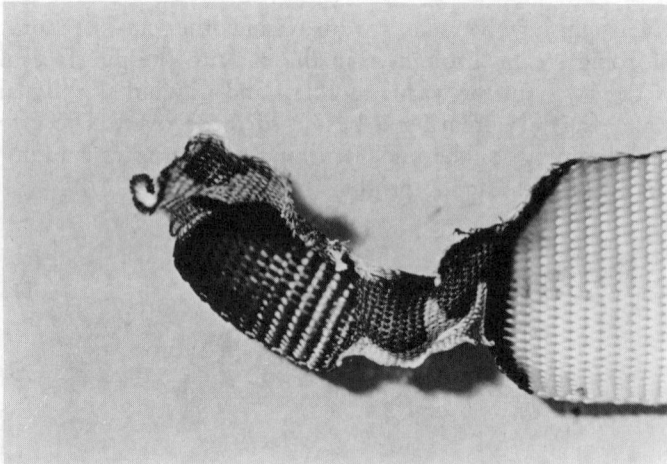

Weighted silk: on burning, the ash retains the fabric structure.

Coating

Fabrics are occasionally coated with resin or a layer of plastic to give them a special look, or to waterproof them. The first coated fabric was oilcloth. Today coating agents range from clear acrylates (for raincoats) to reflective acrylic or metallic coatings (for curtain linings) and expanded vinyl (for leather imitations).

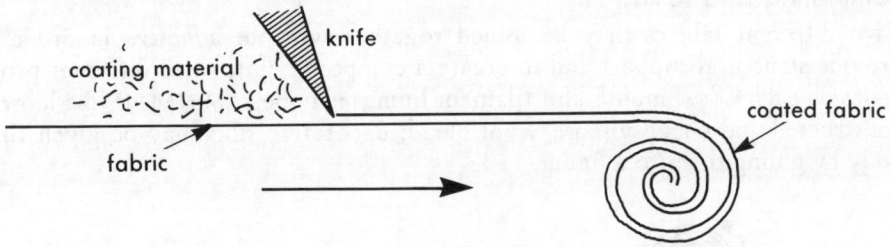

Knife coating.

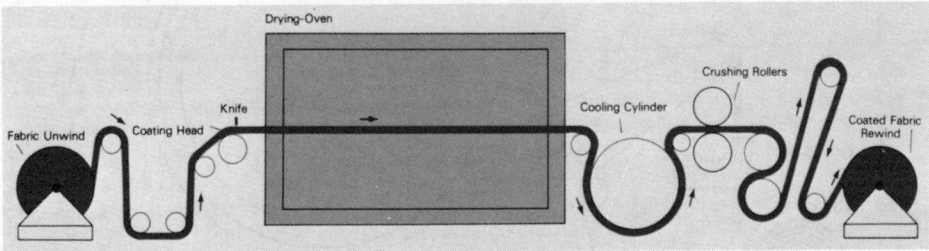

Coating, curing and crushing acrylic finishes on curtain fabrics. Crushing the foam gives a cork-like structure with excellent insulation properties.
(Bruck (Australia) Ltd, Wangaratta)

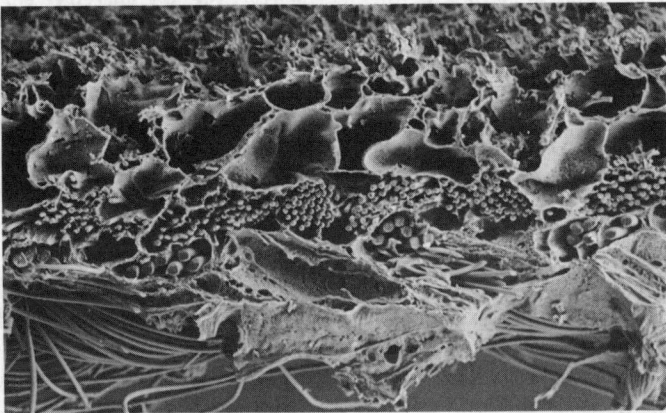

Synthetic suede. Note the polyurethane bubbles on the fabric base.
(Vivian Robinson)

The coating is usually done by knife spreading, followed by curing. Other methods include bonding an extruded or cast sheet to the ground fabric with a suitable adhesive. A foaming agent is added to vinyl or polyurethane so they will expand on curing.

Laminating and fusing

Two different fabrics may be joined together with an *adhesive* in order to provide structural support and to create a composite cloth with different properties on each side. Smooth knit filament linings may be LAMINATED to a loosely constructed and cheap viscose/wool blend; decorative silks may be given firm body by gluing them to a lining.

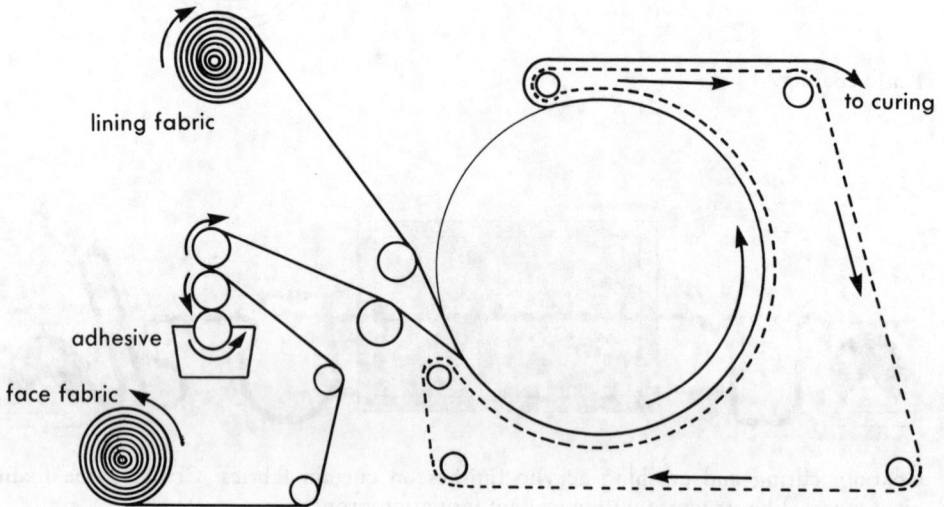

Fabric lamination.

Many of the properties of the laminated fabrics depend on the properties of the adhesive. The very act of gluing the two fabrics restricts the movement of the fibres and yarns in both – this reduced freedom of movement creates *stiffness and stability*. When the adhesive fails, however, the two fabrics separate and cannot be glued together again. Such failure appears as unsightly blisters on the fabric surface after incorrect dry-cleaning or laundering, or faulty manufacture.

An example of fabric laminating is the application during garment manufacture of *fusible interlinings*. The adhesive used for these is a thermoplastic resin which melts during pressing, and bonds the lining to the face fabric.

A recent development is the fusing of a transparent film of polyurethane to the fabric surface. This process is used to make easy-care and attractive good quality plastic-coated aprons and tablecloths. The coating may crack if folded repeatedly, and it is heat sensitive. Objects straight from the oven cannot be placed on plastic tablecloths, and sparks from barbecues will make holes in plastic aprons. For many consumers, however, the wipe-down ease of cleaning of these items outweighs the disadvantages.

A similar coating technique is used to make 'breathing' waterproof wound dressings and sportswear. Teflon or polyurethane film containing millions of tiny pores is applied to the fabric surface. Water vapour can pass through the pores, but liquid water cannot. The films are elastic, and can be made washable

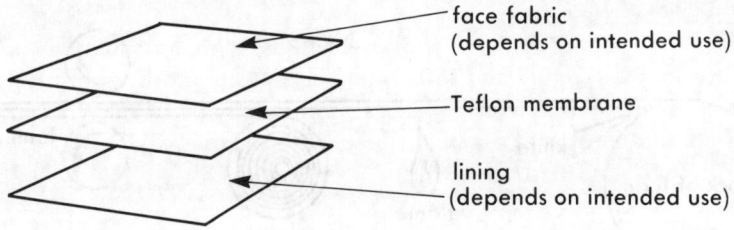

The structure of Gore-Tex fabrics. The choice of lining and face fabric depends on the intended use. Newer developments have improved the initially stiff handle.

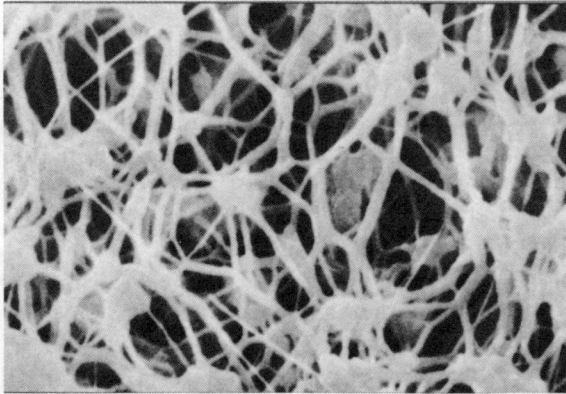

Gore-Tex membrane under electron microscope. Gore-Tex, one of the better-known micropore finishes, sandwiches a Teflon membrane between two layers of fabric. The pores in the membrane allow water vapour to pass through, but are too small to allow either liquid water or bacteria to pass.
(W.L. Gore and Associates Inc.)

and fire-retardant. The process is still quite expensive, but has found commercial acceptance, particularly in ski wear.

Foam laminating

In FOAM LAMINATING, a thin layer of polyurethane foam is peeled off a solid block with a sharp knife. The thin layer of foam is then passed over a gas flame and, while still molten, pressed onto the fabric to be laminated.

Occasionally, the foam is *sandwiched* between two fabrics, as a lining feels less unpleasant against the skin. The foam layer can, however, give the fabric an unpleasant, rubbery handle.

The foam is an *excellent insulator.* It gives fabrics a *stiff drape:* garments made from laminated fabrics were particularly popular in the late 1960s, for the 'geometric look'.

Care can create problems. If exposed to high temperatures or to sunlight, the foam may yellow and crumble. Dry-cleaning is generally not recommended, because delamination is irreversible. Gentle hand washing, with the article allowed to drip dry, or a light sponging of the fabric surface, are the recommended approaches.

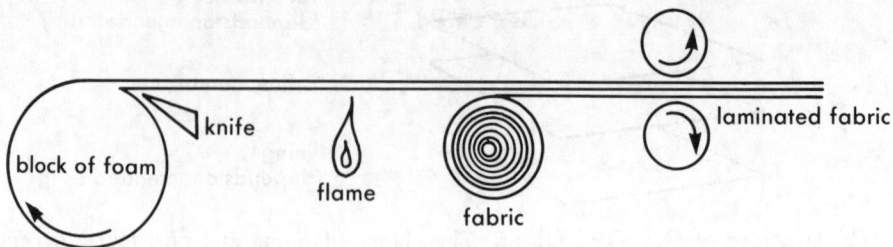

Foam lamination.

Flocking

FLOCK is the name given to very short textile fibres. These can be *glued* to various fabrics to form pile-like surfaces. Because of the low cost of the process, flocking is often used to imitate other, more expensive, textiles such as dotted Swiss voile, suede, or even velour.

The adhesive is first applied to the fabric base, the flock is shaken onto it from sieves and the excess (which is not set in the glue) is then vacuumed off. An electrostatic field may be used to charge the fibres of the flock. The charged fibres repel one another and stand upright. Once the adhesive sets, the uniform, deep, velour-type pile remains.

A knitted fabric, when coated with vinyl resin, heated (this expands the vinyl into a soft spongy structure) and sprinkled with nylon flock, will give a convincing, serviceable, cheap imitation of a suede, though problems may occur with cleaning such fabrics.

Another application is in nonwoven blankets, made of polyurethane foam bonded to two sides of a strong nylon net. Both sides of the net are coated with acrylic adhesive, and flocked with 1 cm nylon fibres. These blankets are completely machine-washable and dry-cleanable, and have the appearance of a very light and warm velvet.

Flocked design on a nylon novelty handkerchief.

Flocking can be used as decoration: the fabric is printed in a pattern with adhesive, and then sprinkled with flock. Any flock which is not held by the glue is vacuumed off, leaving a raised decorative effect.

The adhesive used will dictate many properties of the finished textile, so it must have good drape, flexibility, wash fastness, and durability. *Acrylic resins* perform quite well.

Heat reflective finishes

Curtain linings are often required to act as insulators: to keep heat out of a room during summer, and to trap heat in the room during winter. *Metallic finishes* act as reflectors for both heat and light and prevent the degradation of the base fabric. The lining (usually cotton sateen) is coated on one side with metallic flakes held in a transparent resin binder. Unless the fabric is very tightly woven, the finish may flake off in wear and during cleaning.

It is important that a resin used on curtain linings be able to resist the effect of weathering by sunlight and heat. *Acrylics* have excellent sunlight resistance. When applied as an expanded foam coating, both heat insulation (because of the air spaces in the foam) and light and heat reflection (because of the white pigment included) is effected.

Vinyl resins with white (for reflective) or black (for black-out) pigments are also used for *opaque* linings. Vinyl linings can be satisfactorily dry-cleaned in petroleum solvents, but expanded acrylic linings require special care in handling and cleaning.

Special decorative effects

Calendering

Calendering is rather like continuous ironing: it consists of passing the fabric in open width between a series of rotating, heated, weighted rollers. Different types of calender are used for different effects.

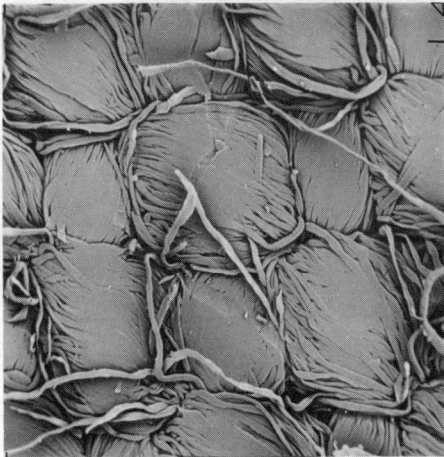

Chintz, face (left) and reverse (right). Note the flattening of the fibres on the face of the fabric.
(Vivian Robinson)

In FRICTION CALENDERING, the weighted top roller rotates much faster than the fabric, and effectively polishes the top surface of the cloth passing under it.

Friction calendering is used to create the shiny surface of *chintzes*. The cotton fibres are squashed flat, which makes the surface lustrous, but the fabric is liable to rip easily.

Glossy, leather-like surfaces may be produced by CIRÉING. The fabric is impregnated with wax and hot friction calendered. The effect is not permanent.

SCHREINER CALENDERS are embossed with very fine, parallel diagonal lines. When the heated heavy roller passes over the fabric surface, the fine lines imprinted on it give the fabric a soft lustre. *Satins* and *polished cottons* are often finished in this way.

A similar process creates the watermark effect of MOIRÉ. Traditionally, two ribbed fabrics were pressed together in a calender, but the modern technique uses an engraved roller.

Moiré effects do not last very well on rayons. They last longer on silk, but repeated laundering and ironing will destroy them. Cottons with resin finishes hold the patterns quite well, and they are permanent on the thermoplastic materials (acetate, polyester).

Moiré.
(David Bailey)

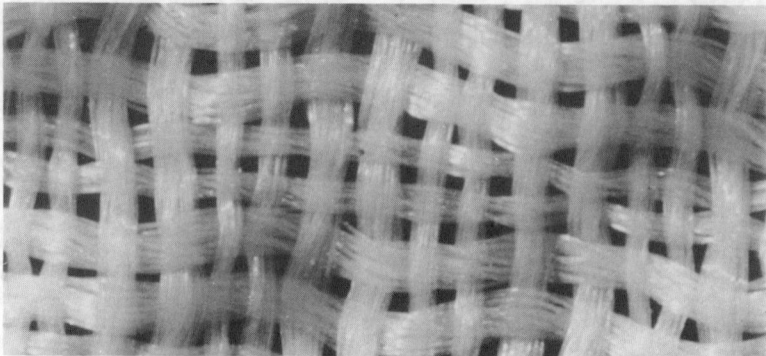

Heat embossed nylon plissé.

Embossing

Embossing creates three-dimensional effects in fabrics. The heated weighted top roller is deeply engraved with the design, and the fabric is supported on a paper roller. For flat embossing (such as used for brocades and velvets), the bottom roller is left flat. For a three-dimensional effect, the paper roll is wetted, and run without any fabric until the top roller imprints its exact reverse design into the softened surface. Once both rollers are ready, the fabric is fed through for embossing.

Thermoplastic fabrics are permanently embossed by means of heat only; cottons need to be treated with special resins to accept and hold the three-dimensional design.

Polyester knit, heat embossed to give a dull surface.
(Vivian Robinson)

Milling

Fulling is an older name for this process, which is carried out on *woollen fabrics* only. Taking advantage of the felting properties of wool, MILLING creates a soft, *fuzzy surface,* bringing the yarns in the fabric closer together to produce a *full handle.*

The rotary milling machine repeatedly compresses and releases the fabric until sufficient felting has taken place. Milling is carried out in the presence of detergents, in either acid or alkaline conditions. Dyes used for woollens that are to be milled should not run or wash out under such harsh conditions.

Milling.
(Australian Wool Corporation)

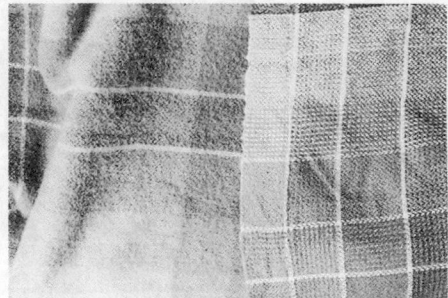

Milled and unmilled blanket.
(Australian Wool Corporation)

Raising and sueding

A fabric may be given a NAP or raised surface by *brushing*. Cylinders covered with flexible wires brush against the surface of a fabric with loosely twisted yarns. Some of the fibres are caught in the wires and partly lifted from the fabric, forming a fuzzy surface. Knitted *tracksuit material* has special thick, soft-twist yarn loops laid into its structure, which makes it particularly well suited to such a raising operation. *Flannelette* is another popular example of a brushed finish.

If, instead of wire brushes, abrasive rollers are used, the raised effect will be finer, even suede-like. Close examination of *sueded denim* shows the warp-faced twill surface to be a tangled mass of broken cotton fibres. These contribute to both the soft feel and the warmth of sueded denim.

Imitation suede may be produced from warp-knitted multifilament yarns. Yarn floats on the back of the fabric are brushed to break some of the fine filaments, giving a short, fibrous, dense, suede-like surface.

Nap raised by brushing.
(Australian Wool Corporation)

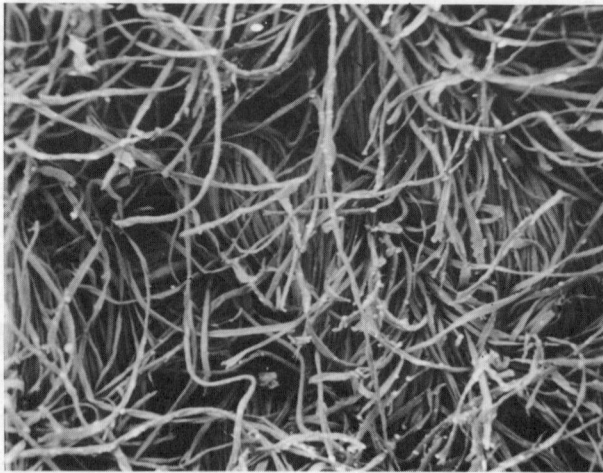

Twill weave polyester/cotton denim with sueded finish. Note the low layer of damaged fibres on the fabric surface.
(Vivian Robinson)

444

Translucent effects

All *cotton* fabrics can have a degree of translucency imparted to them by a carefully controlled dip in sulfuric acid, quickly followed by neutralization. This treatment produces a stiff fabric. *Organdie,* first developed for use in ballerinas' tutus, is made this way. Patterning may be achieved by printing the fabric with a resist paste before the acid immersion.

A DEVORE effect of lacy patterning is achieved on polyester and wool blends when the fabric is printed with alkali. This destroys the wool in the desired pattern, but leaves the supporting polyester fibres intact.

Blends of cotton and polyester can also be printed with acid in a predetermined pattern. The acid destroys the cotton, leaving some transparent areas. In a rayon/nylon fabric, phenol may be used to burn out the nylon. In rayon/acetate blends the acetate can be removed by printing the fabric with acetone.

Devore. The wool wrapping the polyester filament core yarn has been burnt out with alkali to give a translucent effect.
(Vivian Robinson)

Finishes for durability and ease of care

Fabrics and garments may become unserviceable before they are worn out, their use limited by unsightly changes – snagging, holes at the seams, moth holes, pills, fraying. Most of these problems can be minimized by suitable fabric finishes.

Mildew-proof and anti-bacterial finishes

Bacteria and fungi (mildew) are especially likely to grow on hydrophilic fibres under moist, warm conditions. In order to stop their growth, BACTERIOSTATS are applied, usually during dyeing.

The principal types of anti-bacterial chemicals are chlorinated phenols, organic mercury compounds, and some quaternary ammonium compounds. These are all highly toxic to the small organisms they are designed to kill, but also to humans. They are therefore bonded tightly to the fabric to prevent risk to humans.

Fungus growing on a nylon shower curtain.
(Vivian Robinson)

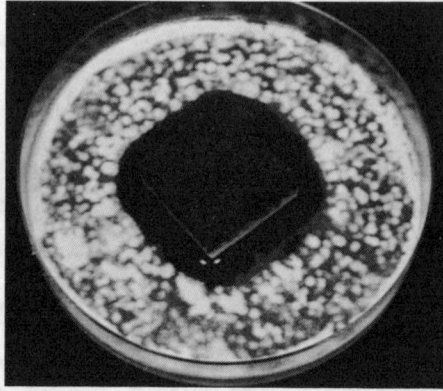

The effect of biostat treatment. The growth of the micro-organisms is inhibited close
to the treated wool square.
(*Bayer Farben Revue* No. 17, 1969, by permission of the editors. *Bayer Farben Revue*
is published by Bayer AG, Leverkusen, West Germany.)

For *carpets,* Sandoz and Ciba-Geigy have collaborated in the development
of a product which resists multiplication of bacteria, build-up of static electricity,
and soiling. It has been claimed that such a sanitized finish is especially useful
in hospitals, where it reduces the risk of cross-infections being carried by feet
from ward to ward.

Anti-bacterial finishes can also be applied to socks. Their presence prevents
the multiplication of bacteria which feed on organic matter such as perspiration
absorbed by the sock.

Moth-proofing is described on pp. 97–8.

Anti-pill finish

PILLING occurs because strong fibres are partly pulled out of the structure of
the fabric, and hold small, broken fibres anchored to the surface. To reduce
pilling, it is necessary to make it harder to pull fibres out of the structure.

The strong fibres can also be weakened so they break away from the fabric rather than form unsightly pills on it.

Singeing helps to prevent fibre ends from protruding. For *polyester/cotton blends,* a special alkali treatment followed by heating can weaken the polyester fibres at points where they are already under some tension in the yarn. This is sufficient to allow pulled fibres to break off, without significantly affecting the strength of the fabric. This treatment is not widely used commercially.

Snags, slips, and abrasion

SNAGGING occurs when fibres or yarns are pulled out of the fabric. It is a problem mainly in knits. Polyacrylate and polyurethane resins can be used to bind the fibres in the yarn without too great an adverse effect on the handle.

Smooth, multifilament yarns may be prone to *slip* when loosely constructed fabrics are subjected to tension. Seams may slip off the edge of the fabric, or distort, or there may be excessive fraying of cut edges. To prevent these problems, the fabric may be coated with a resin before entering the stenter. When the resin is cured, it will bond the fibres at some points where the yarns cross, and so give the fabric the necessary stability.

Thermoplastic resins may be applied to give *abrasion resistance* to such vulnerable items as pocket linings.

Lubricants for ease of sewing

When knit-goods are sewn with needles that are too large or too sharp, the yarns may be cut. High-speed sewing may generate enough heat to weaken thermoplastic yarns. When the fabric is subjected to tension during use, it develops *holes* in the path of the needles. As the holes develop into runs, these faults are a frequent cause of consumer complaints.

If a finish is applied which makes the yarns slip over one another during sewing, the yarns can slide away in front of the needle without being damaged. Later, the yarns return to their original position, and the hole made by the *round*-tipped needle becomes smaller.

Polysiloxanes and aqueous polyethylene dispersions are frequently used. Polysiloxanes, because of their greater heat stability, are particularly suitable for lubricating sewing threads.

Soil-resistant finishes

Silicones and fluorocarbons have very little attraction for other types of molecules. They are neither hydrophilic nor oleophilic (oil-loving), and so can form an excellent barrier against oily and water-borne stains. The more expensive fluorocarbons have been more successful in this role.

Such finishes may be applied to synthetic filaments during manufacture, or to the fabric after construction is completed. Their only short-coming – apart from cost – is that, due to the low affinity of fluorocarbons for other substances, they *wear off* in use.

Effectiveness of a soil-resistant finish.
(3M Australia)

TRY YOUR HAND

You will need:
9 samples, each 5 cm × 5 cm, of each of a variety of white fabrics: polyester, cotton, polyester/cotton blend, durable press all-cotton fabric such as elephant crepe
can of Scotchgard or other stain-repellent spray
red oily stain, prepared by mixing paprika with vegetable oil
watery stain, prepared by adding blue food dye to water

What you do:
Leave one of the samples for reference for each fabric.
Spray 4 others with the stain repellent.
Now put a drop of oily stain in the middle of two of the sprayed samples, and a drop of watery stain on the other two.
Repeat the staining process for the unsprayed fabrics. Be careful to label each piece indelibly for identification.
Repeat for each set of 9 pieces of the other fabrics.
First wipe, then wash, one of each identical pair of stained fabrics.
Compare the ease with which the stain is removed from the treated and untreated samples.

1   Was the spray equally effective against oily and water-borne stains?
2   What was the difference between the different types of untreated fabrics in their affinities for the different stains? Did the durable press fabric behave the same way as pure cotton?
3   Was the protective action of the spray the same in each case?
4   How would you apply your findings to furnishing fabrics? to children's clothes?
5   How could you compare the effectiveness of various home-use stain repellents available on the market?

448

Special-purpose finishes

Water repellent finishes

Fabrics may be *waterproofed* by a coating of polymer emulsion which forms a film on curing, or by bonding to a preformed plastic (usually polyurethane) film (pp. 436–7). Micropore coatings, which allow the passage of water *vapour*, may also be applied. With both these approaches, garment construction – seams and closures – can be as important as the fabric finish.

For many uses, *water resistance* and *comfort in wear* are more desirable than perfect water-proofing. Such water resistance may be achieved by careful fabric construction combined with suitable finishes. Hydrophilic fibres such as cotton swell when wet. If the fabric is tightly woven, this swelling may be enough to close any gaps between the yarns. A water-repellent finish, such as a wax or metallic soap which coats the fibres, will make the fabric more water-resistant but allow space for the passage of air and water vapour. A very heavy shower may penetrate the fabric, but under most conditions it will perform satisfactorily.

The distinctive garb of the mountain cattlemen employs treated cotton for the coats. (Dale Mann/Retrospect)

TRY YOUR HAND

You will need:

shower testing apparatus, made from 250 mL funnel, shower rose, and plug, as shown

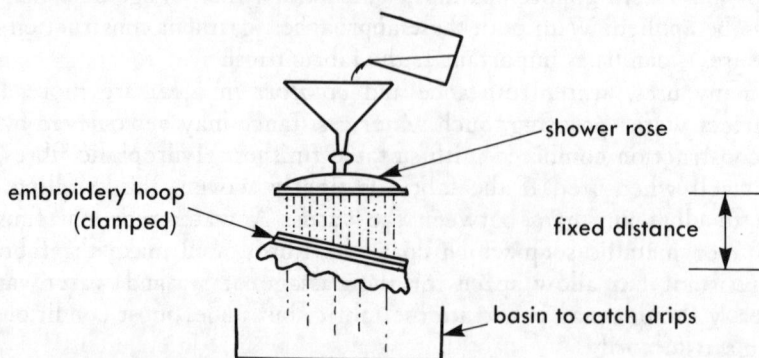

10 cm embroidery hoop, mounted at 45° angle

basin to catch the drips

6 samples of each of an assortment of fabrics, each 15 cm × 15 cm. Include plain cotton – poplin or sailcloth

beaker each of soap solution and magnesium or calcium chloride solution

detergent

furniture wax in spray pack

non-stick food spray

stain-repellent spray

What you do:

Mark the samples clearly. Keep one piece of each fabric type as reference.

Soak the first sample of each fabric first in the soap solution, then in the calcium or magnesium chloride solution. This forms insoluble metal salts of the soap in the fabric. Rinse and dry.

Spray fresh samples of each fabric with each of the remaining reagents.

What effect does each reagent have on the feel of each fabric?

Mount each sample in turn in the shower test apparatus; pour 250 mL of water through, and observe the extent of wetting and droplet formation.

Tabulate your observations.

1   What was the effect of each reagent on ease of wetting of each fabric? Why?

2   What fabric type and treatment would you recommend for a showerproof coat? Justify your choice.

3   In what other situations might some of the other fabric and treatment combinations be used?

Fire retardant finishes

Most textiles can catch fire. Some will burn with a strong, steady flame, others extinguish quickly, yet others melt and drip fire. There are only a few specially produced fibres which will not burn – these are used for high fire hazard situations, firefighters' clothing, safety curtains and such.

Fabrics made from other fibres need to be *treated* if they are to resist fire. There are various consumer applications which require fabrics that present a low fire hazard. These are *children's nightwear, curtains* and *mattressing.* Each winter some children suffer serious burns because of poorly designed heaters setting fire to flammable, poorly designed nightwear. Flammable curtains can propagate a small fire rapidly and provide enough heat to begin combustion of window surrounds and other building components. Mattressing may catch alight when a burning cigarette falls from the hand of a sleeping person.

Cotton is often involved in such fire hazard situations. Since it is a highly flammable fibre, it needs to be treated with chemicals to make it flameproof. In 1735 Obadiah Wyld was granted a patent for flame-proofing 'paper, cotton, lenen and such like substances' with 'a mixture of alum, borax and copperis'. Although it is effective, the borax treatment is not washfast and is therefore not suitable for the modern consumer.

Modern washfast flame-proofing agents contain compounds of phosphorus which are able to form bonds with the hydroxyl group of cellulose. The halogens (chlorine, bromine) are also effective anti-flame agents.

In the PROBAN process, tetrakis(hydroxymethyl)phosphonium chloride (THPC), modified to improve fabric handle and durability, is padded onto cotton or rayon fabrics and allowed to form an insoluble, flame-resistant polymer inside the fibres.

For industrial purposes, such as awnings and tarpaulins, the fire retardants must be *weather-proof.* Antimony and titanium oxides in the presence of chlorinated high molecular weight hydrocarbons are often suitable. The fire retardant for any situation must be chosen only after the efficiency and durability required have been accurately assessed.

TRY YOUR HAND

You will need:

3 samples each of cotton, wool, and acrylic fibres, each 5 cm × 25 cm
experimental set-up as shown in the diagram

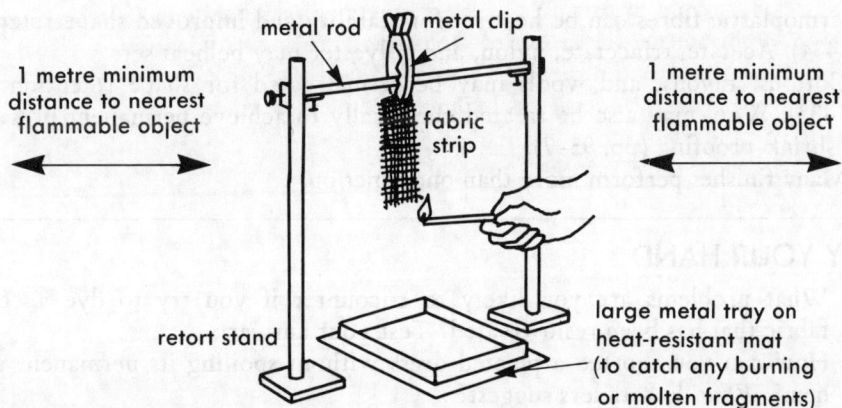

451

70 mL boric acid made into a paste with a minimum of water, mixed with 30 mL of borax solution (solution A)

equal parts borax and diammonium phosphate dissolved in water (solution B)

What you do:

Leave one sample strip of each fabric untreated, as reference.

Dip a piece of each fabric into solution A, and a piece into solution B. Mark them clearly.

Dry your samples, and observe the effect of the reagents on their handle and appearance. Now check their behaviour after a lit match is held for 2 seconds at the base of each hanging strip.

Record the colour and intensity of the flame, the speed with which the sample is consumed, any melting, dripping, afterglow, or self-extinguishing characteristics.

1   Were the flame-proofing agents equally effective for all types of fabric?
2   Which reagent least affected the handle of the fabrics?
3   Which would be your choice of treatment for each of the three fabrics?
4   For what applications would you consider using any of these flame-proofing methods? When would these treatments not be effective?
5   Compare the handle of fire-retardant-treated nightwear available in shops to that of untreated nightwear. Compare their prices as well, and comment on the consumer significance of fire-retardant textiles. This issue is discussed further in Chapter 12.

Other finishes

Anti-static

Static is a problem mainly in synthetics. To overcome this, anti-static agents may be added to the polymer before extrusion (p. 128), carbon or metallic fibres may be added to blends (pp. 178, 180), a conductive film containing charged molecules (usually quaternary ammonium ions) may be applied to the fabric during finishing (p. 439), or cationic fabric conditioners may be added to the rinse after each wash (p. 522).

Shape retention

Thermoplastic fibres can be heat set for stability and improved shape retention (p. 434). Acetate, triacetate, nylon, and polyester may be heat set.

Cottons, rayons, and wools may be resin-treated for shape retention (pp. 33, 433). Wool may also be treated chemically to achieve permanent press and for shrink-proofing (pp. 95–7).

Many finishes perform more than one function.

TRY YOUR HAND

1   What problems are you likely to encounter if you try to dye a cotton fabric that has been resin-treated? Test your answer.
2   How can you shorten a pleated dress without spoiling its permanent press hem? What do retailers suggest?

3 How would you be able to tell if a piece of fabric is suitable for tie-dyeing?

Consumer questions answered

Q1 I bought some linen tea-towels with a shiny, smooth finish. The finish disappeared after a couple of washes. The tea-towels then looked coarser but were more absorbent. Why?

A The shiny finish on the fabric was a starch filling which was removed during wear and care, leaving the absorbent flax fibres to play a more effective role.

Q2 How can one tell whether a fabric – such as sheeting – is high quality or whether it has had fillers added to improve its appearance?

A Hold the fabric firmly between the thumb and forefinger of both hands, leaving about 1 cm space between your hands. Rub the thumbs vigorously together, flexing the fabric thoroughly. Now hold the fabric up against the light – if it is more transparent in the flexed portion, then appreciable amounts of filler have been added to it during finishing. High quality fabric should be unaltered by this treatment.

Q3 Can socks really eat foot odour?

A Well – not actually. Foot odour is a result of the *bacterial* breakdown of perspiration and sebum (the natural oil of the skin). Without bacteria, these products of the skin are nearly odourless. If the socks are treated with bacteriostatic agents, then the bacteria will not multiply, and the foot perspiration will not be broken down. A quality deodorizing finish should be incorporated into a *washfast* resin.

There are certain *shoe liners* which are claimed to absorb foot odours. These are porous foam pads which contain carbon *(activated charcoal)*. The fine surface of the carbon particles is excellent at holding on to molecules which, if released into the air, would be noticed as having a bad odour.

It should be noted that the deodorizing finish only works on the perspiration and sebum *absorbed by the sock*. Any sweat which is left on the *skin* will be free to decompose to produce odorous compounds.

Q4 I bought a cheap singlet. After a couple of washes it has completely shrunk out of shape. How can I avoid such a bad buy next time?

A Knitted fabrics are stretched considerably during manufacturing. If the loops are not compressed to a stable form before laundering, the warm, wet treatment will allow the structure to relax to its natural shape. This can cause a shrinkage of up to 30 per cent in length of the garment. Some of the relaxation is also taken up by a widening of the loops, and hence a widthwise stretching of the fabric.

It is not possible to predict shrinkage visually at the point of sale – the consumer must rely on the reputation (or lack of it) of the manufacturer.

Q5 I made a dress from polyester jersey, and after I had worn it a few times I noticed that the fabric was breaking into holes at the seams. Can I do anything about it?

A   It would seem that the fabric was damaged during sewing. During high speed sewing, friction between the needle and the fabric can generate enough heat to melt the thermoplastic fibres. Sharp needles may also cut the fine yarns. So knitted goods need to be sewn with blunt needles which push the yarns apart. To help ease of sewing, fabrics and sewing threads are often lubricated with polysiloxanes which help the yarns slide over one another during sewing. This allows them to move out of the way of the needle and then to move back and close the hole formed by the stitch.

To prevent complete seam failure, your garment will need another row of stitching inside the old seam, using ball-point needles this time!

To help with your next project, you can spray a silicone lubricant (such as Ezy Glide) along the seam before stitching. This will ensure that the yarns slip over one another during sewing to make room for the needle.

Q6   I saw a raincoat made of woven fabric which seemed to be coated with a layer of plastic on one side. I commented how hot it must be to wear but the salesperson assured me it was comfortable and allowed the body to 'breathe'. Can this be true?

A   Micropore coatings are a recent development for textile applications. Although the film used for coating the back of the fabric appears solid, it is filled with millions of tiny, interconnected cells called 'micropores'. These cells are too small to allow liquid water – or wind – to penetrate, but they do allow water vapour to pass through. In this way, heat generated by the body during gentle exercise can still be lost through insensible perspiration, and the wearer can remain dry and comfortable.

Q7   Does reflective curtain lining really work as an insulator?

A   The pigments incorporated in the reflective lining – titanium dioxide or metallic particles – reflect light and heat quite effectively. However, curtains are used on the *inside* of windows, and so some of the reflected heat is trapped between the curtain and the glass. Reflective surfaces on the outside are more effective in keeping out heat. Nevertheless, curtains with coating are far more effective than curtains without coating. The acrylic polymer used in the process has excellent sunlight resistance and therefore can be expected to protect the curtain fabric and remain serviceable for a long time.

Q8   Why does one find holes in the selvedges of most fabrics?

A   These holes are the pin marks left by the stenter which is used to stretch the fabric to its required size during drying in the finishing processing.

The pins are also called tenterhooks – hence the phrase 'on tenterhooks', for tense anticipation.

Q9   Some small fabric shops have a strange fishy smell which makes my eyes water. What is it?

A   The vapour is formaldehyde, released from drip-dry cottons and rayons as some of the resin used for crease-resist finishes breaks down under the hot humid conditions.

Q10   Someone told me that if I treat my lounge chairs with a stain repellent spray they will end up looking worse than without the protection. Is this possible?

A   The layer of fluorocarbon that you spray on acts as an excellent soil repellent, and will save your chair from absorbing water-borne or oily stains. Some parts of the chair, however, receive more abrasive wear than other parts. The protective fluorocarbon layer may be worn off these areas more quickly. Those parts which no longer have protection collect soil more readily than do the parts on which the spray coating is still intact. So the contrast between the 'clean' and 'dirty' parts is greater than if the whole chair is allowed to get dirty fairly evenly.

To prevent this effect, the chair should be cleaned and re-treated with the fluorocarbon spray at regular intervals.

Q11   How can I tell whether a flannelette sheet is good quality or not?

A   The raised nap on fabrics gives a soft, pleasant handle and is warm to the touch. The nap may also be used to hide poor fabric and yarn construction which in the long run will make the fabric unserviceable.

The flannelette sheet which best withstands wear and laundering needs a closely woven structure, with a twill weave to allow flexibility and effective brushing up of the fibres.

A loosely woven fabric will soon lose most of the brushed-up fibres – the few fibres left in the nap will matt and flatten down.

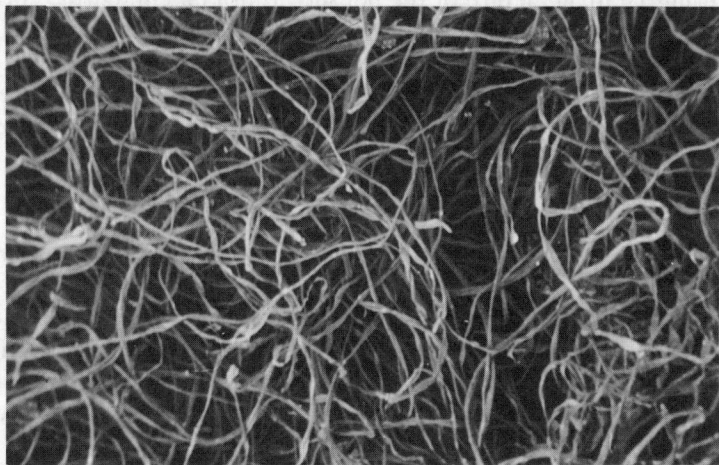

Raised nap on brushed cotton sheeting. The loosely twisted yarns in this sample were loosely woven, so the sheets cannot be expected to wear well.
(Vivian Robinson)

Q12   What is the difference between 'genuine' watermark taffeta and imitations?

A   Genuine watermarking is achieved by lining up two layers of ribbed fabric so that the ribs are at a slightly skew angle. The fabrics are then pressed (under 10 tonne pressure) in a heated calender. The pressure of the ribs against one another causes them to flatten in places. Irregularities in the weave create pleasing changes in the moiré pattern.

Watermark patterns can be heat embossed onto thermoplastic fabrics with engraved rollers. These 'imitation' moiré fabrics show a more repetitious watermark design than true moiré.

Q13 I have an old cotton dressing gown with a pretty embossed pattern on it. The embossing has flattened out during wear and care. Are there any embossed fabrics which keep their design for the lifetime of the garment?

A Yes. When fabrics made from thermoplastic fibres are embossed, the design can be damaged only if the fabric is exposed during use to temperatures higher than those used for embossing.

Q14 I have a tartan jacket which is made from a foam laminate. I handwashed and tumble-dried it – now the foam is crumbly and yellow. Why?

A Excessive heating can readily degrade the polyurethane foam used between your facing and lining fabrics.

Q15 My shirt collar has developed small blisters. Can I do anything about it?

A The blisters are due to the laminated interlining separating from the surface fabric. Once delamination has occurred, it is irreversible. If you have followed the care instructions, complain to the manufacturer.

Q16 I heard that however attractive chintz upholstery covers are, they tend to rip easily if caught by a sharp object. Is this true?

A During the manufacture of the polished finish on chintz, the cotton fibres are flattened by the polishing rollers. This distortion of their natural shape weakens the fibres. Also, since there is greater cohesion between the squashed fibres than in an ordinary fabric, chintz behaves a little like paper (even to the point of having a papery rustle) and does tend to rip if snagged.

Q17 My new jeans are guaranteed to shrink. When can I expect them to stop shrinking?

A Shrink-proof fabrics have a residual shrinkage of no more than one per cent. Most of the shrinkage is in the warp direction. If the fabric has not been shrink-proofed, the warp yarns which have been stretched during weaving can be expected to relax back to their original length – up to as much as 15 per cent. Full relaxation will occur over a number of washes, after which no more shrinkage in the warp direction should be expected.

Q18 My jeans are always tight fitting after a wash – then they 'grow' in size until the next wash. Is this normal?

A Since jeans are a tight-fitting garment, during wear there is considerable tension exerted on the fabric in both the warp and the weft directions.

To allow for some of this stretching, the crimp in the weave of the fabric flattens out; the weft yarns at the seat of the pants and the warp yarns at the knee straighten. This increases the size of the fabric in the direction of the stretch. During laundering, the fabric relaxes back to its original structure. Since no fibre slippage is involved, this is a reversible change.

Q19 I have hand knitted a jumper on giant needles, with lots of space between the yarns. The jumper now stretches and moves every which way. Yet my open knit synthetic curtain fabric has a stable structure. Why? How can I stabilize my knitted garment?

A   Open knitted structures are unstable because the yarns do not support each other in all directions. The curtain fabric, however, was heat set. High temperatures softened the thermoplastic fibres, and on cooling they took as their preferred form the shape of each open loop. After such setting it takes some effort to distort the loops, and on release of the stress the fibres will tend to move back into the set shape.

Acrylics cannot be heat set, but you can partly set your knit by steam pressing it into the desired shape. For *thermoplastic* fibres, use an ironing cloth and no steam. *Hydrophilic* fibres will lose their steam-set shape after hot laundering.

Q20   I tried to make my own bonded blind by ironing a fabric onto a fusible lining. However hard I tried, I could not avoid bubbles forming. Why?

A   No doubt you were using an iron to heat the adhesive of the lining and melt it into the fabric. As you moved the iron back and forth on the soft ironing board, the surface of the fabric *stretched* ever so slightly. Since parts of the top fabric were now actually larger than the lining, it was no longer possible to fit the two perfectly together.

The secret of successful bonding is to *press* rather than 'iron'. If no large press is available, it is best just to press the iron down on the fabric, leave to heat through, then lift and press in the next position.

Q21   I live in a warm, humid place. I am concerned about my clothes getting mildewed. How can I prevent this?

A   Bacteria and fungi are rarely able to exist on materials which do not contain contaminants with varied nutrients. This means that clean clothes do not get mildewed. 'Clean' in this case also means no *soap* residues in the fabrics. Therefore the best prevention against moulds and bacteria is the frequent washing and airing of clothes. Even moth larvae will die of malnutrition on perfectly clean wool!

Q22   I had a showerproof overcoat. I hand-washed it gently with the mildest detergent, but it nevertheless lost its resistance to water. Why? Can I do anything about it?

A   Your coat was treated with a waxy, hydrophobic finishing agent which repelled the droplets of water falling on it. The detergent attached itself to these hydrophobic molecules, and changed the surface to a hydrophilic one – remember one of the main functions of a detergent is 'wetting'. Even if you rinsed the garment carefully, some detergent still clung to the fibres. Therefore, the fabric no longer repelled water.

You can have your coat re-proofed at a professional dry-cleaners, who will apply a fresh emulsion of waxy material to coat the fibres. Alternatively, you may apply a waterproof finish – polysiloxane or fluorocarbon – available in a spray pack.

Q23   Every scout knows that one should never touch the inside of the tent canvas when it is raining. Why?

A   The canvas is woven in plain weave, with spaces between the yarns to help air circulation. The cotton yarns are coated with a water-repellent substance, which allows water to form droplets on the surface. If you touch your finger to the inside of fabric while there is a drop on its outside, the capillary space created between your finger and the yarns will suck the water through to the inside and so wet the canvas.

12 Textiles in use

The satisfaction you gain from using textiles depends on your ability to make appropriate choices. In many instances, people select textiles for various purposes without really attaching much importance to the task: the more familiar the problem to be solved, the less it is thought about in a critical way. There are times, however, when the consumer needs to work through a problem consciously and systematically. For example, if you have never been to the snow before, what sort of clothing is appropriate? If you are responsible for solving erosion problems along an embankment, would a geotextile be of use? Can an existing textile solve a given problem, or is a new product going to render traditional solutions obsolete?

Solving a problem in textiles choice, as in all other situations, is a matter of asking questions: to clarify the problem; to seek creative solutions to the problem; to judge which is the most suitable solution; and to evaluate that solution.

This chapter brings together all the factors which contribute to textile properties, and examines them in the context of consumer choice. It suggests answers to such questions as:

- Why do apparently similar textiles vary so much in price, and is the price difference reflected in performance?
- How can I check if a jumper will tend to form unsightly pills during use?
- How can I choose a carpet which will look good for a long time?
- What type of material is best for sails?
- Why are some sleeping bags warmer than others?
- What is the most suitable seam to use with sheer fabrics?
- Which is warmer, two thin jumpers or one thick one?
- What does the comfort of active sportswear depend on?
- Why do some fabrics resist staining, while others stain easily?
- When should a garment be dry-cleaned, and when washed by hand?
- How can fabrics be used in making roads?

Choosing a textile

The choice of a textile can be broken down into four stages.

The first stage is to identify and clarify the problem: to set a goal. What is the problem? What do I need to know? What will the textile item be used for? Are there any special requirements, or special properties which are desirable? Are there any other constraints or considerations, such as price?

The second stage collects and organizes information and ideas: it identifies resources. What do I know already? What extra information do I need, and where can I find it? Have I gathered enough information and ideas in defining the problem?

The third stage analyses and evaluates the possible solutions: it sets the standard of what is acceptable. What range of options do I have? What would be the likely consequences of each option? After consideration of any constraints which I have identified, which option most closely matches my needs?

The fourth stage is, of course, making the decision and assessing its validity. Which decision can I defend as the most appropriate at the given time? If I act on this decision, how can I evaluate the outcomes?

These stages do not necessarily form a sequence. They overlap, and the person making a decision moves backwards and forwards between them.

All the steps involve attitudes and values. The solutions which you find satisfactory are those which allow you to satisfy your needs according to your own attitudes and values, taking account of the resources which you see as being available to you, within any constraints which you identify.

Other people may see different needs, and arrive at different solutions. Generally, there is no right answer: different answers are appropriate for different people. Similarly, different people will have different standards, different ways of assessing whether or not they have achieved their goal.

TRY YOUR HAND

Choose an example of a textile use with which you are not familiar, perhaps clothing for a sport which you have not played. Go through the decision-making process outlined above to choose a textile item (and possibly also design) appropriate to that use.

Swap your work with someone else, and read the other person's work carefully. Would you have defined and worked through the problem in the same way?

In what ways would you have differed? Were there differences in goals, in standards, or in the resources which you would consider useful?

When you make decisions about textiles, the decisions are a resource which you use to achieve the goals you have set. Depending on your goals, you will make use of other resources in conjunction with your textiles. For example, if you have decided to spin some wool to knit a jumper, you will need to draw on a range of resources to carry out your decision. You will need to use your *knowledge* (or someone else's) about what wool is appropriate. You may need to buy the wool, in which case you will need *money*. You will need *time:* both spinning and knitting are time-consuming; and you will need the *skill* and *equipment* these tasks require. These resources are all necessary to achieve your goal. What if you cannot spin? You may make the resource 'skill in spinning' available by spending time and effort in learning how to spin, or you may substitute the resource 'money' by paying someone else to do the spinning for you. Your satisfaction with the solution depends on your ability to see the range of resources you have available, and to use them effectively.

TRY YOUR HAND

Max and Ian are both tertiary students. Ian lives in an old house which he shares with other students. His room is cold and rather draughty. Max lives with his parents, and has a warm, comfortable room: this is just as well, as he feels the cold. Both love to go skiing when they can afford it.

What special needs does each have when he is studying? skiing?

Using this book as one resource, draw up a list of specifications for clothing to meet the special needs of each student in each situation. Base this list on the above information only. Write up your list as a design brief.

Swap your design brief with a partner, and work from your partner's brief.

Following the brief literally, recommend suitable clothing for each student.

Does the brief provide enough specific information for you to be able to solve the problem?

Swap the work back, and look critically at your partner's solution to the problem. Which solution is the more satisfactory? Why?

The case study provided only limited information. What other factors could be relevant in providing a more satisfactory solution for each problem?

The role of the consumer

The textile manufacturer produces a range of commodities, and the consumer must choose between them. The consumer and the supplier are interdependent: each has responsibilities to the other. The consumer must have a clear idea of her or his needs, and must be able to communicate these needs to the supplier. In return, the supplier must provide the information necessary to enable the consumer to make an informed choice.

TRY YOUR HAND

Work with a partner, one to be a consumer, the other a supplier.

The consumer invents an unusual textile item, and describes it without naming it or using technical information.

The supplier asks questions, and tries to work out what the consumer really wants.

How effective is this process?

Now, write up a design brief which communicates precisely what is wanted.

Information required by a supplier includes:
(i)   What does the consumer want?
(ii) How much is the consumer prepared to spend?
(iii) What standards will the consumer apply to the item?
(iv) In what form is the textile item wanted, for example as fabric or as made up items?
(v)   Where are the consumers who want this item?

(vi) Where do these consumers purchase their textile needs?

Few suppliers sell direct to the private consumer. A big consumer may deal directly with a supplier, but the individual buyer is usually separated by a chain of intermediaries, including importers, wholesalers, and retailers. Communication in this case is through the intermediaries, or through consumer demand for or rejection of a product.

Choosing a fabric

At the point of purchase, a big consumer may demand samples of textiles and test them to determine their suitability for the intended end-use. The individual consumer cannot do this. However, some knowledge of properties and performance will help the consumer choose.

Follow instructions

Each ready-made garment is provided with labels that describe fibre content and give instructions for care and laundering.

Each pattern recommends suitable fabrics.

Rolls of cloth are provided with fibre content labelling, and occasionally with warning labels regarding fire hazard. Care instructions are also provided.

It is the responsibility of the manufacturer to provide correct information about the product to the consumer.

It is the *responsibility of the consumer* to read and consider the information provided, to check all relevant facts about the article, and to make sure that it is well-suited for its intended purpose.

Shop-floor tests

It is very tempting to choose a garment or fabric at first glance, on the basis of colour, texture, and shape. There are, however, a few simple tests a consumer can make on the shop floor.

Close attention should be paid to whether the *colour* of a garment or fabric compliments the wearer's complexion. Hold the fabric or garment near your face, and check the effect in a mirror. Try to seek natural light if possible. Beware of spotlights and coloured boutique lighting: the fabric may look entirely different by the light of day.

Crushability could be a problem during wear. Never buy a fabric or garment without the 'hot fist crush test'. Squash a fistfull of the fabric for a count of ten, release, and observe how – and whether – the fabric recovers from the trauma.

Drape and *stretch* are important in determining the final appearance of a garment. To check the drape, allow the fabric to drape over your hand. Place a length of it over your shoulder and allow it to fall to your front – gather it at the waistline with your hands. Observe carefully the ease with which it falls into folds. If the pattern has gathers, gather some of the fabric in your hand, and check whether it is too bulky or too limp to produce the desired effect.

For a stretch fabric, both the direction and the extent of stretch are important. Pull the fabric in all directions, and check its extensibility against the minimum stretch recommended on the pattern. Remember that not all knits are stretch fabrics – some can be quite inextensible.

Cover is important for some applications. A translucent light-coloured skirt will look most unattractive with a dark blouse tucked into it. Always place a hand behind a light-coloured fabric before purchase to check just how transparent the finished garment will be.

Some knits are prone to *pilling*, as are some loosely woven and some brushed fabrics. The pilling potential of a garment or fabric can be roughly gauged by rubbing two parts of it together with a brisk circular movement. If ten or so rubs disturb the fabric surface, then pilling could be a problem in use.

To see whether seams in a fabric are likely to give way because of *yarn slippage*, check the ease with which yarns can be removed from the fabric. If a cut edge frays readily, the fabric should not be used in garments that have a tight fit or that are subject to stresses at the seams.

Dark colours may give rise to problems with *staining* and *colour transfer*. Check crocking fastness with a white handkerchief. Wrap the handkerchief round your fingertip, moisten, and rub on the suspect fabric. If the white cloth is stained, the dark fabric has a low crocking fastness and could give rise to problems throughout the life of the garment.

Consumer protection

In Australia, the Commonwealth Trade Practices Act protects the consumer from false claims and defective or dangerous goods. Since 1978, importers, manufacturers, and sellers of consumer goods have been bound by law to stand behind the goods and services they provide.

The consumer buying second-hand goods is not afforded the protection given the consumer buying from a commercial retailer.
(Gary Tregaskis)

In return, it is the consumer's responsibility to shop wisely, to make considered choices, and to follow up any genuine causes of complaint through established legal channels. The consumer must examine goods carefully before buying; make sure the goods received match the chosen sample or description; check with the sales staff that the goods are appropriate for their intended use; and keep all contracts, receipts and other records for future reference.

If the consumer is dissatisfied, the first step is to take the complaint to the retailer. If the retailer is unwilling to accept responsibility or to pass the complaint on to the manufacturer or importer, then the consumer should contact the local consumer affairs agency.

Each State in Australia has a separate agency guarding the rights of consumers. In New South Wales it is called the Department of Consumer Affairs; in Victoria, Queensland, and the A.C.T., it is the Consumer Affairs Bureau; in South Australia the Department of Public and Consumer Affairs; in Western Australia the Bureau of Consumer Affairs.

TRY YOUR HAND

Write to your local consumer affairs agency to find out what steps you would need to take if you wished to complain about an unsatisfactory textile article. Whose responsibility is it if:
1 a garment stretches out of shape during laundry?
2 a dry-cleaner ruins a pair of trousers while trying to remove a stain?
3 a shop displaying a 'no refunds, no exchanges' sign sells a faulty garment?
4 you buy a garment labelled 'seconds' and it is faulty or damaged?
5 you change your mind about a purchase?

TRY YOUR HAND

Look at some magazine advertisements published at the turn of the century. Facsimiles are sometimes available. Carefully note the wording of the claims.
How do these advertisements differ from the kind of advertising we have today?
Are the old and the modern advertising equally credible and reliable?
Research 'the snail in the bottle': your legal studies teacher, or a local solicitor, may be able to help.

Textiles labelling

Content labelling

The Textile Products Labelling Act, 1970 requires textile products to be labelled to show the fibre composition, and also country of manufacture if the goods are imported. Where more than one fibre is present, the label must list them in descending order of percentage in the blend.

Textile products containing 95 per cent or more wool can be labelled 'pure wool' or 'all wool'. Those containing from 5 to 95 per cent wool must state the percentage of wool and list the other fibres present.

Artificial fibres must be listed among twelve categories, which were established to simplify the naming of fibres and to avoid the confusion which could arise from the use of brand names. The categories are: acetate; acrylic; chlorofibre;

elastomeric; glass; metallic; paper yarn; polyamide or nylon; polyester; polyolefin; polyvinyl alcohol; rayon.

These labels must be securely attached and in a position where they are easily noticed. Rolls of cloth may be labelled with a swing tag. Traders who sell unlabelled goods may be prosecuted; consumers can assist by refusing to buy unlabelled goods and by drawing the attention of the retailer to the existence of unlabelled products.

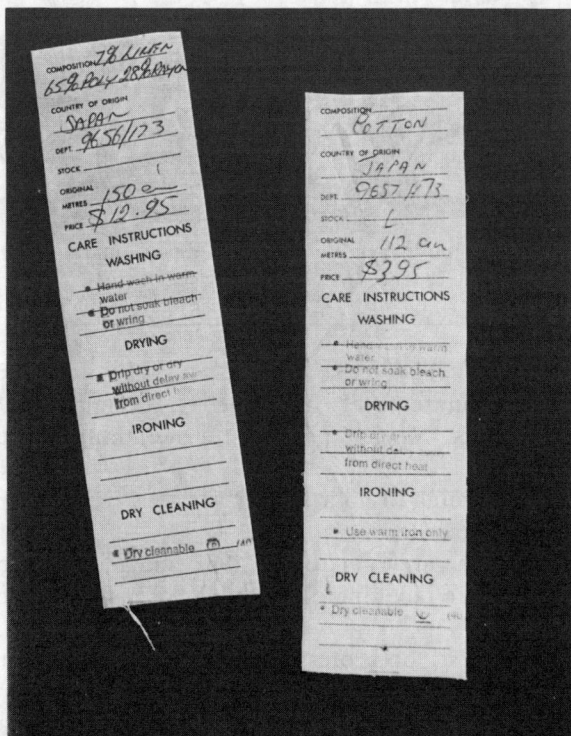

Care labels

Since 1980, all clothing, household textiles, furnishings, piece goods, and yarns have had to be labelled with instructions for their care. The care label should include instructions for washing, drying, ironing, and dry-cleaning. The wording on the label should be clearly legible – at least 1.5 mm high – and the label must be permanently attached.

There are clearly defined phrases and symbols which may be used in the labelling. This is designed to make it easier for the consumer to understand the care instructions. Special schedules to this Act list exemptions (handkerchiefs, sewing threads, floor coverings) and modified requirements for some items.

The manufacturer must make sure that the care instructions are correct, and that they apply to the whole article, including buttons and trims. If the consumer for any reason removes the care labels after purchase, she or he has the responsibility to ensure that correct instructions are passed on to the launderer or dry-cleaner. If the article is cleaned according to the instructions and becomes damaged, the responsibility for the damage lies with the manufacturer.

The consumer can assist by buying only goods that are adequately labelled.
(David Bailey)

TRY YOUR HAND

You are a garment manufacturer. You need to design correct fibre content and care labels for three of your products: a swimming costume, a suit, and a shirt.

What would be the wording of your labels?

Where would they be attached?

Check your answers by referring to the 1982 amended version of Australian Standard 1957–1978 'Care labelling of clothing', obtainable from the Australian Standards Association.

Why is it in the interests of professional dry-cleaners to educate consumers regarding the use and importance of care labelling?

What other sectors of the community benefit from consumer awareness of the care of textiles?

Safety labelling

Flammability of textiles is discussed in detail on pages 485–93.

Children's nightwear is the only textile item for which flammability performance regulations apply nationally. Other potentially flammable textiles – carpets, furnishings, upholstery, bedding – are regulated on a local or State basis. State authorities, supported by manufacturers, are negotiating to introduce uniform labelling and performance requirements in all products, not just textiles.

Fabric design

At the start of fabric production, the manufacturer needs to give clear and accurate specifications for each stage in the manufacture process.

The *yarn manufacturer* needs to know what types of yarn are required: fibre;

thickness; twist, structure – plied, fancy, simple; and what variations in quality will be acceptable.

The *weaver* needs to know the warp and weft counts, the width of the fabric, the structure, and whether coloured or fancy yarns are to be used and in what pattern.

The *dyer and finisher* need to know whether to bleach, dye, or print; what width the finished, relaxed fabric should be; whether it is to be shrink-proofed; and any other special finishing required.

A record of these specifications is kept with a sample of the fabric so it can be manufactured again if desired. The yarn manufacturer keeps a record of the yarn specifications as well as details of the fibre purchased, the equipment used and the machine settings, so the weaver can be supplied with the identical yarn if the need arises.

Dress fabrics

When dress fabrics are designed, the designer usually looks at the last year's successes and modifies them slightly if necessary.

If a *lighter* fabric is required, finer yarns may be specified, or a looser SETT (lower count). For knitteds, a lower gauge machine will give a softer, more pliable, less stable fabric construction.

If *softness* needs to be increased, finer fibres or filaments can be selected for the same yarn thickness.

Extra crimping or texturing gives fabric greater *bulk*. Unless the structure is loosened, the cloth may become rather stiff and graceless. To improve *drape and crispness,* an increase in the level of twist is used in the yarns. *Any* change at any point in the yarn or fabric manufacturing processes will result in changes in the properties of the fabric.

Trial samples may be knitted or woven and their potential evaluated. Even if only one type of fibre is being used, many different kinds of fabrics can be created by using different fibre preparation and yarn and fabric manufacturing processes.

TRY YOUR HAND

Collect a variety of fabrics made from 100 per cent polyester and compare their properties.

Use a binocular microscope and compare:

the structure of the fabrics

the thickness of the yarns

the cover (how much space there is between yarns)

the structure of the yarns (twist if any, staple or filament, textured or smooth fibres, ply, etc.)

the thickness of individual fibres

the lustre of the fibres

What is the contribution of each of the above properties to the final look and feel of each 100 per cent polyester fabric?

TRY YOUR HAND

You are now asked to design an all-polyester outfit – suit, shirt, and cape or overcoat.

Describe in detail the fabric properties needed for each garment and consider the contribution that yarn will make to each.

How will each fabric be constructed?

What types of yarns will you use – staple, filament, textured?

Will you use the same thickness of individual filaments for each?

The same amount of delustrant for the polyester?

What spinning systems would give the correct textures for your yarns?

How would you achieve the correct drape, weight and surface effect for each fabric?

Give detailed manufacturing instructions for your fabrics.

TRY YOUR HAND

Go into a fabric shop and examine the rolls of cloth that are labelled 'linen'.

Are all these fabrics made of flax? What other fibres are used to create linen-look fabrics?

Do all these 'linens' have the same properties? What are the characteristics they have in common? How are they different?

Suggest some techniques the manufacturer can use to make linen-look fabrics.

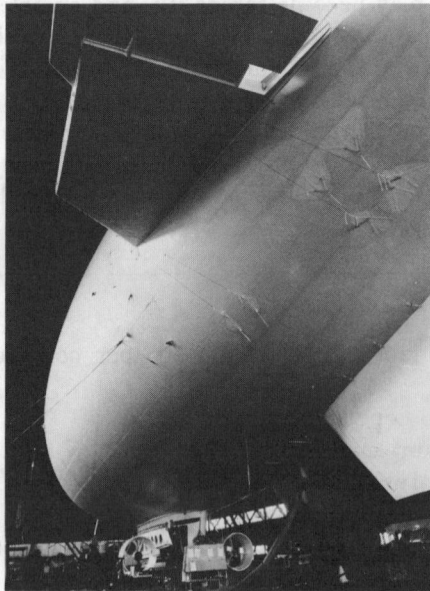

The principal criteria for the fabrics used in airship construction are light weight and strength and resilience. Because the design requirements could be met by synthetic materials, there is very little metal in the modern airship. This has created a new market: with very little metal, the airships do not show up on radar, and so are to be employed by the United States coastguard for coastal surveillance.
(Design Council/Colin Curwood)

Engineering of industrial fabrics

For dress fabrics, the most important attributes are those that are difficult to define accurately – aesthetics, handle, drape, lustre, 'feel'. For industrial applications, however, other properties are important, and these can often be precisely stated. Strength, extensibility, weight, porosity, thickness, are all attributes which can be measured.

The user will not care whether the industrial filter is made from woven or nonwoven fabric. He will, however, make sure to specify how much air the filter is to allow through each minute, what size particles are to be trapped in it, and whether it needs to be resistant to acids or not.

Honeycomb of Nomex. Because of their great strength, fire resistance, and light weight, aramids are increasingly being used as construction materials in aircraft and boats.
(Du Pont)

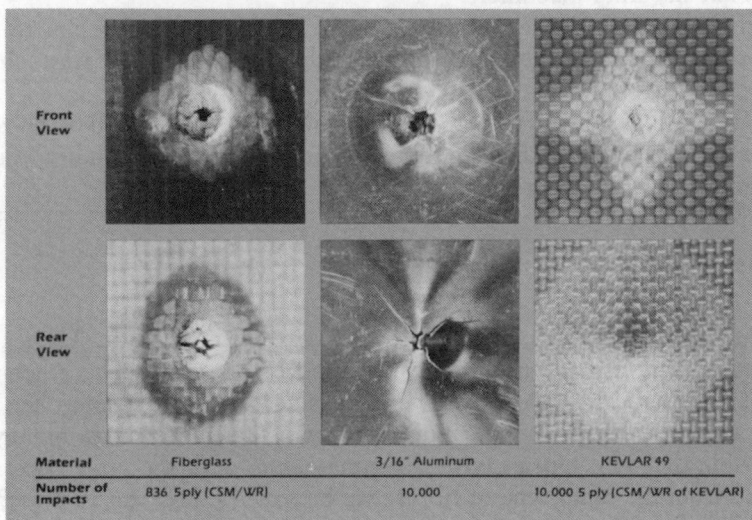

Impact tests on fibreglass, aluminium and Kevlar 49 demonstrate the aramid's greater ability to withstand repeated impacts. The laminates tested were of equivalent structure, and had the same stiffness as the aluminium plate. Kevlar's ability to withstand impacts is so great that 5 centimetre-thick Kevlar laminate was used as the secondary dust shield on the European spacecraft Giotto, which was sent close to the nucleus of comet Halley in 1986.
(Du Pont)

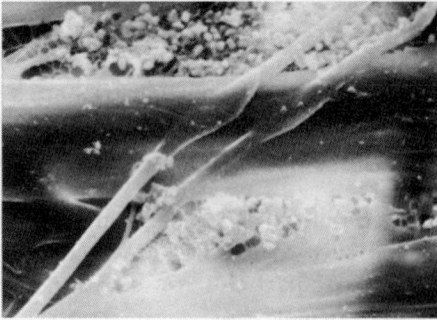
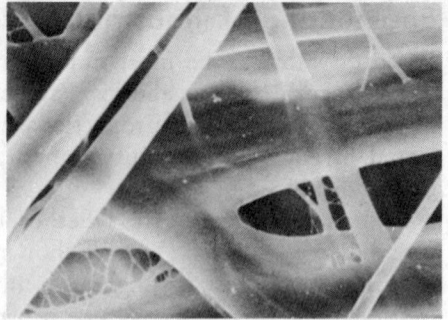

Packaging for medical supplies must be impermeable to bacteria and bacterial spores. Here, electron micrographs show bacteria on the outside surface of Du Pont's Tyvek (a spun-bonded polyethylene); they are unable to pass through the membrane to the inner surface. The tough, water-resistant polyolefin has proved resistant to bacterial penetration over many years of testing.
(Du Pont)

TRY YOUR HAND

You are a manufacturer of nonwoven fabrics. One of your products is needled felt used as air filters. Your customer says that he needs a new filter which allows more air through it.

How will you modify your fabric to make it more porous?

Will you use finer or coarser fibres? Thicker or thinner web? More or fewer passes of the needling machine?

Can you think of any other modification to fabric or filter design to achieve the same result? Which do you think would involve the greatest change in the cost?

Will your new filter still trap the same amount of dust? Will it let some of the finer dust through? If the consumer complains, can you modify your filter further to suit his requirements?

The cost of quality

What price quality?

On the shelf in a fabric shop there are two bolts of cloth. They are both satin cotton. One is made in Japan, and costs $4.50 a metre. The other is Swiss cotton, and costs $25.00 a metre. The expensive cotton feels smoother, silkier, and has a slightly better lustre – but is it worth that much extra money? How is it possible that two fabrics with the same name, with the same construction, and made from the same type of fibres, can be so very different in price?

Would the more expensive fabric perform better than the cheap one? Let's analyse the reasons for the difference in prices.

• The Swiss fabric was made from more expensive raw material than the Japanese cotton was. The cotton fibres used were both finer and longer. The finer each individual fibre is, the softer the handle of the fabric. Since the more expensive fibres of cotton were also longer, the yarn spun from them was smoother and more uniform; this contributed to the final lustre of the Swiss cloth.

- The Swiss fabric was woven from combed cotton yarns. Combing is a slow and expensive process which removes the short fibres before the yarn is spun. In the Japanese cotton the short fibres were left in. The short fibres stick out from the yarns of the cheaper fabric and form a slight fuzz on the fabric surface. This detracts from the sheen that is desirable in a cotton satin fabric.

  If the loose fibres are not held securely in the yarn, they are further loosened during the agitation of washing and tumble drying. During agitation, short fibres fall out. The more fibres are lost, the weaker the yarns become, and the more easily the rest of the fibres are lost. So, the more short fibres there are in a yarn, the more quickly the fabric will deteriorate during washing: loss of strength, loss of lustre, and even a loss of handle.

  Sometimes in poorer quality satin cottons the finished fabric surface is covered with a layer of resin, which is then 'ironed' on to create a relatively temporary lustrous finish.

- During the spinning process, a great deal of care has been taken to ensure that the yarns of the Swiss cotton become uniform and with well-aligned, parallel fibres. This means more processes of drawing. The amount of twist is not much higher but, as the fibres are more parallel, the Swiss yarns are more compact in structure than the Japanese ones. This compact and uniform arrangement of the fibres yields a much smoother yarn, and a lustrous fabric. Since the fibres are securely held in the yarn, the fabric is likely to keep its lustre and firm but soft handle, even after repeated washing and drying cycles.

- Not only was the spinning process of the Swiss cotton more complicated and more expensive, but the weaving was more costly as well. Where only 12 coarser yarns were used for the warp of the cheaper fabric, 18 yarns of the same length were needed to be spun for the more expensive one: an increase of 50 per cent in spinning costs alone. Once the loom is set up, the density of the warp threads does not add greatly to the cost of the fabric. Although there are more yarns that could break during weaving, causing faults and machine down-time, the better yarns are also stronger and more uniform in strength and so are less likely to break than the cheaper yarns.

- The rate of insertion of weft yarns has a profound effect on the costs of weaving. Since a loom will work at a certain rate of inserting picks, the more weft threads that are needed to make a metre of fabric, the longer it will take to carry out the weaving. The fine fabric had ten weft threads in the same space that the cheaper had seven: a great difference in the cost of weaving.

- Finishing costs were also different for the two fabrics. The Japanese fabric was scoured, bleached, set (stentered) and printed by roller printing. The Swiss fabric was scoured, bleached, mercerized and preshrunk, and printed in an exclusive design in limited quantities by screen printing.

- There is also a dollar or two for 'name'.

What are the advantages of all this extra processing? The *mercerization* makes the fabric both stronger and more lustrous. This lustre is a permanent effect, which will not be removed during the wash. *Preshrinking* ensures that the garment made from the fabric will not change in size during laundering. Note that the cheaper fabric may not be preshrunk, but careful *stentering* can ensure that subsequent relaxation shrinkage is no more than 5 per cent and the fabric is still serviceable. The superior strength and stability of the Swiss cotton means it is better suited than the Japanese for shaped and fitted garments. The *exclusive* design is purely an aesthetic factor.

So it seems that the more expensive fabric cost the manufacturers more to produce. It also appears that the more expensive fabric is better able to retain its superior aesthetic properties during extended use. Not only does it look and feel better to start with, it will keep on looking good for longer. Swiss cottons have a well-justified reputation, it seems.

It takes a lot of skill to make fabrics of superior quality. Not every country or city has a reserve of skilled workers to cover every aspect of textile manufacture. From the earliest times certain cities have specialized in the production of special types of cloth (Manchester for cotton goods and sheeting, Calcutta for calico, Worstead for worsted wools).

Even today different countries specialize in the types of fabrics they produce for the local and international markets. Japan manufactures quantities of medium quality fine cottons, and a lot of good polyester/cotton blends. Australia does not manufacture fine quality cottons, though it does weave most of its coarser cotton requirements.

Quality control remains labour-intensive and therefore expensive. Here, blankets are being mended by hand before milling.
(Australian Wool Corporation)

Buying quality

How can the consumer determine the quality of a fabric at the time of purchase?

Evenness of yarns and of weave are always good guides. Holding the fabric up to the light will show variations in evenness and in the density of weave. The hand-squash method of determining crushability and resilience is indispensible. Draping the fabric over the hand, and holding a length of it against the body, will reveal many aesthetic qualities.

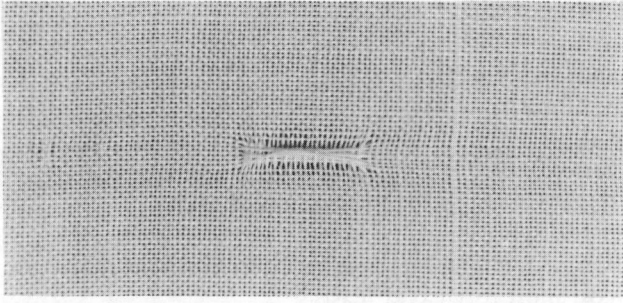

Weaving faults in a cotton fabric.

Clear labelling, including suitable reporting of shrink resistance, other performance characteristics, and country of origin, is intended to give consumers a better chance to make the decision that is right for them.

TRY YOUR HAND

Here are photographs of three cotton voiles. Look at them carefully.

Check the following points:

evenness of yarns

twist (singles and/or plying)

amount of loose fibres

density of threads

coarseness of yarns

List the processes involved in the production of each.

Which cost most to spin? least?

Would the different spinning methods result in different fabric properties?

Which fabric cost most to weave? least?

Would all fabrics have the same handle and appearance? If not, what properties could you predict for each?

Go into a shop and examine a range of voiles. Take a magnifying glass or pick glass to look at them closely. Can you relate the prices of the fabrics to their handle and/or their construction?

Special requirements

Often, a consumer will require characteristics in a textile which result from a combination of fibre type, yarn structure, fabric construction, and finish. For example, clothing must be appropriate for the conditions under which it will be worn. This encompasses not only weather conditions, but comfort and flammability. Considerations such as thermal resistance are important for blanketing as well as for clothing; wind resistance is important for flags and sails as well as for outerwear. Flammability is important not only in children's clothes, but also for furnishings.

The design of special-purpose clothing must take into account all components of the garment and their performance in the expected conditions of use. Seams and closures can be as important as the choice of fabrics. Can you suggest why army trousers have buttons instead of zip fasteners?
(Army Public Relations)

Thermal resistance

Heat loss from the body

The surface temperature of the human body must be maintained within a few degrees of 37°C, yet people live in places where the air temperature is commonly as low as –40°C, or as high as 40°C. The body needs to stay cool in the hot conditions, but loss of heat must be prevented in cold climates.

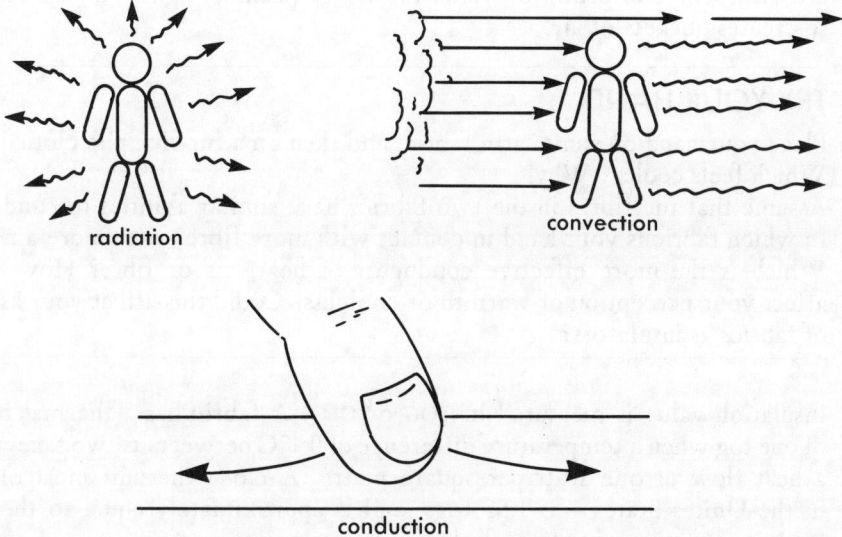

radiation

convection

conduction

Mechanisms of heat loss from the body.

Loss of heat from the body can occur by radiation, convection, and conduction.

RADIANT HEAT is heat in the form of infrared radiation: long wavelength, low energy electromagnetic waves. The heat from the sun reaches us as radiant heat, independent of any carrier medium. Anything which is warmer than its surroundings can lose heat through radiation. Although white clothing can reflect some of the sun's heat away from the body on a hot summer day, radiation is not a very important factor in the thermal functioning of clothing. It is important, however, in reflective curtain linings.

CONVECTION is the carriage of heat by moving gas or liquid. This form of heat loss is especially important for outdoor clothing and windy conditions. Weather reports for cold-climate countries often quote *wind-chill factors,* to warn of the extra cooling effect of the wind.

CONDUCTION is the process of heat transfer from the warm side of a material to the cooler side. Textile fibres differ in their ability to conduct heat, but it is the thermal resistance of trapped pockets of air, not the conductivity of the fibres, which determines the thermal properties of the textile. This is because the average textile contains as much as 75 per cent air. In a blanket, the volume of fibre can be as little as 10 per cent of the volume of the blanket: a wad of steel wool (steel has a high heat conductivity) can be nearly as good an insulator as a wad of pure wool of similar low density. This insulation value depends on the trapped air remaining still, a condition which may not be met in windy weather.

Fabrics as insulators

There are many fabric structures which can trap still pockets of air and so act as insulators under still air conditions. These may be pile fabrics, milled, or quilted, for use as blanketing; or thinner and more flexible knitted and brushed fabrics for clothing: *the more air trapped in a textile, the higher its insulation value.* To increase the amount of air space within a fabric, synthetic fibres are *crimped*. The crimp prevents the fibres packing tightly in the fabric, and so creates pockets of air.

TRY YOUR HAND

Place your hand on some satin fabric, and then on a fur-like pile cloth.
Which feels cooler? Why?
Assume that the fibres in the two fabrics have similar abilities to conduct heat. In which fabric is your hand in contact with more fibre? with more air?
Which is the more effective conductor of heat: air or fibre? How does this affect your perception of warmth or coolness? Could this affect your assessment of fabrics as insulators?

Insulation value is measured in TOG or CLO. A fabric has a thermal resistance of one tog when a temperature difference of 0.1°C between its two faces produces a heat flow of one watt per square metre. A clo – the unit most often used in the Unites States – is 1.55 togs, and is approximately equal to the thermal resistance of normal indoor clothing.

TRY YOUR HAND

Study the table of insulation values.
The tog values for the different fabrics are very similar. Why? (Remember that they contain about 90 per cent air.)
In what applications have you found each of these textile materials?
Comment on your own experience with these materials. Which do you think is warmer, a fluffy acrylic blanket or a polyester quilt?
Look at the warmth/mass column. Is your perception of the warmth of these two blanketings supported by the data? If not, how do you explain your perceptions?
Which material is the best insulator (has the highest warmth/mass)?

Insulation values of selected materials

Material	Insulation value (tog/cm)	Warmth/mass
50/50 down and feather mix	2.6	70
Polyester staple quilted fabric	2.4	40
Sliver knit acrylic high pile fabric	2.3	22
Closed cell expanded polyethylene	2.4	19

Hollofil is polyester fibres with a tubular cross-section. Ten per cent of each fibre is replaced by an air space. It is claimed to have a better insulation value than solid polyester fibre. But 90 per cent of a standard polyester quilt is air anyway; only 10 per cent is fibre, so the extra air in Hollofil increases the total trapped air by only 1 per cent. Therefore, the value of Hollofil is not in the increased amount of trapped air, but in the 10 per cent reduction in the mass of quilt needed to produce a given insulation value.

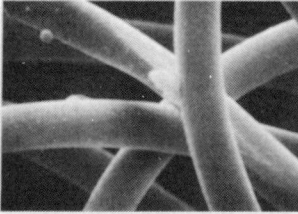

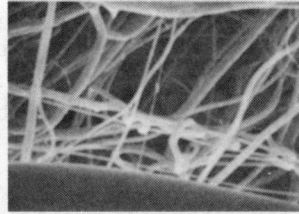

One problem with traditional insulating materials is their bulk. 3M's Thinsulate achieves almost double the insulation values of down and fibrefill of equivalent thickness, by virtue of its microfibre construction. The electron micrographs compare fibrefill and Thinsulate. Because it is 65 per cent polyolefin and 35 per cent polyester, Thinsulate does not absorb water, and so does not lose its insulating properties in damp conditions. (3M Australia)

TRY YOUR HAND

You will need:
samples of various fabrics 2 cm × 5 cm. Compare thin nylon tricot with woollens, cotton sheeting, bulked polyester jersey, and other fabrics including blanketing, fur fabric, quilt and felt
thermometer
needle and thread
beaker, water, and heat source
tissues
stop-watch

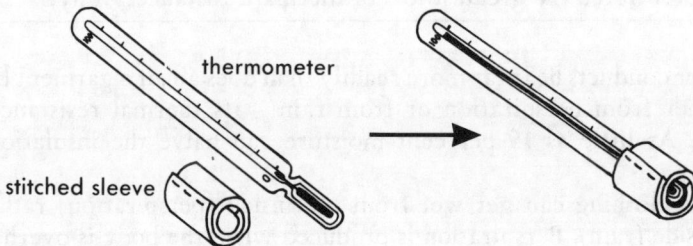

What you do:
Take a strip of fabric, and roll it around the end of the thermometer to form a double layer. Secure this close-fitting sleeve with a few stitches so that it will not unroll.
Label it and set aside.

Do this for each fabric type.

Make another set of sleeves with a single layer of fabric only.

Place the thermometer in boiling water and leave until it shows a steady 100°C.

Remove the thermometer from the water, quickly wipe it with a tissue, and place a single-layer fabric sleeve on it.

Start the stop-watch and record the time taken for the temperature to drop to 50°C.

Repeat for all the single-layer and then all the double-layer sleeves.

Tabulate the time taken for the temperature to drop for single and double layers of each fibre type and fabric structure.

1   Which fabrics were the best insulators?
2   What is the effect of doubling the layer of insulation? Does it double the time for the temperature drop, or is the time more than doubled, or is the effect less than doubled?
3   How do you explain the effect of doubling the insulation layer?
4   Which would you expect to be warmer, two thin jumpers or one thick one?

The effect of water on thermal resistance

Substances conduct heat by transferring the energy directly from molecule to molecule. It follows that a substance like air, with its molecules far apart, will conduct heat far less effectively than a substance like water, which has its molecules much closer together. In other words, air is a better insulator than water.

TRY YOUR HAND

According to the theory above, wet fabrics should be poor insulators. You can test this statement.

Wet and blot some thermometer sleeves from the previous exercise, and check the time taken for the temperature to drop to 50°C.

Compare your results to the times found for dry fabrics.

Which fabrics suffered the greatest loss of thermal resistance? Why?

Because water conducts hear far more readily than does air, if a garment becomes wet – whether from perspiration or from rain – its thermal resistance drops dramatically. As little as 15 per cent moisture can halve the insulation value of clothing.

Sometimes, clothing can get wet from the inside (perspiration) rather than from the outside (rain). Perspiration is produced when the body is overheated.

To prevent perspiration from wetting the insulating fabric the wearer should not overheat. Therefore, people doing vigorous exercise should remove some layers of insulating clothing to maintain comfort. When exertion stops, and the body starts to cool down, the layers of insulation should be gradually increased. Rushing about in cold weather perspiring in one's clothing is a sure way to catch a chill.

In the wind and the rain

It is the *stillness* of the air trapped in clothing which gives it the ability to prevent heat from being conducted away from the body. If wind blows through the garment, fresh *cold* air replaces the insulating layers and heat is rapidly lost. For clothing to insulate efficiently in cold and windy conditions, a *windproof* layer is necessary.

While wind should not be allowed to penetrate the outer layer, water vapour – produced as insensible perspiration by the body – should be able to escape. The most suitable fabrics for such an effect are tightly woven from natural fibres. The water vapour can pass through the hydrophilic fibres, while wind and rain are kept out by the tight fabric construction.

As the fabric is wetted – say by a light shower – the fibres swell and provide a still tighter fabric structure. This serves as a barrier to the penetration of water. Such fabrics, even if coated with a waxy film, are not completely waterproof – they are SHOWER REPELLENT.

Waterproof fabrics are not permeable to water vapour. Under waterproof garments, insensible perspiration readily condenses into liquid and wets the clothes. Micropore coatings were developed to overcome this problem, but even they have much lower water vapour permeability than do shower-proof fabrics.

Waterproof garments should be worn only while it is raining. After the rain has stopped, the impermeable garment must be removed to allow the escape of water vapour generated by the body.

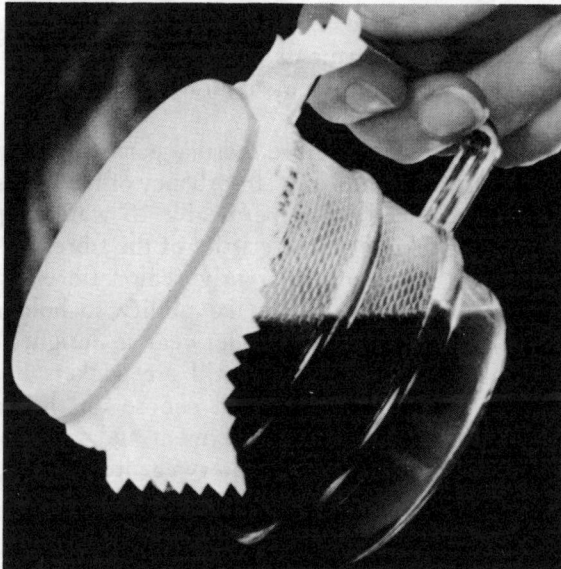

A dramatic demonstration of the Gore-Tex membrane: permeable to water vapour but not to liquid water.
(W.L. Gore and Associates Inc.)

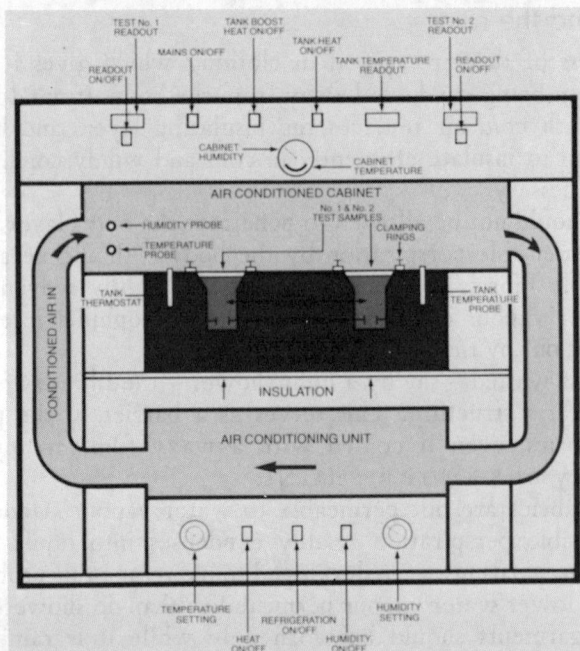

Apparatus developed by Peter Storm to test water vapour permeability. With many manufacturers developing vapour-permeable waterproof finishes and fabrics (including Peter Storm's own finishes), an accurate measure of permeability is necessary.
(Peter Storm)

Cold-weather clothing

There are a number of special garments on the market sold as 'thermal'. Some of these undergarments are made from natural fibres, others from synthetics (mainly chlorofibre, polyester, nylon).

The manufacturers of the natural-fibre thermal garments claim extra comfort for their products because of the moisture absorbency of the fibres. The synthetic-fibre garments are also claimed to be permeable to water vapour because of their open structure and wicking characteristics of the fibres.

While different fibres perform differently (chlorofibres have the highest thermal resistance while wool has particular ability to hold air still because of its scale structure), all such thermal underwear is designed to be excellent insulation. All the fabrics hold plenty of still air pockets in a light-weight, pliable, soft garment. The consumer should choose on the basis of personal preference for the handle and feel of the garment and, most of all, be careful to wear it with a suitable combination of outerwear items.

We have already seen that the thermal resistence of two layers of fabric is greater than the sum of the resistances of the two fabrics. This is because of the air trapped between the layers. Therefore it makes sense to wear a number of thin, light-weight, flexible insulating garments rather than one heavier weight garment. To prevent wind from penetrating the clothing and to help trap still air, the garments should be designed to have some close fitting points, such as cuffs and waistband.

Kim Logan (left) and Peter Hillary (right) in the western cwm of Everest (6600 metres). They are wearing wool beanies, and thermolactyl suits under Gore-Tex windsuits. On their feet they have Gore-Tex/Thinsulate 'Super-Gaitors' over plastic double boots (Koflach; Aveolite inner boots), wool socks, and neoprene vapour barrier socks. (Roderick Mackenzie)

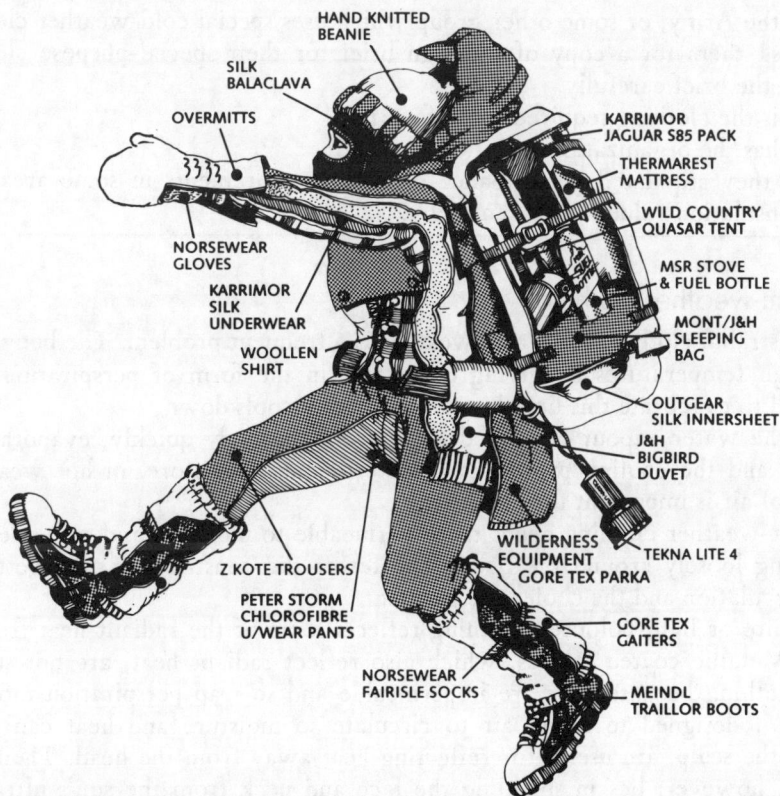

HAND KNITTED BEANIE

SILK BALACLAVA

OVERMITTS

KARRIMOR JAGUAR S85 PACK

THERMAREST MATTRESS

NORSEWEAR GLOVES

WILD COUNTRY QUASAR TENT

KARRIMOR SILK UNDERWEAR

MSR STOVE & FUEL BOTTLE

WOOLLEN SHIRT

MONT/J&H SLEEPING BAG

OUTGEAR SILK INNERSHEET

J&H BIGBIRD DUVET

WILDERNESS EQUIPMENT GORE TEX PARKA

TEKNA LITE 4

Z KOTE TROUSERS

PETER STORM CHLOROFIBRE U/WEAR PANTS

GORE TEX GAITERS

NORSEWEAR FAIRISLE SOCKS

MEINDL TRAILLOR BOOTS

The well-equipped bushwalker.
(Gear for Wild Places, Sydney)

481

Because the scalp is very rich in blood vessels, it can radiate much heat from the body. Indeed, at 15°C, the heat loss from the unprotected head can be as much as three-quarters of the total heat loss from the body. Therefore warm headwear is absolutely essential to protect the body in extreme cold conditions.

The other extremities – hands, feet – also contribute to heat loss. If the temperature drops, the body conserves its heat by reducing the flow of blood to the fingers and toes. This lowers the temperature there, but conserves heat for the vital organs. If the reduction in blood supply to the extremities is severe or prolonged, it can lead to frost-bite, the death of skin and tissues. Both warm footwear and protection for the hands are necessary, therefore, to prevent frost-bite. Mittens are more effective than gloves in keeping the hands warm, as they allow each finger to warm its neighbours. In extreme conditions, electrically heated gloves are necessary to maintain circulation.

Gloves and hats are a convenient way to regulate body temperature during outdoor activity in cold weather: you can take them off as you start to feel hot, and put them back on when the temperature drops below comfort level.

TRY YOUR HAND

Write to the Antarctic Division of the Department of Science and Technology, or to the Army, or some other group which uses special cold-weather clothing, and ask them for a copy of a design brief for their special-purpose clothing. Study the brief carefully.
How is the clothing required to perform?
How has the organization specified its needs?
Have they stipulated performance or other requirements in some area other than insulation value? What, and why?

Warm-weather clothing

In Australia, clothing for warm weather is a frequent problem. The body reacts to high temperatures by giving out water in the form of perspiration. Heat is used to evaporate this liquid, and so the body cools down.

If the water vapour is carried away from the body quickly, evaporation is faster and the cooling process is more efficient. Therefore, in hot weather a flow of air is important for comfort.

Hot-weather clothing needs to be permeable to moisture and to be designed to hang loosely around the body. The fewer the constrictions, the better the air circulation and the cooler the garment.

White or light-coloured clothing reflects some of the radiant heat from the sun. Metallic coated fabrics, which also reflect radiant heat, are not suitable for clothing because they are impermeable and so trap perspiration moisture. Hats, if designed to allow air to circulate so moisture and heat can escape from the scalp, are useful for reflecting heat away from the head. Their main value, however, lies in shielding the face and neck from the sun's ultraviolet rays.

482

Wind resistance
The wind resistance of fabrics

TRY YOUR HAND
You will need:
15 cm × 20 cm samples of several different fabrics: thick, thin, woven, knitted, felted, pile, loose and dense structures
large cardboard box
scissors
tape
hairdryer
lighted candle
measuring tape

What you do:
Cut a window 10 cm × 15 cm in the side of the box.
Tape one of the samples over it, and arrange as shown.

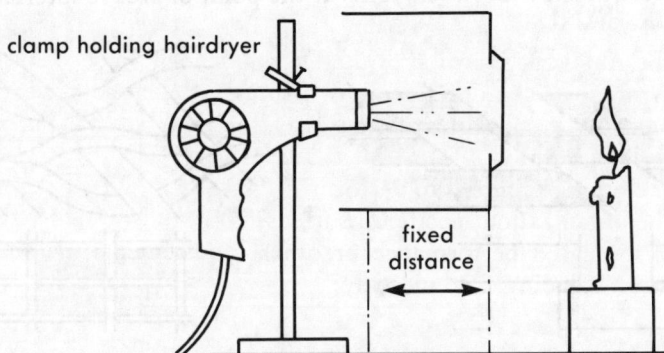

Switch on the hairdryer, and measure how close to the fabric you can move the candle before it starts to flicker.
List your results in a table.
Examine the fabrics under a pick glass or a microscope to assess the amount of space between the yarns.
How does the ratio of yarn to space (SETT) affect the wind resistance of the fabrics you examined?
What kind of structure would you expect sailcloth to have?
Are high-twist or low-twist yarns best for wind-resistant fabrics?

In order to make sails both light weight and wind resistant, they are made from untwisted filaments woven in a very tight structure. Sails are mostly made of polyester, spinnakers from nylon. This is because nylon stretches more easily, and so can absorb the impact of any sudden gust of wind which catches the large spinnaker.

The importance of light weight and high wind resistance.

Cover

The COVER of a fabric refers to its lack of transparency. Cover depends on yarn thickness and yarn density (number of yarns per centimetre). A *flat* yarn can achieve a better effect than a thick yarn. The flatter the yarn cross-section, the less space will show between yarns at the point of weave interlacings.

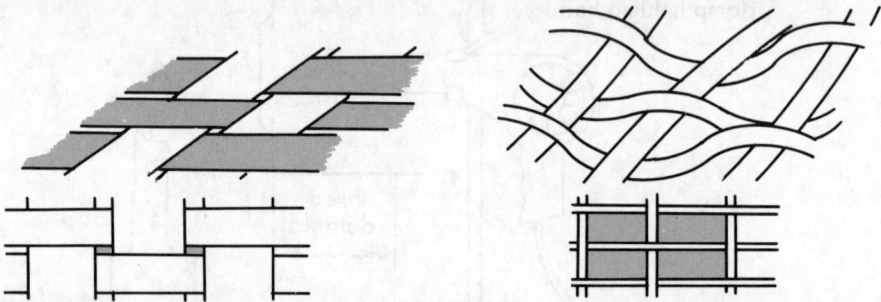

Flat yarns provide higher cover than do round yarns, because less space shows at the point of weave interlacings.

The ratio of yarn diameter to yarn spacing (d/p) expresses the fraction of space covered by either warp or weft yarns. d/p is often used as a measure of cover, and is called FRACTIONAL COVER. Warp and weft are quoted separately. Where $d/p = 1$, the fabric is 'jammed' – it has no space available between its yarns.

Some 'jammed' cloths are sailcloth (for sails), duck canvas, typewriter ribbon, and many fabrics made from twistless yarns. Some fabrics with low cover are cheesecloth, voile, organdie, chiffon, muslin and gauze.

For historical reasons, a different cover factor, K, is widely used. This measures yarn diameter in terms of cotton count, N, and yarn spacing in 'ends per inch', n.

$$K = \frac{n}{N}$$ Where n = number of yarns per inch

N = cotton count of yarn

Total cloth cover equals K(warp) + K(weft). If $K = 28$, then complete jamming occurs, whilst K = 7 is a very loose fabric construction. This concept is much less meaningful than fractional cover (d/p).

TRY YOUR HAND

Under a pick glass or a binocular microscope, measure the cover factors d/p in both the warp and weft directions of the fabrics you used in the previous activity.

How does cover relate to wind resistance?

What cross-section shape provides the highest cover, flat or round? low twist or high twist yarns?

How would you design a fabric for use as a sailcloth?

Would you use a similar fabric for a banner to be carried in a street procession or to be hung as a street decoration? What about a flag?

Find out what fabrics are commonly used for these purposes, and why.

Comfort and handle

Much consumer satisfaction has to do with the aesthetic qualities of textiles. Colour and pattern are the most obvious of these, but the feel, texture, handle and softness of textiles also add to their sensuous appeal.

What governs the sensations involved in the handle of textiles?

Softness is determined not only by the inherent *pliability* of the fibres used, but also by their *diameter*. Acetate is an amorphous fibre, much softer than nylon, yet it is possible to produce ultrafine nylon filaments which are soft, and coarse acetate filaments (such as Estron) which have a harsh, crisp handle. A yarn made from fifty fine multifilaments is much softer than a yarn of equal thickness made from ten coarser filaments. The coarse wool used for carpets is rougher than lambswool, which in turn can be rougher on the skin than ultrafine Merino wool.

The *cross-sectional shape* of the fibres also affects the handle of the fabrics. Think of the harsh, almost scratchy, feel of metallic (Lurex) yarns. They have a rectangular cross-section, and often need to be twisted together with other fibres to give a softer handle.

Processes used in *yarn manufacture* affect the comfort of the wearer also. A highly twisted crepe yarn is less soft to the touch than is a low-twist yarn. Fabrics made from yarns with a hairy surface feel entirely different from fabrics made from smooth-surfaced yarns. The spinning technique used can alter fabric handle. The harsher, more compact yarns made by open end spinning yield a fabric which feels different from the full, soft fabrics made from ring-spun yarns.

Fabric construction and *finishing* processes can also alter the aesthetic properties of fabrics.

Flammability

Fabrics burn, and their ability to burn is a significant cause of a number of fatal accidents, and a very large amount of property damage, each year.

Despite this, consumer interest in flame-resistant fabrics is small. There are many ways in which textiles can be made flame-resistant, but many of these processes alter the handle of the fabrics, and are expensive. Consumers have not been prepared to pay to remove a hazard which they feel does not concern them. Textiles treated with Proban and other flame-resistant finishes have proved a financial disaster for their manufacturers.

In Australia, the greatest number of burns are caused by flammable liquids – often carelessly thrown on a barbecue or backyard incinerator. Some years ago, a large proportion of the fatal burns were to children dressed in flammable nightwear who stood too close to heaters or open fires. Now, commercially available children's nightwear is labelled to show the fire hazard, and the most dangerous fabrics and styles are banned. *Daywear* is more commonly ignited.

Fire injuries in Australia

Substance first ignited	Victims under 16 (per cent)	Victims over 16 (per cent)
Flammable liquid	49	49
Clothing	21	12.5
Flammable liquid and clothing	5	5
Gas	2.5	10.5
Bedding, carpet, curtain, furniture	2	4.5
Cooking oil	–	4
Other	17	11.5
None	4	3

Source: *ANZBA Journal*, August 1982

Agent causing flame burn	Victims under 16 (per cent)	Victims over 16 (per cent)
Flammable liquid	37	41.5
Clothing	33	22
Flammable liquid and clothing	5	5.5
Bedding, carpet, curtain, furniture	1.5	3.5
Gas	2	10
Cooking oil	–	3.5
Unknown	1.5	2.5
Other	20	12

Source: *ANZBA Journal*, August 1982

Clothing involved in thermal injury

Clothing	Percentage	Ratio male:female
Adults' daywear	58	3.5:1
Children's daywear	29	2.3:1
Adults' nightwear	8	1:1.7
Children's nightwear	5	1:1.25

Source: *ANZBA Journal*, August 1982

Classification	Description	Label	Colour
Category 1	Wool, fire-proofed (e.g. with Proban), or other low flammability fabrics	LOW FIRE DANGER	Red writing on white background
Category 2	Nightwear designed to reduce fire hazard: e.g. ski pyjamas or other close-fitting or short style	STYLED TO REDUCE FIRE DANGER	Red writing on white background
Category 3	Flammable fabric, unsafe design, or both (loose, long, frilled designs catch fire easily and burn rapidly)	WARNING HIGH FIRE DANGER KEEP AWAY FROM FIRE	Black writing on red background. Carries fire hazard symbol
Restricted fabrics	Fabrics containing more than 50 per cent cotton, rayon and/or acetate are banned from sale unless they are flame-proofed or styled to fit into Category 2. For example, flannelette ski pyjamas are permitted, but a flannelette dressing-gown is not; cotton baby-doll pyjamas are permitted, but a cotton ankle-length nightdress is not.		
Banned	Cotton chenille and molleton		

FR wool 348 g/m² broke open after 6 seconds

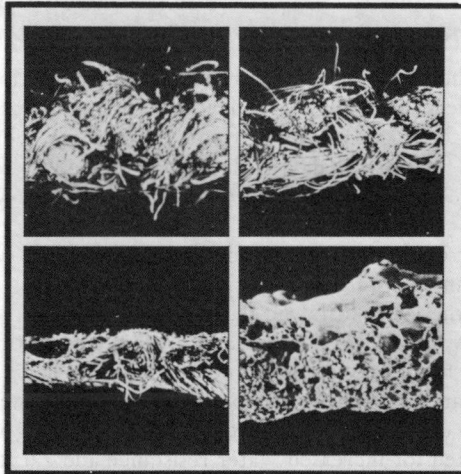

Non-improved aramid 190 g/m² broke open after 3 seconds

FR cotton 332 g/m² broke open after 10 seconds

Nomex III 265 g/m² no break-open after 20 seconds

In addition to flammability, break-open is an important consideration in protective garments. If a garment does not burn itself, but breaks open to expose underlying skin or garments to the heat, its protective value is limited.
(Du Pont)

For nearly 90 years the outfit worn by firefighters of the Metropolitan Fire Brigade, Melbourne, changed very little (right). A lancer style tunic was made from 390 g/m² wool serge lined with 270 g/m² wool flannel. Trousers were of 270 g/m² pure wool, and were used for normal day-to-day wear as well as for firefighting duties. The only change since the 1890s is the 1983 addition of lime-yellow reflective tape to the jacket.

Since 1984 the Metropolitan Fire Brigades Board has been working on the design and development of an integrated protective clothing system, incorporating modern design principles and the latest in fabric technology (left). The new Firemark uniform is a combination of waisted overtrousers and a medium-length tunic, which ensures that the firefighter is totally covered and comfortable. It has a number of variations which enable the wearer to adjust or modify the outfit to suit his or her personal comfort, without loss of protection. The tunic fabric is 360 g/m² pure wool cavalry twill, Scotchgard treated for increased water resistance and Zirpo treated for increased fire resistance. The removable inner lining is 220 g/m² wool flannel. Water-resistant yellow shoulder protectors are added, and the collar is black corduroy. The trousers are a hard-wearing resilient aramid called Tejin Conex, in a 270 g/m² twill weave treated for water resistance. Thigh glove pockets give added protection. The lime-yellow colour was chosen for maximum visibility. All fabrics are dry-cleanable, and Tejin Conex may be laundered.

(Information and photographs courtesy Metropolitan Fire Brigades Board, Melbourne)

488

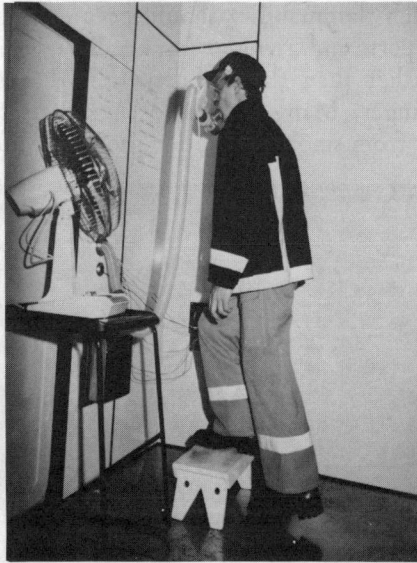

Part of the environmental heat chamber tests conducted at Footscray Institute of Technology.
(Metropolitan Fire Brigades Board, Melbourne)

Unexposed FR cotton
300 g/m²

Unexposed Nomex III
265 g/m²

Exposed to flame
FR cotton 300 g/m²

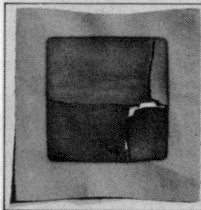

Exposed to flame
Nomex III 265 g/m²

The effect of flame on flame-retardant-treated cotton and on Nomex III: the aramid thickens on exposure to intense heat, and so continues to insulate.

In the United States, 5 per cent of all building fires were caused by burning curtains. Lined curtains are more dangerous than unlined curtains, as the light cotton lining burns more rapidly than the heavier curtain fabric. For half a minute, no flame may be visible, yet another 5 seconds later the flames reach the ceiling and burning fabric pieces fall to the floor.

In Britain, the single largest cause of fatal house fires is cigarettes left to smoulder on upholstered furniture, or trapped between cushions.

Australian legislation sets flammablity standards for mattress covers and upholstery, but there is no legislation covering what fabrics people may use or wear in their homes.

Other areas in which flammability should be a consideration for consumers include tents for campers, and awnings (especially in bushfire-prone areas). In some applications, rather than flammability it is the *toxicity of combustion products* that is a danger. Many victims of house fires die from the effects of poisonous fumes or from smoke inhalation, not from burns.

In only 12 seconds, the cotton pyjamas are flaming, but the wool bootee is only slightly charred.

After 50 seconds, the dòll dressed in cotton is engulfed in flames, but the wool pyjamas are barely smouldering.
(Australian Wool Corporation)

For normal clothing requirements and for furnishing, wool offers excellent fire resistance and comfort. The aramids are still too expensive to compete in other than high-risk areas, and cannot match wool for comfort.

490

Safety legislation

The safety legislation developed in Australia ranks among the most advanced in the world. Rather than testing and labelling fabrics as to the extent of their flammability, the design of the whole garment is considered. Certain design and fabric combinations are banned, others are labelled as dangerous, and yet others labelled as having low fire hazard.

In 1972, out of 90 cases of clothing fire accidents in Melbourne, 11 involved cotton chenille, flannelette, or cotton molleton. These fabrics surface-flash: flames travel extremely rapidly up the raised fibres in their napped surface. After the 1978 legislation which banned the use of these fabrics in children's nightwear, few such accidents have been reported.

Studies of complete garments have shown that some styles encourage the burning process, while others slow it down. Trims and frills can make a great difference.

Low (left) and high risk garment designs (right). The looser the garment construction and fabric weave, the more air is made available to feed the fire.

TRY YOUR HAND

Go to a department store and check the children's nightwear for sale.

Do all the garments comply with the regulations?

Are all the garments labelled correctly?

What is the current wording on Category 2 garments? Why do you think the words 'styled to reduce fire danger' are proposed instead?

If possible, examine the garments for handle, and check their prices. How do garments in the different categories compare?

Flammability testing

The flammability of fabrics and garments depends to a great extent on how much air is allowed to come into contact with the flame during burning.

A fabric which is loosely constructed will always burn more readily than a tightly woven fabric made from highly twisted yarns. A strip of fabric hanging vertically will propagate a flame more easily than one laid flat or at a 45°

angle. Therefore, to test fabrics for different applications, different kinds of flammability tests have been devised. For example, carpets and furnishings are tested by placing a red hot piece of metal on them, to imitate the conditions that occur if a glowing cigarette is dropped.

Different fabrics burn in different ways – some melt, some burn rapidly, some give off toxic fumes, others are self-extinguishing. The following activity will help you examine the burning behaviour of different materials.

TRY YOUR HAND

You will need:
2 retort stands
2 clamps
a rod
some alligator clips
heat-resistant mat
stopwatch
matches
fabric samples 5 cm × 20 cm: include cotton voile, cotton denim, towelling, nylon tricot, acrylic knit, woollen, worsted, polyester, brushed nylon, flannelette, vyella, and any other fabric of your choice. You could also investigate a strip of foam, PVC, and other furnishing materials

What you do:
Set up your equipment as shown.

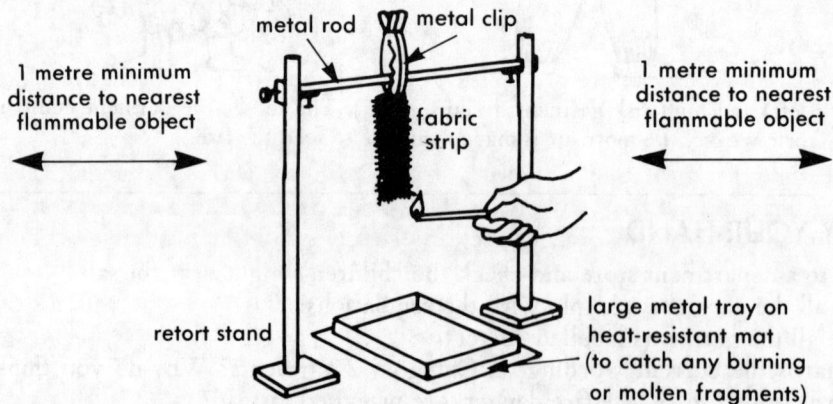

metal rod   metal clip

1 metre minimum distance to nearest flammable object

fabric strip

1 metre minimum distance to nearest flammable object

retort stand

large metal tray on heat-resistant mat (to catch any burning or molten fragments)

Apply the flame to the lower right hand corner of the sample for 2 seconds exactly.
Time how long it takes for the flame to reach the 10 cm mark on the fabric.
Observe the sample as it burns.
1   How quickly does the flame travel?
2   Does any molten fabric drip down?
3   How much smoke and fumes are produced?
4   What factors affect the rate of burning? Consider fabric surface, fabric weight, and fibre composition.

5   What are typical applications of the fabrics you have tested? Do you consider
     that their flammability performance, as found in this exercise, is appropriate
     to those uses?

Imitating nature
Silk

Silk has always been a high-priced, luxury fibre. The drive to produce an artificial
silk began in the nineteenth century, and every synthetic fibre, no matter what
its current use, began as an attempt to make a better artificial silk.

The first 'art silk' was rayon. Tricot rayon filaments are still used in swami
for lustrous silky underwear. Although rayon is more absorbent than silk, it
creases awfully. Other fabrics which were once made of silk, and where rayon
is now often used as a substitute, are taffeta, shantung, lining fabrics, and brocades.

Although silk stockings were also a most desirable – and expensive – fashion
item, rayon was not strong enough to be a substitute in this application. When
nylon burst onto the market after the Second World War, silk was completely
displaced in hosiery, underwear, and nightwear. Other traditional silk fabrics
now made from nylon are satin, organza, and chiffon.

Polyesters and *acrylics,* as partly delustered, fine filaments, have also
encroached on the silk market. Polyester has been most successful in chiffons
and georgettes: light-weight, silk-like fabrics for shirts and dresses. Orlon 81,
an acrylic filament especially engineered to imitate silk, was pushed out of
the market by the success of polyester. Its production has now been discontinued.

Leather and suede
Leather

Real leather and suede are products of the same industry – prepared from tanned
hides by different methods of finishing. To copy them the synthetic textile
industry had to resort to various manufacturing techniques.

Leathers are imitated by fabrics coated with a layer of polymer. The base
fabric may be knitted, woven or nonwoven. It is usually made from staple
yarns which hold the coating securely. Early imitation leathers were just a
thin layer of PVC spread on woven fabric, and lacked the convincing handle
of modern artificial leathers. The coatings most often used today are expanded
vinyl or polyurethane. Both these materials feel soft and pliable because of
the air bubbles they contain.

The surface of simulated leathers can be embossed to imitate a wide variety
of leather grains. Such materials are used for upholstery, car interiors, bags,
shoes, accessories and fabric trims.

TRY YOUR HAND

Collect a variety of vinyl and leather products. If they are expendable, cut
a thin slice from them and examine the cross-section under the microscope
or with a magnifying glass. Use the chlorine and nitrogen tests (p. 196) to
decide whether the coating is PVC or polyurethane.

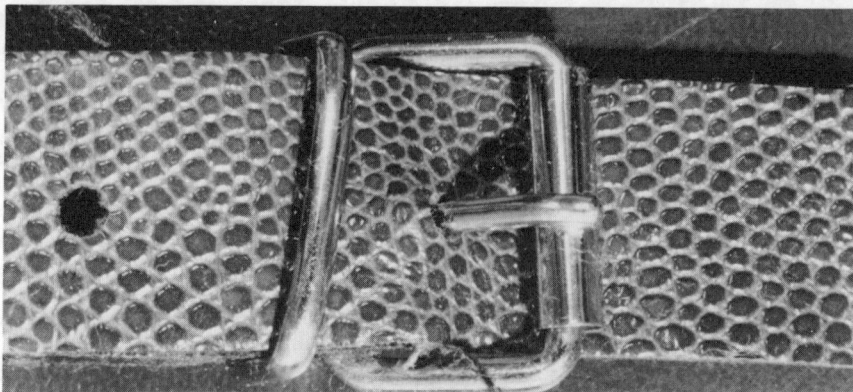

Embossed vinyls can be crude imitations of leather, as here, or almost indistinguishable from the real thing.

TRY YOUR HAND

Make a trip to a shoeshop and a handbag shop. Compare the look, feel, smell and price of similar items made from real and simulated leather. Can you identify which is real and which is imitation? What would be your reasons for and against buying each?

Imitation leathers can be kept clean easily with soap and water, except for a few stains – ink, dye – which are almost impossible to remove. Vinyl-coated fabrics must never by dry-cleaned. This is because the vinyl is kept soft by special solvent-soluble agents. Dry-cleaning can remove these softeners and the coating is likely to shrink and crack.

Real leather is not thermoplastic. Also, it is rarely bonded to fabrics – though this is done in some footwear applications.

The major difference between real leather and vinyl or urethane imitations is that real leather 'breathes'. It is protein and therefore hydrophilic – hence it cannot be wet-cleaned – and it allows the passage of moisture through. Shoes made from real leather are less likely to give owners sweaty feet than are vinyl shoes. For other applications, this property is less important.

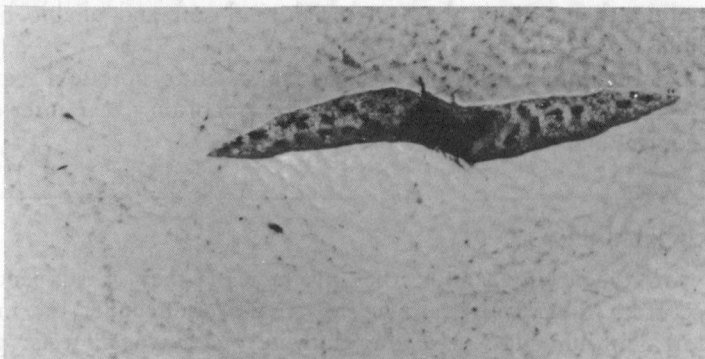

Shrinkage and cracking was a problem particularly with the early vinyl coatings.

494

Suede

Real suede is made from split hide brushed to give an even, fuzzy surface.

The first fabrics to imitate suede were woven flocked fabrics. Although the surface appeared similar to suede, the drape and handle – because of the adhesives used to stick the flock to the fabrics – were very different from the real thing. Recently, adhesives with much improved flexibility and durability have been developed.

Other approaches to simulate suede have included applying a layer of expanded polyurethane to a knitted or woven fabric, and then brushing its surface. This method gives a good suede-like handle and surface, but with abrasion during wear the coating is liable to come off the backing – especially at the collar or cuffs, where the oil content of the skin weakens the polyurethane.

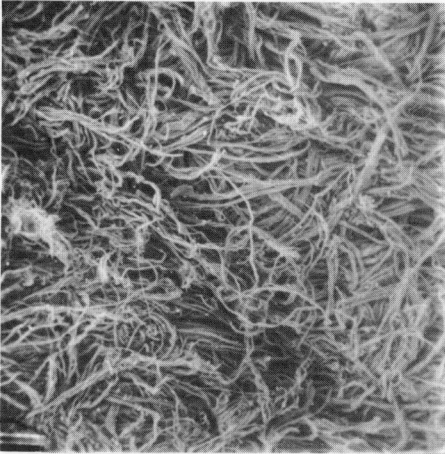

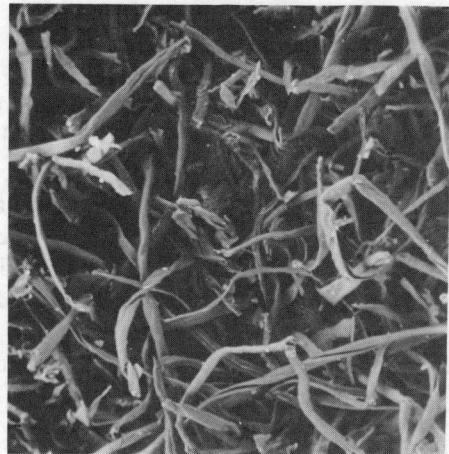

Surface of genuine (left) and flocked imitation suede (right), at the same magnification. (Vivian Robinson)

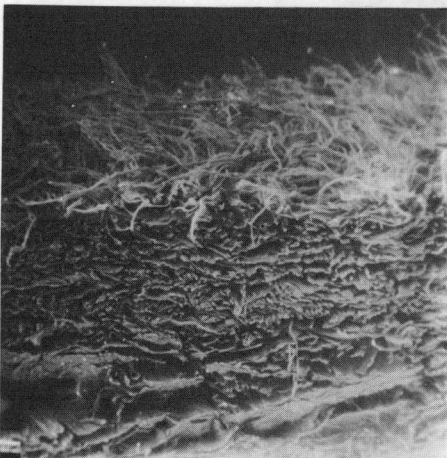

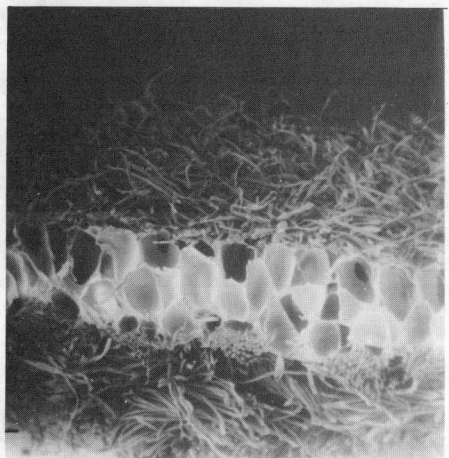

Cross-sections of genuine (left) and synthetic suede (right). (Vivian Robinson)

Washable suede is a nonwoven fabric made from nylon and polyester which is impregnated with polyurethane for a resilient handle. In order to make the fabric 'breathe', the nylon filaments are burnt out with hydrochloric acid to leave narrow channels for air and moisture. Such hydrophobic fabrics are easy to wash and dry, and retain their appearance for a long time. Their main disadvantage is an affinity for oily stains, though real suede stains far more easily. Washable suede is marketed under the Ultrasuede and Amara labels.

Other suede-like fabrics are made by using abrasive rollers to brush the floats on the back of specially constructed polyester warp-knit fabrics. These fabrics are rather lighter than real suede, but they retain their appearance and are fully washable.

Furs

Awareness of the need for wild-life conservation, and the resulting limits or bans on the hunting and trapping of wild animals, have limited the supply of real furs. Furs are not only becoming more and more expensive, they are also becoming an undesirable status symbol to many people.

The appearance of mink, persian lamb, or sheepskin can be imitated quite readily – screen printing can reproduce the leopard's spots or the tiger's stripes.

Although the natural variability, handle and lustre of authentic pelts cannot be perfectly imitated, modern technology has devised fur fabrics which have other advantages over the genuine article.

Many different techniques are used to produce fake furs. They include pile weaving, laid-in yarns, tufting, and sliver knitting. The yarns used may be filaments textured by the knit-deknit process for persian lamb, acrylic slivers for synthetic sheepskin, or blended from high- and low-shrinkage modacrylic fibres for the soft undercoat and sleek guard hairs of fake mink.

Soft toys are an important market for fur-like fabrics.
(Dale Mann/Retrospect)

The advantages of synthetic pile fabrics over real furs are their lightness, their resistance to insect attack, and their washability. Real fur coats – though less comfortable because of their greater weight – are windproof as well as being better insulators.

Much of the pile fabric produced today is used not only for apparel but also for soft toys and furnishing.

Silver and gold

Because clothing is so much more than just protection from the elements, craftsmen have for centuries embellished clothes with fine drawn or hammered wires of silver and gold. Such garments were very expensive, and were heavy from the weight of metal worked into them. Eventually, the silver tarnished, though gold threads retained their magnificence long after the royal garment was worn pale and ragged.

Silver and gold are no longer exclusive to royalty. Light-weight threads with a metallic appearance are available in splendid colours which never tarnish. They are no longer restricted to ceremonial occasions – their use is limited only by the vagaries of fashion.

Metallic yarns such as Lurex are made from extruded sheets of polyester. These, because of their sharp edges, are rather uncomfortable to wear. For many applications the metallic yarn is twisted with other softer threads which allow it to gleam through the fabric structure at intervals.

TRY YOUR HAND

Collect a variety of fabrics which contain metallic threads and investigate their structure under the binocular microscope. Discuss how many different ways such yarns can be used in fabrics.

What is the effect of twist?

For a solid metallic effect, is the fabric woven or knitted?

Where the Lurex yarns are untwisted would you expect them to be the warp or weft of the fabric?

Use a magnifying glass to investigate metallic embroidery threads and metallic braids. How is the gilded effect achieved in each?

Invisible textiles

Dress and shoe linings

Anyone familiar with dressmaking will be aware that many notions go into the making of a garment, in addition to the original fabric. Lining, interlining, and interfacing are all fabrics which are not seen yet which show their presence by the drape and shape of the garment.

These invisible textiles are used to support or stiffen the surface fabric. Some linings are needed to prevent transparency, or to provide a pleasant, smooth surface next to the skin.

The amount of stiffening required in interlinings depends both on the style of the garment and on the nature of the visible fabric used. It is generally recommended that a soft fabric have soft lining and interlining; stiffer interlining should be used for heavier materials. The invisible fabrics may be woven, knitted, or nonwoven. Whichever is used, its laundering properties must be compatible with those of the main fabric of the garment.

In order to choose the correct lining, it is best to drape the lining, covered with the top fabric, over the back of the hand. This will show the final effect.

Nonwoven fabrics, even if thin, are rather inelastic. For interlining garments made of knits, nonwovens have slits cut into them which allow them to stretch and move with the main fabric of the garment.

Fusible interlinings are coated or printed on one side with a low melting-point polymer. When heat is applied, the polymer melts and adheres to the top fabric. When fusing the two fabrics together, it is important not to move them relative to each other. This is best achieved by *pressing* rather than using a smoothing action with the iron. If the fabrics are imperfectly joined, unsightly bubbling may form which cannot be corrected later. This sometimes happens in shirt collars and is a fault in their manufacture.

In shoes, the lining of the upper and the interlining of the sole can often determine comfort in wear, though mostly the consumer chooses according to the outer designs.

Interlinings and linings for shoes are mainly made from needle-punched, random-laid, or spun-bonded materials. These allow water vapour to pass through, but give good dimensional stability to the shoe. Occasionally, the nonwoven base fabric – usually made of polypropylene, on account of its lower cost and greater rigidity – is impregnated with polyurethane foam. This gives the fabric a suede-like texture while still allowing for permeability. For insoles and other leather-like effects, the impregnated nonwoven may have a further coating of polyurethane on the surface.

Invisible textiles can make dramatic differences in the performance of a garment. (Vilene Australia)

TRY YOUR HAND

Take some old shoes and cut them apart. Investigate component fabrics under the microscope. How many different materials are used in each? What kinds of fabrics did you find? What were their roles? What was written on the sole or inside of the shoes to describe their content?

Threads

TRY YOUR HAND

Collect as many different kinds of sewing and embroidery threads as you can. Carefully note any advertising or other information that is available for each at the point of sale. Now note the information that is written on each spool or hank.

What do the numbers mean?

Why is the fibre content important?

How should the size of the needle in the sewing machine relate to the thickness of the thread?

Unravel each thread, and note its structure and composition. Look carefully at each under the microscope.

What is the reason for a plied structure?

When are core yarns used?

How do elastic sewing threads differ from ordinary threads?

Are all sewing threads made from staple fibres?

What would happen if the sewing thread had a very slippery surface?

What if it was uneven and did not slide smoothly through the needle or fabric during sewing?

What is the difference in evenness and in strength between cheap and expensive threads?

How do such differences affect results of the sewing?

In order to ensure evenness, all threads are manufactured as a number of singles yarns doubled. The plying twist is in the opposite direction from the twist in the singles yarn, so as to produce a balanced twist structure in the final thread. If the twist is not balanced, the thread will cause problems by snarling and twisting during sewing.

The plied structure stops the staple fibre ends from sticking out of the yarn. The smoother the yarn surface, the easier it is to sew with.

All sewing threads made from cotton are mercerized – this increases both their strength and their sheen. In addition, they are finished with a coating of protective lubricant. The lubricant helps the thread pass smoothly through the eye of the sewing needle. It must be stable at high temperatures, because friction between needle and thread in high-speed sewing can be considerable – sometimes enough to melt thermoplastic threads! If such threads are effectively coated with lubricant, the friction – and therefore the heat generated – is less, and sewing becomes trouble-free. Samples of all sewing threads are tested at very high speeds as a quality control measure.

Industrial textiles

Filters

Filters are important for many industrial applications. Some are used to provide clean, dust-free air for processes, others are in chimneys to make sure that no harmful products are spewed into the atmosphere. Filters for vacuum lines, air-conditioners and foundry exhausts all have to remove fine particles from air. They need to be permeable to air, and they should have a slippery surface to allow the filtered particles to slide off the filter and not clog it. Frequent shakedowns help this process of self-cleaning. If the filter is designed to remove fine particles it can have only small air spaces – this however slows down the flow of air. For large volume or high-speed filtration, large filter bags are needed.

TRY YOUR HAND

Repeat the wind resistance investigation on p. 483 using as large a variety of nonwoven fabrics as you can find.
Which allows for the best flow of air? Which would trap the most dust?
Which would last longest in use?
Check the filters in home appliances, such as range hoods, air-conditioners, vacuum cleaners, the family car. How do they differ? Why?

For many air-cleaning applications in foundries or in chimney stacks, the filters need to be resistant to acids and to organic molecules. For this reason, spun-bonded polypropylene is rapidly gaining popularity. Other filter materials are made of cellulose, polyester, acrylic, or viscose, by random-laying, heat-bonding, or needle-punching. Polyurethane foam is also used for some air-conditioning applications.

Filters for liquid media have different requirements. The filter in a car, for example, is expected to trap the particles of dirt in the oil. When the filter is full, it needs to be removed and discarded. Filters for liquids tend to be thicker than air filters, and are often made from needle-punched nonwovens.

Tyrecords and belts

Tyres need to withstand both lengthwise and sideways distortion – stresses which change rapidly and which often occur at high temperatures. If tyres were made of vulcanized rubber alone, they would not be strong enough to keep their shape and withstand the rigours of braking or cornering at high speeds.

In order to increase the dimensional stability of the tyre, a web of cord or strong woven fabric is embedded between the layers of vulcanized rubber. This fabric may be made from high-strength rayon, from nylon, or even from fine steel.

Another application of textiles where strength is of great importance is industrial belting. Belting made from double and triple woven narrow fabric is used for seatbelts in cars. Its use is gradually spreading to replace the heavier ropes and steel chains for cargo handling.

The advantage of using belts rather than ropes for cargo handling is that a knotted join in a rope reduces its strength, whereas the seam used to join two parts of a webbing belt has hardly any effect on strength. Steel chains tend to rust, and are both heavy and expensive compared to textile webbing.

The belts are made from thick multifilament yarns of polyester, nylon or polypropylene.

Ropes and nets

In spite of the growing popularity of belts for cargo handling, ropes are still holding their own for marine applications. Originally ropes were made with a corded, plied structure from strong natural fibres such as manila, hemp, and sisal.

The strength of ropes can be of vital importance.

Ropes made from Kevlar are replacing steel in many applications. They are light, strong, and do not corrode. At the same breakstrength, Kevlar is one-fifth the weight of steel. (Du Pont)

501

The materials mostly used for ropes today are not rotted by water or by micro-organisms, and are generally lighter and stronger than natural fibres. Polypropylene ropes even float on water, so they can't be lost at sea – though they can cause problems by getting tangled in ships' propellers!

Nylon or polyester filaments, each about 25–30 microns diameter, are twisted or braided into ropes. Polypropylene ropes are made from fibrillated film directly twisted into rope form, or from staple fibres cut from extruded monofilaments. Nylon ropes are more extensible than polyester or polypropylene; they can absorb the shock of a sudden heavy load by stretching.

Four different types of rope construction are used today. The traditional three-strand rope is twisted from three twisted strands of yarn. A squareline rope is braided from eight strands, four of which have S and four Z twists. Because the final structure is not twisted, the load distribution is particularly even, and the squareline is less likely than the three-strand rope to be distorted during careless storage.

Qualities of synthetic ropes

Fibre	Strength	Extensibility
nylon	highest	great
polyester	medium	medium
polypropylene	lowest	lowest
Construction		
three-strand	medium	medium
squareline	medium	greatest
parafil	lowest	least
braidline	highest	very low

Cargo-handling nets are still widely used in smaller ports and for carrying loads beneath helicopters.
(Department of Agriculture and Rural Affairs, Victoria)

Fibre type and construction of fishing nets depends on whether they are to be used in rivers, as in traditional Thailand, or for deep-sea trawling.
(Anna Janca)

Parafil is a fairly heavy and inextensible construction – the sheath adds only weight, not strength, to the rope. It is mostly used to anchor fixed structures: it competes successfully with wire ropes for aerial stays.

Braidline is the strongest construction and has low extensibility. It is used in diameters of 300 mm as tow ropes for supermovers. Such ropes have a stable structure, are easily spliced, and can carry 1500 tonnes of load.

Nets are made from similar raw materials. They are knotted into large and intricate structures which vary according to final application. Some fishing nets are required to float, others to sink and dredge the ocean floor. The diameter of the yarns and filaments, as well as the size of the holes in the net, are determined by the type of fishing for which the nets will be used. Nets are also used for cargo handling, for example to support loads transported by helicopter.

Geotextiles

Roads, railway lines, dams and embankments need to be well drained in order to withstand heavy loads under wet conditions. The wear on roads depends not only on how many vehicles pass over it, but also on how heavy they are – a single 8 tonne truck is equivalent to 30 000 one-tonne cars! If a road gets waterlogged it cannot support the weight of the vehicles passing over it, and the surface breaks into potholes.

Traditionally, the drainage for such applications has been achieved by layers of carefully graded stones and gravel. The spaces between the packed stones allow the water to flow away.

Finely graded gravel is expensive and often difficult to obtain in the right qualities. If unsuitable stones are chosen and layered, they may sink into the subsoil or be eroded away from the embankment. *Textiles* can act as excellent filters. If they are used as a barrier between the soil and the top layers of gravel, they help to distribute the load and also save on the amount of fine-grade gravel needed. Fabrics which are used for such construction projects are called GEOTEXTILES.

Use of a geotextile in road-making. Here, Du Pont's Typar polypropylene nonwoven is employed.
(Du Pont)

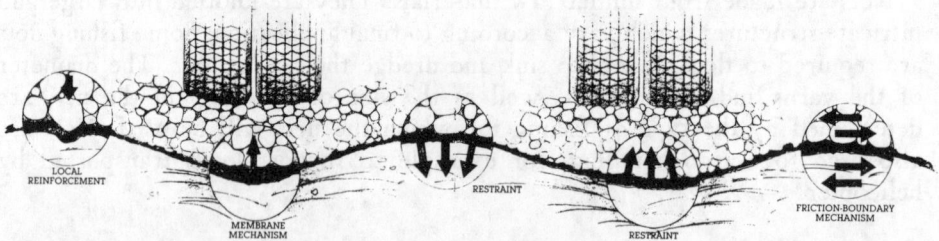

Load-supporting mechanism of a geotextile.
(Du Pont)

Geotextiles may be woven or nonwoven fabrics. They are mostly made from low cost, rot-resistant polypropylene. Generally, they are thick and stiff constructions made from spun-bonded, fuse-bonded, or needle-punched fibres.

Geotextiles are frequently used today as a cost-efficient solution to building roads, drains, dams and embankments. Unsealed or sealed access roads can be built both faster and cheaper, even in tropical climates, than by conventional methods. The geotextiles separate the subsoil from the road surface, allow water to drain away quickly, and distribute loads evenly.

Geotextiles have even been used to reclaim land from the sea – the sludge of water and sand is pumped behind barriers constructed of stones and geotextiles. The water drains away through the barrier, leaving the soil behind in the area being reclaimed.

504

Car interiors

Textiles are used in cars in many different ways – in the tyres as reinforcement, under the bonnet as filters for air and oil, in the fan belt, in rubber-coated hoses, as insulation between the leads in electrical wiring.

TRY YOUR HAND

Next time you sit in a car, make a list of all the different textiles that are used in its interior. Don't forget window seals or plastic-coated trim. Look in the boot too! What were the different construction methods needed to produce all these fabrics?

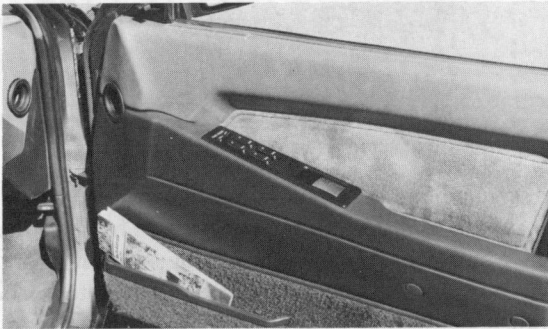

Mitsubishi Magna Elite. Where else in the car can you find textiles used? (Mitsubishi Australia)

The earliest cars had leather upholstery, after the style of the elegant carriages which preceded them. Leather was replaced by durable vinyl seating in the 1950s. Current trends have been to fabric upholstery with comfortable loop or velvet pile, with flat woven fabrics being introduced as the next fashion direction. The following table shows the use of different fibres and constructions in Western European cars in 1985.

Car upholstery usage, Europe, 1985 (tonnes)

Construction	Fibre				Percentage share
	Wool	Nylon	Acrylic	Polyester	
Woven velour	1180	460	700	110	11
Flat woven	450	1600	500	8600	51
Double raschel	110	–	1660	2470	20
Knit	–	–	–	3960	18
Percentage share	8	9	13	70	100

Textile fabrics have advantages over coated fabrics for car upholstery because of their greater ability to transmit water vapour and because their lower thermal conductivity gives a more pleasant 'feel' in hot and cold conditions. They are also more durable: the coated fabrics had a tendency to crack and peel along crease lines. Wool, because of its excellent comfort properties, is gaining over acrylics. It has proved itself in public transport as a serviceable upholstery fibre with excellent appearance retention.

Fibres and constructions other than the traditional poly/cotton stretch terry are now being explored for stretch carseat covers.

Textiles in medicine

Textile fabrics perform many different roles in the field of medicine.

The most spectacular applications are in surgery and prostheses, where precision-knitted tubes have been used as artificial arteries and heart valves, and where fabrics form an important part of artificial limbs and orthopaedic supports. Bandages, sutures and swabs of many different kinds are used for surgery and after care.

Woven sheets, blankets, drapes and other textiles in general hospital use need to be kept as free from infection as possible. Exacting laundry procedures are essential, often using strong bleaches, alkali, and high temperatures. Such harsh – if hygienic – treatment of hospital sheets rapidly degrades the fabrics.

Reusable nonwoven sheeting of lower cost is gradually gaining the attention of hospital administrators. Disposable, single-use sheets have been found cost effective mainly in operating theatres rather than general hospital use. In the operating theatre, where risk of wound infection remains a concern, low-cost nonwovens are used in disposable surgical drapes, gowns and face masks. This has reduced the possibility of cross-infection from one surgical case to another.

Various absorbent nonwovens form comfortable wet sheets and incontinence pads, and help reduce the need for laundering. Such pads are usually absorbent air-laid expanded cellulose contained between a backing of impermeable polyethylene film and a permeable, soft, hydrophobic nonwoven face fabric of heat-bonded polyester or adhesive-bonded viscose. The structure of feminine hygiene pads and disposable baby nappies is basically the same. Such pads – because of the hydrophobic surface layer and the super-absorbent filling – actually keep the surface dryer than do conventional fabric nappies.

Although not of direct application in medicine, floor coverings have an important role to play in hospitals. Carpets are often passive transmitters of infection, as soles of shoes may carry bacteria from one area to another. If conditions are suitable for the multiplication of these bacteria, there is real danger of epidemics through the hospital. Many modern carpets for hospital use are treated with biostats – chemicals which, while not killing all bacteria, prevent their multiplication. Mats made from layers of nonwoven fabrics impregnated with stronger antiseptics can be useful in doorways to prevent the carrying of infection from one area to another.

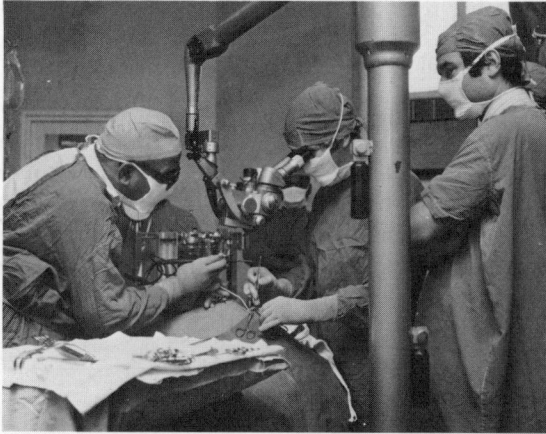

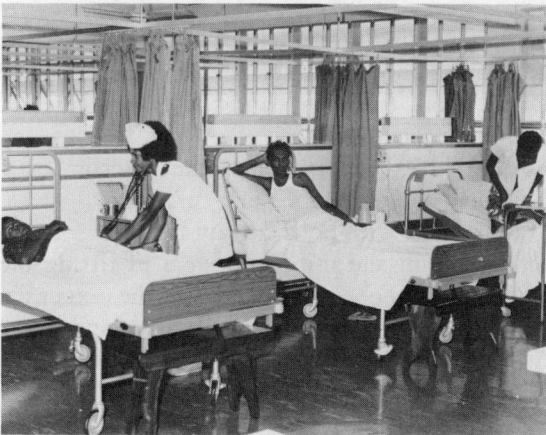

Textiles used in hospitals must be easily sterilized and able to withstand repeated laundering.
(Public Relations Office, Fiji)

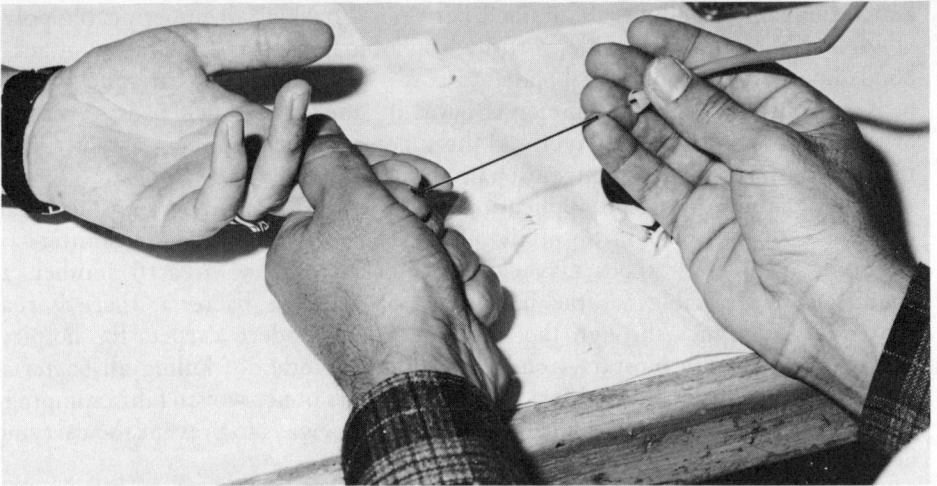

In some areas, disposable textiles are an economical solution, but while it may be reasonable to discard swabs, disposable sheets become very expensive.
(Robert Madden, Audiovisual Services, Royal Women's Hospital)

Carpets and other furnishings

While textiles form an important part of our daily environment, they seldom involve the consumer in major financial outlay. Expenditure on fabrics mostly involves small amounts of money, except where the purchase of furnishings is concerned. Curtains, carpets, and furniture covers not only need to look attractive but, because of their high cost, they need to provide satisfactory service for a long time.

TRY YOUR HAND

You have a flat which you wish to furnish, and you have a choice of carpets and curtain fabrics.

The cheapest broadloom carpet that you like costs $80 a metre; the most expensive is $190 a metre. You will not even consider the carpet at $240 a metre! You have two rooms and a hall to carpet, and the carpet layer says you will need 10 metres of broadloom carpet.

The curtains could be $4.50 a metre for a cheap scrim, or $12.99 for a more elegant fabric. You need 35 metres for your curtains.

Both carpets will look good with both curtain fabrics.

1   Calculate the maximum and minimum costs of fitting the flat with carpet and curtains. What is the cost if you buy the expensive carpet and the cheap curtains? What if you buy the cheap carpet and the expensive curtains?
2   What reasons can you think of for the differences in the prices?
3   Where could you find information on differences in quality between the items?
4   What factors other than cost should you consider when you choose which to buy? ('Quality' is too vague: list specific aspects of performance, and consider your own special requirements.)

Carpets

Many factors are involved in estimating the serviceability of carpets. A consumer – though willing to spend a considerable amount of time and energy – cannot become an expert at evaluating each carpet style.

There are many tests available for carpet performance. Recovery from static loading, pilling test, abrasion test, light fastness, trapped-turf adhesion, and flammability are some of the tests routinely carried out on carpet samples. Regrettably, none of these tests shows exactly how a carpet will behave when actually in use. Some predictions can be made, of course, but it is very difficult to compare the expected performance of two carpets that are made from different fibres and by different techniques.

The Australian carpet mark was developed in order to help the consumer make decisions. The Carpet Manufacturers' Federation of Australia has set up a scheme which gives carpets a grade from 1 to 5: Grade 1 is suitable only for light domestic use; Grade 5 is for heavy contract use such as offices and hotel corridors. A separate luxury classification caters for those carpets that are expensive but may not withstand heavy wear. The grades are given to carpets according to the verdict of a panel of experts who take into consideration factors such as pile fibre, weight, density, and construction.

The International Wool Secretariat has worked out a different grading scheme, but theirs is valid only for pure wool or wool blend carpets. These carpets are graded by scientifically combining the results of 17 separate tests and measurements that are carried out on each carpet sample. The grades range from light duty to medium, heavy and extra heavy duty, as well as a separate 'decorative' category.

Although grading of carpets is not compulsory, more and more carpet manufacturers support the scheme in order to build consumer confidence. Often, however, low quality local carpets and imported lines are not registered for grading. In such cases it is buyer beware!

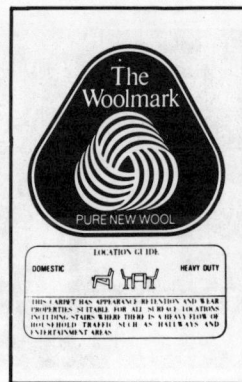

Extra Heavy Duty:

Qualifies for outstanding service in all heavy traffic areas, including stairs. Also suitable for use in some commercial offices and shops.

Heavy Duty:

Qualifies for use in hallways and entertainment areas where there's a heavy flow of household traffic. Suitable for stairs when indicated.

Informative labelling aids consumer choice.

TRY YOUR HAND

Obtain some samples of carpets. Note their gradings and price.

If you can, find out the WEIGHT – usually quoted in ounces per square yard – of the pile. If you cannot find the weight, carefully shave the pile from a 10 cm square piece, and measure the pile mass in grams. Measure the height of the pile above the backing fabric. Calculate pile density from the approximation

$$\text{pile density} = \frac{\text{mass of pile}}{\text{pile height}}$$

Tabulate your results, listing carpet name and construction (woven, tufted), price, pile weight, pile density, grading, and fibre type.

1 Do the carpets with the denser pile have the higher gradings?
2 Does the cost of the carpets correspond to their quality as indicated by the pile density and grading?
3 Does cost correspond to pile weight?
4 How do the costs compare for carpets made from different fibre types?
5 How do you explain the differences in price between your carpet samples?
6 Which carpet do you think would keep its looks the longest, and why?

Fibre content

Throughout history, carpets have been produced with pile yarns made of wool. Not every kind of wool fibre is suitable for carpeting, however. In order to withstand the flexing and abrasion dealt out to the carpet, the wool has to be particularly thick and strong. The wool yarns for today's carpets are blended from the fleece of sheep specially bred in the U.K., Pakistan, India and New Zealand. Most of the outstanding coarse wool of the sheep herded by Iran's nomadic tribes is used for hand-knotted rugs.

Wool has unparalleled resilience as a carpet fibre, and an ability to retain an 'as new' appearance. This involves both recovery from distortion – e.g. under furniture – and resistance to soiling. Wool, however, is far less abrasion resistant than some of the other fibres used in carpet piles.

A display of wool carpets. Wool and wool blends remain the first choice for resilience, appearance retention, and comfort for many carpet buyers.
(Australian Wool Corporation)

Fourth-generation nylon carpeting is designed to rival wool's resilience and is vastly superior to the first nylon carpets that appeared on the market.
(Fibremakers)

Nylon pioneered the way for synthetic fibres. Because of its great strength (which, actually is not a great advantage) and abrasion resistance, nylon carpets were made using a much lower weight of pile than was traditional for wool. These cheap carpets brought wall-to-wall carpeting within the budget of the average family. At the same time, however, nylon developed a reputation for being cheap and nasty. Today's nylon manufacturers are emphasizing that the so-called fourth-generation nylons are a far cry from the early nylon fibres used in carpeting.

A number of the coarser cellulosic fibres can be used in tough, hard-wearing floor coverings.
(The Natural Floor Covering Centre, Stanmore, N.S.W.)

Fourth-generation nylons have delustrants and specially designed cross-sectional shapes to scatter light and 'hide' soil optically. Some nylon filaments have been made to conduct electricity by the inclusion of carbon or other conductive materials in the melt. These electrically conductive filaments – usually round in cross-section – have been blended with the other nylon fibres in the multifilament which is twisted into the pile yarn for the carpet. In order to improve the rather poor soil resistance, the freshly extruded nylon filaments are coated with a fluorocarbon compound that repels dirt and stains.

While fourth-generation nylons perform far better than the early nylons, their cost is almost on a par with that of carpet wool. Not all nylon carpets are made from such expensive fibres, and hence it is rather difficult to generalize about the performance of nylon carpets.

Polyester is an attractive carpet fibre with a soft handle and good durability. It is less prone to static than other synthetics, but is difficult to clean of oil-based stains and dirt.

Polypropylene is mostly used in carpet backings, or as pile for indoor/outdoor carpets and artificial turf. It has excellent abrasion resistance but tends to flatten quickly, and it needs special protection against damage by sunshine.

The use of *acrylics* – which most resemble the look and feel of wool – has gradually declined over the years because of their ease of soiling.

Yarns

Carpet pile may be made from staple or from continuous filament yarns. Staple yarns made on the woollen system are used for velvet and Berber carpets. Those made on the worsted and semi-worsted systems make tight level-loop and high-twist pile carpets. Felted wool yarns are used for special tuft definition. Continuous filaments twisted into ply yarns are often used for velvet or other cut-loop styles where the 'synthetic' look is part of the carpet design. In the cheaper end of the range, continuous filaments are used for high/low loop carpets and kitchen carpets. Their only disadvantage is that if the edge of a stiletto heel gets caught in the pile, the owner of the shoe can be tripped up by the strong filaments.

Which yarns are the best for carpet durability? Those that hold the fibres tight. Hence, high twist and felted yarns look better longer than do the loose woollen yarns that produce such an excellent 'bloom' in level-pile velvets. High twist yarns have more resilience than soft twist yarns. Loop pile carpets are less likely to flatten than are cut pile carpets.

In carved or high/low designs the pile will tend to lean over from the edges of the high areas.

Finishes

Wool carpets release dust and soil readily, but they need to be treated to protect them against attack by carpet beetles and the larvae of clothes moths. The insecticides used are Eulon, Mitin and Perigen, which all form a reasonably long-lived bond to the wool fibre.

Synthetics are not attacked by insects. However they need to be treated to repel dirt and stains. The finish for this purpose is usually a fluorocarbon coating of the pile fibres. Such a treatment – e.g. Scotchgard – is applied to the top of the pile after the carpet is dyed and backed. The coating repels soil effectively, but it rarely reaches all the way to the backing and it is often worn off during the life of the carpet. Where the finish wears off in the traffic lines, the carpet in fact looks dirtier in comparison to the rest of the (protected)

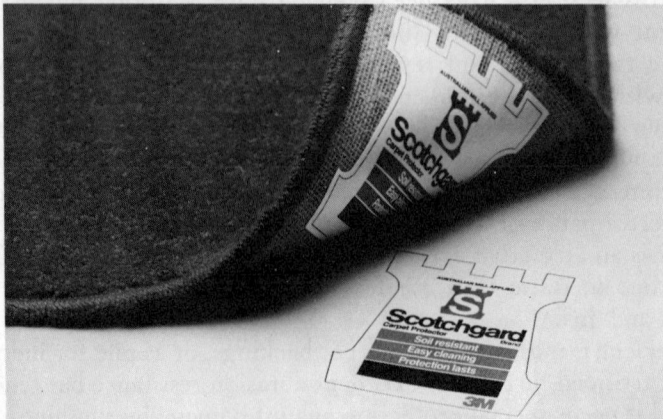

Soil-resistant finishes applied during manufacture provide lasting protection. (3M Australia)

512

pile. A somewhat more effective technique is to apply the fluorocarbon finish to the freshly extruded filaments at the end of melt spinning. This, followed by repeated heating processes, ensures a stronger bond between the coating and the fibre, and makes a more durable soil repellent finish.

Cleaning

The greatest enemy of carpets is dust and sand, which settle deep in the pile and cut the fibres as the weight of the traffic presses them down. Regular vacuuming, therefore, can actually lengthen the life of a carpet.

Spot cleaners and shampoos used on carpets must contain detergents which turn into powder when dry. This powder – with the attached dirt – can be neatly removed by vacuuming. If the detergent used on a carpet is not a powder, it may coat the fibres with a layer of sticky, hard-to-remove surfactant which clings to dirt and dust. After incorrect shampooing the carpet may soil more readily than before.

Even if correct methods are used, the first wet cleaning may cause some fibres to lose whatever resistance to soiling they may have had. Acrylic carpets have developed a poor reputation mainly because they are so easily soiled again after cleaning.

Staining is a separate problem. In general, hydrophobic fibres hold on to greasy or oily stains, while fruit juice and other natural colorants are harder to remove from natural fibres than from synthetics.

The best thing is to soak up any spilt liquids with absorbent towelling as soon as possible. Rather than rub the pile – and risk matting the fibres – it is best to throw several layers of towelling on the wet stain, and tread on them. The weight of the foot presses the cotton cloth into close contact with the liquid stain and allows the liquid to be readily absorbed.

Carpet laying

Much of the final satisfaction with a wall-to-wall carpet depends on correct laying and the use of suitable underfelt. During laying, care must be taken that the carpet is correctly cut, that it is neither stretched nor laid loose and wrinkled, and that patterns and pile directions are matched accurately at the joins. Carpet underlay must be chosen carefully. A soft, bouncy underlay, topped by a deep, plush carpet, may wobble uncomfortably under the heels of users. On the other hand, used with a low-pile carpet such an underlay can create a more luxurious impression, and even lengthen the life of the carpet by absorbing some of the impact of the footsteps.

Furnishing fabrics

In decorating their homes, people use the whole range of textiles available. Fabrics on walls, floors, canopied ceilings, seats, tablecovers – everywhere. Wovens, knits, laces, nonwovens – the products of all fabric manufacturing techniques – find their way into interior furnishings.

The three important considerations for most furnishing fabrics are abrasion resistance, light fastness, and stain resistance/soil repellancy. Comfort is important in upholstery fabrics. Which properties are most relevant, and therefore which textile is most appropriate, depends on the intended use.

TRY YOUR HAND

Select an application of furnishing fabrics for a case study, for example a lounge chair.

Find out what fabrics are used in this application – an upholsterer or manufacturer may be able to help.

Collect as many different samples of these fabrics as you can. Test them for light fastness, soil resistance, abrasion resistance.

Tabulate your results. How do the different fabric structures, fibre types, price ranges compare? How appropriate is each for the particular end use?

What other considerations may be important to the consumer choosing between these fabrics?

Textiles in art

Textile art is probably as old as the use of textiles. Exquisite tapestries, rugs, laces and embroideries of the last 2000 years form a treasured part of the collections in many art museums around the world. Jewellery, mosaics, paintings, sculpture and literature from more ancient times all indicate a rich and varied tradition of textile arts in many cultures. The tradition continues with the textile artists and craft workers of today.

In addition to art forms such as macrame, appliqué, patchwork, knitting, crochet, embroidery, tapestry, rug-making, lace and weaving, new avenues are opening up for the textile artist. Soft sculptures, knitted spaces, fabric collage, draping of buildings and interiors, masks, woven three-dimensional hangings all make use of the properties of textile materials to achieve artistic expression.

The textile artist is, of course, a consumer of textiles, and is as much concerned with the properties of textiles as is the manufacturer of industrial filters – although usually they are different properties. The commercial techniques of textile production and modification provide the textile artist with many of the raw materials; in many cases, these commercial techniques are derived from traditional practices still employed by textile artists.

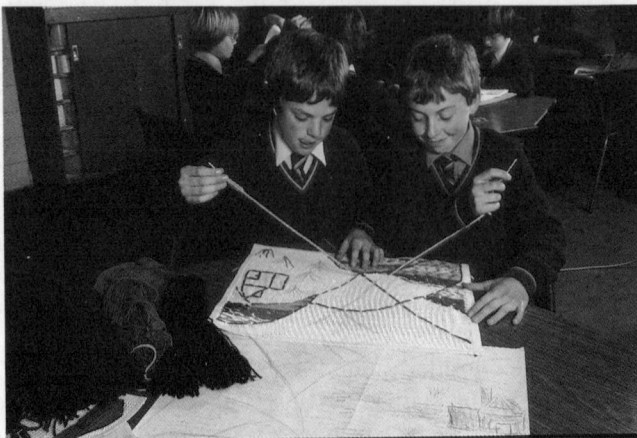

(Australian Wool Corporation)

514

TRY YOUR HAND

Before you consider the work of textile artists, think about how you relate to textiles. Your experience of, sensitivity to, and knowledge of textiles is a very personal thing.

Describe your favourite textile.

What words did you use? What senses are you employing as you characterize this textile?

Sight and touch are obvious. What about sound? Consider taffeta. And smell? Consider a greasy wool jumper, or the smell of wet linen.

When you described your favourite textile you did several things. You were able to describe the fabric because you had sensed differences between this fabric and others: you were able to discriminate, you had analysed certain characteristics of the fabric. Having done this, you were able to interpret this information: this fabric appeals to you, it is the one you like best.

Your past experiences influence your appreciation of textiles. If you have never heard taffeta, you cannot be expected to have feelings about its sound. Similarly, if you have never felt silk in all its forms, from fine smooth silk to rough raw silk, your appreciation of silk will not be the same as that of someone who has. Your values – ideas about what is important – and your cultural background can also influence appreciation. If you believe that animals should not be killed for their fur, this will influence your appreciation of fur coats. If you are Danish, you may be particularly fond of traditional Danish cross-stitch patterns (or you may hate them!). These influences, this process of describing, analysing, interpreting and judging, are important to your perception and appreciation of textiles.

What you know and feel about textiles is an important part of your appreciation of the work of textile artists and designers. You can look at a work of art and simply say that you do, or do not, like it; or you can apply some knowledge and skill to describe, analyse, interpret, and judge the work. And, of course, this process should take account of the artist: it should recognize that the artist would have wished to create a particular impression or satisfy some specific function.

TRY YOUR HAND

Choose a particular piece of textile art.

1 What is the function of the piece – what impression did the artist wish to create?
2 How has the artist achieved this purpose? What structural design and surface enrichment is employed? How have the principles of design been applied?
3 How do you respond to the piece? Has the artist succeeded in creating the desired impression? What attitudes, background, and experiences of yours influence your perception of the work?

Contact a local arts group and talk to the artists about their work. Find out what they do and why.

Use the questions above to help you examine their work. Explore the links their chosen art forms have with the past.

4   What is your own cultural heritage? What textile art forms form part of that heritage?

Visit a major art gallery or an exhibition devoted to textile art. Look for textiles in public buildings – tapestries, wall hangings, rugs.

5   What techniques have the artists used?
6   What is the contribution of the chosen fabrics and yarns to the total effect?
7   Are they appropriate? Why?

Fabric care

Caring for textiles appears at first sight to be a rather complex area, often best left to experts. If something cannot be thrown straight into the washing machine and come out perfect, it is regarded as too difficult to handle.

Many manufacturers cover against possible recriminations by instructing consumers to 'dry-clean only'. Professional dry-cleaners offer 'all care but no responsibility'. Most consumers who have cared for textiles have experienced disasters such as dyes bleeding, fabrics shrinking, linings pulling, garments wrinkling.

However, garments must, by law, carry detailed instructions for laundering and care. In spite of the bewildering variety of textile fibres and fabric structures there are some principles that can guide the consumer to care correctly for textiles.

Ease of care is an important consideration in clothing choice.
(David Beal)

516

Will it wash?
(Cumberland Newspapers, Victoria)

Many wool garments can now be washed with care, a development precipitated by competition from synthetics.
(C.S.I.R.O.)

Stain resistance and removal

Hydrophobic fibres repel water-soluble stains, but are readily stained by oily and other hydrophobic colorants. It is very difficult to remove such stains. Hydrophilic fibres on the other hand may be stained by both oily and water-soluble stains. Since hydrophilic fibres are swelled by water, many stains can be removed by either clean water or by detergent and water.

Dry-cleaning fluids can remove oily stains from fabrics because they act as solvents for the staining material. The same solvents however may not be effective for water-soluble stains. Commercial dry-cleaners always include some water and a little detergent in their 'dry' cleaning solvent blends, in order to attack both oily and water-soluble stains.

TRY YOUR HAND

Make a stain from red paprika dissolved in hot oil, cooled and blended with a little mayonnaise and with some instant coffee dissolved in a little hot water. In this emulsion, the oily component is red, the watery part is brown; the mayonnaise acts as an emulsifier.

Stain three samples each of a number of different fabrics – polyester, nylon, rayon, cotton, wool. Rub the stain in, and remove excess.

Wash one set of samples in warm water with detergent added.

Soak the second set of samples in dry-cleaning fluid.

The third set should be treated with a dry-cleaning spray – be sure to start spraying in a circle around the stain and then spray towards the centre – this will avoid any ring marks after drying. Brush powder off when dry.

Mount your dried, cleaned samples on a board and compare the effectiveness of the different stain removing processes.

Record the colour of the stain left on the fabrics by each of the processes.

Which fibre held most tenaciously onto the red? onto the brown?

Which method was most effective in removing all stains? oily stains? water stains?

There are special methods recommended for the removal of particular stains from particular fibres; and some stains simply won't come out! Not to mention that often the source of the stain is unknown. In such cases, it is important to start with the simplest method and progress to the more involved ones. Unless the fabric is likely to be affected by moisture (e.g. non-washable crepe), always try clean water on the unknown stain first!

An appendix at the end of the book lists recommended methods for the removal of particular stains from particular fibres.

Laundry practice

The consumer is easily overwhelmed by the huge variety of washing powders, laundry liquids, detergents, soaps, presoakers, water softeners, fabric softeners, enzyme soakers and special cleaning mixes available on the supermarket shelves. How to choose between them? Which one to use when?

TRY YOUR HAND

Check on the following properties of detergents – solubility, cleaning power, brightening power, value for money.

You will need:

a universal stain made from instant coffee powder dissolved in *minimum* amount of hot water, and mixed with real mayonnaise. The stain has a water-soluble dye (coffee), an oily component (oil in mayonnaise) and a protein component (egg).

9 beakers (500 mL)

7 bunsen burners and tripods, or a bain-marie on a stove to hold beakers

9 stirrers

thermometer
measuring spoon
dryer
iron
laundry marker

Fabric samples:
20 samples 10 cm × 10 cm of white cotton calico fabric
20 samples 10 cm × 10 cm of white polyester fabric
20 samples 10 cm × 10 cm of white wool fabric

Detergents:
1   cold-water detergent powder
2   hot-water detergent powder
3   soap flakes
4   liquid detergent
5   wool detergent
6   hair shampoo
7   bleach
8   enzyme soaker
9   water only

What you do:
First take one sample of each fabric and mark it UU (for unstained, unwashed) with the marking pen, and set aside as reference.

Next take 10 samples of each fabric and rub one drop of stain into the centre of each. Keep one stained sample of each as reference, marked SU for stained, unwashed.

Number the fabric samples U1 to U9 for unstained, and S1 to S9 for stained.

Number the beakers 1 to 9 with marker.

Prepare detergent solutions using 1 teaspoon of each number detergent to 400 mL of cold water. Stir, and note readiness to dissolve.

Heat detergent solutions to 55°C (hand hot). (Do not heat the bleach or soaker.) Note any undissolved detergent.

While the solutions are being heated, tabulate the recommended quantities and price of the detergents – calculate the cost for one wash load.

Into each beaker place samples, stained and unstained, of each fabric type. For bleach and soaker, soak for 1 hour, then rinse. Keep the 6 beakers of detergent and the beaker of water warm for 5 minutes and stir their contents. Then give the samples 10 'standard' squeezes each in their liquids, and rinse.

Dry the samples and iron smooth.

Tabulate the effects of each cleaning medium on the stained and unstained fabric samples.

1   Were any methods less effective in removing the stain than washing in water? the same? better? much better? completely effective?
2   Did the washing agents work with the same effectiveness on each fabric type?
3   Were all the residual stains the same in appearance?

4   Could you explain any of the results if you considered the affinities of hydrophobic stains to hydrophobic fibres, and of hydrophilic fibres to small dye molecules?
5   What other colour effects did you observe?
6   What do you think the effect of the soaker is? Look at the packet description for clues. Do soakers work on all kinds of stain?
7   Did all the detergents dissolve to give a clear solution? What do you think might cause any cloudiness? How do you think the solubility of the detergent changes its effectiveness?
8   Take the samples into a dark room, and illuminate them with black light. Did any of the detergents contain optical brightening agents?
9   Compare the washed unstained and unwashed unstained samples. Did any of the treatments affect the textile in any way? If so, could these effects be tolerated in laundry practice?
10 Find out what mechanics' overalls are made of. Is priority given to ease of washing, strength, abrasion resistance, or some other factor? What is used for uniforms worn in a fast food outlet? What fabric property is given priority? Explain why you agree or disagree with the fabric choices in each instance.

Action of detergents

Detergents are large, hydrophobic molecules with a charged or highly polar functional group attached.

The hydrophobic part of the molecule attaches itself to dirt, which is mostly made of hydrophobic particles or oily components. The hydrophilic part of the detergent is attracted to the water molecules, and helps to carry dirt away from the fibre and into the water.

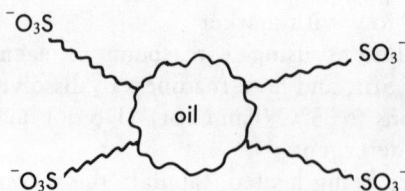

Different detergents have different compositions. Cold water detergents, for example, are slightly more soluble in cold than in hot water. The detergents themselves may be anionic (e.g. soap), cationic (e.g. fabric softener), or non-ionic. Cationic and anionic detergents cannot be used together – their charges would cancel each other out. Cationic detergents can also act as mild germicides. Non-ionic detergents are compatible with either anionics or cationics, but are all liquids which do not foam well in water.

Hard and soft water

Soaps are ionic detergents which are sodium salts of fatty acids, made by boiling mutton fat and other oils with alkali. Hard water contains calcium and magnesium ions. These can replace the soluble sodium in the salt and form *insoluble* calcium and other salts of the fatty acids. The insoluble salt separates as a white scum which floats on the surface of the water. The ionic soap has been effectively removed from its solution – it is no longer a soluble cleaning agent.

Functions of detergent additives

Ingredient	Function
Surfactant (e.g. soap or synthetic detergent)	Removes greasy dirt – detergency
Builder (e.g. Calgon or washing soda)	Removes hardness from water and improves detergency
Soil suspender (e.g. carboxymethyl cellulose, CMC)	Prevents dirt from being redeposited on clothes during the wash
Bleach (e.g. sodium perborate)	Whitens clothes in warm water
Optical Brightening Agent (OBA)	Makes fabric appear brighter and whiter
Filler (e.g. sodium sulfate)	Dilutes synthetic detergents and keeps the powder crisp and free-flowing
Foam stabilizer	Unnecessary, but added to synthetic detergents to give high foam preferred by consumers
Enzyme	Removes any protein stains
Dye, perfume	Makes product more attractive

TRY YOUR HAND

You will need:
some soap solution
4 beakers
a solution of calcium chloride
some Calgon or other water softener
egg beater (not to be used for eggs again!)

What you do:
Half fill three beakers with tap water, one with rain water or distilled water.
To the first two add the same amounts of calcium chloride. Add some Calgon to the second beaker, and equal amounts of soap solution to all four.
Use an egg beater to test the ability of the three solutions to produce foam.
What is the effect of the calcium on the ability of the soap to foam?
What role does the water softener play?
Is the soap more effective in rain water than in tap water?
Is the water supply in your area 'hard'?

The water softener in Calgon and other commercial preparations combines with the calcium and magnesium ions of hard water and removes them from solution. In this way the soap remains as effective as it was in soft water.

It is possible to wash effectively in hard water using synthetic *non-ionic* detergents instead of soap. Since these detergents do not form salts with calcium, they are not precipitated out in hard water.

Washing soda (sodium carbonate) is an alkali which can be added to hard water to precipitate the calcium ions as insoluble calcium carbonate. If soap is added after this reaction is complete it cannot form calcium salts, and acts as efficiently as in soft water. The *disadvantage* is that the sodium carbonate solution is quite strongly alkaline. Washing soda – although cheap and often used in commercial laundries – cannot be used with protein fibres.

Special-purpose products

TRY YOUR HAND

Make a survey of all the laundry agents available on your supermarket shelves. Note the instructions for use listed on each. Classify them into groups. Can you predict their composition from the information presented on the packages?

In addition to detergents, prewash *soakers* generally contain enzymes which attack any protein stains on the fabric. Naturally these must not be used on wool or silk fabrics.

Prewash sprays have solvents to attack oily, greasy stains.

Fabric softeners have cationic surfactants which act as mild germicides and which also leave a charge on the fibres. The charged fibres repel one another and so the dried fabric becomes fluffy and soft.

Bleaches are mainly chlorine types, unsuitable for wool or silk, or perborates which are safer for coloureds but not safe for wool.

Starches may be either synthetic or conventional. They are added to the wash or sprayed on before ironing. Both types form a coating on the fibres and create a temporary crisp finish on the fabric.

Professional care

Difficult-to-handle fabrics, luxury and formal garments, tailored clothes and heavy outerwear are usually entrusted to professional cleaners. The professional then has the responsibility to clean the garment, and to take all possible care not to damage it. If the garment is damaged, the professional cleaner may be held liable. Large dry-cleaning firms employ advocates to negotiate on claims for compensation from dissatisfied customers. Smaller firms find this too expensive, and so run the risk of punishing damages.

Ideally, of course, every garment has a proper care label and a label giving fibre content, and all the dry-cleaner needs to do is follow instructions. In practice, however, labelling is often inadequate. Garments made before labelling became compulsory will often carry insufficient information; labels on more recently bought garments may not comply with the legislation, or may be inaccurate; consumers may remove the labels for many reasons (they are often located where they will cause the most discomfort); home-made garments simply are not labelled, and many home dress-makers do not take sufficient note of fibre content or care requirements when they choose their fabrics, and so are unable to give the professional cleaner the necessary information. It is up to the consumer to insist that manufacturers provide relevant care instructions, and to refuse to buy articles which are not labelled properly.

Another concern of the professional cleaner is to ensure that employees are working under healthy conditions. Many dry-cleaning solvents are toxic, and care is needed in their use. Laundry services have to ensure that their effluent meets the requirements of local sewage authorities.

TRY YOUR HAND

Talk to a local dry-cleaner about the problems she or he encounters. What options are available to him or her to tackle them? Is your dry-cleaner an independent tradesperson, or an agent for a chain of dry-cleaners? How does this affect how these problems are approached?

Sewing seams

To be of use for garments, pieces of fabric have to be joined. The type and quality of the join or seam contributes greatly to the quality of, and consumer satisfaction with, the finished garment.

The sewing machine

Most sewing machines – whether industrial or home – employ the lockstitch mechanism. The seam is formed by two interlocked threads, one passing through the needle, the other supplied from the bobbin underneath. Ideally, the stitches should be perfectly balanced, but changes in yarn tension can cause various inperfections in the finished seam.

Different sewing machines have different controls. Some can move the needle from side to side and so do zigzag stitches; others are equipped with cam mechanisms for embroidery. All machines can have control over two very important features. One is the tension at which the top thread is fed into the needle. The other is the size of the stitches made: this is controlled by the motion of the feeding ratchet under the fabrics.

These two variables can affect the suitability of stitching for many different fabrics.

TRY YOUR HAND

Using high tension, then low tension, then small stitches, then large stitches, stitch four lines 5 cm long in a doubled piece of denim.
Now unravel the stitching, and measure the length of the top and bottom threads used.
1   Which line of stitching used the most thread? the least?
2   Were any of the seams balanced?
3   What setting would you suggest for a balanced seam in denim?
4   Which was fastest to sew? Which was slowest?

5   If you were a garment manufacturer, what considerations would affect your choice of machine settings for tension and stitch size?

The ideal seam

An ideal seam does not gather the fabric, nor is it so bulky that it distorts the garment shape. When the two parts of the fabric which it joins are pulled apart, the seam should not allow its threads to 'grin' through.

TRY YOUR HAND

You will need:

a sewing machine
several pairs of identical fabric pieces, each 10 cm × 10 cm

What you do:

Stitch the pairs of fabric pieces together with a 1 cm seam allowance, using different tensions or stitch lengths for each seam.
Pin the fabrics to wooden coat hanger, as shown.

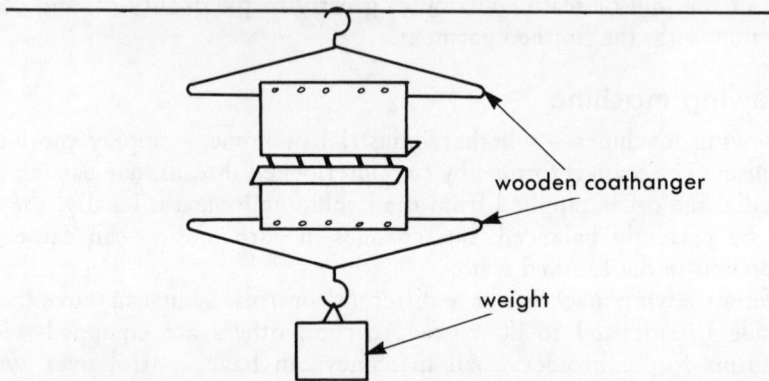

Hang on a hook and attach weights to the lower half.
Note at which loading the seam grin becomes noticeable, and at which loading it is over 1 mm wide.
1   How did yarn tension affect seam grin?
2   How did stitch length affect seam grin?
Repeat the exercise, using fabrics of a variety of thicknesses.
3   Would you recommend the same machine settings for all fabrics?
4   Which needed smaller stitches, sheer fabrics or thick fabrics?
5   Which needed higher tensions, sheer fabrics or thick fabrics?
6   What problems could occur with incorrect settings?
7   How could you ensure that you have chosen the correct stitch length and tension for a given fabric?

Sewing books recommend many effective and attractive seams and seam finishes. Not all these seams are suitable for all fabrics.

TRY YOUR HAND

Select four different seams or seam finishes, and try them on six different fabrics.
1   Are any of the seams too bulky for any of the fabrics? too flimsy, or unsuitable in any other way?
2   Did any of the finishes fray too easily? This will show most after machine washing.
3   Which seams and finishes appeared best for which fabric types?
4   Would the design of the garment – curved or straight seams – have any effect on which is the most suitable choice?

Needle and thread

The appearance of a seam can be affected by the choice of needle and thread.

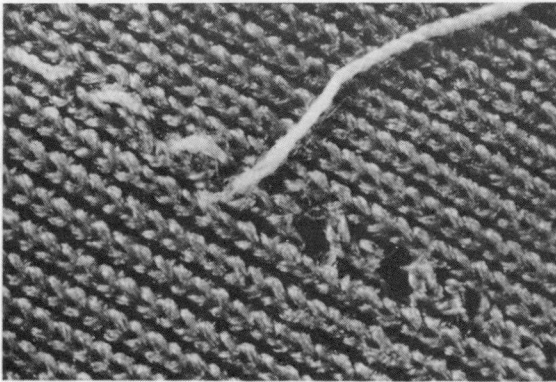

Damage caused by inappropriate needle choice may be concealed by the stitching. Here, the problem was a worn needle.
(J. and P. Coats Limited, Glasgow, Scotland)

TRY YOUR HAND

You will need:
a selection of fabric pieces of assorted weights and constructions: include some knits
a selection of different sewing machine needles: shape (diamond), ball point, thick, fine; and a sewing machine
fine and coarse yarn

What you do:
On each fabric make seams about 10 cm long, 2 cm apart, with each needle and yarn combination (eight seams per fabric).
1   How did the needles and threads affect the appearance of the seams?
2   Did any combinations cause thread breakage?
3   Unravel some stitching. Did the shape needle cut any yarns in the fabric?
4   Would any of the combinations create problems in sewing knits? fine fabrics? leathers and suedes? felts? velvets?

Seam pucker

Because stitching introduces an extra yarn into the fabric structure, all seams can be expected to produce some disturbance in the fabric. In closely woven fabrics this disturbance is often seen as unslightly puckering.

The pucker is worst when the seam is competing directly for space with the warp and weft threads, and is least when the seam is on the diagonal, or at least 15° off grain.

Other reasons for seams puckering can be excessive yarn tension, or the sewing thread shrinking after laundering. Care must be taken during sewing to make sure that the top and bottom threads are fed to the fabric at the same rates. If two different fabrics are being joined, the more extensible fabric should be placed underneath the other, to prevent pucker due to the fabric stretching during stitching.

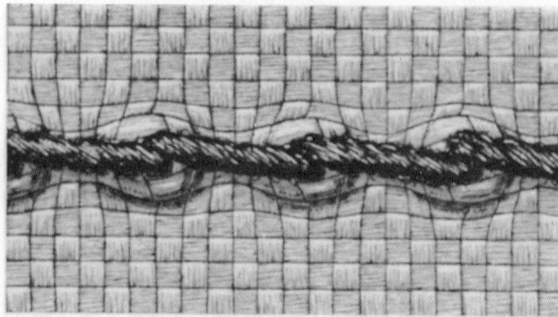

The cause of seam pucker: yarns in the fabric are displaced to make room for the sewing thread.
(J. and P. Coats Limited, Glasgow, Scotland)

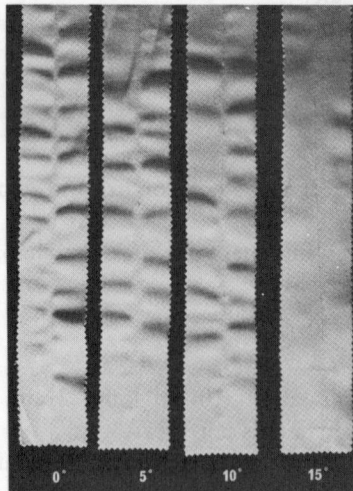

These samples indicate the advantage of stitching seams at a bias angle. The seam on the left was stitched parallel to the warp (0°), the other three at increasing angles.
(J. and P. Coats Limited, Glasgow, Scotland)

Consumer questions answered

Q1   Which is the warmest type of blanketing?

A   The warmest blankets are those which trap the greatest amount of still air for a given weight of blanket, and which are resilient enough to keep their original efficiency for a long time during use.

Pure down quilt fits these specifications much better than any other insulating material. It is followed by feather quilts, by polyester staple quilts and then by loose-structured double woven 'mohair' blanketing, acrylic pile fabrics, and so on.

Q2   Which is the best way to dress in cold weather – one thick jumper or two thin ones?

A   Weight for weight, two thin jumpers offer both greater flexibility and better insulation because of the layer of still air trapped between them.

Q3   Can thermal underwear cure rheumatism?

A   In a sense, all underwear is 'thermal'. Some types are especially produced to trap small pockets of air in a thin, flexible structure. Thermal underwear made from natural fibres, because of their absorbency and their permeability to water vapour, tends to be more comfortable for active wearers.

Sufferers from rheumatism feel more comfortable when the affected joints are kept warm. Underwear made from PVC is a particularly effective insulator because of the high thermal resistance of PVC fibres. However, no textile material can claim to be a cure for any illness.

Q4   Which are warmer – gloves or mittens?

A   Mittens are warmer than gloves, because they allow better blood circulation with greater freedom of movement for the fingers, and because the fingers can warm each other instead of each radiating heat to the cold air. Down mittens are the warmest, but even they need to be put on while the hands are warm. Once the extremities have cooled and the blood vessels have contracted to prevent heat loss from the body, it is very difficult to warm them again just by providing insulation.

Q5   What is the difference between sailcloth and the cloth sails are made of?

A   Sailcloth is a fairly tightly woven, fairly thick, slightly ribbed cotton fabric. It may well be like the fabric used in the days of heavy, cumbersome linen sails.

Modern sails are made from untwisted, closely woven filaments of polyester or nylon. For different wind strengths different sails are designed with varying amounts of 'give' and of wind resistance. Sails make a rusty, papery sound because the fibres and yarns in them are jammed very tight together.

Q6   I used a dry-cleaning fluid to remove a stain from a suit, and now I have a large ring around the original stain. What can I do?

A   You should have dripped the dry-cleaning fluid in a circle *around* the stain. That way, the clean fluid would move towards the stain from the outside, and dissolve and concentrate it in an even smaller area. You can absorb the fluid containing the dissolved stain by holding some cotton wool at the centre of the soiled spot. If you drip the solvent directly on the stain, the dirt will be carried outward in a circle of soil. Now you will need to reverse the process by patiently treating the whole patch as a large stain, and starting with clean fluid on its outer perimeter. Talcum powder sprinkled on the whole area can act as an effective absorber of soil and solvent.

Q7 I have noticed that the detergent I use does not quite dissolve in water. How can such a mucky solution clean my clothes?

A Detergents often have additives which stay suspended in water in fine dispersion. This insoluble material, carboxymethyl cellulose (CMC), provides a surface which will attract the dirt that is removed from the washing. CMC prevents the soil from redepositing on other parts of the clothes, and ensures that a thorough rinsing is successful in removing all soil and detergent from the washload.

Q8 When would you recommend the use of washing soda?

A Washing soda is an alkali which is used to remove magnesium and calcium ions from hard water. Because it is alkaline it must never be used for washing delicate materials such as wool, and it need not be used at all unless the water is 'hard', i.e. if soap does not lather but forms a scum. It is possible to avoid the use of washing soda in hard water by using non-ionic detergents instead of soap.

Q9 Is water softener the same thing as fabric softener?

A No. Water softeners – such as Calgon or washing soda – remove the calcium and magnesium ions from hard water to allow soap to function effectively. Fabric softeners are cationic detergents which coat fibres with a positive charge. They are added to the rinse water, and make fabrics softer and fluffier and help to combat the build-up of static electricity in clothes.

Q10 Should I add detergent and fabric softener to the wash load at the same time?

A No. Soaps and many synthetic detergents are anionic – they carry a negative charge. If the cationic fabric softener is mixed with these, the two oppositely charged molecules will neutralize each other, and both will lose their effectiveness. Softeners should be added only after the soap has been rinsed out. Non-ionic detergents – often used for delicate fabrics – tolerate cationic softeners without loss of effectiveness.

Q11 When should a garment be dry-cleaned, and when washed by hand?

A Each garment and each roll of fabric for sale is labelled with instructions for care. In the case of a garment, the instructions apply to all notions – buttons, zippers, lining, trim. It is in the consumer's interest to treat labelled garments according to care instructions.

   The label showing fibre content can be used as a guide for care in the absence of other information. It must be remembered, however, that while a fibre may be washable, the dye or finish used in the fabric may not be, and that caution must be exercised to prevent shrinkage, pucker, fading, staining, and other disasters.

Q12 When choosing fabrics for a garment, can I take the price as a guide to quality?

A   Sometimes it is possible to find excellent quality fabrics at bargain prices – there is usually a marketing rationale behind such a 'find'. Sometimes, for fashion and marketing reasons, a poor quality fabric may be unreasonably highly priced. On the whole, however, it is more expensive to produce high quality fabrics than ordinary lines. Costs are greater at every stage of processing: raw materials, yarn and fabric manufacture, finishing, quality control, handling. If in doubt, a check of drape, handle and structure should indicate differences between better and poorer quality fabrics.

Q13   In what way is a polyester georgette blouse different from a pure silk georgette blouse?

A   The polyester georgette is likely to be less expensive than the pure silk blouse – particularly if the latter is made from a top quality silk fabric. (Yes, there are differences in performance among silks too!) In general, the silk georgette drapes better than the fine polyester, and will recover from creasing more effectively. This is not only because of the difference between the fibres – after all, polyester is very resilient – but because silk georgette is made from high-twist yarns, while in polyester georgette the multifilament yarns are textured rather than twisted. Both garments can be carefully hand-washed, though there is some danger that the silk will shrink. Manufacturers are likely to protect themselves against rough handling of both delicate fabrics by recommending dry-cleaning. The polyester is not likely to lose any dye during washing, but careful tests must be made on the silk garment to check washfastness before wet treatment.

      Polyester is more hydrophobic than silk, though they are both rather 'warm' fibres. Silk can be ironed with a steam iron, while a dry iron is effective for the polyester. Durability is about the same, though the abrasion resistance of weighted silk is much less than that of polyester.

Q14   Why do my feet get hot and ache in vinyl shoes, while they feel quite comfortable in leather shoes?

A   Expanded PVC is a hydrophobic non-permeable material. If it is coated on a synthetic nonwoven fabric – as it often is for shoe uppers – the composite is likely to have very little 'give'. The tight-fitting shoes will not stretch as readily as real leather, which stretches to accept the shape of the feet. Because real leather is more hydrophilic, it allows the feet to 'breathe', reducing local overheating and perspiration wetness.

Q15   What is the best way to clean vinyl? leather?

A   Soap and water – without abrasives – will work well on vinyl. Fresh inkspots and rust can be treated with water and vinegar, but dry-cleaning solvents should not be used on vinyls.

      Because leathers and suedes are treated with a variety of oils during their manufacture, any attempt to remove greasy stains from them will alter the finish (pp. 110–12). Do not use either detergent or solvents: send them to a professional dry-cleaner instead.

      Washable – synthetic – suede may be both washed and treated with solvents to remove oil-based stains.

Q16   Are all synthetic sewing threads the same?

A   No. There are a number of different yarns structures made from *polyester* for use in sewing.

Textured filament core wrapped with cotton gives a strong, highly extensible yarn suited for sewing stretch fabrics.

Staple polyester yarn, used for sewing synthetic fabrics and knits, is slightly more extensible than cotton.

Air-jet textured – taslanized – yarn is strong but inextensible. It is used for woven fabrics.

Monofilament yarns (usually *nylon*) are used for invisible hemming and other finishes.

Q17   I have been told that I should not vacuum newly laid carpet for the first few weeks, or I may make the pile go thin. Is this true?

A   If there are any short, loose fibres in the carpet, they will be removed by the vacuum cleaner whether in the first few weeks or later. With wool carpets a certain amount of felting occurs during wear, and this may anchor fibres which otherwise could have been sucked away by the vacuum. In either case, however, vacuum cleaning should not be able to remove large quantities of loose fibre from the carpet. If shedding continues for a long time after the carpet is laid down, it is likely that the pile fibres have been damaged at some point during manufacture and are in fact breaking under the strain of traffic. It is possible to check before buying whether the pile has been made from inferior short fibres, by firmly stroking the carpet surface and seeing whether any fibres are released.

Q18   What are the most important things to look for when choosing carpets?

A   The carpet must first of all please you, and suit the room it is to grace. However,
*   A mottled or patterned carpet will show dirt much less than will plain carpets.
*   A dark plain carpet looks dirty just as quickly as a very light coloured one – medium brownish shades hide dirt best.
*   Loop pile is more resilient than cut pile.
*   High-twist yarns are not flattened as readily as velvet-pile low-twist yarns.
*   Short dense pile is far more durable than deep shaggy pile of the same weight.
*   The most important factor for appearance retention and serviceability is the density of the carpet pile.

Q19   What causes puckering in seams?

A   Thick needles, coarse thread, uneven fabric feed, excessive thread tension, sewing in a direction parallel to warp and weft yarns – these are all causes of seam pucker which are easy to eliminate.

Q20   I have sewn a patch pocket on a dress, and however careful I was, the topstitching seam puckered badly.

A   Cut new pockets, this time cutting on the bias or at at least 15° to the direction of the grain. You will find that the topstitching will pucker much less.

Q21 I have bought a one-way stretch fabric. How should I lay out the pattern pieces on it?

A Always consider in which direction the fabric will need the greatest stretch: in a bodice – across the shoulders; for sleeves – along the length, to bend with the elbow; in pants – from waist to ankle, for ease of movement at seat and knees; in a skirt – from side to side; waistband – cut in non-stretch direction.

Many one-way stretch jeans, however, stretch in the side-to-side rather than in the waist-to-ankle direction. While this is not ideal for fit and comfort, it is necessary because denim is a warp-faced fabric. Fabric construction is easier if the stretch is in the weft direction; for the denim to look attractive, the garment is cut on the grain as usual, and finishes up with the horizontal stretch.

Q22 Why do pattern instructions recommend that more fabric be bought if the garment is to be made from napped fabric?

A A fabric which has no nap appears the same in both directions up and down the grain, so when laying out the pattern space can be saved by judicious positioning of the various pieces. A napped fabric appears lighter 'down' the nap, and darker when the observer looks into the pile. Therefore care must be taken that all pattern pieces face the same way on the fabric. This usually takes about half a metre of extra fabric, depending on the design.

Q23 Sheer fabrics and laces are considered to need special skills in handling. What problems can an unsuspecting sewer meet, and how can these be overcome?

A The two main problems in handling sheers are that they are often slippery, and that they may catch and pull or pucker easily.

To avoid slippage during cutting, the fabric should be pinned with a lot of pins to a firm cutting surface. An old sheet can be used as a backing – the scissors can be slipped between the sheet and the sheer fabric during cutting.

Since markings may show through the fabric, snip markings or tailor's tacking should be used.

For best results in machining, use 5–6 stitches per cm. If necessary, tissue paper can be placed over and under the seamline during stitching. This gives extra support to the fabric and keeps it from puckering. The tissue paper can be removed later by pulling it away from the seam. During sewing the fabric should be held taut, and ball point needles with polyester thread are recommended.

Never use fusible interfacings with sheer fabrics – the melted adhesive will show through most unattractively.

Q24 Is there a special way I should treat stretch fabrics during sewing?

A In order to allow the seam to stretch with the fabric, stitches should be short, preferably using stretch threads. Stretch the fabric slightly during stitching – this will allow the seam to take up more yarn. When the garment is stretched, the seam will have more give, and won't break as readily.

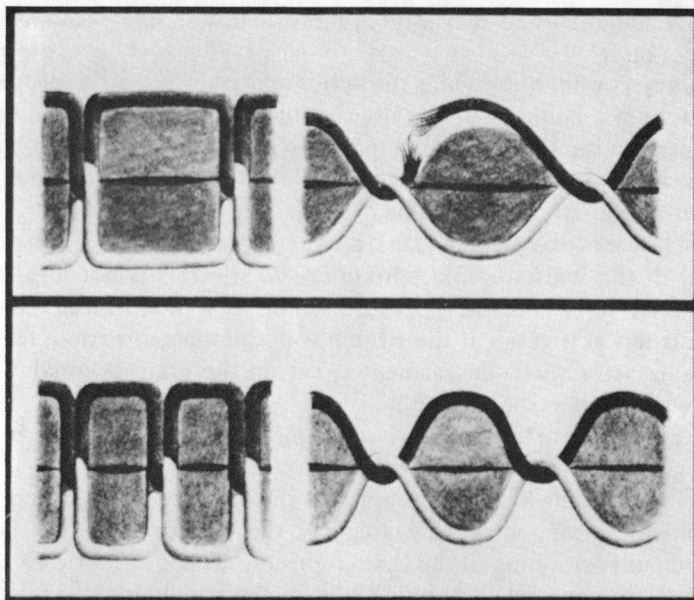

A simplified model demonstrating why seams stitched with short stitches have more 'give'.
(J. and P. Coats Limited, Glasgow, Scotland)

Q25 I have been sewing a garment from synthetic leather, and am having problems with making the seams lie flat. Can I iron the seams flat, as I would for woven fabrics?

A No. Vinyl is a thermoplastic material and would be badly marked if pressed. To flatten the seams, apply an adhesive to the seam allowance after sewing, and press flat with a cold iron or other heavy weight.

Q26 How can I iron velvets without crushing the pile?

A Ideally, a pin board should be used. In its absence, however, the velvet should be placed face down on a thick towel, and lightly steamed – not pressed! This method also works for corduroy and other pile fabrics.

An alternative is to use a piece of the same pile fabric as a pressing cloth, place napped sides together, and steam iron lightly.

Light brushing after a thorough steaming can revive an accidentally crushed pile. Hanging velvet garments in steaming bathrooms is another proven method of eliminating wrinkles.

Q27 In the days of sophisticated steam irons, pressing cloths should surely be considered out-of-date!

A Not at all. Pressing clothes are not only used as wet fabrics placed under an old-fashioned iron to generate steam. A dry pressing cloth placed between the steam iron and the fabric prevents the development of shine in the fabric being pressed. This is because the ridged weave of the pressing cloth presents an uneven surface to the fabric being ironed, and so allows general flattening of the fabric without its surface being squashed into a smooth, shiny replica of the iron's sole plate.

Q28 I have heard that if I put my silk blouse into the freezer for half an hour or so it will be much easier to iron. Is this possible?

A Yes. And not only silk – the idea applies to other hydrophilic fibres as well, but it only works if the air is fairly humid. In the refrigerator the fabric cools down. When you take it out, moisture from the air condenses in an even film on its cold surface, and the moistened fabric will respond better to ironing.

Q29 Should garments made from light-weight stretch fabrics be lined?

A No. In general, a sewn-in lining is likely to cause puckering as the stretch fabric is fed over the lining during sewing. If you must line a stretch fabric, use a tricot lining, and leave it loose: join only at the waist band, collar, and cuffs. Buttonholes should be backed to prevent stretching of the fabric during stitching.

3 Clothing and fashion

This chapter looks at why people wear the clothes they do: the reasons given by the wearers, and the reasons given by the sociologists.

Fashion, of course, relates to choices in cars, in sport and leisure activities, to holidays, to hairstyles, to all aspects of our lives. In this chapter, various theories are applied to *clothing* fashions as the most obvious example in textiles, but, to be valid, a comprehensive theory of fashion must be applicable to all these other areas.

Because fashions change, no text can claim to give rules on what will or will not be fashionable. However, through this chapter the consumer is encouraged to consider some of the factors behind clothing choices, and to understand how the forces of fashion affect our lives.

The function of clothing

We take it for granted that all of us wear clothes.

The reasons are complex. In all societies the body is clothed – adorned, festooned, decorated – in some way. Occasionally it is the climate which makes some body covering necessary, yet even where the climate would permit complete nakedness, people wear some body covering.

TRY YOUR HAND

Find out what the traditional forms of clothing are (or were) for as many cold-climate cultures as you can: Lapp, Inuit (Eskimo), Mongol, Tierra del Fuegan, Tasmanian Aboriginal, Tibetan.

Each of these cultures has used available resources to solve the problem of dressing for the harsh climate.

For each culture, identify aspects of the traditional dress which do not contribute to protection from the weather.

Does traditional costume vary with the age, sex, or social standing of the members of these cultures?

There is enormous variation between cultures in styles of dress. Only part of this variation can be attributed to differences in problems to be solved and available resources. Over and above the need for protection from the elements,

534

there seems to be a need to *decorate*. This may be expressed in hairstyles, jewellery, scar patterns and tattoos, or in clothing.

Within a culture, there is again enormous variation in dress and costume. Forms of clothing act as symbols, indicating the roles that an individual plays within the society. Clothes *signify the status and role* of the wearer. As these symbols are culturally determined, they may be difficult to interpret for people from outside the culture, but they are easily interpreted from within. The group to which we belong is usually well equipped to receive the messages inherent in our clothing; other members of our society can identify the group to which we belong.

Tattoos, Marquesas Islands, 1813.
(Mitchell Library, State Library of New South Wales)

Tattoos, Melbourne, 1986.
(Dale Mann/Retrospect)

Clothing symbolism. The ceremonial head-dresses worn by these young women at a wedding in Tanjung Karang, Sumatra, Indonesia, indicate that they are unmarried and that they stand in a particular relation to the bride.
(Photographic Archives of the Royal Tropical Institute, Amsterdam)

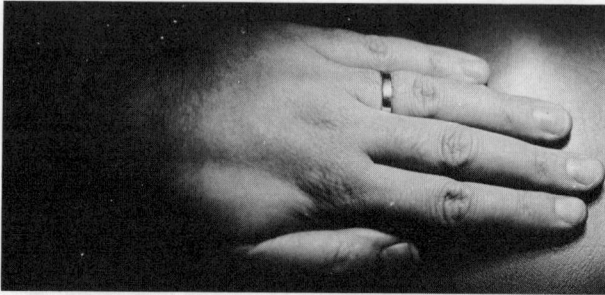

Clothing symbolism. The ring is sufficient cue to tell most Australians that this man is married.

TRY YOUR HAND

How much information can you derive from these photographs? List everything you can discover about each person, and the cues which the photographs provide.

(Dale Mann/Retrospect)

536

(Stephen Dattner, Furrier)

(Dale Mann/Retrospect)

(This exercise is, of course, asking you to treat the people photographed as stereotypes.)

Within a society, one person may have many different roles. Often the changes in role are accompanied by changes in clothing. These changes may be made several times in a day, or a particular costume may be worn once only in a lifetime, like the traditional graduation robe, white wedding dress, or christening robe.

The clothing worn within a wider society by members of a given sex, social and economic class, age group, religion, and occupation *identify the wearers as members of that particular group.* If the group does not acknowledge individuality – if it demands that the members identify completely with the group – then usually it will allow very little individual variation in clothing. A soldier on parade, for example, will have his hair cut the same way as his neighbour, his hat at the same angle, his tie tied in just the same way. A sports team or school may also demand identification with the group by means of a uniform, but will not demand total identification: there may be differences in hair style, in shoes, in minor details of dress. These differences may be argued to differentiate members of a group, to be an *affirmation of individuality*.

Clothing and roles. These work clothes are unlikely to be worn at breakfast, for a night out on the town, for a day at the beach, or for a casual social visit.

In the Fijian band, the only variations in the uniforms are the insignia of rank. The 1862 cricket team, however, exhibits considerable variety while maintaining a group identity.
(Nitin Lal, Fiji Visitors Bureau) (Herald and Weekly Times Ltd)

TRY YOUR HAND

Conduct a survey of an identifiable group, such as family, friends, classmates.
Define your group carefully.
1 List the clothes each person is wearing at a particular time on a particular day.
2 Ask each person to provide a list of reasons for the clothes she or he is wearing, and to indicate which is the most important reason.
3 Classify the reasons given for clothing choice. From the data you have gathered, suggest some factors which influence these people in making their clothing choices.
4 How much freedom of choice does each person feel she or he has?

TRY YOUR HAND

Maria, straight out of University, joined a law firm. On her first day she appeared in the office wearing the jeans, T-shirt and sneakers that she used to wear to lectures.
1 Do you think Maria's colleagues were dressed in the same way?
2 If you were a client of that law firm, would you be surprised to see Maria's outfit? Would you trust her to handle your case? What would be the reasons for your misgivings, if any?
3 Would you expect Maria to change to wearing suits or dresses if she continued in her present employment? Why, or why not?
You are invited to a party.
4 How would you feel if you wore jeans and everyone else was 'dressed up'? If you were dressed up and everyone else was in casual clothes?
5 How would you feel if you, and all the other guests, wore jeans? if you all wore dressy clothes?
6 How would you feel if someone else wore *exactly* the same clothes as you?
7 How can you explain your answers to the above questions?

When asked to carry out an assignment on why people wear clothes, a group of high school students gave these reasons: *protection* (from the sun or the wind); *beauty* (personal likes); *modesty* (to hide the naked body); *attraction* (to please the opposite sex); *showing off or display* (to look more important than others); and *belonging to a group* (distinctive uniform). University students gave the same reasons, but with fancier labels: physical/physiological; cultural/aesthetic; behavioural/ornamental; sexual/biological; psychological/assertion of individuality; sociological/status.

TRY YOUR HAND

Look at the pictures, and consider:
1 How is each of the following people influenced in his or her clothing choices by each of the factors above?
2 Do the same influences operate on each?

(Herald and Weekly Times Ltd)

(Army Public Relations)

(The Age)

3   If so, why do the same influences have different outcomes in each case?
    If not, what influences do operate, and why do they give rise to different
    outcomes?
In answering these questions you are, of course, treating each of the people
in the photographs as a stereotype, not as a real person.

The above lists are not the only way to classify the many influences on clothing choices. Nor do they tell us anything about *what* people wear; about which clothes will satisfy each of these needs in a given culture or setting.

Theories of fashion

There are many possible definitions of fashion. A useful definition is:

'Fashion is simply the modal [most commonly found] style of a particular group at a particular time. It is the style which is considered appropriate or desirable.'

Anne Hollander, *Seeing through clothes,* Viking, New York, 1978, p. 350.

This definition is about what is popular or well-accepted by a particular group. It is not about *haute couture* or high fashion.

A satisfactory theory of fashion must take into account all the reasons for people's clothing choices; it must cover popular fashion and *haute couture;* and it must explain why fashion changes.

Functional explanations

Fashion as a search for meaning and/or identity

An individual may select a particular style or be coerced into wearing a particular style. In either case, the mode of dress will have an impact on that individual's identity and the meaning of his or her existence.

Fashion has a dual role, *conformity and individualization*. It is used to identify members of a particular group (they conform), and at the same time it differentiates members from those outside the group.

Fashion is used by people to *enhance themselves and their attractiveness* to others, either personally or as a member of a group.

Fashion *establishes identity*.

Fashion *defines roles and role distance*. An individual adopts the style of dress of a particular group, but at the same time may manipulate or change the clothing slightly to indicate that commitment to the group is not total.

Group identity. The basic costume is jeans and nylon windbreakers, but there is great freedom in colours and manner of wearing the clothes.
(Education Department of Victoria)

TRY YOUR HAND

1   Is there any particular form of clothing which you and your friends all like to wear? List the most popular examples.
2   Identify another group of people about your age but with whom you feel you have very little in common. Is their clothing identical to yours? Is it similar? Explain any differences.
3   Would you wear the clothing of an entirely different age group, your parents', for example? your grandparents'? What are the similarities and differences here?
4   Think of a time when you took extra trouble to look your best. Why did you do this? What did you do?
5   How do you choose your clothing? What influences you? What clothing makes you feel good?
6   Does your clothing always exactly match what you and your friends think of as fashionable, or do you sometimes like to be a little different?

Fashion and the struggle for identity

According to this theory, fashion has a social significance that enables the individual or group to manipulate dress in order to maintain or change *status* (social or sexual).

Competition

Individuals use fashion as a means of competing for status, for example to display rank, wealth, power, or sexual appeal.

Clothing and status. The badge worn by this man from Ming, near Mt Hagen in Papua New Guinea, identifies him as a luluai, or local police officer under the Australian colonial government. The shells and feathers identify him as a man of considerable status and wealth in traditional terms: they perform the same function as the diamonds on a duchess.

(W. Brindle. Australian News and Information Bureau)

Does wearing a hero's number impart sporting prowess?
(Dale Mann/Retrospect)

Imitation

The clothing worn by a revered person or group is copied by individuals. The intention is that by looking the same as those they imitate, the imitators will gain equality with their model. This is known as the 'trickle down' theory.

TRY YOUR HAND

1   Are there times when you, or others you know, flaunt wealth, sexiness, or some other form of superiority, through choice of clothing?
2   Can you think of a time when you have tried to copy the clothing style of someone you admire?
3   Can you identify any people who seem to be copying the style of dress of some sports star or pop idol?

There are two opinions on the outcome of this theory of fashion. One possible outcome is that imitation breaks down class distinctions because people look very much the same. The second outcome is that it serves to enhance class distinctions because the wealthy can afford to change their clothing style as often as necessary to maintain a distinct appearance.

Economics of fashion

'Unlike any other kind of demand . . . fashion means that the consumer pays – willingly – for impermanence, for the stamp of recency upon goods.'
R. Lauer and J. Lauer, *Fashion Power: the meaning of fashion in American society*, Prentice Hall, Englewood Cliffs, 1981.

Demonstration of affluence

People who use clothing for this purpose often choose expensive clothing which is quite impractical for doing any form of work.

Fashion means jobs

Employment is created by the fashion industry. If fashions did not change, there would be less employment in the industry.

Some fashion garments are designed to be worn at most once or twice.
(Herald and Weekly Times Ltd)

TRY YOUR HAND

1 Have you, or someone you know, ever been tempted to buy something totally impractical, or been tempted by some wonderful creation in a magazine?
2 Estimate how much of your clothing you discard before it is quite worn out. Be honest, you are not justifying a new purchase.
   a Why do you discard this clothing?
   b How much longer could you have worn each garment before it was worn out?
   c For the sake of argument, assume that on average people obtain only 70 per cent of the possible wear from their clothing. If everyone obtained all possible use from their clothes, what impact would this have on the clothing industry?

Fashion as erotic

At different times, different parts of the male and female anatomy are seen as erotic, and become the focus of attention. According to this theory, clothing is not a means of attaining modesty; rather, modesty is the effect of clothing.

Portrait of Henry VIII after Hans Holbein the Younger.
(Walker Art Gallery, Liverpool)

Yali man, Irian Jaya.
(Robert Mitton, *The Lost World of Irian Jaya*. Oxford University Press, Melbourne, 1983)

Both these photographs illustrate the dangers inherent in interpreting clothing symbolism from other societies and times. Although both costumes emphasize male genitals, in neither case can the intention be described as erotic. In both societies (sixteenth century England and present-day Irian Jaya), maleness is associated with power and status, and so emphasis on the male genitals identifies both men as powerful figures. Henry's massively padded shoulders extend the symbolism.

TRY YOUR HAND

1 Use a fashion magazine to identify parts of the female and male body which are highlighted, and parts which are hidden by current fashion.
2 Are there any styles which are currently thought of as 'sexy'? How do these differ from other clothes?
3 Look at pictures from other times. Can you identify parts of the body which are highlighted? Be very careful if you choose costumes from a culture you are not familiar with, as clothing symbols will differ between cultures.

Dynamic explanations
Fashion as diffusion

The adoption of fashion innovations depends on: the characteristics of the fashion and how these compare with existing values and needs; the way the fashion is communicated; the activity of change agents (such as advertisers) and opinion leaders; and the reaction of early adopters (fashion leaders from all classes) who accept or reject a new fashion.

New designs must find acceptance before they can become fashion.
(Dale Mann/Retrospect)

TRY YOUR HAND

1 What is the most recent fashion you have adopted? Consider new colours as well as styles.
2 What influences were brought to bear on your decision to adopt this new fashion? Did you feel you needed something like this? How did you find out about the fashion? Who were the first people to wear it?

Recurring cycles of back fullness (left), tubular (centre) and bell-shaped contours (right). (Agnes Brooks Young, *Recurring Cycles of Fashion, 1760–1937*. Cooper Square Publishers, New York, 1966.)

Cycles of fashion

It has been suggested that fashion is *cyclic*. Basic shapes in clothing reappear through the ages. Attempts to break away from the 'look' before it has run its course will fail, and even world events, such as a war, do not have any real impact on the cycle.

TRY YOUR HAND

1   Can you identify a basic shape for the current fashion? (Women's clothes are probably easier to study here than men's.) For example, does it have back fullness like the bustle dress, is it tubular, or bell-shaped?
2   Can you identify another age when this shape was the fashion?

Fashion as a sign of the times
A reflection of wider cultural concerns

Fashion, like every facet of culture, is a reflection of the ZEITGEIST, the spirit of the times. Thus an understanding of the changing society is a necessary prerequisite for understanding fashion. This is a very broad idea, and can be applied together with other theories.

TRY YOUR HAND

Is there something unique about the present time which is reflected in current fashions? Consider sportswear for leisure clothing. Does this reflect a current concern with fitness? Find other examples where clothes seem to reflect current concerns.

The above exercise looks at popular clothing as a reflection of the Zeitgeist. *Haute couture,* too, can be studied in this way.

In the mid-1960s, when all was peace, love, and back-to-nature, loose-fitting clothes and bare feet symbolized the (temporary) rejection of high technology.
(Herald and Weekly Times Ltd)

Thai dancer performing a sacred dance. The costume and the temple architecture both symbolize God.
(Thai Airlines International Ltd) (Anna Janca)

The conception of what is beautiful changes in different ages and different cultures. Collect information on the grand architecture (churches, public buildings, palaces) and on the clothes worn by the elite (the rich and powerful have the means to change their clothing styles) for the following periods: Gothic, renaissance, Edwardian (art nouveau), 1960s. Each of these periods shows distinct styles in architecture. Are the same concepts of what is beautiful reflected in high fashion clothing?

Fashion and ideology

Fashion as a tool of change

The Shah of Iran attempted to Westernize the economy of his country, and to introduce many social changes. Amongst the changes he promoted was the participation of women in public life. At the same time that women were being encouraged to take a more active role in public life, they were encouraged to adopt a more Western style of dress.

The revolution which deposed the Shah led to a restoration of traditional fundamentalist Muslim values. Women were once more excluded from public life, and returned to wearing traditional dress. (However, in other countries in which Muslim women are returning to traditional dress, there has not been the same erosion of women's rights.)

Textile designs in the Soviet Union in the late 1920s served as propaganda for the industrialization of the country.

Industry
Cotton print, 1930
Designed by D. Preobrazhenskaya
Russian Museum, Leningrad
Acquired in 1931

In China during the Cultural Revolution, the notion of equality was reinforced by the wearing of Mao suits, which became a sort of uniform for the population. Now the country is developing closer ties with the West, it is also allowing more Western styles of dress to be worn.

Cardin to cater for Peking trendies

From TONY WALKER

PEKING, 8 Nov. – 'But Madame Wong, you'll look simply adorable in this little black dress at the next Great Hall tea party with the comrades . . . and what about these slinky leather pants for those dreadful inspection tours of the provinces. . .'

Paris chic is coming to Peking – without compromises – according to Pierre Crey, public relations director for Pierre Cardin.

This month the Cardin invasion will begin when the French designer opens a showroom in Peking to exhibit a range of fashions from haute-couture to his ready-to-wear lines, and he hopes that young Chinese women will drop by.

'The showroom will not be directed at Western tourists visiting Peking,' Cardin official Elizabeth de Molliens told reporters in Paris recently. 'We hope the Chinese population will visit too.'

The Chinese will be able to shop at a Cardin boutique in the heart of Peking, from next year, the first time a Western fashion house has been able to sell its wares in China since the communist takeover in 1949.

Cardin is perhaps being a little optimistic if he expects Chinese to shop at his boutique, after all the average Peking wage is around $A26 a month, hardly enough to buy a cheap scarf from his normal range.

Young Chinese try hard to emulate Western fashions, but the effect is often 20–30 years out of date, rather like leafing through an old and faded fashion magazine.

Fashion, like most other things in China is political. Thus, no self-respecting and politically ambitious Chinese matron is too adventurous in the clothes she wears.

Dresses in summer . . . yes, but nothing that reveals too much. The treatment meted out to Wang Guangmei, the widow of the late President Liu Shaoqi, is a too recent reminder for most Chinese of the hazards of an individual style of dress.

Madame Liu committed an unpardonable sin in the eyes of the Maoists: On a visit to Indonesia with her husband she wore a colourful cheongsam (close-fitting dress with a slit up the leg), something for which she was pilloried during the Cultural Revolution.

Even Mao's widow, the hated Jiang Qing, who has been given a suspended death sentence, compromised in her style of dress, according to her biographer, Roxanne Whitke, who recorded that Madame Mao: '. . . always dressed for our meetings in a pastel crepe-de-Chine dress of simple Western cut, sheer nylon anklets and white plastic sandals . . .'

Age, 9 November 1981

TRY YOUR HAND

Aboriginal Australians have a history stretching back some 40 000 years. An important recent influence on their history is European settlement in Australia, which has affected their whole way of life.

1   What factors affected the clothing worn in traditional Aboriginal societies? Remember, *all* of Australia was Aboriginal land, and that Aboriginal cultures showed considerable diversity as well as having many features in common.

Choose one part of Australia for which you can obtain information: northern and central Australia will be the most accessible, but a group project could discover some interesting material on other regions. Superficial answers are worthless.

2   What body decorations were traditional in this region? What purpose did they serve?

3   Do any people still wear these body decorations? Who, and under what circumstances? If not, try to find some of the reasons why use of these body decorations has discontinued.

4   How did early European contact influence the clothing worn by the Aboriginal population? What were the main points of contact (military, exploring, gold rush, farming, missionary, trade, or other)?

5   What clothing is now worn by the Aboriginal population of your chosen area? Why?

1890: the bustle.

Around the turn of the century, women's demands for higher education and the right to participate in public life were becoming more forceful, and some concessions were beginning to be made. The 'new women' were determined to show themselves the equal of men; and the fashion was for tailored jackets, shirt-style blouses, and ties.

However, if this fashion served to seal the change of women's entry into public life, what was the effect of the fashion which followed?

1900: tailor-made.   1910: hobble skirt and big hat.

(*Argus*, 26 November 1910. La Trobe Library, Victoria)

Forces for change

The American researchers Carole Robenstine and Eleanor Kelley set out to study the connection between social change and fashion change *(Home Economics Research Journal,* September 1981, Vol. 10, No. 1). Two conflicting theories of fashion change are often expressed: (i) that changes in fashion occur inexorably and are not determined by external events; and (ii) that fashion, a social behaviour, is related to the setting in which it occurs.

Robenstine and Kelly selected France in the eighteenth and nineteenth century for their study. In that time, France underwent considerable political change (there were changes in who held power), and considerable social change (there were changes throughout the social fabric, in religious, economic, educational, familial, and other aspects of everyday life). These changes did not always go together. Portraits and fashion plates were used as sources of information on the clothing styles of the elite (those with the means to change their clothing styles), and various features of the clothing style – such as sleeve shape – were selected to allow fashion changes to be studied statistically.

The two hypotheses Robenstine and Kelley tested were (i) that patterns of stability and change in dress fashions will coincide with patterns of institutional [social] stability and institutional change; and (ii) that patterns of stability and change in dress fashions will coincide with patterns of political stability and political change. At the end of all their analysis, they found that these hypotheses were *not* supported by their data, but they were careful to point out that these conclusions do not necessarily apply to other periods, countries, and cultures.

Forces for social change influencing clothing styles. A mission school room in Papua New Guinea in the days before independence.
(Department of Territories)

Evaluating theories of fashion

Each of the theories of fashion seeks to explain fashion in clothing; to offer a theory about clothing. Two American sociologists, Robert and Jeanette Lauer, made a thorough study of American clothing of the last 200 years, and of the various theories of fashion. They suggested that although each of these theories contributes to an understanding of fashion, there are two main criticisms.

(i) Each of the analyses is unable to account for all of the evidence, and so must be considered incomplete.

(ii) The link between the individual and society is ambiguous or unrealistic. The *interdependence* of the individual and society is not recognized.

Fashion and the individual

When the Lauers took all the analyses of fashion into account, they concluded that *fashion is outside the control of and greater than individuals, that it is coercive upon individuals.*

As W. G. Sumner wrote in his 1906 book *Folkways* (quoted in R. & J. Lauer, *Fashion Power*), 'Fashion is by no means trivial. It is a form of the dominance of the group over the individual, and it is quite as often harmful as beneficial. There is no arguing with fashion... The authority of fashion is imperative as to everything it touches. The sanctions are ridicule and powerlessness. The dissenter hurts himself; he never affects the fashion.'

The Lauers go on to assert, 'Essentially, fashion is a process of collective definition in which a particular alternative in a set of possibilities is selected as appropriate. All phenomena may vary or change in a number of different ways over time. A particular way is selected and becomes the fashion as the result of collective definition. The definition is the outcome of ideological evaluation... which is made along two lines: whether fashion is consistent with values and roles and whether it is useful for reaffirming or establishing an individual's identity and/or status.'

Where other theories focus on social control or social change, both of which fail to explain all the evidence, the Lauers also include the *meaning* of fashion to the individual. In other words, their theory is meant to include all the reasons which people give for their choice of clothes. The Lauers place *fashion as a form of control in a changing society. Fashion is the identification of those new directions which are collectively described as appropriate.*

Every fashion change must be ideologically acceptable; it must not flout current ideology. In every society there are contradictory ideologies, so the fashion suited to one set of values is not necessarily approved of by people holding a different set of values. Acceptance or rejection of a fashion, as a result, can lead to acceptance or rejection of a person. Consider an employer's assessment of a 'scruffy' person as unreliable; the acceptance or rejection of a newcomer to a group on the basis of appearance; the attribution of rape or sexual harrassment to the way the victim dressed: 'they were looking for it'.

A currently topical fashion issue is the wearing of furs and leather. Although most furs are obtained from animals bred for that purpose, rather than from endangered wild species, the use of furs is sometimes criticized as unjustified exploitation of animals.

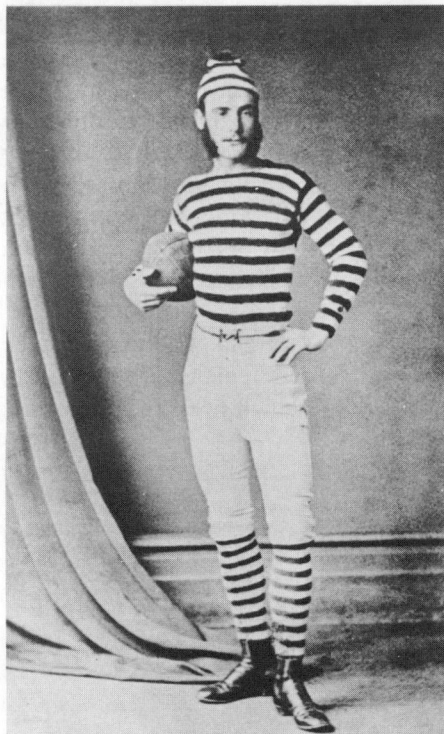

(Herald and Weekly Times Ltd)

Football uniforms in 1879 and the 1930s. How do they compare to the uniforms of today? How can you account for the differences?

TRY YOUR HAND

1 Talk to the careers teacher in your school about the appropriate way to dress for a job interview. Ask the teacher to explain why this is important. What meaning and ideology underlies the explanation?

2 Study the figures of fire accidents on p. 486. How can you account for the greater number of girls suffering burns from nightwear? boys and daywear? Do these figures relate to clothing styles? If so, assuming parents do not, in fact, want their children to be burnt, how can you explain the acceptance of some of the more dangerous styles of clothing?

3 Research the history of bathing costumes in Australia, finding the arguments put forward by the proponents and opponents of each fashion change. What ideologies were involved in each change? Did the introduction of each style necessarily mean that the opponents' ideology had lost ground? Did the ideology of the proponents necessarily gain wide acceptance?

4 Survey literature on popular forms of clothing and its impact on people's lives. Some feminist literature, in particular, has studied harmful clothing practices.

5 Consider current fashion trends and their impact on your life. Make a list of positive and negative aspects to identify the underlying ideology. Does identifying the ideology affect your acceptance or rejection of the fashion?

Marketing fashion

The commercial fashion designer does not necessarily consciously evaluate each new style in terms of all the theories of fashion. Instead, she or he may study the direction of change in recent fashions; look at fashion trends in minority groups and subcultures which may be adopted by the majority; and study new ideas and attitudes and interests in other areas which could be translated into clothing design. The designer for popular clothes will look at the work of the *haute couture* designers, and perhaps adapt it to suit her or his perception of the market. These processes take account of the explanations of fashion which are offered by the various theories.

One problem faced by the designer is the *real–ideal* gap. Clothes must be displayed to their advantage, and so models, male and female, are tall, slim, beautiful, elegant. Few consumers are this size: clothes which stress the unrealistic ideal may be rejected.

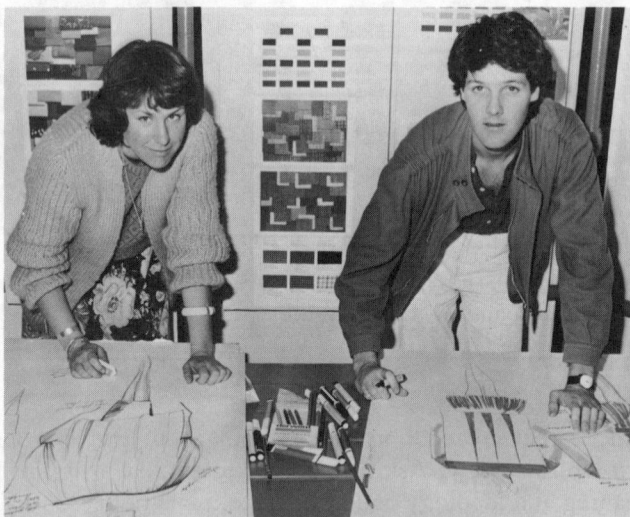

Fashion designers at work.

TRY YOUR HAND

Read the following extract:

A new target for wool fashion

Australian woolgrowers will need to 'chase' the younger generations in the USA, Western Europe and Japan if their industry is to remain viable in the future, Dr John McPhee CText FTI, deputy managing director of the International Wool Secretariat and President of the Textile Institute said in Canberra in October.

The young people will have a relatively high level of disposable income and above-average interest in fashion and clothes. They will pay high prices for products commanding their interest, and it is essential that wool should be among those products if wool producing is to remain a viable industry in Australia.

'The challenge for those directing future wool promotion activities is that these same young people are not particularly interested in the classical, traditional products in which wool currently enjoys its highest market share', Dr McPhee said.

... Dr McPhee said that wool's research and development programmes must be closely orientated to market needs.

... Dr McPhee said that especially for the younger consumer there was intense competition between textile and non-textile products. He added that although superficially the 'casual' clothing styles of today's young people do not appear to offer good possibilities for wool, in practice this market has expanded and developed its own high-priced segment for people who want to demonstrate that they can dress with 'style'. People in this part of the market are likely to purchase wool products if they are of the 'right' design and styling and their performance provides satisfaction.

Textile Horizons, December 1983.

(Australian Wool Corporation)

Plan a marketing campaign designed to make wool more desirable to the younger consumer. Use the Lauer theory as a guide to what your campaign must try to achieve.

TRY YOUR HAND

Develop a proposal for a new fashion. Design something new and different yourself, or use an avant-garde fashion magazine, a currently controversial fashion idea, or a future fashion book such as *Fashion 2001*.

1　Try to gauge the potential of your proposal; that is, try to find evidence of collective acceptance or rejection of the proposal by a group of people representative of your target market. For this, you will need a clear idea of who your market is to be, in terms of sex, age, interests and so on.

2　What meaning do your potential customers give the clothing? Identify attitudes, roles, identity (self-image), and status requirements in the responses. A questionnaire may help, but will need to be tested on a small sample before you try it on all your group. Consider carefully how you will analyse the data you collect.

TRY YOUR HAND

1　Survey two department stores.

  a　Do they cater for (i) early adopters of a fashion (trend setters), (ii) fashion followers, (iii) obsolescents (those who do not take up fashion trends as quickly as the rest of the population)?

  b　Do both stores cater for the same types of customer? Where does each store see its market? If you were the manager of a store, which market would you try to 'catch'? Why?

  c　If you were the buyer of the 'under twenties' section of the store, what criteria (style, price, size, colour, fashion currency, etc.) would you set for your purchases? Why? How would your criteria change if you were buying for the 'elegant' market? for the 'little dollar' customer? for the older woman or man?

2　Collect advertisements for various styles and prices of clothing. What models of fashion and clothing use underlie each advertisement? What need is each item claimed to meet?

14. Issues in textiles production

This chapter looks at some of the issues which have affected textiles production and consumption in the past, and which continue to affect the industry and consumer today.

Issues of conservation, pollution control and automation, concerns which are foremost in the minds of machinery manufacturers and mill managers, are examined in detail. The economics of the textile industry: resources, trade, and government influence, are also studied.

Questions of concern to the textiles consumer are considered, questions such as:

- What problems face the Australian garment manufacturer?
- Why do some countries specialize in making particular types of textile products?
- What is it like to be a textile worker in different parts of the world?
- What are the implications of innovations in the textile industry?
- Why doesn't the 'cheapest' country produce all the world's textiles?
- What role do hand-crafted goods play in a technological world?

Consumption and production

Trade in textiles has an ancient history. The emergence of trade routes was an outcome of trade development; of these, the Silk Road between China and Europe is the most famous.

Put crudely, trade occurs when a manufacturer can produce an item for less than someone else is prepared to pay for it. Before the industrial revolution, the textile and clothing needs of most of the population would have been met through local production: trade would have occurred between neighbouring towns and villages. Trade over longer distances and between countries required more organization and involved greater risks: it was most profitable in luxury goods.

The technological developments of the eighteenth and nineteenth centuries lowered the costs of processing fibres. The lowered costs meant more people could buy more goods, and this in turn increased the demand for raw materials. Commerce in luxury goods such as tobacco, tea, silk, spices and gold had led to improvements in navigation and the opening of reliable trade routes; these routes were now employed to transport raw materials to the mills of England,

and to distribute the relatively low-cost textiles to the rest of the world. Textiles for the common people had become an important item of world trade.

Transporting wool, 1920s. Getting products to the markets has always been a significant problem and expense, in all fields of production.
(Department of Agriculture and Rural Affairs, Victoria)

Packing wool for transport, 1980s. Container transport has helped control shipping costs.
(Australian Wool Corporation)

Hyderabad cotton waiting for transport, India, 1880s. British industry's demand for raw materials led to the development of the Indian railway system. However, British industry was jealous of competition, and so efforts were made to stifle the production of manufactured goods in India: the country was to remain a source of raw materials only.

Per capita consumption

The effect of increased production can be shown in the per capita consumption of textiles.

World fibre production

Period	Average annual world fibre production (million kg)							Civilian consumption United States (estimated) (kg/head/year)
	Synthetic	Rayon	Cotton	Wool	Flax	Silk	Total	
1920–30			4 500	900	620	70	6 000	
1930–40			5 900	900	620	70	7 500	
1946–50			5 900	1 000	540	16	9 000	
1950–55	70	1 600	8 000	1 100	810	23	13 000	18
1955–60	185	2 200	9 500	1 300	620	32	14 800	16
1961–65	2 500	2 900	10 800	1 450	650	32	17 500	19
1965–70	3 500	3 500	11 000	1 590	700	36	20 000	23
1971–75	5 800	3 500	11 200	1 000	620	40	24 000	25
1976–78	9 500	3 300	13 000	1 450	n.a.	50	27 500	26
1981–83	11 000	3 100	14 900	1 610	n.a.	56	30 800	n.a.

n.a. = not available

TRY YOUR HAND

1 By what percentage has total fibre production increased since 1930? By what percentage between 1950 and 1980?

2 In 1950 the world population was 2.5 billion; in 1980 it was 4.5 billion, an increase of 80 per cent. Consider the increase in fibre production for the same period. Was more or less textile fibre available per head?

3 The last column in the table shows that in the United States, civilian consumption per head increased by 45 per cent between 1950 and 1980. How does this relate to the increase or decrease world-wide?

4 What type of fibres contributed most to the increased fibre production in the world?

5 How has the percentage share of each fibre type changed over the period 1950–1980? How can you account for these changes?

6 Based on the current trends, discuss what you would expect to happen by the year 2000. (A graph will help here).

The history of hosiery illustrates some factors which made the increased consumption possible.

Development of hosiery

Date	Development
700 BC	Greek poet Hesiod mentions piloi – matted animal hair used for lining shoes. Employs felting properties of wool.
200 BC	Romans wrapped their feet and legs in strips of leather or woven cloth. This warm but bulky garment was used in the Balkans and Russia into the nineteenth century.
200 AD	Udones – Roman hose – were cut and sewn from fabric, felt, or leather. These fitted better than the leg wrappings they replaced, but still lacked elasticity.
300	Knit socks were used in Egypt. They were elastic, and conformed to the shape of the foot.

Date	Development
1400	Cut and sewn woven cloth hose, each leg separately attached to the waist of the doublet, were worn with breeches by European men. These hose did not fit the body accurately. Bulky hand-knitted hose were also worn.
1589	Fully fashioned machine-knit stockings made on flat-bed knitting machine, then sewn into a tube. Production six times faster than hand knitting: patent refused to William Lee in England, so invention taken to France.
1764	Fully fashioned seamed stockings with top welt and shaped heel made from wool, cotton, or silk, on William Cotton's power-driven flat-bed knitting machine. Production was much faster and the garments were form-fitting. Silk stockings could be afforded only by the rich.
1816	Tube-like cheaper stockings which were not very form-fitting made on Marc Brunel's circular knitting machine; improved by Peter Clausan in 1845.
1900	Long-lasting, thick stockings worn under long skirts by women. Market shares: 88 per cent cotton, 11 per cent wool, 1 per cent silk.
1920–1935	Rayon and silk increase their market share at the expense of 'sensible' stockings. The invention of viscose rayon made the silk-like look available to the middle classes; fashions with shorter skirts dictated shiny sheer stockings and drew attention to the legs. Stockings were usually fully fashioned and seamed.
1940	Sheer nylon stockings replaced silk amost immediately on introduction to the market. Although expensive at first, the price dropped rapidly. Within five years, a pair of stockings cost the same as 10 kilograms of bread.
1950s	Seamless form-fitting stockings with formed heel knitted from a continuous nylon filament on a circular knitting machine, then heat-set into shape of the leg. Heels were formed by reciprocating action of the machine.
1960s	Improvements in technology of circular knitting made micromesh stockings which did not ladder easily: ladders would run upwards only.
late 1960s	The creation of textured nylon made possible sheer clinging stockings which had excellent form-fitting properties. Mini skirts made sheer-to-waist pantyhose a necessity: in 1968 they held 8 per cent of the market; in 1972 they held 35 per cent.
1970s	Pantyhose hold 95 per cent of the market by 1980: the fully automated circular knitting machine makes a complete three-dimensional garment without the need for seams. Anti-static finishes and many different styles become available. At start of decade, a pair of pantyhose cost the same as 1 kilogram of bread; by 1980, they cost the same as half a kilogram of bread.

The textile pipeline

There are many different manufacturers involved in producing the textile garments bought by consumers. Fibre producers, yarn and fabric manufacturers, distributors, garment manufacturers and retailers all need to make decisions about what – and where and how much – to produce, who to buy from, who to sell to and for how much.

Each member of the textile pipeline – from fibre producer to consumer – depends on the decisions made by governments and by others in the pipeline, both in the same country and overseas.

What are the options open to each of the sectors of the Australian textile pipeline?

Fibre producers

Fibre producers sell their product to the yarn manufacturers, not to the consumers. Generally, the fibre producers appeal directly to the fabric and garment manufacturers. They offer key persons (or companies) in the industry various inducements – cheaper prices, technical help, designs, assistance with promotions, etc. – to win them over to use their products at the expense of other fibres. This is called merchandizing, and is aimed at getting particular products onto the retailers' shelves.

Nevertheless fibre producers have often tried by advertising and other marketing techniques to influence the fibre choices that consumers make. For polyester fibres Du Pont has promoted the brand name of Dacron, ICI that of Terylene, Hoechst that of Trevira, Toyo Rayon that of Tetoron and so on. After a brand-name is established in the public consciousness, the expensive promotion is usually stopped.

(The Campaign Palace, Melbourne, for Sheridan Textiles)

During the early 1970s most of the patents held by major synthetic-fibre manufacturers expired. In the face of increasing competition, synthetics today are sold on the basis of who can offer the lowest price. On the other hand, the International Wool Secretariat and other allied organizations continually campaign to make consumers aware of the advantages of the wool fibre. By clearly establishing the image of wool in fashion parades and advertisements, they represent the interests of the wool growers and encourage consumers in

Knitting wool promotion, Japan.
(Australian Wool Corporation)

Wool promotion in Great Britain.
(Australian Wool Corporation)

the higher-cost market to consider fibre content rather than just shape and colour when choosing textiles.

Yarn manufacturers

This segment of the industry has the least direct effect on consumer choices. Yarn manufacturers tend to specialize and seek their market among the fabric manufacturers. Some fibre manufacturers have even taken over the role of the throwster, and often topmakers and spinners are tied to the requirements of a few selected outlets. They spin yarn as ordered by the fabric manufacturer, and are often part of their client's company.

Fabric manufacturers

They employ designers, collaborate closely with both fibre and garment manufacturers, and are an important link in the carefully timed chain of events that produces a fashion item in its correct season for the retailers' shelves.

Dyers and finishers are totally dependent on the decisions of fabric manufacturers. Often they are an integral part of the company. Most dyeing and finishing innovations are researched and developed not by commission dyehouses but by their suppliers – machinery manufacturers and dye-producing chemical companies.

Most fabric manufacturers specialize in a particular segment of the market, and limit their merchandizing activity to selected garment manufacturers and wholesalers. Fabric manufacturers – other than in carpets or blanketing – rarely have direct contact with the consumer by way of trade names or advertising.

Fashions change rapidly; this necessitates only small orders for any one fabric style. Other markets – uniforms, underwear – can derive benefits from long production runs and large orders. If the designer or garment manufacturer intends to use fabrics produced overseas, he or she must allow for delays in delivery, or rely on stocks held by the wholesaler or importer. In this way, the *fabric distributor* also plays an important part in the fashion chain.

Garment manufacturers

Sometimes garment manufacturers act as their own retailers in boutiques carrying their brand-names, but mostly distribution occurs through independent retail outlets. The price structure, design, and variety of products – even the amount of fabric used or the quality of design employed – depend on the kind of retailer the garment manufacturer is aiming to sell his goods to. The merchandizing of the garment by the manufacturer is closely related to what he perceives to be the needs and wants of the particular target segment of the consumer market.

Retailer

It is not easy to identify just who determines what fashions will be available for the consumer at the start of each season. During the garment manufacturers' showings, retailers influence the consumers' choice by ordering what they think the consumer will buy. As the season progresses, reorders confirm or correct the retailers' original decision.

Great fashion houses and fashion magazines play a direct role by influencing both the garment designer (and hence the fabric manufacturer) and the consumer. Consumers, however, cannot be dictated to – witness the disastrous failure of the maxiskirts in 1971. Yet five years later they were all the rage!

Not all retail outlets are sensitive to the vagaries of fashion. Many discount and variety stores still sell the same types of garments they sold several years ago. Consumers with different priorities choose retail outlets that stock items most suited to their needs and means.

Consumers

The 'sharp' end of the long chain of textile marketing, the consumer is by no means a uniform entity. Different groups of consumers have different needs, according to income level, ethnic background, and many other social factors. Retailers are aware of this diversity and cater to it to the best of their ability. The role of *government* as a significant consumer of textile materials should not be overlooked when discussing the forces that forge the character of a country's textile industry.

The timetabling of fashion

It takes time to produce a garment, and so a new fashion must be planned well ahead. Manufacturers need to allow for delivery time as well as production time: this can be a deciding factor in the choice between local and overseas purchases. The reliability of the supplier, in meeting deadlines and in meeting quality requirements, can be crucial; so can the ability of the supplier to provide more stock at short notice.

TRY YOUR HAND

1   At the start of the new season, go to a variety of clothes shops, and check how many entirely new fabrics have 'hit the market'. (How will you define what is 'new'?)
    a In what kind of shops did you find these fabrics?
    b Did all the shops you visited stock the 'newest'?
    c Were there many new styles that obviously made from last year's fabrics?

d Were there many 'old' fabrics that were being used in a new range of colours? For whom would this strategy be an advantage?

2 Choose a new style garment. Again, how will you decide what is 'new'? Visit a boutique, a department store, a variety store and a discount store, and find designs similar to your chosen garment. Discuss the differences among the items in terms of price, quality, fabric used, brand-names, amount of intricate sewing required, cleverness of design details, quantity of fabric used, colours used, colour range available, fit, final effect. Since this could prove complicated, give grades of one to five for each of the above qualities. How did each garment score? What can you say about the type of consumer that the retailer is aiming to attract in each case?

The timetabling of fashion

Activity	Activity commenced
Weaving design, including selection of yarns and finishes	October of Year 1
Knitting design, including selection of yarns and finishes	February of Year 2 (knitting is faster than weaving)
Fabric sample production	May/June of Year 2
Fabric sample ranges shown to garment manufacturers	
Fabrics are ordered and produced	
Garment samples designed and produced	
Garment ranges shown to retailer	September of Year 2
Garments are ordered: production	
Retail release of garments	March of Year 3
	Total 18 months

Raw materials
Wool in Australia

From 1890 to 1910, wool was the principal industry in Australia, and everything which affected the wool industry had a resounding effect on every part of the community. The 700 million kilograms (five million bales) of fine wool exported annually supplies half the world's needs, and provides 12 per cent of Australia's export income. Nine per cent of the clip remains in Australia, and coarse wools for carpet manufacture are imported from New Zealand and India.

Because wool is so important to the Australian economy, the federal government established the Australian Wool Corporation in 1971. This body finances wool research, markets the wool clip (through an auction system), and sets a reserve price. If there is a slump in the price of wool, the Corporation steps in and buys stock to prop up prices and protect producers. When prices rise, the Corporation releases some of its stockpile and recovers its expenses.

The International Wool Secretariat, based in the United Kingdom, promotes wool on the world market. It is funded mainly by Australia, New Zealand, and South Africa.

Wool baler in operation. The bales are left partially open for easy inspection before sale.
(Department of Agriculture and Rural Affairs, Victoria)

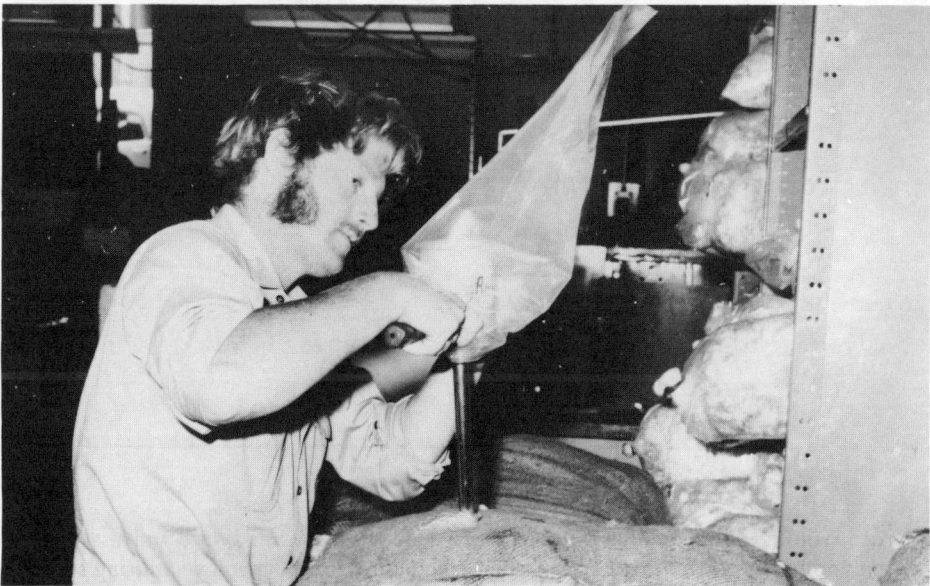

Taking a sample for analysis from the centre of a wool bale. For Australia's credibility as a supplier, it is essential to ensure that bale contents match their description.
(Department of Agriculture and Rural Affairs, Victoria)

Wool samples displayed for auction. Buyers must be able to be sure that the samples are representative of the contents of the bales.
(Department of Agriculture and Rural Affairs, Victoria)

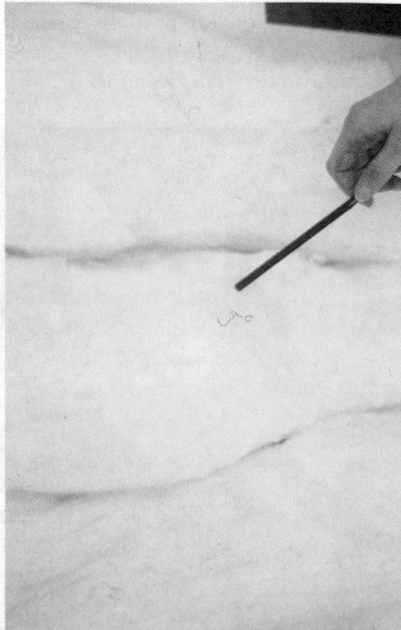

Carded wool contaminated by polypropylene baler twine. Unlike jute and sisal twines, the polypropylene cannot be destroyed by carbonizing. It is not removed during scouring and carding, and its presence may pass unnoticed until the fabric is dyed. Contamination of this sort drastically reduces the value of the wool, and is a severe problem for wool producers.
(Australian Wool Corporation)

Cotton in Australia

Cotton grown in Queensland was important briefly last century during the American civil war. Once the American industry recovered, cotton in Australia went into a decline from which it recovered only slowly.

In 1951 the Australian government started a five-year system of bounty payments to cotton growers in an attempt to stimulate cultivation. Today, the most extensive cotton growing areas are in north-western New South Wales (Namoi and Macquarie rivers), which yields 70 per cent of Australia's cotton crop, in the Murrumbidgee, and in southern Queensland. Current production is around 75 million kilograms, of which 50 million kilograms is exported. The remainder satisfies Australia's needs for medium-quality cotton, but finer grades are imported.

Much of Australia's cotton industry is based on American plant stock, American expertise, and American capital.

Cotton is a hot-climate plant, and because it requires irrigation the decision to grow it involves government as well as the farmer. During growth of the crop it requires inter-row weed control, which is usually by means of herbicides. It is subject to damaging pests which are resistant to insecticides: the boll weevil destroys about one-eighth of the crop. For political reasons – an attempt to stimulate development in the sparsely populated north of Western Australia – in the late 1960s and early 1970s massive amounts of government money were poured into cotton growing in the Ord River irrigation area of the Kimberleys, but the crop failed repeatedly due to damage from insect pests. The attempt has now been abandoned.

Flax in Australia

Until 1964, Australia produced about 600 tonnes of flax per year for the domestic market. This was not enough to meet local demand and costs were high. Flax-growing for fibre was discontinued after this date. Small quantities of flax are still grown for linseed.

Scutching flax in Victoria in the 1920s: a primary industry abandoned.
(Department of Agriculture and Rural Affairs, Victoria)

Cotton

Soviet Union (21%)
United States (18%)
China (16%)
India (9%)
Pakistan (5%)
Africa South America Australia

Wool

Australia (30%)
Soviet Union (18%)
New Zealand (11%)
Argentina (7%)
South Africa and others

Flax

Soviet Union (75%)
Eastern Europe
France
Belgium
Ireland
Japan
and others

Jute

India (40%)
Bangladesh (30%)
others

Silk

China (32%)
Japan (35%)
India
Soviet Union
Italy
Thailand
and others

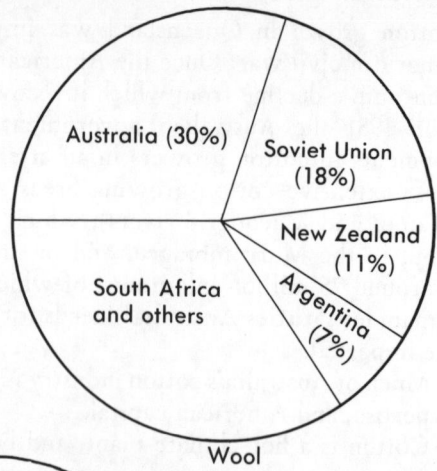

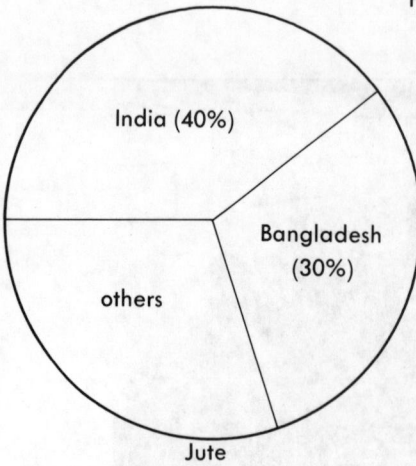

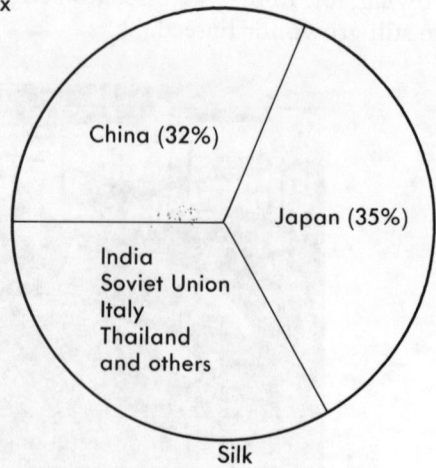

World distribution of natural fibre production. Natural fibre production depends largely on climatic conditions; the synthetics, on the other hand, are produced mainly in the most heavily industrialized countries (the United States, Britain, Japan, West Germany, South Korea).

Synthetic fibres in Australia

Rayon and cellulose acetate were first produced in Australia by Courtaulds Limited in Newcastle in 1933. This, the only rayon manufacturing plant in Australia, was closed in 1976, due to rising labour costs in Australia and competition from cheaper imports.

The only manufacturer of synthetic fibres left in Australia is Fibremakers. This firm began manufacturing nylon in Bayswater, Victoria, in 1958, and has been producing polyester as well since 1964. Much of their output is used for carpet yarns. Another company, Allied Chemicals (Australia), which manufactured nylon at Penrith in New South Wales, has been acquired by Fibremakers, and high tenacity polyester for tyre cords and seat belts is manufactured at their plant.

Australia produces all its industry requirements of nylon and polyester, but acrylics, rayon, acetates and elastomerics are all wholly imported.

World fibre production

To produce a natural fibre, a country above all needs the appropriate climate and conditions; that is why, for example, Australia produces mainly fine wools and New Zealand produces mainly carpet wools. Almost as important, however, is political and economic support for the industry: at the same time as bounties were introduced to stimulate Australia's cotton industry, the bounties on flax production were being withdrawn.

Similarly, the production of synthetic fibres is a political and economic issue. Would Australia's rayon industry have closed if it had been given the support that the wool industry receives? The raw material for rayon production is wood pulp, yet 14 per cent of the world's rayon is produced in Japan, a country which has almost no forestry industry. The raw materials for other synthetics are petrochemicals, yet Japan, which imports all its petroleum needs, manufactures a large share of the world production of these fibres as well. Why Japan?

How can countries such as Japan, Switzerland, the United Kingdom, afford to import textile fibres and re-export finished textiles? Why does Australia sell so much of its cotton and wool to countries like Japan and then re-import the finished textiles to meet local demand?

The issue of the location of industry is extremely complex. It is a result of government support and economic policy, of labour costs and available technology, of geography and historical accident. In turn, it affects employment, the environment, and political relations between nations.

TRY YOUR HAND

1   Journals such as *Australasian Textiles, Textile Horizons,* and *Textiles Asia* carry many articles about the textile industry in Australia and in Australia's trading partners. Articles also appear in newspapers from time to time. Collect any material you can find on the manufacture of synthetic fibres. What issues do the writers of these articles see as important in the manufacture and marketing of synthetic fibres? How does this relate to whether the fibres are manufactured in Australia, or imported?

2 The results of wool auctions are reported in the newspapers. Use these and any other sources of information to discover the current price of wool, which gradings sell best, and who the most important customers are. In what form do the customers buy the wool? What do they use it for? Is the same grading of wool always preferred? What implications does this have for the Australian wool grower?

3 Using textile magazines, Department of Trade publications, and *Year Book Australia* and any other sources of information, find out what the projected development plans are for the Australian cotton industry. Where are our major markets? Is the amount of cotton grown to be increased? Will new cotton gins be required? Are there plans to grow different qualities of cotton?

Labour and technology

There are three kinds of resources needed for the production of textiles: raw materials; technology; and labour. A considerable input of labour and technology is required to turn raw materials into textile products.

Different countries have different access to these resources. Some have abundant raw materials: wool, cotton, oil, a chemical industry; others have the money and the expertise to employ advanced technology; some have a large population which can be employed as a cheap labour source.

Trade between countries with different resources allows each to make the best use of its resources and production capacity.

The interdependence of national industries

The following sketch gives an extremely simplified picture of world trade, but it serves to illustrate the interdependence of the different countries.

The parable of the three lands

There are three countries. One is called Baaland. This country has a lot of space for sheep and growing crops. There are not many people in Baaland so the cost of their labour is high. With their high wages they can afford to be choosy. The few people of Baaland each want to buy a little of a large variety of things. This country produces a lot of raw materials – mainly food and natural fibres – more, in fact, than its population can use.

Populand's chief resource is people. It has so many that it needs all its lands to produce enough food for them. Not all the people are needed for food production, however, and there are many spare hands to work in factories. Labour in this country is cheap, with so many willing workers – in fact, there are not enough jobs for all. Those who are employed earn only low wages, and so cannot buy many varied or expensive textile products.

Clickland has been industrialized for a long time. For many generations its inhabitants have worked in factories and they often make changes and improvements to the machines they understand so well. Although it costs a lot of money to build the complex machines that Clickland people invent, a few machines can do the work of many people. Because each machine produces goods so quickly, there are large quantities of products made in Clickland. Because the machines work so efficiently few workers are needed to supervise them. These workers, however, earn well, and are able to buy large amounts of many different things.

Baaland's people need goods in exchange for their surplus food and raw materials. Populand's people need more food, but have no more land to produce it. They are happy to exchange their labour for food and for some of the things that the Clickland people make. The people of Clickland need raw materials to feed their machines. They also like some of the hand-made things that the cheap labour of Populand can produce.

Trade allows each country to dispose of its surpluses, and to receive what it lacks. So the three countries trade with one another. They cannot simply exchange goods – that would be far too complicated – so they use money as a medium of exchange.

Sometimes it is the government of the country which handles the money, sometimes it is individual traders. Each country and each individual tries to get as much of this precious trading resource as possible. It is quite a scramble, and as the competition goes on each country's situation changes slightly.

TRY YOUR HAND

1 Name some countries which are like Baaland and Clickland. What is the importance of the size and character of the *internal* markets in each of these countries?
2 What would happen if Clickland technology were introduced in Populand:
    a to the people of Populand?
    b to the cost and availability of goods everywhere?
    c to the people of Clickland?
3 What would happen if Baaland adopted Clickland technology:
    a to the goods produced?
    b to the market for the goods?
4 If Populand had Clickland technology, where would be the first market for its goods?
5 How can Clickland find *new* markets for all the goods its machines produce?
6 What effect does unemployment have on the selling of goods?
7 In the long run, Clickland technology will be available to all three countries. What would you expect to be the long-term result for all three countries and their people?

The cost of labour

Many aspects of the textile industry are labour-intensive; that is, many working hours are needed to produce textile products. The cost of labour is therefore an important part of the cost of textiles production. Manufacturers, of course, want to produce their goods as cheaply as possible, and so will, if they can, produce them in the countries with the cheapest labour.

The following case study on the clothing industry illustrates some of the issues involved.

The six machinists

Margaret is a machinist living in Australia. She is a member of the Textile Workers Union, and clears $200 for a 35-hour week. She is the second wage earner in her family, and her mother takes care of her daughter after school. The family of four live in a three-bedroom house in a suburb of Wollongong. Margaret is a keen supporter of the 'Altogether Australian' trademark scheme. 'We will all be out of a job unless we support the local manufacturing industry,' she says.

Ean works as a machinist for an Australian-owned company in Malaysia. Her company does much of the design, planning, and purchasing work in Australia, then delivers fabrics and cut pieces to the Kuala Lumpur factory for assembly and finishing. It is worth the company's while to ship goods back and forth between Malaysia and Australia because Ean earns only $27 for her 60-hour week – including overtime. Her wages are needed to help support her family of nine, of whom only three are working.

Wingyan works for a Hong Kong manufacturer. She earns $80 a week, and is proud of the effort she puts into her work. 'If I were less conscientious, or slower, I might lose my job. But if I work hard, I will be able to ensure a good education for my children. As long as I have a job we are able to enjoy good living standards, even though we have no insurance against illness, and there are no old-age pensions. My children's success will be enough insurance for my old age.'

Mariko is a Japanese machinist. She works for a small company which does contract work for one of Japan's largest textile mills. Although the company is small, it has benefited in recent years from many technological improvements, and Mariko has been trained to use the semi-automatic machinery that produces shirts accurately and quickly. Her wages are only about $100 a week, but she knows that her job is secure, and she has a great deal of loyalty to her company and is proud of the high quality work she produces.

Salida finished school in Sri Lanka with excellent marks, but she had no hope of finding employment until a garment factory was established in her town to manufacture low-cost shirts. Although she makes only $6 a week, she is glad to have a job, as she is the only income earner in her family. Six brothers and sisters live with their parents in a single room. She is worried about their future. 'Our industry was established with assistance from the World Bank. It was necessary to provide jobs for so many people. We cannot all work on the land or on the tea plantations. If other countries refuse to buy our products, if they impose high tariffs on our goods, then many of us will lose our jobs and starve.'

Janet is a machinist in Manchester in the United Kingdom. Her family has worked in the textile industry for three generations. Today she is worried about the growing unemployment in Britain. 'If we did not import so many cheap shirts from Asia, our factories would not be in danger of closing down. A third of the garments on British bodies are made in another country. It is a large market here, and we should protect it, instead of giving it to other countries where people are willing to work for a few pennies. We must not lose our traditional industries and skills.'

How realistic is this type of marketing exercise in a world of interconnecting and interdependent economies?
(Dale Mann/Retrospect)

TRY YOUR HAND

1 Which of the six countries appears to have the most secure textile industry? Which has the least secure industry?
2 What changes have happened in the textile industries in each country? Is it likely that conditions will change further in the near future? What are the possible changes, and what will be their effect on the six workers?
3 Where is the market for the shirts produced by each of the six countries? How does the industry in any one country affect the industry in the others?
4 What is a multinational company? How could a multinational company benefit from trade between the six countries?
5 If the cost of labour was the only consideration, which country would manufacture most of the shirts?

This picture is, of course, extremely simplistic.

The Japanese company could not decide overnight to shift all its manufacturing to Sri Lanka: it employs a skilled labour force which it had to train and which is not easily replaced; and it has expensive technology, which makes the cost of setting up a factory very high. The Malaysian industry survives because its savings in labour costs more than offset the transport costs between Malaysia and Australia: a large increase in transport costs could change this balance. Sri Lanka has to import all its raw materials, and so is vulnerable to shipping costs both ways, and to the cost of these raw materials. All the other countries produce the polyester used in their shirts, but Australia is the only country which produces its own cotton. Each country imports the raw materials for its polyester, so each is affected by oil prices. Nor is the cost of labour the only cost after raw materials and transport: the factories need a source of energy, and energy costs vary enormously between countries. Energy in Australia is relatively cheap, due to hydroelectricity, coal, and oil and natural gas. Britain has a large coal industry and oil to supply its energy needs; Sri Lanka is energy-poor, importing its oil and lacking the electricity-generating capacity needed by a large industry base. Governments try to encourage their own country's industries at the expense of foreign competition, and so the highly-taxed Hong Kong shirts on the Australian market may not be much cheaper than the subsidized Australian product. Consumers like Margaret in Australia or Janet in Britain may deliberately choose not to buy the cheapest product ... There are many other factors involved as well.

Polypropylene (left) versus jute sacking (right). The one is a product of an advanced petrochemical industry; the other an agricultural product of one of the world's poorest countries.

The wages of the workers in the various countries reflect only their cost to the manufacturer. The figures are a result of the cost of *money* in the various countries. What they mean is that for the cost of employing Ean in Malaysia you can buy only an eighth of the oil, gold, or some other international commodity that you could buy for the cost of employing Margaret in Australia. Margaret may cost the manufacturer more, but that does not mean she is overpaid: wages must be compared to the cost of living (food and housing) before their value to the workers can mean anything. In those terms, Mariko in Japan is not paid very well, as the cost of living in Yokohama is quite high. She does have security of employment, however, and this may be more valuable to her than a few more dollars.

In the British and Australian examples, the wages of employees may not reflect the average labour costs. The clothing industry often employs *outworkers,* people who take the cut-out garments and assemble them in their homes for so much a garment (piece rates). The manufacturer does not have to pay for factory space or energy costs, or any of the overheads associated with employing staff. Although it is not always the case, some unscrupulous manufacturers exploit vulnerable sections of the workforce, such as migrant women with little English, by employing them as outworkers at appallingly low piece rates.

Labour-saving devices

In the textile industry, robots lift bales and rolls of fabric and help prepare the warp for looms. Automatic knotters and yarn splicers keep production running smoothly without the need for constant human supervision. In cotton processing, measuring, weighing, and fibre distribution are automatic. In the dyehouse, computers guarantee the accurate weighing of dyestuffs, the correct heating of the dye liquors, and precise timing of the dyeing process. In garment manufacture, computerized cutters cut garment pieces from up to seventy layers of fabric at a time. The results of automation are better quality yarns and fabrics, lower production costs, and greater consumer satisfaction – as well as increased unemployment.

The cost of unemployment

The cost of an employee in Australia is roughly double the employee's wages. There is payroll tax, workers' compensation insurance, sick pay and holiday pay, the cost of providing facilities at the work-place. These costs are borne directly by the manufacturer, and passed on to the consumer as part of the price of the goods. If the employee is replaced by a machine, what does it cost, and to whom?

There is the cost of the machine, and the cost of running it. These costs will still be passed on to the consumer as part of the price of the goods. The displaced employee will receive unemployment benefits, which will be paid for by the consumer's taxes. The ex-employee will no longer be paying taxes her/himself, and so the tax burden on the rest of the community rises to make up the lack. Instead of paying the Medicare levy, the unemployed person on social security receives free medical cover, and possibly also rent assistance and other support. These costs are not borne by the manufacturer who bought the machine, and so do not appear in the price of the goods, but they are borne by the community. In addition, because the person now has a much lower income, she or he has less buying power and can buy fewer goods: the reduced demand means the manufacturer sells less product, which is reflected in higher unit costs and fewer jobs.

This sketch is the worst possible case. In times of economic growth and high employment, the employee's chances of finding alternative work are good, and so the cost savings from using the machine benefit the community. However, if the economic climate is such that many people are out of work for a long time, the savings to the consumer will be eroded by the cost of supporting the unemployed, although the savings to the manufacturer may not be affected in the short-term.

TRY YOUR HAND

The tweed industry accounts for 45 per cent of the £33 million annual turnover of the crafts industry in Scotland, and the economy of the Outer Hebrides is largely dependent on it. Eighty per cent of the tweed is exported. The textile is claimed to be an indigenous product of a discrete culture, and this is part of its appeal to the consumer. To use the name Harris Tweed, the cloth must be made from 100 per cent pure virgin Scottish wool, dyed, spun, hand woven, and finished, in the Outer Hebrides. However, faster hand looms have increased each worker's output (the number of weavers fell from 1300 in 1970 to 650 in 1983); and the trade mark permits industrialized spinning, dyeing, and finishing. These changes mean that the product is less 'authentic', especially as some of the new dyes bear no relation to the colours traditionally obtained from local plants and lichens. The industrialization of these processes also has caused social changes, as the jobs replaced by machinery are the traditional female roles.

1   If much of the market value of tweed is due to its appeal as a traditional craft product, what is the possible effect of increasing production and introducing modern technology in some steps of the manufacture?
2   How would the mechanization and increased production affect the economy of the islands?

Milling tweed: a labour-intensive community activity. The Harris Tweed trade mark allows finishing to be industrialized, removing this source of employment.
(Harris Tweed Association)

3   Does the restricted use of the Harris Tweed name protect the industry, or does the use of non-traditional methods, dyes and colours risk devaluation of the name?
4   There is no suggestion that the mechanized product does not perform as well as the traditional product. Why should it matter how the tweed is made?
5   Do you see the future of the indusry as lying in increased production, with the risks that entails, or in keeping the product as close to its craft origins as possible? Justify your answer.
6   Do you see any worth in trying to maintain a genuine craft industry, or is it just romatic nonsense? Justify your answer.

The role of government

In countries under a centralized economic system, the government decides how the resources of the country should be utilized, how much to produce of different items, who to trade with, and so on. In Western-style economies the government has only an indirect effect on the country's production and consumption patterns. Industry is in the hands of individual decision-makers who regulate production in line with what they expect will be the demand and in a way which will produce as much profit as possible for themselves.

Nevertheless, it is the responsibility of all governments to regulate the country's trade and industry to the greatest benefit of the whole population. Each country contains many different industries with opposing interests. The people in all these groups lobby the government to take steps to favour their particular industry. Fibre manufacturers would like bounties and export incentives, and a ban on the import of foreign fibres. Yarn manufacturers want to prohibit the import of yarns. Fabric manufacturers want free yarn imports so that they can produce cloth at competitive prices. They also want to restrict the import of foreign fabrics. Garment manufacturers need freedom of choice and low fabric prices, so they are at loggerheads with the interests of the fabric producers.

Somehow, it appears that somewhere else in the world, there is always someone who makes yarns and fabrics more cheaply than the local industry can. Yet if there are no buyers for the products of local industry, the local industry will die. This situation applies to most countries in the developed world, and even to some less developed countries.

Governments use a number of different tactics to regulate imports and exports. The three main approaches are through *tariffs, quotas,* and *bounties.* To these can be added export incentives, taxation incentives, trade restrictions, research and development grants, manufacturing standards, local content requirements, and incentives to industry to decentralize.

Tariffs

A TARIFF is the money that an importer must pay to the government's customs office when goods are brought into the country.

Customs houses – Wodonga – Victoria. Passing the customs officers at Wodonga. (*Australasian Sketcher*, 13 August 1881. National Library of Australia)

Tariffs are determined for each category of goods – e.g. fabric of a certain weight and construction – so that its imported cost is brought into line with the cost of manufacturing it locally. This helps local industry by discouraging imports. Effective tariffs ensure employment and satisfactory living standards for local workers. If a certain type of goods is not manufactured in the country, a tariff on imports would mean that the consumer was paying higher prices without helping local industry.

TRY YOUR HAND

The tariff on the import of a woollen fabric is $2.00 per metre plus 10 per cent of the imported cost per metre. You are a fabric wholesaler, stocking both local and imported fabrics. Since the local labour costs and taxes are higher than overseas, the landed cost (cost before duty) of a metre of woollen from Japan is $5.00 while the local equivalent fabric is $8.50.

1   Calculate how much the Japanese cloth will cost per metre after the tariff has been paid.
2   What risks does a wholesaler run in using imported rather than local products? List all the factors you can think of which will affect the cost to the wholesaler of importing goods, or which may jeopardize supplies.
3   What are the advantages to a wholesaler of stocking some imported and some local goods?

Quotas

In order to stop particular goods entering the country in unlimited quantities, the government can issue a QUOTA on the amount that can be imported. An absolute amount of goods (say some many thousands of square metres of woven fabric of a particular quality or fibre content) is specified as the maximum that can be imported.

To establish the quota figure, historical precedent, lobby pressures and government planning are considered. The quota applies to the whole industry. No single trader could possibly want to import all the fabric included in the quota, so individual traders place bids in an auction system for a certain share of the quota. The quota is allocated among the highest bidders. Traders may further sell or buy quotas among themselves. All this adds to the final cost of the imported item, even though the quota system is less inflationary (less likely to force up prices) than the tariff system.

Bounties

In order to encourage local industries, governments may offer BOUNTIES (cash grants) to producers. Such bounties are distributed according to the amount and quality of the goods produced by each manufacturer or grower.

Bounties are mostly given to newly established industries, to encourage their growth and competitiveness through the early years. The cotton industry in Australia was started on a five-year bounty plan, during which $20 million was paid out to cotton growers. This allowed the industry to establish itself successfully: Australia is now a cotton exporter.

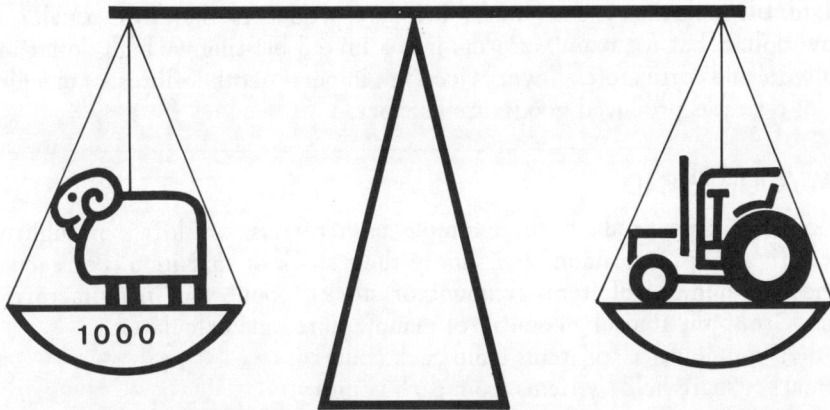

An item of machinery may have the same value on the international market as 1000 merino fleeces. If the dollar is weak, this value may correspond to $25 000. If the dollar is strong, this may correspond to only $15 000. In other words, a weak dollar means more dollars for the primary producer, but more dollars paid out by the importer. A strong dollar means fewer dollars for the primary producer, and fewer dollars paid out by the importer.

Other tactics

Governments can also have *indirect* effects on an industry. Regulations concerning waste disposal or performance standards can drastically affect the viability of manufacturing practices. For example, Japan has very strict laws concerning water pollution and waste disposal. Wool scouring produces a great deal of waste material, and so many Japanese worsted manufacturers have decided to establish wool scouring and top making plants in Australia, and to import clean wool tops, rather than build costly new factories in Japan that conform to the stringent anti-pollution regulations.

Exchange rates

The EXCHANGE RATE of a currency is its value in terms of the currency of another country, or its value in terms of some international commodity. When the value of the dollar is high, few dollars are needed to buy goods from other countries, but Australian goods are expensive in terms of the currencies of those other countries. When the dollar is weak, many dollars are needed to buy from overseas, but Australian goods are relatively cheaper on international markets.

This means that a *high value of the dollar favours importers* and disadvantages local producers and exporters, but a *weak dollar favours exporters and local producers* and disadvantages importers.

Governments try to stabilize the value of their currency in terms of their perception of the needs of the whole economy. Regulation of interest rates, and controls on the amount of money entering or leaving the country, are two of the tactics they use.

In relation to textiles production, a low dollar means that wool or cotton growers receive more dollars for their produce, but imported fibres, garments, and textile machinery cost more. Primary producers therefore usually want a low dollar, but for manufacturers it is a mixed blessing. A high dollar means that wool and cotton fetch lower prices, but imports of other fibres, of machinery, and of overseas produced goods, are cheaper.

TRY YOUR HAND

Choose a textile product, for example bath towels, or shirts, or nightwear. Select five shops at random and survey their stock of that item. For each item found, list number of items (amount of stock), country of manufacture, and price. Group the items by country of manufacture, and calculate:
(i) the average price for items from each country;
(ii) market share held by items from each country.

$$\text{market share} = \frac{\text{total number of items from that country}}{\text{total number of items in whole survey}} \times 100\%$$

1   How does price compare with market share?
2   Discuss the role that imports play in the Australian market-place for your chosen item. How do Australian-produced goods perform?
3   How do your results compare with those of students studying other items? Explain some of the differences you found.
4   Find out from the tariff board or a customs agent what the import tariff is for your chosen item. How does this influence the cost of imports relative to the Australian product? How does this affect the choice and price of goods available to the consumer?

The following table illustrates, in simplified form, some of the ways government policies have influenced the wool industry in various times and countries.

Effects of government intervention in the wool industry

Year	Action taken	Effect
57 BC	Romans under Julius Caesar invade Belgian Gaul; find Flemish and Walloon tribes are producing fine wools. Cloth ordered for the legions. (Government orders products from local industry; direct spending of government money in return for goods.)	By 300 AD Flemish wool cloths are famous, and are too expensive for all but the Roman nobility. By the tenth century, the Flemish are textile craftsmen rather than farmers. (Industry stimulated and skills created. *Danger:* guaranteed market with no price and quality controls can be counterproductive.)
1337	Edward III of England prohibits export of greasy longwools to encourage local processing.	In England, increased cloth production. Increased government revenue due to tax on exports. In Flanders, weavers bound by guild rules to use English wools are impoverished. Non-guild rural industry develops 'French' comb to use the shorter Spanish Merino wool.

		(Retention of raw materials stimulates local manufacturing skills. Traditional export market turns to other suppliers and develops new skills: competitors adapt to new conditions.)
1610	James I of England pays off large personal debt to Alderman Cockayne by granting him a monopoly to finish all broadcloths before their export from England. Previously, the raw exports were expertly finished in Flanders. (Foreign politics outside industry control.)	This cloth finishing is of poor quality; the cloth finds no export market. By 1616, half the looms in England are idle. By 1641, this and the Thirty Years War in Europe combine to ruin the English wool and linen industries. (Local manufacturing skills stimulated *only* if price and quality are competitive; otherwise both primary and secondary industries are damaged.)
1825	Colonial wools allowed to enter England duty-free. (Lifting import duty on raw materials leads to loss of tariff revenue, but income tax revenue increases.)	Australia's wool production increases enormously. (Raw material producer (Australia) benefits; user industries (Britain) are stimulated.)
late 1850s	Victoria's gold rush is over, with many unemployed and no money to buy goods.	Textile factories supplying diggers close down: more unemployment.
1878	Victorian Tariff Bill with over 300 categories including livestock and manufactured goods from other Australian colonies. (Tariff protection of local products against cheaper imports. Some small revenue gain to government from tariffs, but offset by loss of export revenue: inflation.)	Tariffs raise the price of imports to more than that of local products: industry re-established, economy stimulated. Duty paid on all goods entering Victoria. (High wages and standards of living assured. Industry protected, and therefore jobs. No ability to compete on export market means costs are transferred to local consumers: inflation.)
1973	Australian Wool Corporation established to operate a minimum reserve price scheme for all wool sold at auction. Funds (up to $150 million) provided by government, the balance from growers by 5 per cent levy on shorn wool. (Large initial capital outlay, recouped as stock is sold later.)	The minimum reserve price does not allow the value of wool to drop, even when markets are over-supplied: the Australian Wool Corporation buys the excess. This provides a stable income for growers. (Protects profitability of producers, but useless if alternative sources of the product are available elsewhere in the world.)
	Research funded at considerable cost to government: recouped only when industry implements research findings, and products can be sold profitably on local and world markets.	Industry often needs direct incentives to employ the results of research.

TRY YOUR HAND

Read the following articles about the textile industry in various countries. These articles show the conflict between the need for local employment and world trade, and illustrate the difficulties governments and economic planners have in striking a fair balance.

Trade commentary: Swiss clothing makers

Following the favourable results achieved in 1980, the important Swiss clothing industry entered rough water during 1981. During the first quarter all indices were below the corresponding figures for 1980 and this unsatisfactory situation was not improved in the second quarter: although production rose 2.4% and the nominal order influx by 7.6%, real turnover continued to decline by 4 to 5%.

The decline in orders booked could be compensated for by a reduction of work in hand and by re-stocking during the first six months, but an adjustment of production to the changed market conditions became necessary during the second half of the year, aggravated by the continuous increase in overheads which is affecting earnings.

In addition to sales resistance in the home market, exports during the first half of the current year fell by 1.4% to SF329M but the volume shrinkage of exports totalled 7.5%. Due to the economic recession, export reductions were as high as 29.2% in Sweden, 14.9% in Austria, 3.4% in Japan and 1.6% in the UK. Sales to Germany BRD were practically unchanged, but shipments to the USA rose 68.3% and those to France by 12.6%.

Protectionist policies and various restrictive practices, combined with imports from Asiatic producers (which rose 37.6% in the first half of 1981) are causing increasing anxiety. Much attention has, therefore, been paid to a recent market investigation which indicated that, in contrast to the textile manufacturing industry, the competitiveness of the Swiss clothing industry has markedly deteriorated during the last ten years. Although a further period of stagnation is expected in the near future by quarters close to the Swiss clothing industry – which comprises at present 630 firms employing about 28 000 persons – recent developments are bound to result in a renewed effort to improve both quality and marketing and to re-establish its former high level of competitiveness in home and overseas markets.

Textile Horizons, February 1982

Import protest

US textile and apparel companies are enlisting their employees in the battle to reduce imports by asking them to encourage retailers to stock US-made merchandise. Workers are being asked to hand out to retailers thousands of cards which say 'Imported textiles and apparel threaten my job and give it to foreigners. For me to be a customer, I have to have a job. I hope that you will consider this and stock textiles and apparels made in the USA'.

One large Southern chain of 73 discount retail stores has even started its own 'Buy Southern' campaign. Most of the nation's textiles and garments are made in the six south-eastern states. Many manufacturers are sewing 'Crafted with pride in the USA' labels in their garments.

However, some US textile machinery manufacturers are complaining that these same textile and apparel companies are buying almost exclusively foreign-made textile machinery.

Textile Horizons, December 1983

China refutes dumping claim

China has denied that it is 'dumping' textiles on the US market. US textile and apparel trade groups and their unions have accused China of 'unfairly subsidizing exports through the use of dual exchange rates that financially reward textile exporters'. The US Commerce Dept has agreed to investigate. China's Ministry of Foreign Economic Relations and Trade stated that imposition of any new limits on imports from China would endanger economic and trade relations between the two countries. China and the US have only recently worked out a complicated and, in the US unpopular, bi-lateral textile trade agreement after nearly a year of negotiations which were broken off frequently.

The People's Republic of China have also now made a formal request to become a part of the General Agreement on Tariffs and Trade's Multifibre Arrangement (MFA).

Textile Horizons, December 1983

Israeli spinning cut

Israeli cotton spinners have had to reduce production significantly due to a sharp fall in exports (caused partly by the weakness of European currencies and partly by low-cost producers), and due to imports of lower-priced yarn by local factories producing underwear and other garments for export.

In 1980, total cotton yarn output came to 22 000t. Of this, 10 300t were surplus to the needs of weaving and knitting departments of integrated concerns or represented the output of spinning mills not part of a vertical concern. However, 5800t were sold to local factories not having spinning facilities of their own, while 5400t went for export. In 1981, sales abroad fell to less than 2000t (1800t in the first 11 months). Moreover, from the beginning of the year, some local firms began importing cheaper foreign yarn. These were firms working for export, which do not have to pay customs duty.

In view of the continuing low level of prices charged by some countries, eg Brazil and Turkey, the Israeli authorities in November 1981 imposed specific duties (not ad valorem ones) on imports from sources other than the EEC and the US, which in effect bring their costs to the local level. This duty is only being charged on yarn intended for production for the domestic market.

At the same time, production is being cut back further from the 1981 level of 20 000t (itself a reduction of 10% from that of 1980). One spinning mill in the development town of Beit Sha'an, which had been operating at a deficit, was closed down, and other firms have gone over from three-shift to two-shift operations. Clal-tex, the cartel which handles all sales of 'surplus' yarn (ie surplus to own requirements), both on the home market and for export, in September ordered a cut in 'surplus' production from 750t to 500t a month. In view of a slight expansion in home market demand, this figure is now being raised to 600–700t, with 600t intended for the local market and only 100t for export.

Needless to say, a large part of yarn production used by integrated mills and supplies bought by local factories from Clal-tex, finds its way to foreign markets as fabrics and finished garments. But overall exports of yarn, cloth and finished garments in 1981 are not likely to exceed US$365M, as compared with exports of US$455M in 1980. While some of this is due to the weakness of European currencies, it nevertheless represents a drop in the region of 5%.

Textile Horizons, February 1982

1 What were the problems of the Swiss clothing industry?
2 Why did these problems arise?
3 What steps are the Swiss taking to try to improve their position in the world markets?
4 Why did the United States textile manufacturers start their campaign?
5 Are they consistent in their attitude to locally made products?
6 If their campaign is successful, predict some possible outcomes in the local (United States) scene, and in terms of international trade (the trade agreement with China).
7 What is the situation faced by each of the sectors of the Israeli industry – cotton spinners, producers for the export market, local producers?
8 What similarities and differences exist between the needs of the different groups?
9 From the information in the article, does a cut in Israel's spinning output seem the best solution? Why, or why not?
10 To what extent do the writers of these articles see the problems as arising within the international textiles industry, and to what extent do they blame issues such as exchange rates which are beyond the industry's control?

Industry in Australia

Originally, all textiles for use in Australia were imported from England. The first local product was Parramatta cloth, woven by female convicts. The establishment of the wool-growing industry in the early 1800s meant that raw material was available for a local textile manufacturing industry.

Further growth took place in Victoria, where the gold rushes caused a sevenfold increase in the population in the early 1850s. The end of the rush in the late 1850s left a large unemployed workforce, and in 1866 the Victorian government imposed tariffs on imported goods. This was successful in creating employment: in 1865 the Victorian Woollen Cloth Manufacturing Company (later bought by Godfrey Hirst) was founded at Barwon River, Geelong, and by 1875 four woollen mills, producing mainly blankets, operated there.

Australian workers have always enjoyed a high standard of living compared to the rest of the world. Work, other than convict labour, was valued highly in a country with a small population. Because of the high labour costs, fabrics produced in Australia were more expensive than English cloth, even after transport costs were included, but tariffs on the English cloth raised its price above that of the local product. This allowed Australian goods to compete successfully on the local market, but they were too expensive to compete on the open market in the rest of the world. In the protected home environment, the industry thrived and provided jobs for many workers. As the workers earned good wages they were able to buy many different kinds of goods, including their own textile products. Tariff protection therefore ensured jobs and high living standards, but it limited the range of goods available and prevented the export of local products.

There are many countries in the world where labour costs are lower than in Australia, therefore labour-intensive industries can operate more cheaply in those countries than here. If their goods were brought in without any tariff being paid they would be much cheaper to the Australian consumer. It was calculated that in 1977 each Australian family paid $200 for the tariffs that protect jobs in the textiles industry.

In 1973, the tariffs on textiles, clothing and footwear were reduced by 25 per cent. The result was that wholesalers and retailers imported large quantities of these products, giving consumers access to an abundance of cheap goods. Money flowed out of the country to pay for the imports, leading to a money shortage and credit squeeze. Local industries lost business, and many closed down, creating unemployment. In the four years from 1973 to 1977, employment in the clothing industry dropped by more than 70 000 jobs, from 2.8 per cent of the workforce to 1.5 per cent. Some of the workers would have found other employment, or dropped out of the labour market altogether, but as there was a general rise in unemployment at the same time many would have needed to be paid unemployment benefits. So the money consumers saved through lower tariffs and prices had to be paid in social security.

The Industries Assistance Commission, after many years of investigation, and the collection of evidence from all sectors of the clothing and footwear industries, recommended that tariffs be lowered to a point where only the most efficient companies could survive. Other companies were to change to less labour-intensive industries which could compete on world markets. It made no recommendations as to which 'efficient' industries were to re-employ the workers made redundant by the dying textile industry.

The IAC recommendations were not accepted by the government. Instead, in 1980, the industry was given a seven-year protection package. During the operation of this protection package the industry was to prepare actively for a less protected future.

TRY YOUR HAND

1 List the advantages and disadvantages faced by the Australian textiles industry.
2 What steps has government taken in the past to assist the Australian industry? What has been their effect?
3 What is the result of policies which make Australian manufacture less profitable?
4 If tariffs are abolished, what is the effect for the consumer in terms of cost of purchases? in terms of choice available?
5 In the long term, what is the effect of unemployment on consumer behaviour?
6 An industry decides to manufacture in Australia, and asks for tariff protection. It is labour-intensive, and does not make use of the latest technology to improve its efficiency. As a result, its prices rise compared to its overseas

competitors, and it asks for more protection.

a How could a government decide if the industry was efficient?

b How could the industry be encouraged to become more efficient?

c What would happen if tariff protection was removed suddenly? If the government proposed to phase out tariff protection? If instead of tariffs, some other form of protection, such as quotas, was introduced? (Quotas do not increase the price of imported goods as much as tariffs do, but they limit the amount available.)

d How does the size of the market affect the efficiency of an industry?

e What is the importance of having many different manufacturing industries in each country? Consider future contingencies as well as the current situation.

7 If the big glossy advertisements in textiles magazines are inserted by the industries which are doing well, which industries in Australia are the most buoyant at the moment? Survey textiles magazines to find out.

8 Read the following article. What do you predict will be the result of the Industries Assistance Commission's recommendations, if they are implemented? What could be the result of industry uncertainty as to what government policy will be in the future?

Australian assistance scheme begins

Australia: The textile industry here has just begun a new programme of assistance based on tariffs, quotas and bounty. In introducing the programme to parliament in late 1980, the government said that it recognized the special need for protection of the industry but that the overall objective was gradual reduction in protection over the period. The industry initially welcomed the package for the stability it apparently offered, but now is not confident that it will see out its course. The uncertainty is largely caused by an Industries Assistance Commission (IAC) recommendation for across-the-board large-scale reductions in the protection of Australian industry against imported goods.

The commission's recommendations, which it says are not hard and fast, are contained in a discussion paper issued in December. The paper is part of the IAC's deliberations for a government-requested inquiry it is undertaking on the advisability of general reductions in protection. Entitled *A consideration of options,* it examines seven possible methods of reducing the level of protection for manufacturing industry overall, not just textiles, and favours one entailing abolition of quotas and replacement by tariffs which would be reduced over 15 years – and 'preferably ten years' – to a maximum of 20%.

The commission says that the mechanism selected to achieve the desired protection cut should serve three objectives:

It should not only reduce the average level of assistance but should also reduce disparities in assistance;

It should generate a less complex and more stable structure of protection; and

Industries in the program should start the adjustment process as soon as possible.

The options preferred by the IAC to achieve these goals are:

Straight-line adjustment where a target rate of 20% protection is achieved over seven periods;

Combined target and percentage reductions so that all rates over 15% are to be reduced to 15% or by 50%, whichever is the smaller reduction over seven time periods.

When the government's terms of reference for the inquiry were announced last August, there was considerable speculation that they had been framed to exclude all industries on which the government had taken a recent decision on protection – the major ones at that stage being the textile, clothing and footwear (TCF) industries. These industries also lobbied strongly to be excluded. The reference was, however, vague and the commission has forthrightly rejected any such exclusion.

The final IAC report on cuts in protection is required by the government in February, and the discussion paper has been issued for comment at public hearings in January.

Textile Horizons, February 1982

9   Invite a representative from a local textile manufacturer or from a textile union to your school to give the industry or union point of view on trade and protection and the viability of Australian industry.

Concerns of modern industry

Until recently, the main concern of technology has been to produce goods faster and more cheaply. Increased availability of goods at lower prices was supposed to mean more sales to more people; demand and therefore employment could expand indefinitely.

For the first years of the industrial revolution this must have appeared true. However, it soon became clear that demand for goods was not infinite, that the smoke and soot poured into the sky did not vanish in the wide blue yonder, that the sea was not a bottomless lake into which rivers could pour their poisons with impunity. Even the sources of energy would not last forever. The earth was shown to be fragile, and industry – often due to community and government pressure – began to take steps to reduce the damage it was doing to the environment.

Scientists, engineers, and managers of industry today are concerned with more than speed of production. Energy conservation, and air, noise and water pollution, are issues central to modern innovation.

Energy conservation

The production of textiles requires a lot of energy: energy requirements may contribute a third of all production costs. Some of this energy is used for lighting, heating, air-conditioning, and humidity and cleanliness control, but over 60 per cent is used in wet processing.

Efficient heating and lighting systems, and new air circulation equipment that makes use of natural convection currents (in buildings of improved design), all contribute to a significant savings in the energy bill. Automatic energy minder systems ensure that lights and air-conditioning are not left on after working hours. Where possible, round-the-clock shifts minimize wastage due to the warm-up and shut-down of boilers and other equipment.

The processes which use most energy during textile production are those that involve the use of water. Water must be heated, and wet fabric must be dried – therefore many energy conservation projects have centred around the dyehouse. Counterflow scourers, low liquor-ratio jet dyeing machines, padding mangles, thermosol, and foam dyeing and finishing processes are all developments that aim at using a smaller volume of heated water during dyeing. Heat exchangers allow waste water from the completed dyeing to heat the water to be used for the next dyebath.

Yallourn W power station, Victoria. Although electricity is relatively cheap in Australia, efficient use of energy can result in substantial savings to manufacturers.
(State Electricity Commission of Victoria)

Other energy saving devices in the dyehouse are involved in drying. Counterflow hot air dryers direct hot waste gases back to the fabric in the dryer. Efficient microwave, radiofrequency or infrared heating systems are some of the recent innovations used in fabric drying.

Water conservation and energy conservation go hand in hand. The less water needing to be heated, the less energy is used. To encourage the further conservation of water, supply authorities in many countries have special contracts with different industries, fixing higher water rates if the water which is discharged into the sewage system is highly polluted. Since the total payment depends on the volume of water used, this system encourages manufacturers to use as little water as possible. Such incentives have helped some textile companies to invest in new equipment which reduces their water consumption to less than half. The mini-bowl and lo-flo of W.R.O.N.Z. and the C.S.I.R.O. are examples of water saving equipment developed specifically for wool scouring.

Some countries, such as France, have legislated to force their industries to save energy. As a result, the energy consumption of the French textile industry was reduced by 25 per cent in the five years 1975–1980.

Apart from legislation, government incentives such as taxation write-offs for money spent on energy conservation, research and development grants, and audit fee rebates can be used to encourage wise management of energy resources.

Pollution concerns

Water pollution

Like all industries, the textile industry puts waste materials into the environment. Some of the waste pollutes the waterways – detergents, dyes, acids and alkalis, chemicals and oils are disposed of through the sewage system.

Different councils have different regulations as to what can be poured down the drains. Most dyehouses have special tanks in which the hot dye liquors are allowed to cool so they won't kill the bacteria that process the wastes at the sewage treatment works. Any heavy sediments and fibrous fluff is separated at this stage, and acid and alkaline liquors neutralize each other before they are allowed into the municipal sewage system.

Current research concentrates on separating the waste materials from wool scour liquors, and using them as fertilizers. Other anti-pollution measures involve the separation of dyes from waste liquors by flocculating chemicals. The dyes can then be filtered out, leaving a clear solution which will not pollute the waterways. All detergents used in the textile industry must be biodegradable and, in order to allow the bacteria to digest them successfully, all waste liquors must have a certain minimum amount of available oxygen. To reach this level, air is bubbled through the waste in large aeration tanks. After the removal of heavy sediments, and filtration through beds of stone and sand, the water is clear enough to be released into the environment.

Industries forced to comply with stringent pollution control requirements may go off-shore, as the Japanese wool scourers have done, or they may need government assistance to rebuild and relocate factories, as has happened in Victoria.

Air pollution

The waste gases from the boilers used in the textile industry are a potential source of air pollution, the danger varying with the type of fuel. Natural gas is a 'clean' fuel, causing little trouble, but coal, because it often contains sulfur compounds, is a major environmental problem. The sulfur compounds are converted to oxides during burning. These oxides dissolve in water in the atmosphere to create acid rain, and acid rain is killing the forests of Europe and North America. As winds often carry the pollution across national borders, governments have been slow to acknowledge that their country's industries are responsible for their neighbours' dead trees. It is only now that the problem has become a world disaster that steps are being taken to prevent further damage. Strict regulations on the quality of coal that can be burnt, and the use of special filters to trap the harmful combustion products before they leave the chimneys, are among the control measures introduced.

Chimneys in Japan. Pollution is a world-wide problem.

Less disastrous (on a world scale) pollution can arise from the burning of process oils during heat treatments such as drying, stentering, curing, and singeing. Afterburners in the chimneys to ensure complete combustion, and recirculation of the burnt gases, can eliminate most of the problem.

Air pollution of another kind can occur where short staple fibres are processed. As the fine fluff flies about in the air it can cause eye irritations to the workers. If the air is not kept sufficiently clean in the cotton textile industry, the fine fluff which is breathed into the lungs can cause an occupational disease called byssinosis. Moist air conditions and efficient air filtration make modern cotton plants safer than older-style factories.

Noise pollution

Textile mills are very noisy places, so noisy, in fact, that in the past textile workers – particularly weavers – had considered it natural to be hard of hearing. Modern factory management is aware that industrial deafness can be avoided by reducing the noise of machinery on the factory floor.

Legislation has fixed the maximum acceptable noise level in a textile mill. In order to reduce noise levels, old machinery with worn gears and ill-fitting metal parts must be replaced by new, well-maintained machinery designed for quiet running.

Machinery builders are well aware of the need to reduce noise. Smooth-running ballbearings, rubber-padded ring spinners, shuttleless looms, carpet-covered partitions and enclosed machines help create working environments less dangerous to hearing.

TRY YOUR HAND

Read this extract from a British report:

Hearing protection: the law at work

An employee has been prosecuted by the Factory Inspectorate for refusing to wear ear protection while working in a weaving shed where noise levels were approximately 100dB(A). He was charged with 2 offences under Section 7 of the Health and Safety at Work etc Act 1974.

Section 2(1) of the Health and Safety at Work etc Act 1974 places a duty on employers to ensure, so far as is reasonably practicable, the health of his employees while at work. It is now well established that exposure to high levels of noise will cause permanent and irreversible damage to the hearing so, where workers are subject to such noise levels, their employer has a legal responsibility to protect them.

Prior to the prosecution the employers of the worker concerned in the case introduced a hearing conservation programme for those of their employees working in the weaving shed. Following the recommendations of the Code of Practice for Reducing Exposure of Employed Persons to Noise published in 1972 a noise survey was carried out to identify the high noise areas and the employees at risk. The managing director of the company and his weaving manager then discussed the hearing conservation programme with all the workers concerned, informed them of the dangers and described the different types of ear protection available. Instructions were given on how to use this protection and what to do if it became damaged or unsuitable. The company safety policy was amended to include a section on noise and ear protection and suitable notices were posted in relevant parts of the factory to warn employees that they were about to enter a noisy area and that protection should be worn. In addition the weaving manager kept records detailing the type of ear protection selected by each employee and every two or three weeks carried out spot checks in order to find out who was and who was not wearing the protection.

from *Textile Horizons*, February 1984.

Traditionally, it has been employers who have been prosecuted for unsafe working conditions. Where do you think the responsibility of the employer ends and that of the employee begins?

Toxic chemicals

Weeds in growing cotton are treated with herbicides, damaging pests such as the boll weevil are controlled (not very successfully) with insecticides, the cotton plants are treated with defoliants before harvest to make mechanical picking easier. All these chemicals are potential hazards to people living and working in the area, and so must be handled with caution.

Many of the early synthetic dyes are no longer in use because they are now suspected of being carcinogens (cancer causing). All modern dyes are subjected to rigorous safety checks. Efforts are made to use the least hazardous chemicals – solvents, resin components and so on – at all stages of textile manufacture.

Where potentially hazardous chemicals are used, worker exposure is closely monitored to minimize risk.

Conditions of employment

The industrial revolution in England coincided with the rise of humanism, with a growing belief among some influential sectors of the community that animals and other people did not exist merely to be used. For this reason, information abounds on the appalling conditions in the early factories, and, despite the lack of political power of the worst-treated workers, legislation was enacted to limit the abuses.

Today, in the developed countries, legislation sets minimum standards for working conditions, and factory inspectors employed by governments, and shop stewards representing unions, make sure that these standards are upheld. In some less developed countries the need for foreign money and the shortage of employment is such that less stringent controls are enforced, and in some countries unions are actively discouraged.

TRY YOUR HAND

Read the following articles on working conditions.

England, 1833

. . . They who grow cotton are merciful taskmasters in comparison with those who manufacture it. Robert Hildyard (whom you know) told me the other day that Marshall, the Member for Leeds, showed him one of his manufactories, and upon his remarking the extreme delicacy of the children, replied they were consumptive, that a great proportion of them never reached the age of twenty, and that this was owing to the *flew* with which the air was always filled. He spoke of this with as little compunction as a General would calculate the probable consumption of lives in a campaign. A General may do this, under – even a righteous – sense of duty; but I know not where the love of gain appears in more undisguised deformity than in a cotton-mill. The cruelty is never so excessive as it often is in a plantation, but it is more unmitigated; the system is more uniformly and incorrigibly evil. The negroes in a plantation may be rendered happy by kind treatment, and no doubt often are so, but I know not how a cotton-mill can be otherwise than an abomination to God and man.

Robert Southey, letter, 1833.

The decline of the hand-loom weaver, 1835

A very great number of the weavers are unable to provide for themselves and their families a sufficiency of food of the plainest and cheapest kind. They are clothed in rags, and indisposed on this account to go to any place of worship or to send their children to the Sunday schools. They have scarcely anything like furniture in their houses. Their beds and bedding are of the most wretched description and many of them sleep upon straw. Notwithstanding their want of food, clothing, furniture and bedding, they, for the most part, have full employment.

Report of the Hand-loom Weavers' Committee 1835

Shanghai, 1934

'Go and see a bit of life,' said Hers [a Belgian official]. 'Stuck in your university, you think everything can be done by slogans. You will see that the Chinese are worse to their own people than Europeans.'

A small rat-like man with glasses took me to a silk filature, in Chapei or Yangsepoo, quite a way from the International Settlement. It was an area impossibly squalid, a slum of sagging huts, stinking unpaved alleyways. A barn-like structure, with a small courtyard in front, was the filature. Inside, great vats of boiling water, furnaces, and children looking about six but who the rat-faced man told me were all of fourteen, standing round the vats. It was hardly possible to see what they were doing with the steam rising from the vats, and it was suffocatingly hot; but they were plunging silk cocoons in bundles wrapped in fine webs of gauze in the vats. The children's eyes were peculiar, bright red with trachoma, their arms were covered with scalds, and they worked almost naked; the temperature outside was ninety-eight degrees; inside it was one hundred and three or perhaps more. The smell was bad, and I could not stand it and no one wanted me to stay very long. 'These are refugees... we give them work, otherwise they would die of hunger...' The rat-like man told me how kind his manager was to give employment to these children, one hundred and twenty of them. Quickly we went out again. The car of the manager who had arranged this instructive trip for me because Hers had asked him to do so, was waiting beyond the mud lanes, as these were impassable; it whisked me away, back to civilization, in the French Concession... 'Now you have seen the problem,' said Hers. 'Too many poor people in China.' Later I found out the silk filature belonged to a Japanese concern.

... From the famines, the floods, the civil wars, wave after wave of children would arrive at the city with their parents. And they were sold by their parents, or hired out to work, as was done in England. The factories refused to employ men, they would only take women and children. From the refugee camps and the orphanages children would be bought. In crowded lofts, in rickety barns such as this, they were put to work; twelve, fourteen hours a day, no Sundays off; they made flashlight bulbs for the five and ten cent stores; they slept underneath the punch press machines; twice a day they ate gruel, and they died of beri beri, swollen with festering sores, within four months.

[Rewi Alley, New Zealand-born inspector of factories, said] I shall never forget the irrepressible gaiety of dying children, in the lead battery factories... as I went to take their urine... the word child applied to the under twelve. Over twelve they became apprentices till they were eighteen. Apprentices were paid like children.'

'The foreigners always said: It's up to the Chinese administration.' And the Chinese could not do anything even if they wanted to (which they did not) because over seventy per cent of the factories were owned by Japan and Britain.

The foremen in the textile mills were armed with guns and whips. The girls, mostly from the countryside, were brought into the city in batches of thirty, worked fourteen hours a day and slept in lofts on floorboards. The prettiest were sold to the brothels. Rewi saw many a girl working with a child strapped to her back, another child tied to her leg, for there was no place where the children could be left while the mother worked.

The silk filatures of Shanghai had long lines of children, many not more than eight years old, standing twelve hours a day over boiling vats of cocoons with swollen red fingers, many crying from the beating of the foreman who passed up and down behind them with a number eight gauge wire as a whip; their arms were scalded in punishment if they passed a thread incorrectly.

'But we've also been through this, you know, we had it in England during the nineteenth century,' an Englishman loftily said to me when I told him of my gruesome visit. 'And look at the British worker now! Always on strike, the Bolshies, if it hadn't been for Ramsay MacDonald [British Prime Minister, 1924 and 1929–35] they'd have cut our throats!' 'These people would starve if we didn't give them employment.' This was philanthropy.

... Of the textile mills of Shanghai, 77.2 per cent were foreign-owned, 22.8 per cent Chinese. Woman and child labour was near 85 per cent of the total; the monthly earnings were fifteen (silver) dollars for a male, thirteen for a female, eight for a child (under twelve) working twelve to fourteen hours a day for seven days a week. Though the government had passed labour laws in 1931 they were never put into effect. *And all this went on till 1949.*

Han Suyin, *A mortal flower,* Jonathan Cape, London, 1966.

Messrs. Sargood Son and Co.'s establishment, Melbourne.
(*Illustrated Australian and Tasmanian News,* December 1874, p. 3. National Library of Australia)

A modern clothing factory. Many garments can be cut out simultaneously, saving on labour.

Formaldehyde toxicity

Report No. 7 of the Textile Research Institute, Princeton, New Jersey, contains a report on a recent U.S. conference on formaldehyde toxicity. Of particular interest to the textile industry is that, at the conference, no acute or chronic toxicity effects of significance were reported in connection with textile products and processes. In fact, a NIOSH study involving 18 people with an average service of seven years working in a crease-resistant cloth processing and storage plant with exposures of formaldehyde of less than 4.1 p.p.m. reported no effects from the exposure.

Epidemiology studies reported at the conference did not show significant overall increases of carcinogenicity in persons knowingly exposed to formaldehyde when compared with the general population. However, questions were raised at the conference about the significance of these epidemiology studies and further studies are planned.

Australasian Textiles, May 1981

Wage demands unrealistic

The Justice V S Deshpande committee, appointed by the Indian Government in August 1982, to investigate the grievances of the mill workers regarding wages and conditions of service and to make recommendations, has just submitted its report. The high-level tripartite committee comprising representatives of the industry, Government and RMMS, the recognized textile union, had been boycotted by the MGKU which had called the strike by employees in 60 Bombay mills, and by the leftist trade unions. After making a thorough study of the problems referred to it and the conditions in the industry, the majority of the committee has rejected the demand of the workers for higher wages on the ground that the employers are simply not in a position to pay more. It has described even the moderate demands of the RMMS, which was against the strike, as exorbitant and unrealistic as any further increase in wages would financially ruin most of the mills. The committee in its interim report had recommended house-rent allowance which has already become effective.

The committee noted that the minimum total wage of around Rs704 a month was considerably more than the country's minimum wage and every price rise was compensated by the variable cost-of-living allowance being paid to the workers. As for the contention of the MGKU that the engineering and petro-chemical industries pay much higher wages, the committee observed that this was possible in their case because they are capital-intensive, their wage-bill forming only a small part of their total turnover in contrast to the textile industry which is labour-intensive.

However, there were dissenting notes to the committee's report by the leaders representing labour, who felt that there should be some upward revision of the wages.

Textile Horizons, December 1983

1 How are wages and conditions determined in Australia? What controls exist to make sure that both employers and unions observe agreements on wages and conditions?
2 Contact a local textile industry and a textile union and ask about conditions in Australia. Make sure that you hear both sides of any argument.
3 Many unions have contacts with workers in other countries. Find out what you can about working conditions in some countries providing cheap goods for the Australian market. Are conditions as bad as those described here for Shanghai in 1934? Do the workers have the right to form unions? If they do not, what controls are there on working conditions? How do conditions compare with those of Australian workers? (Do not compare wages in dollar terms: it is wages in relation to cost of living which are relevant to working conditions.)
4 Because so much information is available on the conditions during the industrial revolution, there is a temptation to believe that earlier workers lived in some sort of golden age. Consider: in Belgium, lace workers worked with linen threads so fine that they had to work in damp cellars, for the increase in strength that the moist air gave the threads. Light, naturally, was very bad. Blindness and tuberculosis were common. Find out what conditions were like in other trades before the industrial revolution.

You, the consumer

If you have worked through this book, you will have gained some understanding of the technology of textiles production. You will be aware of the properties of each of the generic classes of fibres, and their relation to each of the various techniques of textile making and finishing.

This knowledge is of very little value to you unless you can connect it to your needs as a consumer, to your understanding of the interrelation between technology, aesthetics, culture, economics, politics, environment, fashion, ideology, and yourself.

As a consumer of textiles, you have a responsibility to choose products which satisfy your requirements for quality, price, and design. There is also a *moral* component to your choice: you must choose according to your values. For example, all goods are labelled with their country of origin. If you believe apartheid is wrong, would you buy something manufactured in South Africa? If you are opposed to communism, would you buy Chinese products or goods from the Soviet bloc? If you are opposed to killing wild animals, would you buy leather and furs? These are not questions that can be answered by a book: only you can answer the question, 'How right is my decision?'

Consumer questions answered

Q1   Why doesn't Australia export textiles?

A   Textiles industries have historically been labour-intensive. Because of high labour costs and short runs – due to the small population – Australia has not been able to make textiles that have a competitive price on the world market. Long distances, which cause time delays in the delivery of fashion goods, have also hindered the marketing of Australian textiles overseas.

Q2   Why are there so many imported textile goods in Australia?

A   Australia is a rich country. Its relatively small population can afford to buy a large variety of things. Making small amounts of many different kinds of textiles is not an economical process. So Australian manufacturers make relatively large amounts of fairly simple textiles, and import smaller quantities of a variety of products made elsewhere.

Q3   Why is such high duty levied on imported textiles?

A   The import duty – tariff – raises the cost of the cheap imported textiles to a level where the Australian products can compete with imports on the local market. The amount of tariff levied is determined 'historically' and as a result of lobbying and negotiation between the government and interested parties such as importers and manufacturers.

Q4   If energy is so expensive, why isn't solar power used in the textiles industry?

A   Solar energy is an excellent source of heat where the weather guarantees plenty of sunshine. It is best used for preheating dye liquors and water that is to be converted to steam in boilers. If more of Australia's textile industry were situated in the sunny north or inland, more advantage could be taken of the heat of the sun. As yet, no plant has been designed to depend on solar energy for more than 60 per cent of its energy needs.

Q5   Will robots ever replace people in textile manufacture?

A   Already robots and automated machinery have taken the jobs of many skilled textile workers. However, the fully robot textile mill or garment workshop is still a long way away. Skilled people are still essential for the human input into decision-making in this very people-dependent, fashion-conscious industry.

Q6   What does the textile industry do about preventing the pollution of our waterways?

A   The trend in textile manufacture is to use biodegradable detergents, emulsifiable yarn dressings and process oils, smaller volumes of water, and smaller amounts of chemical additives.

Appendix: Stain removal

General care

If you are to care for your textiles, you should be familiar with the fibre content and care instructions. This information should be provided by the manufacturer: by law it must be included on all manufactured items purchased, and available to you when fabric is purchased. You run a risk of spoiling the item if you purchase it without this information. Once your textile item requires cleaning, it is important to follow the care instructions so that, if the article is damaged, the manufacturer alone is responsible.

Stain treatment

It is always easier to treat known stains. Therefore, if possible, identify the stain. Otherwise try to classify the stain as greasy, non-greasy, or combination. This information is important because incorrect treatment can result in stains being set and therefore more difficult to remove. Care must be taken as treatment for one stain can make others worse.

Stain removal procedures are best carried out before normal cleaning. This ensures that the stain is not set by normal cleaning and that the stain removal agents are removed in the normal cleaning. The sooner you treat the stain, the better. The older the stain, the more difficult it is to remove.

Stain removal agents

It is important to be well informed about the reagents you are using for stain removal and the effects they have on different fibres and fabrics. Some stain removal methods carry a considerable risk; for example oxalic acid treatment of a rust stain is so harsh that it may destroy silk fabrics. Similarly, the use of chlorine bleach on 'cotton' with a polyester component will result in an orange patch where the bleach has reacted with the polyester.

Before beginning any stain removal treatment it is advisable to test the reagent on a scrap of the fabric or a hidden area of the item, e.g. a hem or seam. It is thus possible to avoid a complete disaster. After all, you may prefer a garment with a small stain to one which has a hole burnt in it by a harsh reagent.

Stain removal agents and their uses

Stain removal agent	Example	Use	General instructions
Absorbents	Talcum powder Cornflour Blotting paper Chalk	Non-wash fabrics, for greasy stains and liquids	Spread absorbent powder over the stain. Remove it by brushing, shaking or vacuuming it as it absorbs the stain. Several applications may be necessary. Absorbent agents may be used in conjunction with grease solvents where the use of a solvent alone may leave a ring mark. A relatively dry mixture of the two agents is spread on a grease stain and allowed to dry. This is then brushed off.
Solvents	Cold water	On washable fabrics for water-soluble, colloidal or protein stains	**Non-grease solvents** Solvents other than water are dangerous if inhaled and should be used only in small amounts and with caution. Many are also highly flammable. *Acetone* do not use on acetate, Arnel, Dynel or Verel *Alcohol* Before use, check that it is safe for the dye. Dilute with 2 parts of water for use on acetate. *Amyl acetate* If pure, it can be used on fabrics which are damaged by acetone.
	Organic solvents – dry-cleaning fluids, methylated spirits, turpentine, ether	On greasy and resinous stains and some special stains such as lipstick	**Grease solvents** Available as proprietary lines. Follow care instructions on label. **Method of using solvents** Place article stain side down on a wad of clean absorbent cloth or other material. Dampen a pad of cotton or soft cloth with a small amount of the solvent. Dab stain from the centre out, avoiding rubbing. Change cloths as required to avoid restaining the fabric. Sponge stain irregularly around the edge to help avoid leaving a ring mark. Old paint or tar may require a dampened pad to be left on the stain for the solvent to have sufficient time soften the stain. Solvents may be used in conjunction with absorbent powders in cases where the solvents may leave a ring. This is less satisfactory on dark fabrics which may be marked by the powder.
Movement promoters	Boiling water from a height	Colloidal stains	Should be poured onto the stain from a height to maximize the movement of the stain. Take care when pouring boiling water

Stain removal agent	Example	Use	General instructions
Emulsifying agents	Soap (on washable fabrics) fat	On greasy and resinous stains, such as tar. (Tar can be rubbed with lard to soften it and after 30 minutes scraped and washed)	Soaps and detergents break grease and other soil into small globules which can then be rinsed out. Liquid detergents are often very useful as they are in concentrated form and can also be easily worked into the stain.
Special reagents	Acids (oxalic, lemon juice) Alkalis (ammonia)	Rust, perspiration	*Acetic acid, vinegar* Use 10% acetic acid or vinegar. Keep wet until the stain is removed. Rinse well after use. Note: this treatment is not safe for all dyes. Ammonia may help restore the colour. *Oxalic acid* May cause damage to protein fibres. Dissolve 1 tablespoon of oxalic acid in 1 cup of warm water. Soak stained area in this solution. Rinse thoroughly. A stronger treatment uses the same proportion of acid to water but uses water as hot as the fabric can take. Oxalic acid is toxic so do not use kitchen utensils. *Ammonia* For use with protein fibres, dilute the ammonia with an equal volume of water. For all other fibres, use 10% household ammonia. Soak stain in it until gone.
Bleaches	Hydrogen peroxide, sunlight, cream of tartar, sodium hypochlorite, sodium hydrosulfite	As a last resort for coloured stains	*Hydrogen peroxide* 33% solution is safe for all fibres. It loses strength on storage. Test for colour fastness before use. Use a few drops of hydrogen peroxide on the stain and place in sunlight. A few drops of ammonia added to about 1 tablespoon of hydrogen peroxide makes a stronger treatment. If all else fails, cover stain with a pad of material soaked in the peroxide and heat with an iron. Rinse well. *Chlorine bleach* Do not use on protein fibres, elastomerics, polyester, polyurethane foams or special finishes. A mild treatment consists of 1 tablespoon of bleach in 1 litre of water. Apply to small stains with a medicine dropper, soak large stains. Rinse well. A stronger treatment consists of equal parts of bleach and water. Use as above.

Stain removal agent	Example	Use	General instructions
			Powdered peroxygen bleaches – these include sodium perborate and potassium monopersulfate. Test for colourfastness and do not use in a metal container. Use 1 to 2 tablespoons in ½ litre of water. Mix just before using as the mixture loses strength on standing. A hot mixture provides a stronger treatment but is not suitable for fabrics such as protein fibres sensitive to hot treatment. Always rinse well.

It is worth keeping in mind that professional cleaners have reagents and equipment not available in a normal household. Professional cleaners are not able to remove all stains, but you may find they are able to solve a problem which does not respond to household reagents.

The following table lists common stains and possible removal techniques. Proprietary lines are not discussed, as their use is a matter of personal judgement.

Note: use removal agents according to directions above unless otherwise specified.

Stain removal by type of stain

Stain	Fabric	Method
Acid	All	Rinse the fabric immediately with water, then apply ammonia and rinse again.
Adhesive	All	Harden with ice cubes, scrape carefully with dull edged knife (e.g. table knife) then sponge with kerosene or other grease solvents.
Alcoholic beverages	All	Sponge immediately with cold water. If stain remains, work soap into it then rinse. Red wine stains should be treated with vinegar or lemon juice before soaping.
Alkali	All	Rinse immediately with water, then apply vinegar to area and rinse again.
Antiperspirant	All	Sponge with soap or detergent and warm water, rinse. If stain is still visible, sponge with chlorine or peroxygen bleach.
Blood	All except protein fibres	Treat with enzyme washing powder, soap or detergent and *cold* water. If difficult to remove, a few drops of ammonia and further treatment with detergent may work. Bleach if necessary. Heat-set blood stains will be difficult to remove.
	wool, silk	Sponge with cold water.
Butter or margarine	All	Sponge stain with grease solvent. Dry, repeat. If yellow stain appears, use chlorine or peroxygen bleach. If safe for fabric, use sodium perborate.

Stain	Fabric	Method
Carbon paper	All	Work detergent or soap into stain, rinse. If stain is still visible, place a few drops of ammonia on stain and repeat washing.
Chewing gum	All	Apply ice to harden, and chip off as much chewing gum as possible. Sponge with grease solvent.
Chocolate	Cotton, linen	Use cold water to soak, wash in hypochlorite bleach.
	Wool, silk	Use cold water to soak, wash in hydrogen peroxide.
Chlorine	(no treatment for chlorine stain on wool and silk)	Rinse fabric thoroughly with water. Soak for 30 minutes or longer in a solution of 1 teaspoon sodium thiosulfate to each litre of water (as hot as is safe for the fabric). Bleaching caused by chlorine cannot be removed.
Coffee (black)	Washable	Pour boiling water from a height, then wash. If necessary, treat with perborate.
Coffee (white)		Treat as per milk.
Correction fluid	All except acetates, Arnel, Dynel, Verel	Sponge with acetone.
	Acetates, Arnel, Dynel, Verel	Sponge with amyl acetate.
Cosmetics	Washable	Sponge with undiluted detergent until thick suds appear. Work into fabric until stain disappears. Rinse, repeat if necessary.
	Non-washable	Use grease solvent.
Crayon		As per cosmetics.
Cream	Washable	Sponge or soak in cool water 30 minutes or longer. Work soap or detergent into the stain, rinse thoroughly. Dry. Further treatment with grease solvent may be necessary.
	Non-washable	Spot clean as above.
Dyes	Washable	Sponge or soak in cool water. Treat with soap or detergent. A long soak in soapy water is often effective on fresh stains. Stubborn stains may need treatment with chlorine or peroxygen bleach.
	Non-washable	Spot clean with water and soap or detergent. A final sponge with alcohol helps remove soap. Test fabric with alcohol first.
Egg		As per dye. Enzyme soak may be used on all except protein fibres.
Food colouring		As per dyes

Stain	Fabric	Method
Fruit juice		As per dyes. If safe for the fabric, pour boiling water through the stain. Immediate sponging with cool water is the most effective treatment. Some juices colour after drying and may be difficult to remove.
Furniture polish		As per cream. If the polish contains wood stain, turpentine may help remove it. After using turpentine, work soap or detergent into the area and soak in hot water, leave overnight.
Glue, adhesives	Washable	Airplane glue: as per correction fluid. Casein glue: sponge with cool water, work in soap or detergent and rinse.
	Non-washable	Spot clean as above and finish by sponging with alcohol. (Test fabric first.)
	Washable	Plastic glue: wash with soap or detergent before the glue hardens. Some forms of plastic glue may be removed by soaking stain in hot 10% acetic acid or vinegar. Keep this hot or nearly boiling until stain is removed (about 15 minutes). Rinse.
	All	Rubber cement: scrape gummy glue from fabric and sponge thoroughly with grease solvent.
Grass, flowers	Washable	Work soap or detergent into stain then rinse. If safe for dye, sponge the stain with alcohol. Dilute alcohol with 2 parts of water for acetate.
	Non-washable	As above but use alcohol first if safe for dye.
Gravy, meat juice	Washable	Sponge with cool water or soak for 30 minutes. Work soap or detergent into the stain and rinse. Allow to dry. If a greasy stain remains, sponge with grease solvent.
	Non-washable	Spot clean as above, using alcohol (if safe for dye) to remove soap from article.
Grease	Washable	Rub with soap or detergent and rinse in warm water. For some fabrics it may be necessary to leave soap in it for some hours or overnight. A grease solvent will probably be necessary. If the solvent leaves a stain, sodium perborate solution is effective, but test first.
	Non-washable	Sponge with grease solvent. Several treatments may be necessary. Finish with sodium perborate if safe.
Ice cream		As per cream.
Ink: ballpoint ink	All except acetate, Arnel, Dynel, Verel	Sponge stain immediately with acetone.
	Acetate, Arnel, Dynel, Verel	Sponge immediately with amyl acetate.

Stain	Fabric	Method
drawing ink	Washable	Force water through stain to remove pigment. Wash with soap or detergent. Soak in a solution of dilute ammonia (4 tablespoons/litre)
	Non-washable	Use water as above, or the stain will spread. Sponge with dilute ammonia or a combination of soap and ammonia if necessary. If colour changes, sponge with vinegar.
writing ink	Washable	Sponge with water first then try soap or detergent. Soak if necessary. Further treatment with chlorine or peroxygen bleach if necessary. Inks vary, and the particular stain may require several attempts before the right treatment is found.
	Non-washable	Spot clean with water and/or soap or detergent. Remove soap with alcohol.
Iodine	Washable	Try soaking in water, use soap if necessary. Use 1 tablespoon of sodium thiosulfate per ½ litre of water to soak with if needed. Alcohol may be used; a pad left on the stain for several hours may be necessary.
	Non-washable	Spot clean as above.
Lacquer		As per correction fluid.
Mayonnaise, salad dressing	Washable	Sponge with cool water or soak for 30 minutes or longer. Soap or detergent may be used. Follow with treatment with a grease solvent if a greasy stain remains.
	Non-washable	Spot clean as above.
Medicines		Treat as per butter or other greasy stains if the medicine appears greasy. If syrupy, wash with water. If the medicine is dissolved in alcohol (tincture), sponge with alcohol. A medicine containing iron should be treated as an iron stain. If the medicine is coloured, treat as per dye stain.
Mercurochrome	Washable	Soak in warm water, soap or detergent and ammonia (4 tablespoons/litre).
	Non-washable	If safe to use alcohol, use this to sponge away the stain. Liquid detergent and a drop of ammonia may be used if alcohol is not safe. Rinse well.
Metal		Use acid treatment (p. 603). Do not use bleaches.
Mildew	Washable	Wash thoroughly and dry in the sun. Use chlorine or peroxygen bleach if necessary.
	Non-washable	Send article to the dry-cleaner promptly.
Milk	Washable	As per egg.
Mucus, vomitus		Treat with a lukewarm solution of salt and water (¼ cup salt per litre of water). Sponge stain with solution or soak stain in it. Rinse well. Treatment as per dyes may also be necessary.

Stain	Fabric	Method
Mud		Allow to dry and then brush off. Mud from iron-rich clay will need further treatment as per metal stain.
Mustard	Washable	Dampen stain and rub with soap or detergent, rinse. If stubborn, use hot water and soap or detergent, leave several hours. Further treatment with sodium perborate may be necessary.
	Non-washable	If safe, sponge with alcohol; otherwise spot clean as above.
Nail polish		As per correction fluid or polish remover but test first.
Oil		As per butter or margarine.
Paint, varnish	Washable	Treat immediately with whatever is recommended as a thinner. Follow by washing.
	Non-washable	Spot clean as above, including spot cleaning with detergent.
Pencil		Try using a soft eraser. Further treatment if required may be as per carbon paper.
Perspiration	Washable	Wash in warm water and soap or detergent. Take care with silk because it is damaged by perspiration. Vinegar may be used on old stains; ammonia on fresh stains. Bleaching may be necessary.
	Non-washable	Spot clean as above.
Plastic		Use amyl acetate or trichlorethylene. Follow method for solvents (p. 602).
Rust		Treat with oxalic acid (p. 603). If stain is stubborn, place oxalic crystals directly on the stain and pour boiling water through the stain (this treatment is harsh and may cause damage to some fabrics, e.g. silk). Cream of tartar or lemon juice may be used on less stubborn stains.
Sauces		As per mayonnaise.
Scorch mark		As per dyes. Severe scorch cannot be removed because of damage to the fibre.
Shoe polish		There are many different forms of polish so trial and error is the only way. 1 Treat as per cosmetics. 2 Sponge with alcohol if safe for fabric. 3 Use grease solvent or turpentine (p. 602). Futher treatment with bleach may be necessary.
Soft drinks		As per dyes.
Soot, smoke		As per cosmetics.
Tea		As per coffee.
Tobacco		As per grass.

Stain	Fabric	Method
Unknown stains		If the stain appears greasy, treat as per butter; otherwise treat as per dye.
Urine		As per dye unless colour of fabric has changed. If it has, sponge with ammonia. If this does not work, treat with acetic acid or vinegar.
Wax – floor, furniture, car		As per butter.
Yellowing, brown stains		Treat in the following order as necessary: 1 Wash. 2 Use mild bleach treatment. 3 Use oxalic acid treatment. 4 Use strong bleach treatment.

Glossary of terms and fabrics

acetate hydrophobic fibres made by chemical modification of regenerated cellulose. Usually refers to secondary cellulose acetate, but is sometimes used to designate both secondary acetate and triacetate.

acid dye a dye that is taken up in acid solution. Acid dye molecules are negatively charged, and so dye proteins and polyamides which take a positive charge in acid solution.

acid milling dye an acid dye which remains fast during the rigorous milling of woollens.

acrylic generic name for synthetics based on not less than 85 per cent of the monomer acrylonitrile.

additive primary one of the three colours of light that can be detected by the eye (red, blue, green), so called because they can be added to give white or any other colour of light. Also called light primaries.

adhesive bonding is used to bond fibres in some nonwovens in which fibre–fibre friction alone would not provide the necessary strength.

amino acid the building block of proteins, amino acids contain a basic amino group and an acidic carboxylic acid group.

amorphous amorphous regions do not have an ordered structure. The molecules in the fibre do not pack tightly, and so amorphous regions are weak (less intermolecular bonding) and contain voids (spaces) which can accommodate dye or water molecules.

angora hair of the angora rabbit.

anionic carrying a negative charge.

antistatic agent a small charged or polar molecule added to a synthetic fibre to aid the dissipation of static electric charge.

aramid a polyamide with aromatic components: usually high strength and/ or flame resistant.

aromatic in chemistry, containing the chemically stable benzene ring system.

atlas fabric a warp-knitted fabric with a zig-zag character.

auxochrome a polar group or electronegative atom which intensifies the absorption of light by a dye, and in some cases changes the colour.

Axminster a cut-pile carpet woven on a Jacquard-controlled loom.

azoic dye a dye characterized by a nitrogen–nitrogen double bond that is formed by chemical reaction within the fibre.

610

bacteriostat   a chemical added to a fabric during finishing that inhibits the growth of micro-organisms.

balanced   of a woven fabric, having equal numbers of warp and weft yarns per centimetre.

basic dye   a dye which carries a positive charge in solution, often used to dye acrylics. Despite the name, basic dyes are usually applied under mildly acid conditions. Also called cationic dye.

bast   of fibres, obtained from plant stems.

batiste   a fine, soft, plain woven fabric of cotton or linen.

bave   silk in cocoon form: twin filaments bound by sericin.

beam dyeing   a dyeing procedure in which the fabric is loosely wrapped in open width around a perforated beam through which the dye is pumped.

beating up   pushing a newly laid weft into place in the cloth.

bicomponent fibre   a synthetic fibre extruded from two different polymer solutions, so it has two distinct parts: these usually have differing stretch characteristics, and so give the fibre a natural crimp.

biodegradation   breakdown through biological action: desirable when bacteria are used to digest waste in sewage treatment; undesirable when fabrics rot in damp conditions.

birdseye   a fabric woven with an overall pattern of small spots.

blending   a thorough intermixing of *fibres*.

bouclé   a looped yarn used to produce fabrics with a rough, pebbly surface.

bounty   a cash grant made to producers to stimulate production.

braid   a narrow fabric, either woven on a narrow fabric loom, or produced by interlacing yarns at an angle (no weft).

broadcloth   a cotton cloth used for shirting.

broadloom   a general name for carpets more than two metres wide.

brocade   an elaborately figured Jacquard woven fabric.

brushed knit   a single jersey fabric in which a thick laid-in yarn has its fibres matted by brushing.

brushing   the deliberate damaging of surface fibres by means of rotating wire brushes to create a soft fuzzy surface.

Brussels carpet   a loop pile woven carpet.

calendering   a finishing procedure using heat and pressure to produce surface effects.

calico   a plain, medium-weight cotton fabric, bleached or unbleached.

carbonizing   the treatment of wool with dilute sulfuric acid to destroy any cellulosic contaminants.

carding   the process of partially aligning fibres in a web, either as a preparatory step in yarn manufacture, or to produce an orientated web for nonwoven textiles.

cashmere   hair of the cashmere goat.

cationic   carrying a positive charge.

challis   a soft, fine fabric of cotton, viscose, or very fine wool, either plain or twill weave.

cheesecloth   a plain woven cotton fabric made from highly twisted but hairy yarns in a coarse, loose weave.

chenille a cut-pile fabric made by tufting.

chenille yarn a hairy yarn produced by cutting specially woven fabrics into strips.

chiffon a very light, loose, plain weave fabric, originally of silk.

chintz a plain weave fabric with a glazed finish produced by calendering and resin treatment. It is usually made from cotton and is often printed.

chlorination an early shrink-proofing treatment for wool which employed chlorine to soften and degrade the scales.

chromophore the basic skeleton or system of conjugated bonds in a coloured molecule that is responsible for the absorption of light.

clipspot a fabric with warp pile yarns woven into the ground fabric only at intervals: the long floats are then clipped or cut away to leave a spot pile pattern.

clo a measure of insulation value widely used in the United States, roughly equal to the insulation value of normal indoor clothing.

cloque a woven or knitted fabric with a figured blister effect.

coir coarse, reddish brown fibres from the husk of the coconut, widely used for doormats, but also used for stuffing upholstery and for tough floor coverings.

cold drawing the stretching of synthetic filaments to orientate the crystallites in the direction of fibre length.

combing the removal of short fibres from a staple sliver. Combed yarns are characteristically very smooth.

complementary colours of light, combining to produce the perception of white; of pigments, combining to produce the sensation of black.

conjugated bond system a system of alternating double and single bonds in a molecule which allows smearing or overlap of the bond system. Conjugated bonds absorb strongly in the ultraviolet, and if extensive enough, in the visible region of the spectrum. The benzene ring is a closed conjugated loop.

copolymer a polymer consisting of two or more different monomers.

corduroy a weft-pile fabric in which the cut pile tufts are aligned in lengthwise rows.

core yarn a yarn consisting of a sheath of staple fibres wrapped around a filament core.

cortex the central mass of a hair fibre.

count of fabrics, the number of yarns per centimetre; of yarns or threads, a measure of thickness. Many conflicting systems are in use.

covalent bond a chemical bond in which electrons are shared between atoms.

cover the amount of space occupied by the yarns of a fabric. In fabrics with high cover, there is little space between the yarns; in fabrics with low cover, there is a lot of space between the yarns. Cover may be expressed as a ratio of yarn diameter to yarn spacing.

crepe a woven fabric with a crinkled surface which is achieved either by using highly twisted yarns, or by irregular interlacing of the yarns.

crepe de chine a light-weight crepe fabric made from filament yarns, with untwisted warp and highly twisted alternating S and Z wefts.

crimp the waviness of a fibre.

crocking fastness   the resistance of a dye or pigment to removal by rubbing.

cross dyeing   dyeing a fabric blend in two or more successive operations, one for each component of the blend.

crystallinity   order in structure. In crystalline regions of fibres, molecules are in ordered arrangements and are packed tightly: there is considerable bonding between molecule chains.

crystallite   a small region within a fibre in which the molecules are in a highly ordered (crystalline) arrangement.

cupro   regenerated cellulose fibre (a rayon) made by dissolving cellulose in cuprammonium hydroxide, then precipitating it in an acid coagulating bath.

damask   a patterned fabric in which the ground is sateen weave and the figure is satin weave.

degree of polymerization (D.P.)   average number of monomer units in a polymer chain.

delustrant   white pigment added to synthetic fibres to subdue their lustre by scattering light.

denier   a measurement of filament thickness. By definition, 9000 metres of 1 denier filament has mass of 1 gram.

denim   a warp-faced twill cotton cloth woven with indigo blue warp and white weft.

density   the ratio of mass to volume: in g/mL numerically equal to specific gravity.

detergent   a molecule with both polar or highly charged segments and a long non-polar chain, which therefore has both hydrophilic and oleophilic character: used to remove a break up grease or oil, and to increase wettability.

devoré   a lacy pattern effect in wool/polyester blends achieved by printing the fabric with alkali to destroy the wool.

direct dye   a class of dye, usually long flat molecules with a number of sulfonic acid groups, with particular affinity for cellulosic fibres.

directional frictional effect   the effect of the scales of wool which makes it easier to move the wool fibre in the direction of its root than in the direction of its tip.

discharge printing   a technique of producing small, light-coloured designs on dark fabrics, by printing the dark dyed fabric with a chemical which destroys the dye in the pattern areas.

disperse dye   a class of relatively non-polar dyes, only sparingly soluble in water, used to dye hydrophobic fibres such as polyamides, polyesters, acetates, and acrylics.

disulfide bond   sulfur–sulfur bond linking different parts of a wool protein chain, or joining two different wool protein molecules, which contributes greatly to the strength and resilience of the wool fibre.

dobby   a system for weaving complex patterns through the programmed lifting of up to forty different frames; fabric with a small figured design woven on a dobby loom.

doeskin   a fine warp-faced wool fabric finished to have a handle resembling kid.

dope dyeing the colouring of synthetic filaments by the addition of pigments before extrusion.

double cloth fabric woven with two sets of warp yarns, the weft passing only occasionally from one set to the other. Sometimes used for blanketing and upholstery.

double jersey a weft-knit fabric produced on a knitting machine that has two beds of needles.

double knit the knit equivalent of double cloth.

drape the behaviour of fabric when hung suspended.

drawing the elongation of a staple sliver by passing it through a succession of pairs of rollers, each pair moving faster than the one before.

drill a tough cotton twill fabric.

dry spinning production of filaments by extrusion of their solution into a stream of dry air which evaporates the solvent.

duck a heavy cotton or linen fabric resembling canvas.

duffel (duffle) a heavy woollen fabric, napped on both sides.

dupion an irregular silk produced when two silkworms spin their cocoons together; the fabric made from this.

durable press pleats and creases that are not altered by washing or dry-cleaning.

easy care requiring little or no ironing, and retaining its appearance after washing.

ecru naturally coloured; unbleached.

elasticity the ability of a fibre to recover from extension.

elastomeric a class of synthetic fibres capable of considerable extension with good recovery.

electronegative having a strong attraction for electrons.

electropositive giving up electrons readily.

embossing a calendering process using a deeply engraved roller and a soft paper roller, used to create three-dimensional design effects.

exchange rate the value of the currency of one country expressed in terms of the currency of another country.

fading loss of colour caused by light-initiated chemical breakdown of dye molecules.

faille a fine, soft fabric with a weft rib.

felt a textile composed of a dense mat of fibres.

fibre a fibre is characterized by flexibility, fineness, and a high ratio of length to thickness.

fibril a single spiral within the cotton or other cellulosic fibres; fibrous component of wool and other hair fibres.

fibroin the protein of silk.

filament a fibre of indefinite length.

filet lace lace with a square mesh, produced by Leavers machine, by warp knitting, or by hand knotting.

fingering yarn yarn designed for hand knitting.

finishing collectively, all the processes other than dyeing and colouring to which a textile is subjected after it leaves the loom or knitting frame.

flame resistant denotes a fabric or textile which will burn when flame is applied, but which rapidly self-extinguishes when the flame source is removed.

614

flame retardant   a chemical or treatment used to impart flame-resistant properties to a textile.

flannel   a soft wool fabric with a slightly raised surface.

flannelette   a cotton fabric woven from soft spun yarns and raised to a fluffy nap. Its use in children's wear is restricted because of its high flammability.

flat   of yarns, composed of untwisted filaments.

flax   the bast fibre used to make linen; the plant from which it is obtained.

float   the length of yarn lying across the surface of a knit or woven fabric: usually refers to effect or colour yarns which are worked into the fabric only at intervals.

flock   very short textile fibres obtained by shredding or grinding. These may be glued to a fabric to give a pile effect.

fluorescent   absorbing ultraviolet light and re-emitting it in the visible spectrum.

foam dyeing   a technique of dyeing in which the dye liquor is applied to the fabric surface as a foam: when the foam breaks, the fabric surface is wetted. The process uses concentrated dye solutions, and is economical of solvents and therefore of energy.

foulard   a light-weight silk or synthetic twill fabric, usually printed.

fully fashioned   of knitted garments, shaped during knitting.

fustian   originally, a low cost, hard wearing, weft-faced cloth with linen warp and cotton weft; a weft-faced cotton cloth.

gabardine   a durable, firm, warp-faced twill cloth.

galloon lace   lace with both edges scalloped.

gauze   a light-weight open leno-weave fabric.

generic classification   classification by chemical nature.

geotextile   a textile used as a construction material in roads, embankments, etc., usually designed to be permeable to water but to resist penetration by stones.

gilling   the combing of wool for exact alignment of fibres in the production of worsted yarns.

gimp   wrapped cord, often used as a highlight in laces made from finer threads.

gin   a machine for the removal of seeds and trash from cotton.

gingham   yarn-dyed checked plain woven cloth, usually cotton in white and one colour.

grandrelle   a two-ply yarn twisted from contrasting singles.

greige (grey)   describes woven or knitted fabrics in the condition in which they leave the loom or knitting machine; loomstate.

grosgrain   a shiny weft-ribbed fabric, usually produced as ribbons.

guipure   a heavy embroidered lace, now usually made by embroidery on a water-soluble cloth.

hank   a loose coil of yarn or thread, not wound around a core.

heat setting   the use of heat to break intermolecular bonds, allowing rearrangement of polymer chains in a desired form.

hemp   a fine, light-coloured, lustrous bast fibre.

henequen   fibre from the leaf of the agave plant.

herringbone   two-colour twill woven to produce a zig-zag pattern.

hessian   coarse plain-weave cloth from hemp or jute.

hogget wool   wool from the first shearing of a sheep.

hollow-filament yarn   synthetic filament extruded with a hollow centre: used to reduce mass without loss of bulk for insulation.

honeycomb   woven fabric with a three-dimensional, cellular effect.

hopsack   fabric made with a variation of plain weave in which a number of warp and weft yarns are woven as one.

houndstooth   a small check pattern in plain weave or twill fabrics, produced by using regular groups of contrasting colours in the warp and the weft.

huckaback   a woven fabric with a rough texture produced by warp and weft floats, often used for tea-towels.

hydrogen bond   strong electrostatic attraction between the slightly positively charged hydrogen and the slightly negatively charged oxygen or nitrogen of polar bonds.

hydrophilic   literally, water-loving: attractive to water.

hydrophobic   literally, water-fearing: possessing no attraction to water.

initial modulus   the initial resistance to touch of a fibre.

insulation   resistance to the passage of heat or electricity.

intarsia   a weft-knit fabric with adjacent design areas in different colours.

interlock   a form of double knit in which on one pass every second needle is missed out.

ionic   possessing a whole electric charge.

ionic bonding   very strong electrostatic attractions between oppositely charged atoms or molecules (ions).

Jacquard knit   weft-knit fabric using multiple yarn feeds in different colours to create a complex pattern.

Jacquard loom   a loom which allows the warp yarns to be lifted individually, thus allowing extemely complex patterns to be woven.

jersey   a general name for weft-knit fabrics.

jet dyeing   a dyeing procedure whereby the fabric rope and concentrated dye liquor are moved together in a sealed pressurized dye vat.

jet loom   a loom in which the weft is carried across the shed by a jet of water or air.

jig dyeing   a dyeing procedure in which the fabric is moved back and forth through the dye liquor.

jute   a bast fibre much used for sacking.

kapok   the fine, soft seed hairs of a tropical plant, used for stuffing.

kemp   a coarse, heavily medullated fibre that does not take dye well: occurs naturally in wool, and is now often imitated by synthetics to give subtle colour effects in tweeds.

keratin   the protein of wool and other hair fibres.

knitting   a method of constructing fabrics by interconnecting loops.

knit velour   a pile fabric made by shearing off the tops of the loops of one side of stretch terry. Used for sportswear and upholstery.

laid-in yarns   yarns in knitting which are held in place by the loops but which do not themselves form loops.

lambswool   wool from a sheep less than eight months old.

lanolin   wool fat: the natural grease of wool.

lawn   a fine, plain-weave fabric of cotton or linen.

leno weave weaving with crossing over of adjacent warp yarns working in pairs.

level dyeing dyeing without any streaks or unevenness.

light fast of a dye, not readily decomposed by the action of light.

light primaries those wavelengths of light which together combine to give the sensation 'white' and which cannot be formed by a combination of other wavelengths (red, blue, green). Also called additive primaries.

linen yarns or fabrics made from flax; textile items traditionally made from linen; often used to describe fabrics which imitate the appearance and texture of linen.

lint cleaned cotton fibres.

linters the short fragments of cotton fibre adhering to the seed after ginning. Often used as a raw material for rayon production.

lively of a yarn, with unbalanced torque, and so with a tendency to twist and snarl and curl back on itself. Fabrics woven from lively yarns often have excellent drape.

loftiness lofty fabrics feel full and light, and are not heavy for their thickness.

London shrinking a finishing process used mainly for worsteds, in which the wetted fabric is allowed to dry in a relaxed state.

lumen the central canal of the cotton fibre.

Madras muslin a leno gauze fabric with extra, hand-cut, weft for patterning.

mangle two heavy rollers through which wet textiles are passed to squeeze out liquid.

marl two differently coloured singles yarns or filaments plied together.

marocain a crepe fabric with a weft rib, woven with two highly twisted S wefts alternating with two highly twisted Z yarns, and a closely spaced warp.

marquisette a leno weave or warp-knitted open textured fabric.

medulla the central canal in hair fibres.

melange printing the printing of carded sliver in a striped pattern to produce subtle colour effects in the yarn.

melded fabric nonwoven fabric composed of a mix of fibres with different melting points, or of bicomponent fibres, and bonded by melting and welding.

melt spinning extrusion of a filament from the molten polymer.

mercerization treatment of cotton with alkali to increase strength and improve dye uptake.

metamerism the near identity of two colour samples under some light sources and not others, caused by the different light sources exaggerating or masking differences in the spectra of the samples.

microfibril a small, long bundle of molecules which goes to make up a fibril.

Milanese a warp-knit fabric of double threads with a slight diagonal design.

milling a finishing process in which wool fabrics are pounded in hot soapy water to give them a soft, fuzzy surface.

moiré a finishing procedure in which two layers of a ribbed fabric are passed together between heavy rollers, giving a non-durable watermark effect.

moleskin a heavy-duty cotton fabric with a smooth face, used for trousers and work clothes.

molleton   a heavy reversible cloth with a nap on both sides, originally made from wool.

monomer   a single unit which goes to make a polymer.

moquette   a loop- or cut-pile fabric used for upholstery.

mordant   a metal salt which combines with small dye molecules within a fibre to form a large, insoluble complex, and so give a washfast dyeing.

moss crepe   a crepe weave fabric with a very large repeat and hence a dull, random appearance.

multifeed knitter   a weft-knitting machine which feeds a new yarn to each needle as soon as it has completed a stitch, thus allowing many rows to be knitted at once.

mungo   fibrous material recovered from torn wool cloth and felt waste.

muslin   a light-weight, open, plain-weave cotton fabric.

nap   a fibrous surface on a fabric, often with one-way orientation.

needled felt   a nonwoven textile in which the necessary bonding is obtained by tangling some fibres together by means of barbed needles.

needle loom   the machine for producing needled felts.

needlepoint lace   a hand-made lace worked with needle and thread.

nep   a small tangle of usually immature cotton fibres.

noil   short wool fibres separated from the top during combing.

nonwoven   a class of textile: nonwoven fabrics consist essentially of a web of fibres held together.

nun's veiling   a light-weight, smooth finished plain weave fabric, often worsted or cotton, sometimes silk.

off-grain   describes fabrics in which the warp and weft do not cross at right angles.

oilcloth   plain woven fabric waterproofed by impregnation with linseed oil.

oiled silk   silk or rayon fabrics waterproofed by treatment with oils such as linseed oil.

oleophilic   literally, oil-loving: describes the tendency of hydrophobic fibres to cling tenaciously to oil and grease.

ombre   literally, shaded: fabric in which the colour is graduated from light to dark.

optical brightener   a fluorescent dye used to give 'brighter than white' effects.

organdie   a light-weight plain-weave transparent fabric with a permanently stiff finish.

ottoman   a warp-faced fabric with a bold weft rib, used for upholstery and coatings. Originally made with silk warp and wool weft.

outworkers   people who work at home at piece rates.

Oxford   a plain-weave fabric with two warp ends weaving as one: the yarns are good quality, often cotton, and the fabric is often striped. Used for shirting.

oxidation   in chemistry, addition of oxygen atoms to, or removal of hygrogen atoms from, a molecule.

padding   a dyeing procedure in which the fabric impregnated with dye is squeezed between weighted mangles to recover (and recycle) the excess. It is very economical of water.

paisley   a decorative print featuring the Indian curved pine-cone pattern.

Panama canvas   a plain-weave fabric used for embroidery.

Panama fabric   a plain-weave smooth finished cloth used for men's tropical suitings.

panné velvet   a light-weight velvet with a flattened pile.

parchmentizing   a brief acid treatment of cottons to give a permanent transparent finish (as in organdie) or a crisp linen-like hand.

peau de soie   literally, skin of silk, satin silk-like fabric with a grained or ribbed surface.

percale   a light-weight, fine, closely woven plain-weave fabric, often made from fine Egyptian cotton. The yarns may be combed or carded, and the glazed or unglazed fabric is often used for sheeting.

photodegradation   changes in fibres caused by light-initiated chemical reactions.

pick   a pass of the weft through the warp shed.

pigment primary   one of the three colours of pigments (yellow, cyan, magenta) which can be combined to give all other colours, but which cannot itself be obtained from a mix of other colours. The primary colour of the artist.

pill   a tangle of broken fibres adhering to the surface of a fabric.

piping   a narrow fabric, one of the selvedges of which is wrapped around a core. Used for decorative effects.

piqué   a woven fabric textured by long floating wefts on the reverse side which pull the ground fabric into hills and furrows.

pirn   the small bobbin around which the weft yarn is wound inside a shuttle.

plissé   fabrics with a puckered or crinkled finish.

plush   a fabric with a long but not particularly dense pile, woven, weft knitted, or warp knitted.

plying   the twisting together of individual strands of yarn.

polar   of an atom or covalent bond, possessing a partial separation of electric charges.

polymer   a vast molecule formed by the joining of many small units (monomers), usually in a repeating pattern.

polynosic   a regenerated cellulosic fibre with a fibrillar structure which gives it high wet strength.

polypeptide   a linear polymer of amino acids (protein), so called because the amino acids are joined by the peptide (CONH) link.

power net   an elasticized warp-knit fabric.

primary   of light or colour, one of the three which can be combined to give all other colours. As the light or additive primaries are different from the pigment or subtractive primaries, it must always be specified which are being discussed.

projectile loom   a loom in which single lengths of weft yarn are carried across the shed in the grip of a small, light projectile which is hit across the loom at high speed.

protein   a natural polymer of amino acids, of animal or vegetable origin. Also called polypeptide.

quota   a limit imposed on the quantity of a product that may be imported or produced.

raising the finishing of fabrics by abrasion to produce a surface layer of protruding fibres.

rapier loom a shuttleless loom in which the weft is carried through the shed by a gripper mounted on a long metal strip or rapier.

Raschel lace a warp-knitted lace fabric.

raw silk silk from which the sericin has not been removed.

rayon a regenerated cellulose fibre, the first artifically produced fibre and the first imitation silk.

reactive dye a small coloured molecule containing a reactive group which allows it to react with and bond to the textile fibre.

reduction in chemistry, the addition of hydrogen to or removal of oxygen from a molecule.

reed a comb-like part of a loom, used to beat up the wefts.

regain the degree to which a fibre can absorb atmospheric moisture.

repp a plain weave fabric with a pronounced weft rib.

resilience the ability of a fibre to recover from a slight deformation.

resist printing application to a fabric in a pattern of a substance which prevents dye uptake, resulting in a white or undyed pattern on a dyed ground.

ric-rac a zig-zag braid produced by varying the tensions on the yarns during manufacture.

roller printing application of pigment or dye to fabric in a pattern by means of engraved metal rollers.

sateen weave a weft-faced smooth weave, often used for cottons.

satin weave a warp-faced weave in which the warp floats are arranged to give a very smooth surface, free from diagonal lines.

Schappe silk a spun silk fabric.

scouring the thorough washing of fibres or fabrics to remove oils, dirt, and size.

scrim low quality light-weight plain cloth resembling muslin; semi-transparent loosely constructed fabric of heavier weight used for curtaining.

secondary colour a colour obtained by mixing two primary colours.

seersucker a plain-weave fabric characterized by puckered and relatively flat sections, usually in stripes but sometimes in checks.

semi-worsted yarn spun by carding and gilling then roving fibres of wool staple length.

sericin the gum that holds the twin silk filaments together and binds the cocoon.

shantung a plain-weave silk dress or furnishing fabric with random irregularities in the weft yarns; synthetics imitating this.

sharkskin a woven or warp-knitted fabric with a firm construction and stiff handle.

shed the space created between warp yarns for the passage of the weft.

shepherd's check the small check pattern achieved on plain or twill fabrics by using regular groups of contrasting colours in the warp and weft.

shoddy fibrous material recovered from loosely constructed woollen rags.

shower repellent denotes fabrics treated to resist the passage of liquid water without being completely waterproof.

shuttle a small container which carries a package of weft yarn back and forth through the shed.

silk   a protein filament obtained from the cocoons of a species of moth; loosely used to describe any light, shiny fabric.

silk screen   a stencil used as a resist in printing, once made of mulberry paper on a silk screen, now usually photographically developed on a polyester screen.

single-knit   weft knitting using a single row of needles.

size   starch and other substances added to yarns to protect them from damage during weaving and to make them easier to handle.

sliver   a strip of loose textile fibres after carding.

solubility   the degree to which a substance will dissolve in a solvent.

specific gravity   the density of a substance, expressed as a ratio to the density of water.

spun silk   silk spun from the short fibres from pierced cocoons and the first brushings of the cocoon.

staple   short fibre lengths as opposed to filaments; fibre of cotton, wool, cut filaments, etc.

steam setting   the use of heat and humidity to rearrange the hydrogen bonds and disulfide bonds of wool.

stenter   a device for straightening wefts and easing yarn tensions introduced during fabric manufacture, by passing the wet fabric at controlled speed and tension through a dryer.

stock dyeing   the dyeing of fibres before spinning.

stretch terry   a single-knit loop-pile fabric.

subtractive primary   one of the three colours (magenta, yellow, cyan) produced when light corresponding to a light or additive primary colour is absorbed (or subtracted), and hence one of the three colours of pigment that cannot be obtained by mixing, but which can be combined to give all other pigment colours. Also called pigment primary.

suede   leather with a fuzzy surface produced by brushing the flesh side of the hide with emery coated rollers.

sulfur dyes   a class of dyes which are chemically complex, but which are all sulfur-containing compounds. They give dark, dull, washfast dyeings with poor light fastness.

Superwash   denotes wool that has been shrinkproofed sufficiently to withstand washing without damage.

taffeta   closely woven, shiny, plain-weave crisp fabric with a faint weft rib, used for linings and evening wear.

tanning   the treatment of hides to preserve them and keep them permanently supple.

tappet loom   loom in which the frames are raised and lowered by the action of cams on levers.

tariff   money paid by an importer to the government when goods are brought into a country.

tartan   woollen twill cloth in a check pattern.

tenacity   the breaking strength of a fibre, expressed as the maximum load it can support, usually in g/den or g/tex.

tensile strength   the ability of a fibre to withstand a lengthwise pull, tenacity.

terry   woven fabric with a looped warp pile, used for towelling.

textile textile materials (fibres, yarns, fabrics); textile products (apparel, furnishings, industrial textiles).

texturing of yarns, is the heat setting or distortion of filament yarns to impart bulk and elasticity.

thermoplastic softening reversibly on heating.

thrown yarn a twisted filament yarn.

tog a metric unit for insulation value, equal to the thermal resistance that will allow a heat flow of one watt per square metre from a temperature differential of 0.1°C.

top grain the surface layer of cow hide, yielding the best quality leather.

torque a rotary force.

tow the bundles of fibres produced from many spinnerets; flax, hemp, or other long bast fibre prepared for spinning.

transfer printing a process whereby paper printed with volatile disperse dyes is brought into contact with hydrophobic fibres and then heated, so the volatile dyes pass directly into the fibres in the desired pattern.

tricot a general name for warp-knit fabrics.

tufting the addition of pile to a fabric in a step separate from the manufacture of the ground fabric.

tulle a warp-knit net with hexagonal holes.

tussah silk protein filament obtained from the cocoons of wild moths of a number of species related to the domestic silkworm. Also called wild silk.

tweed a coarse, heavy, weather-resistant woollen cloth.

twill woven fabric with a pronounced diagonal weave design which is produced by the warp (warp-faced twill), or weft (weft-faced twill), or both, passing over more than one yarn at a time, but with the picking pattern moving one or more yarns across with successive passes.

union dyeing the simultaneous dyeing by different means of different fibres in a blend.

van der Waals bonds weak attractions that operate at close range between all types of atom and molecule, and which are the principal source of intermolecular bonding in hydrophobic fibres.

vat dyes a class of dyes with large, flat, insoluble molecules, extremely washfast and with an affinity for cellulosic fibres.

velour a woven pile fabric, made by shearing off the tops of the loops of a terry fabric. Sometimes used for towelling.

velvet a woven fabric with a cut warp pile.

velveteen a woven fabric with a cut weft pile.

viscose regenerated cellulose fibre (a rayon) made by converting cellulose to a viscous xanthate solution then extruding into a neutralizing bath.

voile a light-weight transparent plain-weave fabric, usually of cotton.

warp the yarn that runs lengthwise in a woven fabric.

warp knit knitted fabric made with a separate yarn for each needle going the length of the fabric.

wash fast of a dye, not removed from the fibre even after repeated washing.

weft the yarn that goes from side to side of a woven fabric.

weft knit knitted fabric with the yarn going across the width of the fabric.

weighted silk   silk treated with salts of tin to improve its drape.

whipcord   twill fabric of cotton or worsted yarns, of various qualities.

wild silk   protein filaments from the cocoons of certain species of wild moths: tussah silk. Because it is made from pierced cocoons, wild silk is always spun.

Wilton   a patterned, woven carpet.

winch   a dyeing apparatus consisting of a large vat through which is pulled ropes of fabric sewn in continuous loops.

woollen   a hairy yarn made from wool fibres; fabric made from these.

worsted   a smooth yarn made from closely aligned wool fibres; fabric made from these, which is made smoother still by the cropping of any protruding hairs.

Bibliography

Batterberry M. and Batterberry A., *Mirror, mirror: a social history of fashion.* Holt, Rinehart and Winston, New York, 1977.

Billmeyer F. and Salzman M., *Principles of color technology.* Interscience, New York, n.d.

Boucher F., *20 000 years of fashion.* Abrams, New York, n.d.

Burnham R. W. *et. al., Color: a guide to basic facts and concepts.* John Wiley and Sons, New York, 1967.

Coats J. and P. Limited, *The technology of thread and seams.* Glasgow, n.d. (and *Thread technology* leaflets)

Dixon B., *Resources: a family ecological view.* Rusden State College (now Victoria College Rusden Campus), Melbourne, 1980.

Fineman M., *The inquisitive eye.* Oxford University Press, New York, 1981.

Fourt L. and Hollies N. R. S., *Clothing: comfort and function.* Marcel Dekker Inc., New York, 1970.

Gittinger M., *Splendid symbols: Textiles and tradition in Indonesia.* Oxford University Press, Singapore, 1985.

Gohl E. P. G. and Vilensky L. D., *Investigating textiles.* Longman Cheshire, Melbourne, 1977.

Greenwood K. M. and Murphy M. F., *Fashion innovation and marketing.* Macmillan, New York, 1978.

Hollander A., *Seeing through clothes.* Avon, New York, 1980.

Hollen N. *et. al., Textiles,* 5th edition. Macmillan, New York, 1979.

Horn M. J. and Beeson M., *The second skin: instructor's manual.* Houghton Mifflin, Boston, 1981.

Howard C., *Embroidery and colour.* Batsford, London, 1976. (and any other Howard titles)

Industries Assistance Commission, *Textiles, clothing and footwear. Parts A–H.* Australian Government Publishing Service, Canberra, 1980.

Jarnow J. A. and Judelle B., *Inside the fashion business.* John Wiley and Sons, New York, 1965.

Joseph M. L., *Introductory textile science.* Holt, Rinehart and Winston, New York, 1981.

Khornak L., *Fashion 2001.* Viking Press, New York, 1982.

Kurth H. *Textiles.* Worlds Work Limited, Surry, n.d.

Lauer R. H. and Lauer J. C., *Fashion power*. Prentice Hall, New Jersey, 1981.

Laury J. R. and Aiken J., *Creating body coverings*. Van Nostrand Reinhold, New York, 1973.

Leeder J., *Wool the wonder fibre*. Australasian Textiles Publishers, Melbourne, 1984.

Lyle D. S., *Performance of textiles*. John Wiley and Sons, New York, 1977.

McLaren K., *The colour science of dyes and pigments*. Adam Hilger Limited, Bristol, 1983.

McIntyre J. E. *The chemistry of fibres*. Edward Arnold, London, 1971.

Moncrieff R. W., *Man-made fibres*, 6th ed. Butterworth, London, 1975.

The new encyclopedia of textiles, Prentice Hall, New Jersey, 1980.

Parker R., *The subversive stitch: Embroidery and the making of the feminine*. The Women's Press, 1984.

Robinson A. T. C. and Marks R., *Woven cloth construction*. The Textile Institute, Manchester, 1973.

Schools Council Publication, *Fibres in chemistry*. The English Universities Press, London, 1974.

Simmons M., *Dyes and dyeing*. Van Nostrand Reinhold, New York, 1978.

Spencer D. J., *Knitting technology*. Pergamon Press, Oxford, 1983.

Thomson J., *Laundrywork*. Reed Education, Auckland, 1974.

Trotman E. R., *Dyeing and chemical technology of fibres*, 6th edition. Charles Griffin and Co., High Wycombe, U.K., 1984.

Vogue. *Glossary of fabrics: All you ever wanted to know about fabrics*. Vogue Pattern Service, U.K., 1986.

Wingate I., *Textile fabrics and their selection*. Prentice Hall, New Jersey, 1976.

Woolley M., *Consumer protection*. Pitman, Melbourne, 1978.

Yasinskaya I., *Soviet textile design of the revolutionary period*. Thames and Hudson, London, 1983.

Periodicals

Apparel International. 51 Hillcrest Road, Purley, Surry CR2 2JF, U.K.

Australasian Textiles. Society of Dyers and Colourists of Australia and New Zealand, Box 286, Belmont, Victoria 3216.

Bayer Farben Revue. Bayer AG, Leverkusen, West Germany. Twice yearly, with occasional special issues.

Craft Australia. Crafts Council of Australia, 100 George Street, Sydney 2000.

Fibre Forum. Australian Forum for Textile Arts, P.O. Box 77, University of Queensland, St Lucia, 4067. Three per year.

Fibres and Fabrics. 435 Riley Street, Surry Hills, New South Wales 2010.

Journal of Consumer Studies and Home Economics. Blackwell Scientific Publications, Osney Mead, Oxford OX2 OEL, U.K.

Journal of the Home Economics Association of Australia. P.O. Box 303, Broadway, New South Wales 2007.

Journal of the Textile Institute. 10 Blackfriars Street, Manchester M3 5DR, U.K.

Ragtrader. 435 Riley Street, Surry Hills, New South Wales 2010. Fortnightly.

Textiles. Shirley Institute, Didbury, Manchester M20 8RX, U.K.

Textiles Asia. G.P.O. Box 185, Hong Kong.

Textile Horizons, the international magazine of the Textiles Institute, 10 Black-
     friars Street, Manchester M3 5DR, U.K.

Other references

The Australian Wool Corporation, 369 Royal Parade, Parkville 3052, has numer-
ous videos available for loan. Topics covered include the history of wool, the
Australian sheep industry, the condition of the Australian agricultural sector
in relation to government policy, fashion, marketing, clothing safety, wool
production, and many aspects of textile production. They have for sale slide
kits on Australia's wool industry, on the manufacture of wool blankets, and
on the creative and historical aspects of knitting; and publish many brochures
and booklets providing fibre information, product information, and educational
resource material.

Many manufacturer's brochures designed for the trade are good sources of
technical information.

General advertising material, for clothing and other textile products, often
provides interesting and unexpected information: this is especially true of advertising for special-purpose products.

Acknowledgements

The authors and publishers are grateful to copyright holders for permission
to reproduce copyright material used in the book.

Sources of illustrations and textual extracts are given beneath each extract.

Our thanks also to E.P.G. Gohl for the data in the tables on pages 352 and
385; Jonathan Cape Ltd for the extract from *A Mortal Flower* by Han Suyin
on pages 595–6; *The Age* for the article on page 550; the Textile Institute
for articles from *Textile Horizons*, the international magazine of the Textile
Institute, appearing throughout the book; and the Society of Dyers and Colourists
for the extract from *Australasian Textiles* on pages 302–304.

Disclaimer

Every effort has been made to trace the original source of material contained
in this book. Where the attempt has been unsuccessful, the publishers would
be pleased to hear from copyright holders to rectify any omission.

Index

resin finishes
  care of   39, 70
  cotton   38–9, 427, 433, 443, 454
  diagnostic staining and   192
  glass fibre   175, 417
  glazed chintz   442
  moiré   442
  pigment binding   292
  rayon   50, 52, 433
  shape retention   433
  shrink-proofing   93–4
resists   269, 297, 445
rigidity
  abrasion resistance   45
  crystallinity   16, 42
rods   202–3
roller printing   250, 292–4
roving   337–8
rubber   125, 170, 172
sacking   44–6, 166
salt
  acid dyes and   265
  direct dyes and   259–60
  solubility and   20–21
  use in dyeing   245, 253–4
sanforizing   433
Saran   160
scouring
  cotton   278
  greige goods   429
  synthetics   278
  wool   278, 328–30, 590–1
screen printing
  art   295
  flat bed   294–5
  fur fabrics   496
  glass fibre   175
  origins   294
  rotary   295–6
seams   447, 453–4, 523–6, 530
secondary acetate (acetate)
  care   67
  comfort   59, 67
  dyeing   64, 67
  identification   193, 195
  production   58–9
  properties   58–9, 62–4, 67, 92
secondary colours   204
seersucker   37, 376
selvedges   377, 383, 425–6, 432–3, 454
  leno   377
  neatening   384

tucked   383–4
semi-worsted   334–5
setting
  carpet yarns   97
  flat   97
  garments   97
  heat   58, 67, 138, 320–4, 344, 388
  Si-ro-set   97
  steam   97, 344
sewing threads   36, 325, 499, 525, 529–30
shape of molecules   14
  cellulose   23–4
  dyes   257, 263, 274
sheer fabrics   531
shrink-proofing of wool   90, 93–5, 97
shuttle
  flying   363, 377, 379
  hand-thrown   363
  loom   378–9, 384
  replenishment   378–9
side-groups
  crystallinity   15
  proteins   75
silk
  amino acids in   105–106
  care   107, 109, 116, 529, 533
  dyeing   104, 242, 277
  identification   190
  imitations   46–8, 120, 132, 149, 493, 529
  production   104–108, 570
  properties   104–108, 132, 142
  trade   104, 559
  types   105, 108
  weighting   107–109, 436, 529
silkworm   104
sisal
  identification   190, 193
  properties   45
  production   46
  use   45, 46, 146, 501
size   344, 364
  removal   364, 429
social meaning of textiles   3, 6–7, 110
softening
  chemical   430, 522, 528
  flexing   430
solubility   18–21
  alginates   178
  dyes   251, 262–3, 274–5
  polyvinyl alcohol   161
  salt and   20–21, 253

Notes

Notes

Notes